Complex Biological Systems

Scrivener Publishing
100 Cummings Center, Suite 541J
Beverly, MA 01915-6106

Publishers at Scrivener
Martin Scrivener (martin@scrivenerpublishing.com)
Phillip Carmical (pcarmical@scrivenerpublishing.com)

Complex Biological Systems

Adaptation and Tolerance to Extreme Environments

Edited by
Irina R. Fomina, Karl Y. Biel, and Vladislav G. Soukhovolsky

Scrivener Publishing

WILEY

This edition first published 2018 by John Wiley & Sons, Inc., 111 River Street, Hoboken, NJ 07030, USA and Scrivener Publishing LLC, 100 Cummings Center, Suite 541J, Beverly, MA 01915, USA

For more information about Scrivener publications please visit www.scrivenerpublishing.com.

Wiley Global Headquarters
111 River Street, Hoboken, NJ 07030, USA

For details of our global editorial offices, customer services, and more information about Wiley products visit us at www.wiley.com.

Library of Congress Cataloging-in-Publication Data

ISBN 978-1-119-51040-6

Cover images: Changing Tree - Jan Martin Will | Dreamstime.com
Cover design by: Kris Hackerott

Set in size of 11pt and Minion Pro by Exeter Premedia Services Private Ltd., Chennai, India

10 9 8 7 6 5 4 3 2 1

Dedicated to the memory of Rem G. Khlebopros

Wipe off random features and you'll see –
the world is knowable, – Rem Khlebopros

Contents

Preface

The book you are likely to read was written by a small number of authors who live at great distances from each other (in Asia, Europe and North America) and is a collection of chapters. Each chapter can be read quite independently of the others, but, despite this, they are very much interconnected.

To some extent the book looks like a small pile of smooth pebbles on a sandy beach. They are all either small or large, either reddish or gray, but underlying them there is a common genesis.

The common link uniting the chapters in this book is one of the most ingenious inventions of our Biosphere - namely Photosynthesis. That is, any and all of the issues being discussed here are, in one way or another, related to photosynthesis or at very least are addressed to this natural phenomenon.

Be that as it may, to varying degrees our chapters combine hypotheses that may be tested and present, therefore, concepts for confirmation by strictly scientific protocols.

Responsibility lies with the authors of a given chapter, especially in cases where a hypothesis is expressed for the first time and therefore requires justification. As for the book, taken as a whole, its contents and critiques are addressable to the editors.

Rem G. Khlebopros
President,
International Scientific Centre
for Organism Extreme States Research,
Krasnoyarsk Scientific Centre,
Siberian Branch of the Russian Academy of Sciences,
Krasnoyarsk 660036, Russia

Abstract

The book is a complication of our authors' concepts, hypotheses, and proofs of concepts concerning critical problems of closed and semiclosed biological systems; mechanisms of adaptation to normal and extreme conditions by plants; and the role of photosynthesis and other physiological processes in the activities of the plant itself, in communities, through ecosystems, and in agriculture.

The fourteen chapters of the book are divided into four sections: "Modeling and Approaches", "Hypotheses", "Experiments", and "History and Biography – Tribute". The special section, "History and Biography – Tribute" contains our memories of scientists whose discoveries, ideas and dedication to science have inspired us. In the extraordinary case of Andrew A. Benson, culminating concepts driven by his eponymous Protocol are transferred from science to technology.

We believe that the book will be useful for a wide group of biologists, especially for specialists in ecology, plant physiology, botany, biochemistry, agronomy and genetics; in addition, as a contemporary reference, we expect it to be used in ongoing and future research work in the field. The chapters will serve to educate undergraduates, graduates, postgraduates, educators and all who care about the present and the future of life on our planet.

Contributing Authors

Khlebopros Rem G.,
Ph.D., Dr. Sci., Professor

President,
International Scientific Centre for Organism Extreme States Research, Krasnoyarsk Scientific Centre, Siberian Branch of the Russian Academy of Sciences, Krasnoyarsk 660036, Russia (deceased)

Fomina Irina R., Ph.D.

Institute of Basic Biological Problems
of the Russian Academy of Sciences,
Pushchino, Moscow Region 142290, Russia

E-mail: irafomi@rambler.ru

Vice President of Education,
Biosphere Systems International Foundation, Tucson, Arizona 85755, USA

Website: www.biospheresystems.com

Biel Karl Y., Ph.D., Dr. Sci., Professor

Institute of Basic Biological Problems
of the Russian Academy of Sciences,
Pushchino, Moscow Region 142290, Russia

E-mail: karlbiel@hotmail.com

President,
Biosphere Systems International Foundation, Tucson, Arizona 85755, USA

Website: www.biospheresystems.com

Soukhovolsky Vladislav G., Ph.D., Dr. Sci., Professor

International Scientific Centre for Organism Extreme States Research, Krasnoyarsk Scientific Centre, Siberian Branch of the Russian Academy of Sciences;
Institute of Forest, Siberian Branch of the Russian Academy of Sciences, Krasnoyarsk 660036, Russia

E-mail: soukhovolsky@nm.ru

Balakhnina Tamara I., Ph.D.

Institute of Basic Biological Problems
of the Russian Academy of Sciences,
Pushchino, Moscow Region 142290, Russia

E-mail: tbalakhnina@rambler.ru

Brummel Daniel P.

Department of Agronomy of Iowa State University, 2104 Agronomy Hall, Ames, Iowa 50011-1010, USA

E-mail: DBrummel@iastate.edu

Ivanov Anatoly A., Ph.D.

Institute of Basic Biological Problems of the Russian Academy of Sciences, Pushchino, Moscow Region 142290, Russia

E-mail: demfarm@mail.ru

Kosobryukhov Anatoly A., Ph.D., Dr. Sci.

Institute of Basic Biological Problems of the Russian Academy of Sciences, Pushchino, Moscow Region 142290, Russia

E-mail: kosobr@rambler.ru

Lam Wai-Ki "Frankie", Ph.D.

BRANDT Consolidated, Inc.,
3654 South Willow Ave.,
Fresno, California 93725, USA

E-mail: Frankie.Lam@BRANDT.co

Lauria Allison

Research Agronomist
BRANDT Consolidated, Inc.,
3654 S. Willow Ave. Fresno,
CA 93725, USA

E-mail: Ally.Lauria@BRANDT.co

Makarov Mikhail V., Ph.D., Dr. Sci.

Murmansk Marine Biological Institute Kola Scientific Center of the Russian Academy of Sciences, Murmansk 183010, Russia

E-mail: zelentsy@yandex.ru

Markovskaya Evgeniya F., Ph.D., Dr. Sci., Professor

Petrozavodsk State University,
Petrozavodsk 185910, Russia

E-mail: volev10@mail.ru

Matichenkov Vladimir V., Ph.D., Dr. Sci.

Institute of Basic Biological Problems
of the Russian Academy of Sciences,
Pushchino, Moscow Region 142290, Russia

E-mail: vvmatichenkov@rambler.ru

McBride Michael S.

President

BRANDT iHammer, 479 Village Park
Drive, Powell, Ohio 43065-9178, USA

E-mail: Mike.McBride@BRANDTiHammer.co

Nazarova Galina N., Ph.D.

Institute of Basic Biological Problems
of the Russian Academy of Sciences,
Pushchino, Moscow Region 142290, Russia

E-mail: nazgal@mail.ru

Nishio John N., Ph.D., Professor

Chief Executive Officer,
Biocompatible Plant Research Institute,
Chico, California 95973, USA

E-mail: nishio@biocompatible.org

Nonomura Arthur M., Ph.D.

Carbon Reactions of Photosynthesis Sector
BRANDT iHammer, 479 Village Park
Drive, Powell, Ohio 43065-9178, USA

E-mail: Art.N@BRANDTiHammer.co

Ryzhik Inna V., Ph.D.

Murmansk Marine Biological Institute Kola Scientific Center of the Russian Academy of Sciences, Murmansk 183010, Russia
Murmansk Arctic State University, Murmansk 183036, Russia

E-mail: alaria@yandex.ru

Yensen Nicholas P., Ph.D.

President,
NyPa International, Tucson, Arizona 85705, USA; Biosphere Systems International, Tucson, Arizona 85705, USA (deceased)

Modeling and Approaches

1
Critical Impacts on Complex Biological and Ecological Systems: Basic Principles of Modeling

R.G. Khlebopros[1], V.G. Soukhovolsky[1,2,*]

[1]*International Scientific Centre for Organism Extreme States Research, Krasnoyarsk Scientific Centre, Siberian Branch of the Russian Academy of Sciences, Krasnoyarsk 660036, Russia*
[2]*Institute of Forest, Siberian Branch of the Russian Academy of Sciences, Krasnoyarsk 660036, Russia*

Abstract
In the chapter, methods of description and modeling of complex biological systems, as well as the influences of various environmental factors are considered. A complex system consists of a greater number of components cooperating with each other. Any complex biological system is stable to external impacts below a certain intensity and frequency. This property is a condition for the duration of the existence of such a complex system. This chapter discusses problems that are associated with modeling of such complex systems. Existing methods for macro- and microscopic description of complex biological systems are analyzed. In particular, methods of optimization modeling of complex biological systems are considered. In optimization models it is offered, that the biological system is characterized by the presence of a certain set of goals, which it aspires to achieve. A set of criterion functions, which describe the purposes facing to the system, aspiring to survive in a changing environment is entered into the models. It is supposed that the value of the criterion functions depends on the parameters of the state of the biological system. The process of adaptation of the system is that the system pa-

* *Corresponding author*: soukhovolsky@nm.ru

rameters accept following values, at which sizes of the target achieve a maximum, and the probability of survival of such a system appears maximum under existing environmental conditions. Further in the book optimization approach is used for modeling processes of photosynthesis in the analysis of the stability of closed ecological systems. Here, the problems of the response of complex biological systems to external impacts are discussed. If the intensity and speed of the external impact surpass some critical level, the result of such critical impact is a system transformation. Thus, the state of the components and / or the spatial order in the system is changed. Classification of types of critical impacts on systems and classification of the properties of the systems shown at critical impacts are given. In particular, consider these types of complex biological systems, such as elastic and fragile systems. Methods of modeling of state changes of complex systems are considered at critical impacts. Using the methods of modeling of complex biological systems, the methods being by analogues used in the analysis of critical phenomena in physical systems, is considered in detail.

Keywords: Complex biological systems, Methods of modeling, Optimization models

1.1 Complex Ecological Systems: The Principle of Decomposition, Taking into Account the Characteristic Times of Components

In contemporary science, analysis and prediction of the behavior of complex systems is in great demand. Once, the use of mathematical approaches in different sciences, in natural ones in particular, was limited to the construction and subsequent analysis of model systems such as pendulum, solid-solid interaction, *etc.* Although these model systems were also good for presenting events, we now deal with complex systems and in every particular case we aim to reduce the number of variables describing a system to a minimum, ideally to one. Thus, the main objective now is to examine a complex system and to construct an analogous model system that would have the necessary properties for predicting its dynamics. This is possible if the complex system is defined as a system with a large number of variables. However, not every system with a large number of variables is a complex system. The complex system is the system that exists in reality. This implies that for any of its coordinates there is a domain in which negative feedback dominates. If there are no domains in which for all components there is negative feedback, the system does not exist. This is true for all systems – living, nonliving, economic, social, *etc.*

Let us consider the issue of reducing the number of variables of the analyzed system to the level that cannot be further reduced if we are to stay within the range of correct answers (Tikhonov, 1943; Bogolyubov and Krylov, 2005; Khlebopros *et al.*, 2007). Imagine a complex system and arrange the equations based on characteristic relaxation times (τ) in descending order. If the system analyzed is forest, the longest characteristic times are the times of formation of forested areas – usually several thousand years. Next can be the characteristic time of the existence of tree stands; different plant species growing in the forest have different characteristic times, but these are much shorter times anyway – decades or centuries. The next in the series are forest mammals, whose lifetime is several decades; they are followed by insects with the life cycles lasting 1 or 2 years; then come equations for microorganisms, with their lifetimes of several weeks, days, hours, *etc.*

If we arrange the "scores" of existing components in this way, we can ask, for example, what changes will occur in this forest if insects reach outbreak levels and "eat up" the mature forest. Over the lifetime of one generation of insects, the forest will not actually undergo any changes, but microorganisms will act as a "quick" component. There are very slow components, which for the insect equation will act as continuous, slowly evolving, factors. It will take decades or centuries for the forest composition to change so much that insects will live in the same place but in a different forest, with trees of different sizes and species. Microorganisms, on the other hand, will multiply and vanish several times over the same period, and this component will be averaged over the characteristic times equal to several decades. Then we can only examine the components that have equal characteristic times. These can be insects that consume other insects that, in turn, consume forest trees. Their equations should be retained, while other components will either be present as continuous parameters or become averaged over this time.

Thus, the number of analyzed components will be reduced dramatically. However, this is only possible if we do not have to find out whether insect outbreaks will ruin forests. Forests may be replaced by marshes, and then the characteristic times that we need are characteristic times of the formation of marshes and forests. In this case we deal with changes in the soil under the impact of forest or marsh rather than with insects and trees. These processes last several thousand years, and the effect of all the components with much shorter characteristic times will become averaged. Thus, these equations need not be used in the model.

This mathematical approach can conveniently be used to analyze complex systems because it dramatically reduces the number of equations, components, and variables that have to be taken into account in solving different problems.

So, analysis of a complex system should begin with composing time hierarchy for its components, followed by defining of characteristic times that will be used to examine the system and its dynamics. The system always evolves, but evolutionary processes over very short, short, and long times are different. Analysis of the behavior of the system over these characteristic times will answer the questions about steady and unsteady development, evolution of different complex systems or, vice versa, the risk of catastrophic events (Isaev and Khlebopros, 1973*a*, 1973*b*, 1974; Isaev *et al.*, 1974, 2001).

Another opportunity to reduce the number of coordinates to be taken into account arises when the analyzed system changes from one state to another: either from the current state into a new one or from one stable state into another stable state or from the equilibrium state into evolution around the cycle or along another complex trajectory. All changes have the following property: close to the transition boundary, the number of the components that we have retained is again dramatically reduced. As in this n'-dimensional space ($n' << n$), close to the boundaries the crossing of which will change the state, we reduce the number of coordinates again, the only ones left will be the movement along which is orthogonal to the interface separating the states, while all the coordinates that run almost parallel to this interface can be ignored. This is the second way to reduce the number of coordinates and, thus, to considerably increase the possibility of accurate solution of different extremal problems.

To analyze evolution of the complex system that exists in reality and, hence, has a negative feedback domain for any component, we should construct a phase portrait. The most correct approach is to construct it for the component that has the characteristic time (relaxation time) over which we examine evolution of the system and make forecasts (Volterra, 1931; Kolmogorov, 1936, 1972; Kolmogorov *et al.*, 1937; Poincaré, 1947; Lyapunov, 1954–1959). If, for instance, we examine insect outbreaks that feed on trees and shrubs, we should take into account characteristic times of several generations of these insects.

Prior to constructing the phase portrait by experimental points, we can do the following: plot the number of insects per square meter (tree) on the abscissa and the rate of change in this number on the ordinate (in the case of insects, we should take two successive measurements of insect density and divide the difference between these densities by the characteristic time between these two measurements). Thus, we have the coordinate and its rate of change. There are two types of the phase portrait. One is the narrow phase portrait (Figure 1.1) (Isaev and Khlebopros, 1973 *a*, 1973*b*, 1974).

The narrow phase portrait implies that we can actually use one equation, *i.e.* the rate depends upon coordinate $\dot{x}_i(x_i)$, and all the other coordinates will either be averaged (if they change rapidly) or function as parameters (if they change slowly). Thus, of all the numerous coordinates that describe the forest ecosystem we may need just this coordinate (this insect abundance and the rate of change of this abundance) to describe evolution (behavior) of the forest area. Only one equation is left. The intersection point of this narrow phase portrait with axis X_i is the equilibrium state around which insect abundance will vary. It can certainly evolve over time and have slow trends, but these trends will correlate with other characteristic times such as the time of tree growth and the time of a change in the tree stand. If the rate of change in the insect abundance versus insect abundance is a nonmonotonic function, there may be several intersections (Figure 1.2); in the simple case there will be three intersections and two of them, X_1 and X_2, are stable equilibrium points and X_r is the "divide" point.

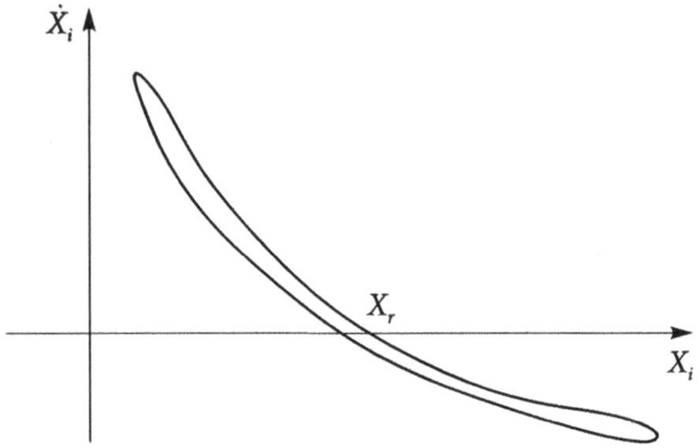

Fig. 1.1. The narrow phase portrait of complex system

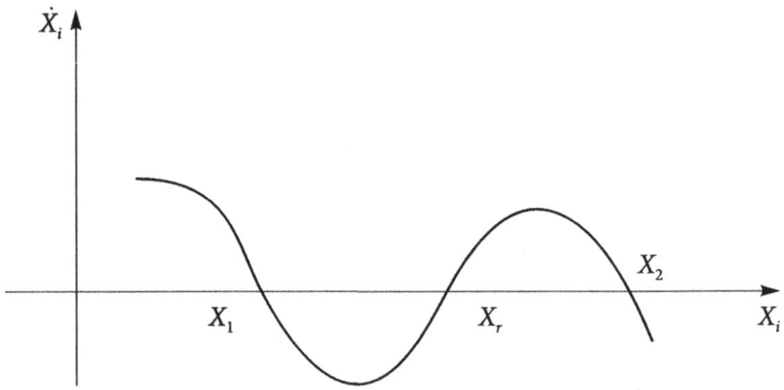

Fig. 1.2. The narrow phase portrait with two stable equilibrium points

If there are more variables (if we take these same insects but with their competitors and enemies that have similar characteristic times), the phase portrait will be broad (Volterra, 1931; Kolmogorov, 1936, 1972; Kolmogorov *et al.*, 1937; Poincaré, 1947; Lyapunov, 1954-1959). That is, for the same X_i values (in the case of insects, for the same densities), the rate will be different depending on the abundance of competitors or enemies – insects or other animals. When the system is delay-free (all enemies and competitors have "agreed" to X_i being constant), this curve can cross the abscissa once (Figure 1.3), *i.e.* there is one equilibrium point, or three times, *i.e.* there are two stable equilibrium points. Although the phase portrait is broad, the behavior of the entire system is sometimes similar to the behavior of the system with the narrow phase portrait (Figure 1.4). The only difference is that in the narrow phase portrait all movements are along the "reproduction curves", whereas in the case just described the movements are above or below rather than along the "reproduction curves" (Isaev and Khlebopros, 1973*a*, 1973*b*, 1974; Isaev *et al.*, 2001).

However, when there is a negative feedback domain (Figure 1.5), there may be positive feedback domains in it. Then, if the curve intersects the abscissa, different variants of the system dynamics can be obtained, depending on the position of points X_1 and X_2 (Figure 1.6) (Isaev and Khlebopros, 1973*a*, 1973*b*, 1974; Isaev *et al.*, 2001; Bazykin *et al.*, 1997).

One of the possible variants is when X_1 and X_2 are in the negative feedback domain, *i.e.* are stable points. Again, we have the situation that has already been examined, one described by the narrow phase portrait, *i.e.* the broad phase portrait whose phase space is entirely the negative feedback domain. If, however, either X_1 or X_2 is out of the "stable" domain, in the positive feedback domain, either outbreak or "anti-outbreak" occurs, *i.e.* there is one stable point from which the system is "ejected" from time to time either towards a dramatic increase or towards a dramatic decrease in insect abundance. An interesting situation occurs when both X_1 and X_2 are in the positive feedback domain. Then we have a very rare event, which can be termed a sustained outbreak, if we deal with insects, or attractor formation, if we analyze any other complex system. The existence of the attractor indicates that the system all the time moves around points X_1 and X_2, without going along the same trajectory twice.

Thus, the detailed analysis of the phytophagous insects – forest area system yielded very interesting trends that can be observed in any complex system. The various behaviors of insects in the forest can be reduced to just six variants:

- one stable equilibrium point;
- one unstable equilibrium point, when the system is far from the equilibrium, constantly moving along an intricate and always changing trajectory of the phase space;
- fixed outbreak, when there are two stable states: one with high abundance and the other with low;
- outbreak;
- anti-outbreak;
- sustained outbreak.

Interestingly, we need not always keep the complex system really complex, *i.e.* we can choose a characteristic time that we are interested in to examine the evolution of the system. Then we can choose the coordinates, the variables whose characteristic times are the same as or close to the characteristic time we are interested in. The next step is to construct a phase portrait for one of these coordinates. A narrow or a broad phase

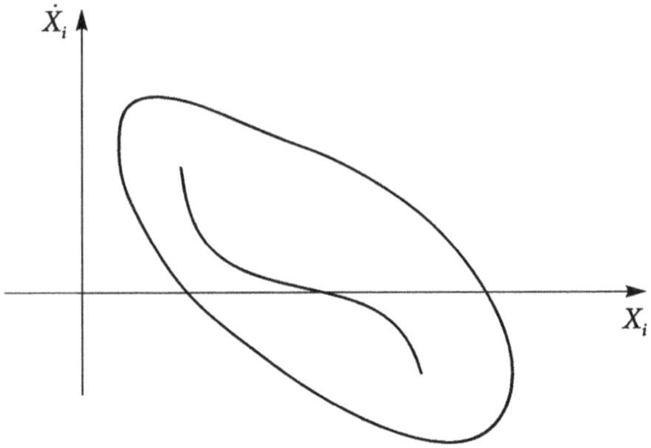

Fig. 1.3. The broad phase portrait with one stable equilibrium point

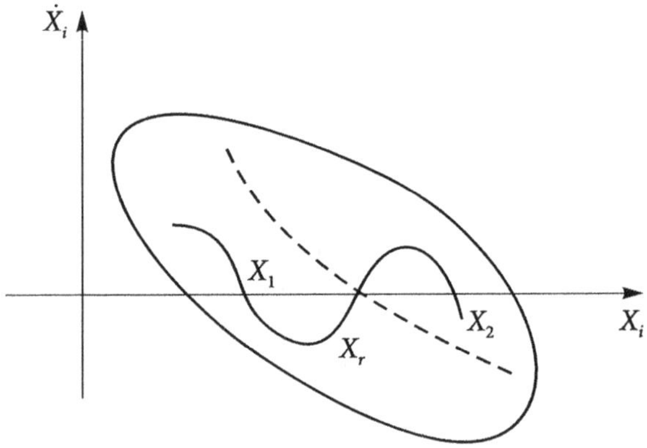

Fig. 1.4. The broad phase portrait with two stable equilibrium points

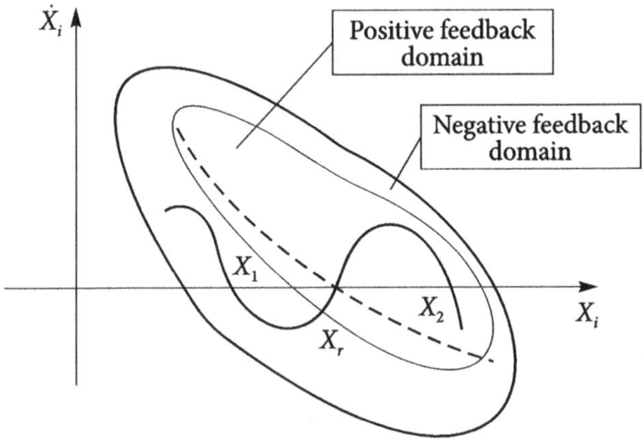

Fig. 1.5. The broad phase portrait with one stable equilibrium point and one unstable equilibrium point

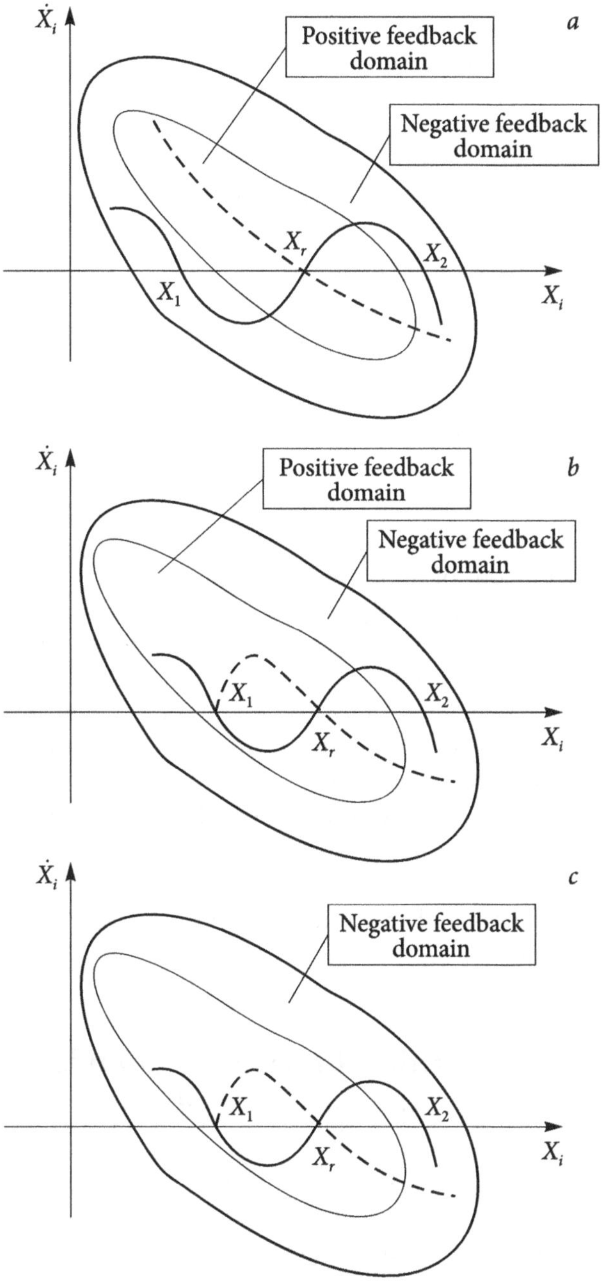

Fig. 1.6. The broad phase portrait with two stable equilibrium points – "fixed outbreak" (*a*), with one stable equilibrium point and one unstable equilibrium point – "anti-outbreak" (*b*), with two unstable equilibrium points – "sustained outbreak" (*c*)

portrait can result. If it is narrow, all depends on how many times the characteristic curve crosses the abscissa: if there is one intersection, there is one stable point; if there are several intersections, there are two, three, four ... equilibrium points. If the resulting phase portrait is broad, we cannot use just one coordinate; the number of coordinates is actually larger than one but much smaller than in the initial system. Then, plotting experimental points in the phase portrait, we can.

First: determine that it is broad;

Second: determine that it is singly connected, that there are no positive feedback domains in it. If this is so, the results are actually the same as in the narrow phase portrait.

If in the phase portrait, within the negative feedback domain there is a positive feedback domain, there may be six variants of the behavior of the system.

The approach that we have demonstrated can be used to examine a complex system step by step. Let us imagine an artist painting a landscape he sees in front of him. This landscape contains mountains, hills, trees, meadows, and, probably, animals. If we scrutinize the landscape closely, we can ask: What may change in a day? The position of the goats may change, but not the position of the trees. The position of the trees will remain unchanged for decades and the positions of the mountains and the hills for several thousand years. One can see "quick" movements in this system and analyze the dynamics based on this; the same system can be analyzed in terms of "slower" movements; and, finally, it can be analyzed in terms of "very-very slow" movements. Anyway, each time the phase portrait is only constructed for the characteristic time that the researcher is interested in to examine the evolution of the entire system.

1.2 Analysis of Critical Impacts on Complex Systems and Extreme Principles of Modeling

In physics, complex systems are described using not only kinetic equations, which account for the dynamics of microscopic variables of the system, but also the so-called extremal principles. In this case, n equations of dynamics are replaced by a certain function relative to which it is assumed that the system tends to reach the state in which the value of this state function is the lowest. Then there are not more than 3–4 independent variables in the model of a physical system, whatever the number of the system components is. This reduction is convenient for finding stable states of the system but not for describing the dynamics of the processes occurring in it (Landau and Lifshits, 1964; Dmitriev, 2004).

Yet, simplification of the models of ecosystems using this metamodel approach seems a promising tactic. There are various approaches to the construction of ecological models, based on using extremal principles (Fursova *et al.*, 2003; Soukhovolsky *et al.*, 2008). In this study we use extremal principles to describe such critical events in forest ecosystems as insect outbreaks, forest fires, *etc.*

1.2.1 Meta-Models of Phase Transitions for Describing Critical Events in Complex Systems

To describe the state of a complex system, physics introduces the notion of phase as an integral state of a certain substance of a physical system (Bruce and Cowley, 1981). A very simple example is three phases of water (solid, liquid, and gaseous). A change in the state of the object or the system is termed phase transition (*e.g.*, ice melting or water evaporation). Phase transitions occur under an external influence (such as a temperature change). It is assumed that the state of the system and phase transitions can be described using a certain thermodynamic potential, G, and condition $G \to min$ is an optimization principle, regulating the process of phase transition (Landau and Lifshits, 1964). The thermodynamic potential of a complex system depends on a large number of factors, and in most cases, the general form of function G is unknown. However, in the case of second-order phase transitions, such as superconductivity, super fluidity, and magnetization, the number of factors that influence the behavior of the system is usually dramatically reduced, suggesting that function G depends on just two variables: the so-called order parameter, q, characterizing certain general properties of the examined object, and an external variable such as temperature T (Landau and Lifshits, 1964). In the theory of phase transitions, the order parameter is regarded as an independent variable. The phase transition begins when the value of the external variable becomes less (or more – depending on the particular object) than a certain critical value.

Using this simplified approach to describing the process of phase transition, the $G = f(T,q)$ relation can be sufficiently accurately represented as an even power series of the order parameter (Landau and Lifshits, 1964).

$$G = G_0 + Aq^2 + bq^4, \tag{1}$$

where b = const, and A is a linear function of external variable T:

$$A = a(T - T_c), \tag{2}$$

where T_c is the critical value of T marking the onset of phase transition.

A characteristic feature of static models of phase transitions is that they are time independent. This assumption significantly simplifies the model: the description of the system only needs stable minimums of function G, which can be derived from the conditions $\frac{dG}{dq} = 0; \frac{d^2G}{dq^2} > 0.$ The value of the order parameter, q, at which the G minimum is reached, will depend on the values of parameters A and b of equations (1) and (2).

The models of phase transitions incorporate the universality principle, in accordance with which processes of phase transitions depend on just a few basic properties of the system, such as dimensionality, the number of components of the order parameter, and distance dependence of interactions in the system (Bruce and Cowley, 1981). The idea of universality of phase transitions in systems with one order parameter and homogeneous spatial structure simplifies and unifies the construction of models of phase transitions. Even without knowing the exact form of function G, one can describe the process of phase transition and find critical values of the external factor and values of the order parameter in stable states of the system.

Below are models describing some critical events in forest ecosystems as second-order phase transitions, based on the universality principle in theory of phase transitions.

1.2.2 A Model of Outbreak as Second-Order Phase Transition

Forest insect populations are characterized by both the density of X individuals (absolute colonization) and the distribution of individuals over the area, specifically, the distribution of individuals on sample units: trees, model branches, *i.e.* In entomological monitoring, the simplest and most frequently used indicator of spatial distribution of insects in sample units is the relative colonization A – the fraction of sample units with individuals of the studied insect species (Isaev *et al.*, 2001).

In accordance with the phenomenological model of population dynamics, an outbreak occurs when the pest population density exceeds a certain critical value, X_r , and the population has colonized all hosts available in the area (Isaev and Khlebopros, 1973*a*, 1973*b*; Isaev *et al.*, 2001).

The relative colonization of the trees in the stand is then close to 1. During the sparse stable population phase the population density is the lowest possible and the insects do not colonize all of the hosts. During the outbreak phase, the abundance of the insects in the population is the highest, and they colonize all suitable host trees. Thus, insect population can be either in the sparse stable phase or in the outbreak phase. Let us hypothesize that insect survival in the population is described by a certain function, G (the function of ecological risk). The function of ecological risk can be represented as the probability of insect death in the population, depending on the effects of various regulating and modifying factors. For every particular insect population, the general form of this dependence is certainly unknown, but one can assume that stable states of the population are characterized by the minimum values of the function of ecological risk, G (Soukhovolsky *et al.*, 2005). Then, to describe the insect outbreak as an ecological analog of phase transitions in physical systems, we should introduce the order parameter, q, to determine the external factor leading to the phase transition, and introduce an equation that would relate these variables.

Let us define the order parameter, q, as the value related to the relative insect colonization, A, of sample units of the tree stand:

$$q = 1 - A = \frac{n-k}{n}, \tag{3}$$

where $A = \frac{k}{n}$ is the relative insect colonization of sample units (trees, plots, *etc.*), n is the number of sample units, and k is the number of sample units colonized by insects. It follows from (3) that $0 \le q \le 1$.

It follows from (3) that the value $q = 0$ of the order parameter corresponds to the outbreak phase, when insects occur in all sample units (trees, plots, *etc.*). Characteristic values of the order parameter do not populate all sample units. As the external parameter, corresponding to the temperature in physical systems, we have chosen population density X. In accordance with the phenomenological theory of forest insect population dynamics, an outbreak (*i.e.*, phase transition) occurs when the population density exceeds a certain critical value, X_r. When the sparse stable population phase is replaced by the outbreak phase, the value of the order parameter is reduced to zero, *i.e.* all trees in the outbreak site are colonized by the insects.

To describe the outbreak as phase transition, we use equation (1). The order parameter (3) will describe the state of the tree stand. The external factor will be population density, X. Coefficient A in (1) will be written as one linearly dependent on population density:

$$A = a(X - X_r). \tag{4}$$

It follows from (4) that if the population density is higher than the critical value $X = X_r$, $A > 0$.

The value of the order parameter for the outbreak phase and the sparse density phase will be found from equation $\frac{\partial G}{\partial q} = 0$, characterizing the minimum value of function G:

$$\frac{\partial G}{\partial q} = 2a(X - X_r)q + 4bq^3 = 0. \tag{5}$$

Equation (5) has two solutions:

$$q = 0 \text{ and } q^2 = \frac{a(X_r - X)}{2b}. \tag{6}$$

Solution $q = 0$ describes the outbreak phase. The second solution is true for the population density below the critical value, X_r. This solution characterizes a sparse population.

At $q = 0$, the value of ecological risk $G_1 = G_0$, and at $q = \sqrt{\frac{a(X_r - X)}{2b}}$, $G_2 = G_0 - \frac{a^2(X_r - X)}{4b}$. Hence, $G_1 > G_2$, and the ecological risk for populations of non-outbreak insect species that have sparse stable densities is lower than the ecological risk for outbreak species. The number of non-outbreak species should be significantly greater than the number of outbreak ones. This conclusion is in good agreement with the data of forest insect surveys. Our investigations in boreal forests of Eastern Siberia showed that of 315 economically important species of forest insects, just 74 were able to reach outbreak levels (23.5%) (Isaev *et al.*, 2001).

Knowing insect population densities provided by insect surveys and relative estimates of insect colonization of tree stands in different phases of insect outbreaks, we can verify the proposed model by determining the relationship between X and q^2 in the $\{X, q^2\}$ plane. Based on model (1), at $q = 0$, points characterizing population density must be located on the abscissa on the right of point X_r. For $X < X_r$, in accordance with (6), between q^2 and density X there must be a negative linear relationship. Point $X = a/2b$ of the intersection of line $q^2 = \frac{a(X_r - X)}{2b}$ and the abscissa gives the value of X_r.

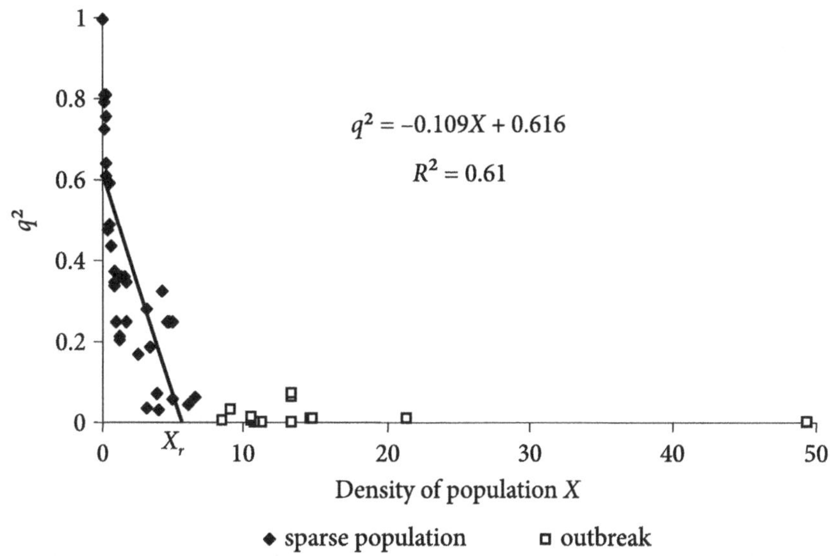

Fig. 1.7. Relationship between parameter q^2 and the density of pine looper *Bupalus piniarius* L. populations in different outbreak phases in Altai, Minusinsk, and North-Kazakhstan pine forests during 1962–1988

Figure 1.7 shows the data of surveys that demonstrate the relation between q^2 and X for populations of the pine looper *Bupalus piniarius* L. in Altai, Minusinsk, and North-Kazakhstan pine forests during 1962–1988.

Bupalus piniarius L. surveys were performed in spring, during the pupal stage, on 1 × 1 m plots (Palnikova *et al.*, 2002). The survey data are clearly in good agreement with the theoretical model (Figure 1.7). For population densities below the critical value, there is a significant linear relationship between population density and the squared order parameter. For population densities higher than the critical value, the order parameter is close to zero. The critical density X_r for the pine looper is 5.65 pupae per plot, and this is in good agreement with the data reported in the literature – 6 pupae per plot (Isaev *et al.*, 2001).

Figure 1.8 shows the relation between parameter q^2 and population densities of the Siberian silk moth *Dendrolimus superans sibiricus* Tschetv. in different phases of the outbreak cycle.

Calculations were done using Y.P. Kondakov's data (Kondakov, 1974). Here we also have a good agreement between survey data and theoretical equation (1). The critical value X_r calculated from equation (1) is equal to 657 larvae per tree, and the number given in the handbook on monitoring of forest pests for the critical density of the Siberian silk moth population is 600 larvae per tree (Isaev *et al.*, 2001).

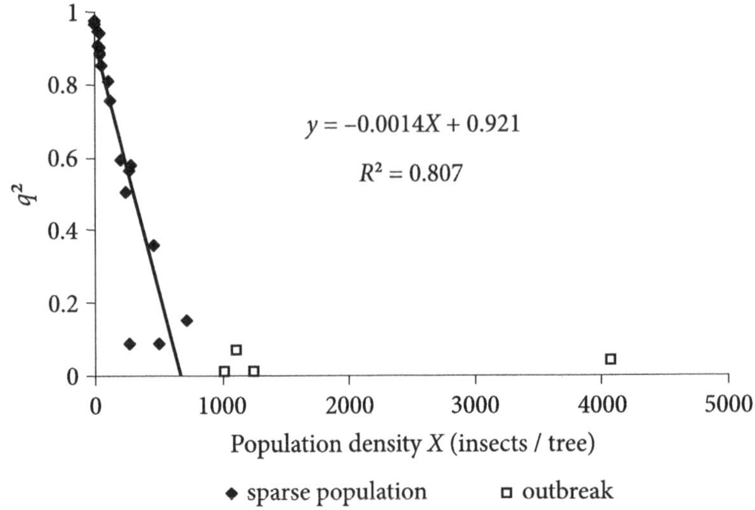

Fig. 1.8. Relationship between parameter q^2 and the density of *Dendrolimus superans sibiricus* Tschetv. populations in different phases of the outbreak cycle

Compare model (1) with the Poisson model describing the relation between the relative and absolute estimates of insect colonization and based on the assumption that all sample units are similar in their properties, individual insects choose their food randomly, and the interaction between individuals is weak if any. In this case, the distribution of insects among sample units can be described by the Poisson model. According to this model, the portion $q = 1 - A$ of sample units without insects (*i.e.* the analog of the order parameter in model (1)) is expressed with the following formula:

$$q = 1 - A = \exp(-X). \tag{7}$$

It follows from (7) that at $X \to 0$, *i.e.* under low population density, $q \to 1$, that is, there are very few sample units with insects in them. At $X \to \infty$, $q \to 0$, *i.e.* almost all sample units are colonized by the insects. It also follows from (7) that the type of relation between relative and absolute colonization must be similar for all insect species, as in (7) there are no species-specific constants.

According to (7), as the population density grows relative colonization of tree stands increases rapidly. At $X = 3$ individuals / sample unit, $q \approx 0.05$ and $A \approx 0.95$. Obviously, if the distribution of insects among sample units is random, at a population density above 3 insects per sample unit there will be very few sample units without insects on the survey plot.

Thus, analysis of survey data on the relation between the relative and absolute colonization by insect populations can provide a basis for choosing between model (1) and model (7). If the results of the analysis suggest the equation of the relation between the relative and absolute colonization is significantly different from the Poisson model (7), this will be indicative of interactions between insects and their host plants as well as between phytophagous insects. One manifestation of these interactions is cooperative colonization of trees. Parameters of these interactions can be estimated from the values of constants b and X_r of model (1). Statistical analysis showed that the survey data disagree with calculations based on model (7). Both survey data and results of calculations based on model (1) suggest that for the Siberian silk moth, population density of 3 insects per tree corresponds to relative colonization of about 4% of the trees and for the pine looper population - about 45% of the trees. Calculations based on model (7) suggest that insects must be present on 95% of the trees.

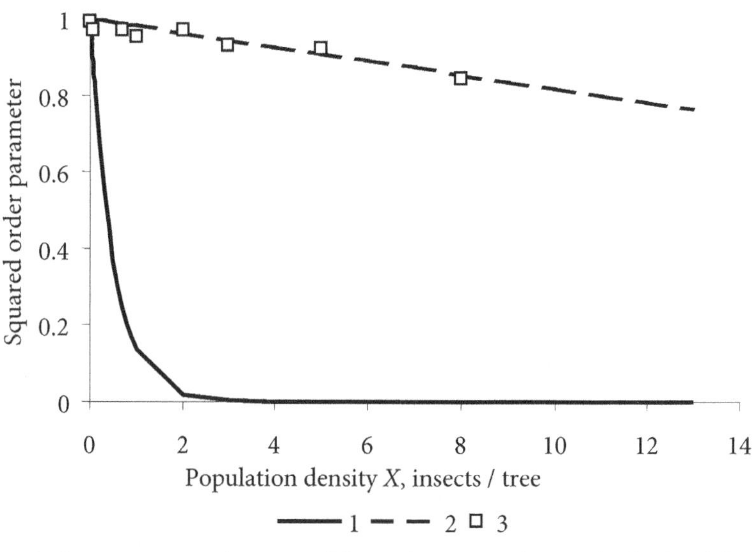

Fig.1.9. Order parameters determined from the survey data and calculated from models (1) and (7) for insects of the order *Homoptera* in the forests of the Yemelyanovskii District of the Krasnoyarsk Territory.
Note: 1, calculations from model (7); 2, calculations from model (1); 3, survey data

Figure 1.9 shows survey data and model results of the squared order parameter calculated from models (1) and (7) to estimate the distribution of *Homoptera* insects (*Lepyronta coleopterata* L., *Philaenus spumarius* L.,

Cinara nuda Mordv., *C. hyperophila* Koch.) on trees of middle-taiga plain forests of the Krasnoyarsk Territory (Bulanova *et al.*, 2008).

Results of the model of phase transitions (1) are in good agreement with the survey data, whereas results of the Poisson model (7) significantly disagree with field measurements. Table 1.1 compares values of relative colonization of the trees found using survey data and calculated from models (1) and (7), based on the highest values, $X(max)$, of absolute colonization by insects observed during surveys.

Table 1.1. Relative colonization of trees estimated using survey data and models (1) and (7)

Order of insects	X(max)	Relative colonization A		
		survey data	model (1)	model (7)
Heteroptera	0.24	0.14	0.139	0.213*
Coleoptera	0.25	0.21	0.219	0.221
Lepidoptera	0.46	0.37	0.371	0.369
Hymenoptera	0.1	0.10	0.101	0.095
Homoptera	8.7	0.08	0.084	1.000*

Note: * the value is significantly ($p = 0.95$) different from the survey data

The survey data and results of calculations using model (1) are very similar. On the other hand, the calculation of relative colonization based on the Poisson model (7) is in some cases significantly different from survey data, and the higher the values of absolute colonization the greater the differences between the calculations of relative colonization based on model (7) and the survey data. At low population densities, calculations from both model (1) and model (7) are in good agreement with the survey data. This good agreement may be accounted for by the fact that at low population densities ($X < 0.5$), the exponent in the right-hand side of (1) can be expanded into Taylor series to within a linear term. Then from (7) we obtain

$$q^2 \approx 1 - 2X. \tag{8}$$

This may suggest that at low population densities, both models predict a linear relationship between relative colonization and population density. As the population density grows, expansion (8) becomes incorrect, and the Poisson model gives too high calculated values of relative colonization, while model (1) remains more correct even at high densities.

It is evident from the data just presented (Figures 1.7–1.9, Table 1.1) that parameters characterizing the relation between the relative and absolute colonization by insects (model (1)) are species specific and vary both between different groups of insects and within one systematic group in different regions. Model (7) is less sophisticated. Our analysis showed that calculations based on interpretation of the insect outbreak as a second-order phase transition suggest a good agreement between the model of phase transitions and field data.

1.2.3 *The Effect of Modifying Factors on the Development of an Outbreak*

A modifying, *i.e.* density-independent, factor, in terms of phase transitions, can be treated as superposition of external field onto the system. For population dynamics and development of outbreaks, the major modifying factor, which significantly affects population increase and development of an outbreak, is ambient temperature. If temperature is regarded as constant time-independent field *T*, we assume that elevation of the environmental temperature will decrease the ecological risk potential. Then, equation (1) can be rewritten as follows (Dmitriev, 2004)

$$G = G_0 + Aq^2 + Bq^4 - Vq\grave{O}, \tag{9}$$

where *V* is a certain constant.

It is clear from (9) that in this case, the transition of the population to the outbreak state will not occur at $q = 0$, *i.e.* during the peak phase not all the trees will be infested by pests.

In this case, no discrete phase transition will occur, and in the external temperature field the transition is diffuse and can take place within a certain interval of pest population densities.

The parameter characterizing the effect of the external field on the order parameter that we use is the sensitivity value $\chi = \frac{\partial q}{\partial h}$. It follows from (9) that

$$\chi = \frac{\partial q}{\partial h} = \frac{k}{2a(X - X_r) + 12Bq^2}. \tag{10}$$

Substituting equilibrium values of the order parameter in the absence of the field into (10), we obtain the expressions for sensitivity to the effect of the external field in the ranges above and below the critical population density X_r:

$$\chi = \frac{k}{4a(X_r - X)}, \quad \textit{if} \quad X < X_r; \quad \chi = \frac{k}{2a(X - X_r)}, \quad \textit{if} \quad X > X_r. \tag{11}$$

It is clear from (11) that on approaching the critical density X_r, sensitivity tends to infinity. At both very low and very high population densities, the effect of the external field on changes in population density is weak; sensitivity increases dramatically close to the critical density X_r. However, according to (11), the sensitivity to the effect of external modifying factors on the sparse stable population is twice lower compared with the population in the outbreak phase.

We assume that the effect of the external temperature field is not significant, *i.e.* the condition $Aq^2 >> VqT$ is fulfilled. Let us find the value of the order parameter q at which the minimum of function (9) is attained:

$$\frac{\partial G}{\partial q} = 2a(X - X_r)q + 4Bq^3 - VT = 0 \qquad (12)$$

or

$$a(X - X_r)q + 2Bq^3 = VT. \qquad (13)$$

At X > X_r and $T \to 0$, the order parameter $q = \overline{q}$, when all trees are colonized by pests and an outbreak occurs, is equal to 0. At $T \neq 0$, the value of $\overline{q}$ is found from cubic equation (13). The solution of (13) is graphically shown in Figure 1.10.

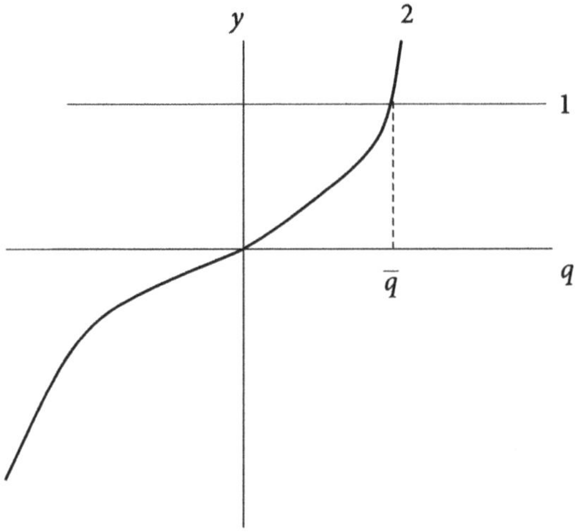

Fig. 1.10. The graphic solution of equation (13)
Note: 1, $y_1 = VT$; 2, $y_2 = a(X - X_r)q + 2Bq^3$

Figure 1.10 shows that the effect of the temperature causes the value of the order parameter q to increase when the outbreak takes place.

At $X < X_r$, we obtain from (13):

$$-a(X - X_r)q + 2Bq^3 = VT. \quad (14)$$

The graphic solution of (14) is shown in Figure 1.11.

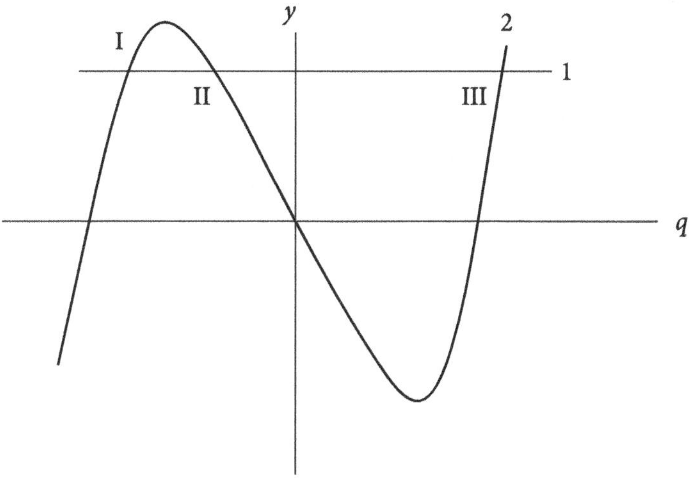

Fig. 1.11. The graphic solution of equation (14)
Note: 1, $y_1 = VT$; 2, $y_2 = -a(X - X_r)q + 2Bq^3$

Figure 1.11 shows that there are three solutions to equation (14). However, the only solution that corresponds to the maximum of function G is II. The positive root of III characterizes colonization of trees by sparse stable population. Note that the condition of the slightness of the effect exerted by the field is necessary for the condition $q \leq 1$ to be fulfilled.

Thus, the effect of the temperature on the insect population in the examined model corresponds to the situation when the transition to the outbreak phase does not result in colonization of all sample units (plots, trees) by pests.

1.2.4 *The Impact of Chemical Compounds on Biological Objects*

A model of a second-order phase transition can be used to evaluate the effect of chemical compounds on biological objects. Typically, these exposures are characterized by a dose of LD_{50} exposure to a chemical substance that kills half the objects, or a dose of LD_{100}, which kills all objects under the influence of LD_{50}.

If we consider the effect of an external factor (chemical substance) from the point of view of the second-order phase transition model, then the relationship between the order parameter q characterizing the state of the biological object and the concentration C of the chemical substance will be described by the following equation:

$$q^2 = \begin{cases} 0, & C > C_r, \\ 1 - BC, & C \leq C_r, \end{cases} \tag{15}$$

where q is the proportion of biological objects (for example, tumor cells) on which the chemical compound is exposed, $C_r = LD_{100}$ is the dose of exposure at which all biological objects affected by the drug die.

The effect of impact of the chemicals on tumor cells in tissue culture is shown in Figures 1.12 and 1.13 based on the data presented in (Severin *et al.*, 2013):

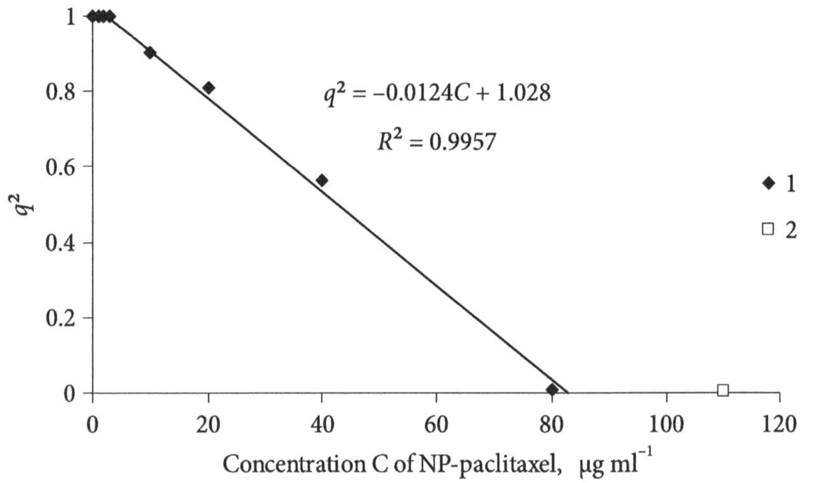

Fig. 1.12. The effect of the chemical preparation paclitaxel included into biodegradable nanoparticles (NP-paclitaxel) on the survival of tumor cells in tissue culture (according to Severin *et al.*, 2013)
Note: NP, nanoparticles; 1, the phase of partial death of tumor cells; 2, the phase of complete death of tumor cells

As can be seen from Figure 1.12, the death of tumor cells under the influence of paclitaxel included into biodegradable nanoparticles (NP-paclitaxel) is very well described by equation (15). The dose above $C_r = LD_{100} = 82.9$ µg ml^{-1} = 1.028 / 0.0124 leads to "overkilling" – an overabundance of tumor cells. The traditional dose of LD_{50} at $q^2 = 0.25$ is easily calculated from equation (15):

$$0.25 = 1.028 - 0.0124\, LD_{50}. \tag{16}$$

Hence LD_{50} = 62.74 μg ml^{-1}. Since the parameters of equation (15) can be found from the data of experiments on the effect of small doses, in assessing the risk of exposure of chemicals to biological objects, it is possible to perform experiments on the effect of small doses on the survival of the studied objects.

One can compare the effect of different preparations on the objects under study by the parameters of equation (15). The relationship between q^2 (quadrate of the proportion of living cells in tissue culture) and the impact of the drug paclitaxel included in the biodegradable nanoparticles associated with recombinant receptor-binding fragment of alpha-fetoprotein (NP-paclitaxel-rAFP-3D), is shown in Figure 1.13. For this chemical preparation LD_{50} = 20.10 μg ml^{-1}, and LD_{100} = 26.84 μg ml^{-1}. Comparing the doses for the two drugs, it can be concluded that the second drug NP-paclitaxel-rAFP-3D has a significantly greater efficacy of exposure than NP-paclitaxel.

Similar results can be obtained by analyzing the impact of herbicides on plants. The results of experiments on the impact of herbicide 2,4-dichlorophenoxyacetic acid (2,4-D) on plant seedlings are shown in Figure 1.14.

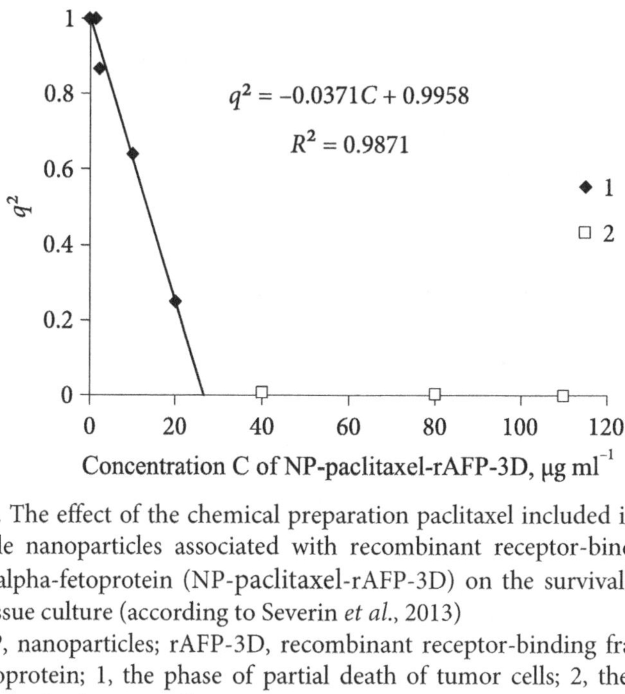

Fig. 1.13. The effect of the chemical preparation paclitaxel included in the biodegradable nanoparticles associated with recombinant receptor-binding fragment of alpha-fetoprotein (NP-paclitaxel-rAFP-3D) on the survival of tumor cells in tissue culture (according to Severin *et al.*, 2013)

Note: NP, nanoparticles; rAFP-3D, recombinant receptor-binding fragment of alpha-fetoprotein; 1, the phase of partial death of tumor cells; 2, the phase of complete death of tumor cells

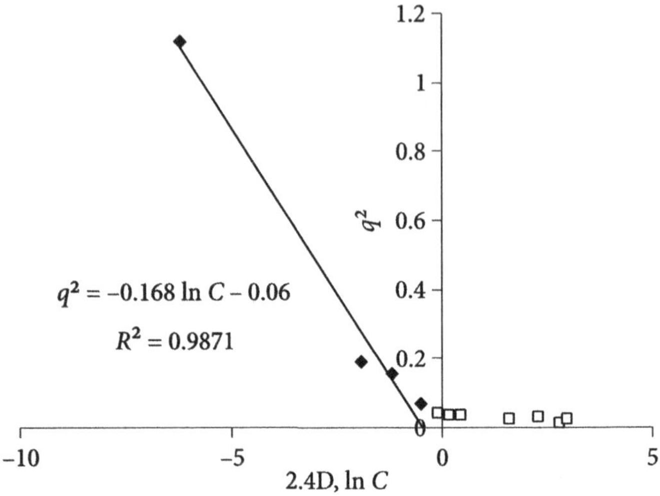

Fig. 1.14. Impact of herbicide 2,4-dichlorophenoxyacetic acid (2,4-D) in concentrations C on the survival rate of q plant seedlings

Apparently, in this case too, the second-order phase transition model describes the impact on biological objects very well. LD_{100} equals 0.70 μg ml^{-1} for the used herbicide.

Thus, the second-order phase transition model can be used to analyze the impact of various chemical compounds on biological objects. It seems that, when evaluating the impact of salts in small concentrations on plant growth, it is possible to obtain species-specific LD_{100} indicators that will characterize the impact on plants.

References

Bazykin A.D., Berezovskaya F.S., Isaev A.S., Khlebopros R.G. Dynamics of forest insect density: bifurcation approach, *J. Theor. Biol.*, 1997, Vol. 186, No. 3, p. 267–278.

Bogolyubov N.N., Krylov N.M. *Nonlinear Mechanics, 1932–1940,* Mitropolsky Yu.A., Sukhanov A.D. eds., *Mathematics and Non-Linear Mechanics,* Vol. 2, Nauka, Moscow, 2005, 828 p. [in Russian]

Bruce A.D., Cowley R.A. *Structural Phase Transitions,* Taylor & Francis Ltd, London, 1981, 376 p.

Bulanova O.S., Borisova E.V., Soukhovolsky V.G., Tarasova O.V. The spatial distribution of insects during formation of young pine forests on fallow lands, *Vestnik KrasGAU (Bull. Krasnoyarsk State Agrar. Univer.*), 2008, Vol. 4, p. 167–172. [in Russian]

Dmitriev A.V. *Fundamentals of Statistical Physics of Materials*, Publ. Moscow State Univer. - Nauka, Moscow, 2004, 668 p. [in Russian]

Freiberg I.A., Ermakova M.V., Tolkach O.V. Phenotypic changes in pine seedlings under the influence of pesticides, *Lesovedenie (Russian Forest Science)*, 1997, No. 3, p. 61–69. [in Russian]

Fursova P.V., Levich A.P., Alexeev V.L. Extremal principles in mathematical biology, *Advances Modern Biol.*, 2003, Vol. 123, No. 2, p. 115–137. [in Russian]

Isaev A.S., Khlebopros R.G. Principle of stability in the population dynamics of forest insects, *Proc. Acad. Sci. USSR, Biol Sci.*, 1973*a*, Vol. 208, Sect. 6, p. 225–227.

Isaev A.S., Khlebopros R.G. Population ecology: Fighting a forest pest, *Nature*, 1973*b*, Vol. 241, No. 5391, p. 504.

Isaev A.S., Khlebopros R.G. Analysis of the dynamics of forest insect numerical population based on the principle of the stability of mobile ecological systems, *J. General Biol.*, 1974, Vol. 35, No. 5, p. 737–745. [in Russian]

Isaev A.S, Khlebopros RG, Kondakov YP Laws of the population dynamics of forest insects, *Russian Forest Sci.*, 1974, Vol. 3, p. 27–42. [in Russian]

Isaev A.S., Khlebopros R.G., Nedorezov L.V., Kondakov Y.P., Kiselev V.V., Soukhovolsky V.G. *Forest Insect Population Dynamics*, Nauka, Moscow, 2001, 374 p. [in Russian]

Khlebopros R., Okhonin V., Fet A. *Catastrophes in Nature and Society: Mathematical Modeling of Complex Systems*, Word Scientific Publishing Co, Singapore, 2007, 336 p. http://www.ebooks-share.net/catastrophes-in-nature-and-society-mathematical-modeling-of-complex-systems

Kolmogorov A. Sulla teoria di Volterra della lotta per l'esistenza, *Giornale dell' Istituto Italiano degli Attuari*, 1936, Vol. 7, No. 1, p. 74–80.

Kolmogorov A.N. 1A qualitative study of mathematical models of population dynamics, *Problems of Cybernetics*, Nauka, Moscow, 1972, Vol. 25, p. 101–106. [in Russian]

Kolmogorov A.N., Petrovskii I.G., Piskunov N.S. The study of the diffusion equation which is connected with an increasing of the amount of matter and its application to a biological problem, *Bull. MSU, Ser. A*, 1937, Vol. 6, p. 1–26. [in Russian]

Kondakov Y.P. Outbreaks of the Siberian silk moth, *Ecology of Populations of Forest Animals in Siberia*, Nauka, Novosibirsk, 1974, 206–265. [in Russian]

Landau L.D., Lifshits E.M. *Statistical Physics,* 1964, Nauka, Moscow: 567 p. [in Russian]

Lyapunov A.M. *Collected Works* (1954–1959*)*, Publ. Acad. Sci. USSR, Vol. 1, Moscow, 1954, 446 p.; Vol. 2, Moscow–Leningrad, 1956, 472 p.; Vol. 3, Moscow, 1959, 374 p. [in Russian]

Palnikova E.N., Sviderskaya I.V., Soukhovolsky V.G. *The Pine Looper in Siberian Forests. Ecology, Population Dynamics, Impact on Tree Stands,* Nauka, Novosibirsk, 2002, 254 p. [in Russian]

Poincaré A. *On Curves Defined by Differential Equations,* Gostekhizdat, Moscow–Leningrad, 1947, 392 p. [Russian translation]

Severin S.E., Moskaleva E.Yu., Posypanova G.A. Target therapy of cancer, *Priroda (Nature)*, 2013, No. 12, p. 71–77. [in Russian]

Soukhovolsky V.G., Iskhakov T.R., Tarasova O.V. *Optimization Modeling of Interactions between Populations,* Nauka, Novosibirsk, 2008, 168 p. [in Russian]

Soukhovolsky V.G., Palnikova E.N., Tarasova O.V., Karlyuk A.Y. A model of forest insect outbreak as a second-order phase transition, *Dokl. Biochem. Biophys.*, 2005, Vol. 403, No. 1–6, p. 297–299.

Tikhonov A.N. About the stability of inverse problems, *Dokl. AN SSSR,* 1943, Vol. 39, No. 5, p. 195–198. [in Russian]

Volterra V. Variations and fluctuations of the number of individuals in animal species living together, *Animal Ecology,* Chaptman R.N. ed., McGraw-Hill, New York, 1931, 409–448.

2
Criticality Concept and Some Principles for Sustainability in Closed Biological Systems and Biospheres

Nicholas P. Yensen[1], Karl Y. Biel[2,3,*]

[1]NyPa International, Tucson, Arizona 85705, USA; Biosphere Systems International,
Tucson, Arizona 85705, USA
[2]Institute of Basic Biological Problems, Russian Academy of Sciences, Pushchino, Moscow Region 142290, Russia
[3]Biosphere Systems International Foundation, Tucson, Arizona 85755, USA

...modeling represents the sole means available to predict possible changes in the biosphere's state resulting from humanity's large-scale influence on it, – Academician Iosif Gitelson (cited by Gitelson *et al.*, 2003, Manmade Closed Ecological Systems)

Abstract
This survey of closed systems offers nomenclature, presents a new construct of criticality, describes the development of closed systems, and suggests future directions in the new science of ecobiospherics. Closed-system life-containing spheres are called biospheres and such systems with two or more biomes are named Biospheres (capitalized). Microbiospheres (smaller than humans) are classified as: biocappillaries, bioculturetubes, biopetris, biobottles, Folsome-Hansen flasks, Ecospheres® and terraspheres, bioboxes, Wardian cases. Macrobiospheres

[*] *Corresponding author*: karlbiel@hotmail.com

(larger than humans) are: Biotubes 1 to 8, CELSS, ALS, BPC, LMLSTP I, II and III, Biohome, Bio-Plex and ALSSIT, LSSIF, MELISSA, CEEF, Test Module, Bios-1, -2, and -3, Biosphere 2 Laboratory, Lunar and Martian bases, Earth [Biosphere 1]. Criticality is defined as a relatively rapid decrease in the ratio of organic material to inorganic forms within a system. Time to criticality is inversely related to initial organic content as a hyperbolic function which asymptotically approaches some minimum positive value. In general, the natural logarithms of time to criticality and volume are positively correlated, *i.e.* microbiospheres of one milliliter with fewer than 100 life forms reach criticality within 48 hours, whereas macrobiospheres may require months to years. Volumes of greater than 10 milliliters (bioculturetubes / biopetris, Ecospheres, *etc.*) may be necessary to achieve perpetual "sustainability". The artificial macrobiospheres tested to date have not demonstrated sustainability for longer than a "qualified" duration of months. Microbiospheres can provide the principles to design sustainable human systems. These principles will provide the basis for sustainable space exploration systems and for the Earth's Biosphere. The Bios systems (Krasnoyarsk, Russia), Biosphere 2 Laboratory (Tucson-Oracle, USA) and similar laboratories provide the best opportunity to understand and regulate sustainability in artificial human systems and to modulate the anthropogenic influenced Earth behaviour such as oxygen depletion and Global Climate Change.

Keywords: Ecobiospherics, Artificial closed biological systems, Biosphere 1, BIOSes, Biosphere 2 Laboratory

2.1 Introduction

Researchers, however, may be able to use small closed ecosystems to push the limits of stability to produce data upon which mathematical models may be tested and to accelerate our acquisition of knowledge in biospherics (Khlebopros *et al.*, 2007). Remarkably, relatively little effort has been expended in this direction, especially compared to the extensive studies on open ecosystems. Even compared to mathematical modeling, small closed-ecosystem testing has received far less attention. Professor Iosif I. Gitelson and colleagues (Gitelson *et al.*, 2003) agree that the use of small closed ecosystems can form a basis for mathematical modeling.

The typical ecosystem study is of an open system which, due to its interconnectedness with other ecosystems, has limited and possibly misleading paradigms for making global predictions. Because of the complexity of the Earth's Biosphere, it is comprehensible that we, as individuals, are all blind to many aspects of the world. Metaphorically, the study of open "global" ecosystems is like three blind men describing the elephant's trunk as a "snake", the leg as a "tree", and the ear as a "leather curtain". But we, as a race, are still remarkably blind to the functional properties of closed ecological systems, not the least of which is the planet Earth's closed ecological system, upon whose function depends the lives of all known species. Although one might think the evidence is clear, scientists disagree about the effects of: increased carbon dioxide, acid rain, the ozone hole, over fishing, and the cutting down of forests, to name a few of the perceived global anthropogenic phenomenon.

The study of small, closed ecosystems is an ideal method to investigate *whole integrated ecosystem interactions* and thus to discover general principles. At present, few principles of these interactions are known even though the concept of a closed ecological system has been around for over 150 years.

The following section considers some aspects of closed system history and descriptions of existing apparatuses. It also offers our suggested classification of closed systems. Further, it provides a "snapshot in time" of the evolving closed system terminology, including our concept of criticality. With respect to planet Earth, and to man-made closed systems, projections and suggestions are made herein as to how humanity may proceed to better understand our potential for sustaining the Earth's Biosphere; and how miniature human-oriented Biospheres and closed systems may be developed for colonizing space, planets and beyond.

2.2 History of Manmade Closed Ecosystems

In 1827 a British physician, Nathaniel Ward, while studying cocoons under glass with soil, serendipitously discovered that ferns and other plants sprouted and grew vigorously under glass. Outside of the glass in the industrially polluted London air, however, these plants could not survive. Dr. N. Ward realized that he had discovered a way to grow plants sensitive to environmental pollutants. After some trials with small glass jars he found the right combinations of light, air, humidity and moisture to support various plants. He then developed "closed-system" cases large enough for houseplants. These became known as ***Wardian cases***. With popularity, the cases became very ornate and elaborate, often mimicking architectural styles. In commerce, it was found that tea plants, rubber plants and exotic plants, when inside the cases, could be shipped to other parts of the world. This discovery resulted in the Indian tea and rubber industries. In the 1970's environmental awareness, as exemplified by Earth Day, Rachel Carson's *Silent Spring*, "flower children", and the hippie back-to-nature movements, brought a renewed interest in Wardian cases leading to the modern terrarium. The 1800's aquarium-sized and bottle-sized Wardian cases represented some of the first "closed" systems, although Wardian-case ecosystems were not generally hermetically sealed and were constructed principally for their attractiveness (see Arthurs, 1975; Michie, 2004).

Dr. R. Warington (1851) was perhaps the first to describe a sealed "*balanced aquatic microecosystem*" type microsphere. But it wasn't until the second half of the 20th century that more serious work began on microcosms. Much of this work, however, still dealt with microcosms of differing degrees of closure for various experiments (Odum and Hoskin, 1957; Myers, 1958; Beyers, 1962*a*, 1962*b*; Golueke and Oswald, 1964; and various others).

Investigations of materially sealed ecosystem microcosms, *i.e.* the truly-closed ***microbiospheres***, *smaller than a human, hermetically-sealed biological systems*, were started and pursued simultaneously and independently in both the USA (Folsome and Hanson, 1986) and in Russia (USSR) (Kovrov, 1992). Some of the sealed microbiospheres created in the 1970's are still alive, including unpublished systems created in 1973 by K. Biel and by N. Yensen (per. obs.). Algal and microorganism-containing microbiospheres were studied by Drs. B.G. Kovrov and G.N. Fishtein (1980) and Drs. C. Folsome and J.A. Hanson (1986). Also, Dr. Bassett Maguire's (1980) experimental studies of macroscopic organism in one-liter microsphere systems produced some of our most rudimentary knowledge of closed systems. In 1985, based on the NASA Folsome-Hansen flask (see below), N. Yensen, D. Harmony and L. Acker developed the synthetic Ecosphere® microcosm by combining species that would not normally occur together in nature into glass spheres to produce a sustainable small shrimp ecosystem (Yensen and Biel, 2005).

The first closed ecosystems were made as curiosities for their aesthetic qualities, but, through the transport of important commercial plants, had a significant impact on civilization. Even today, most closed systems are still constructed by individuals and scientists mostly out of curiosity and as aesthetic objects. It now appears that a myriad of formats is possible, despite the former scientific paradigm that, "*a sustainable closed ecosystem smaller than planet Earth's Biosphere was not possible*".

Today, there are several closed system types to choose from when planning an experiment (Table 2.1). Closed living systems in this classification may be defined according to their closure, size and shape.

2.3 Classification of Closed Biological Systems

2.3.1 *Terminology*

The following terminology is based on the literature, our derivations, and an amalgamation of terms, which previously were at times both redundant and contradictory or used without precision, for instance, *biosphere*, *closed system*, and *microcosm*. We have followed historical precedent and / or the root meanings of the words. Certainly, closed system terminology is not static and will continue to evolve and grow. The omission of deserving terms is not intentional, as the lexicon is far from comprehensive.

"Truly-closed biosystems" are more properly called ***materially closed biological systems.*** These systems *are relatively closed regarding mass, but not to information transfer and / or electromagnetic energy*, such as x-rays, light, heat, or magnetic flux, notwithstanding cosmic rays, neutrinos or virtual particles which, with respect to closure, do not significantly change the mass. For example, at the Biosphere 2 Laboratory a loss of 6–10% of the atmosphere per year was still considered a materially closed system, even though some studies suggest that a 1% change of mass per year can significantly alter a closed ecosystem.

Table 2.1. Classification of closed living systems

SYSTEM CLOSURE / *SYSTEM SIZE*	*MATERIALLY CLOSED*	*SEMI-CLOSED*	
		Periodic	*Organismic*
MICROBIOCOSMS (~MICROECOSYSTEMS smaller than humans)	*MICROBIOSPHERES* Biocappillaries Bioculturetubes Biopetris Biobottles Folsome–Hansen flasks Ecospheres® and terraspheres Bioboxes Wardian cases C.E.B.A.S.	Tide pools Animal guts Aquaria Terraria	Petri dishes Plant bowls Aquaria Terrariums
MACROBIOCOSMS (~MACROECOSYSTEMS larger than humans)	*MACROBIOSPHERES* Biotubes 1-8 CELSS, ALS, BPC LMLSTP I, II, and III Mars Base 0 Biohome Bio-Plex & ALSSIT "LSSIF" MELISSA, CEEF Bios-l, -2, and -3 Test module B2L Lunar and Martian bases Extraterrestrial planetoid Earth (Biosphere 1)	Space stations Arctic stations Antarctic stations Undersea habitats Submarines	Zoos Oceanaria Ponds Lakes Habitat enclosures Reserves Refuges Bionics

***A biological system** may be an open, closed or semi-closed life-containing system.* Semi-closed systems may be defined according to having: extended or periodic closure; single species, multiple species; and / or only closure to specific organisms (organismic closure).

Ecosystems are generally specific sets or subsets of known biological systems that have the capacity to pass materials through various organisms. An important subset of ecosystems is the ***sustainable ecosystem**, which has the capacity to recycle sufficient material to maintain the ecosystem indefinitely.*

We define the term "***microcosm***" *as any system too small for a human to be inside,* and "***macrocosm***" *as any system large enough for one or more humans to be inside.* The terms "***microecocosm***" and "***macro-ecocosm***" *refer to ecological systems smaller and larger than humans respectively.*

Known systems are categorized according to system size (microcosm = smaller than a human; macrocosm = large enough for humans) and with respect to closure (semi-closed = either closure to organisms or periodically open to movement of material and / or organisms; and materially closed = generally considered < 1–10% leak rate per annum, but open to electromagnetic energy and information).

2.3.2 Micro Systems

Microbiocosms *are ecosystems smaller than humans that are organismically closed, wherein the organisms may not leave the confines of the system.* The term microbiocosm applies to closed and semi-closed biological systems smaller than humans. For instance, aquaria habitats may be considered forms of life containing microcosms or microbiocosms. In a microbiocosm energy and material (principally gas and liquids) may (or may not) move into or out of the system without violating organism closure.

1. ***Aquaria*** *are aquatic microcosms or macrocosms which when open to gas exchange form open microbiocosm or macrobiocosm systems.* As such, they may be specially classified according to size, maintenance and the community or ecosystem that they contain.
2. ***Terraria*** *are solid substrate containing microcosms or macrocosms which when open to gas exchange form open microbiocosms or macrobiocosms and range in size from small glass jars to containers large enough for trees.*

Microbiospheres *are materially closed microbiocosms smaller than humans.* Microbiospheres may be constructed from various materials but typically of glass and / or plastic to allow for photosynthetically active radiation light penetration and are named for their principal form or aspect. Microbiospheres are of particular interest due to their potential to statistically elucidate concepts and principles of closed systems within short time frames. For example, 2–6 days in capillary microbiospheres (Photo 2.1) is the equivalent of 300–1000 Earth years (Ortiz, Vender, Adami *et al.*, 2005) with respect to criticality (criticality is the rapid change of organic matter to inorganic matter; see criticality below; see also Yensen and Biel, 2005).

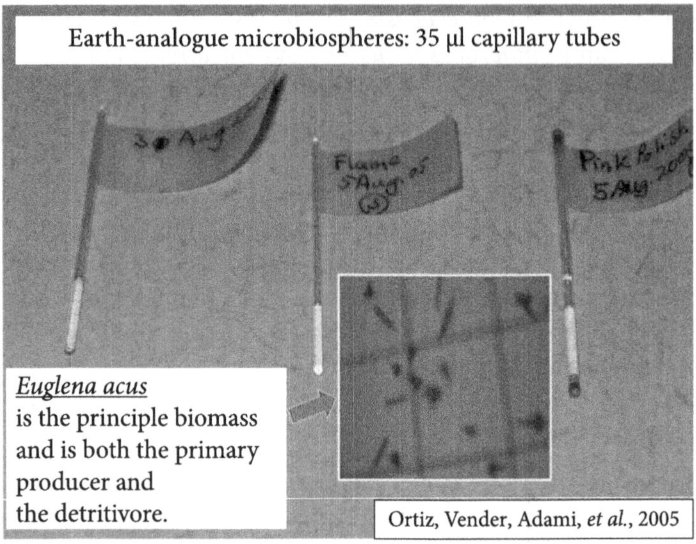

Photo 2.1. Earth-analogue microbiospheres (modified from Ortiz, Vender, Adami, *et al.*, 2005)

The following briefly defines and describes several different kinds of microbiospheres.

1. ***Biocapillaries*** *are microbiospheres made of glass and / or plastic capillary tubes, with microscopic organisms sealed inside.* Biocapillaries may have their ends sealed with a matrix such as clay or the tube ends may be melted closed for extended observations.

2. ***Bioculturetubes*** *are microbiospheres made glass and / or plastic culture tubes, are closed with either a stopper or hermetically sealed by melting, and contain microscopic and / or macroscopic organisms.* Bioculturetubes may be of various diameters and lengths allowing control over volume and varying substrate: gas interface ratios. Neophytes should be forewarned that well-sealed glass systems might develop sufficient internal pressure to destroy the system by explosion due to decomposition of excessive organic material.

3. ***Biopetris*** *are microbiospheres made by sealing Petri dishes.* They have the advantages of: easy replication, high surface to volume ratio, ease for observation, and ease for population counts. Biopetri systems may vary from simple (one species) to complex (hundreds of species), and typically have a limited lifetime depending on the cultured organism(s) and the nutrient substrate, typically an agar media.

4. ***Biobottles*** *are microbiospheres made from lid- or stopper-sealed bottles with micro and / or macroscopic organisms.* Biobottles may be inexpensive canning jars with lids or special ordered bottles, either of which may be hermetically sealed. Again, experimenters should be forewarned that well-sealed glass systems with excessive decomposing organic material might explode.

5. ***Folsome–Hansen flasks*** *are microbiospheres made of natural ecosystems subsets by the sealing of laboratory flasks containing microscopic and / or macroscopic organisms.* Folsome–Hansen flasks are named after Dr. Claire Folsome of the University of Hawaii and Dr. Joe Hansen of NASA's Jet Propulsion Laboratory in Pasadena. These systems were constructed by Drs C. Folsome and J.A. Hanson by going to a tide pool or beach and thereby attempting to acquire enough of the natural ecosystem to result in a sustainable closed ecosystem.

6. ***Ecospheres*** *are microbiospheres made of glass spheres or elongate "pods" containing brackish water and a sustainable, synthetic ecosystem of small shrimp, algae and microbes.* Ecospheres are a commercial product developed in Tucson, Arizona (USA) under a NASA technology transfer of a Folsome–Hansen flask protocol.

7. ***Aquaria*** *are aquatic microcosms or macrocosms which may be sealed to form microbiospheric or macrobiospheric systems.* As such, they may be specially classified according to size, maintenance and the community or ecosystem that they contain.

8. ***Bioboxes*** *are microbiospheres generally of a "rectangular-solid" shape.* Bioboxes provide an ideal system for investigating short-term effects of system dynamics, such as atmospheric changes, water recycling and growth rates for crop or other plants.

9. ***Terraspheres*** *are solid substrate containing systems which are sealed to form microbiospheric or macrobiospheric systems.* Terraspheres may range in size from small glass spheres to large enclosures containing soil, sand, rocks and / or terrestrial plants and animals in a closed, ecosystem typically of algae, mosses, ferns, bryophytes, other small plants, fungi and microbes.

10. ***Wardian cases*** *are terraspheres which may not be tightly sealed and range in size from small glass jars to containers large enough for trees.* Wardian cases were developed by Nathaniel Ward in 1827 to protect plants from ambient air contamination and stress.

2.3.3 *Macro Systems*

Macrobiocosms *are ecosystems larger than humans that are organismically closed, wherein specified organisms may not leave the confines of the system.* For instance, Zoos may be considered forms of macrocosms. The following delimits some of the characteristics of macrocosms and macrobiospheres.

1. ***Zoos*** and ***live animal "museums"*** *are terrestrial and / or aquatic macrocosms.* As such, they may be specially classified according to size, maintenance and the community or ecosystem that they contain. They are usually intended for a single or few species with an external food source, they may be considered semiautecological macrocosms which do not have the support of a larger synecological system. And further, because they typically allow the exchange of insects and other organisms, they are semiorganismically-closed.

2. ***Aquaria*** *are aquatic macrocosms or microcosms which may be sealed to form microbiospheric or macrobiospheric systems.* As such, they may be specially classified according to size, maintenance and the community or ecosystem that they contain.

3. ***Terraria*** *are solid substrate containing macrocosms or microcosms which range in size from small glass jars to containers large enough for trees.*

Macrobiospheres *are ecosystems larger than humans that are organismically closed, wherein specified organisms may not leave the confines of the system.* For instance, Zoos may be considered forms of macrocosms. The following delimits some of the characteristics of macrocosms and macrobiospheres.

1. ***Wardian cases*** *are solid substrate containing macrocosms or microcosms which may be sealed to form microbiospheres or macrobiospheres and range in size from small glass jars to containers large enough for trees.* Wardian cases were developed by Nathaniel Ward in 1827.

2. ***Terraspheres*** *are solid substrate containing systems which are sealed to form macrobiospheric or microbiospheric systems.* Terraspheres may range in size from small glass spheres to large enclosures containing soil, sand, rocks and / or terrestrial plants and animals in a closed, ecosystem typically of algae, mosses, ferns, bryophytes, other small plants, fungi and microbes.

3. ***Humanobiospheres*** *are macrobiospheres constructed with the intent of supporting humans.* Their principle conceptual function is for providing life support in the colonization of space, submarines and Antarctic habitats. They have been constructed in various countries. Some examples are: Biotubes 1-8; CELSS, ALS, BPC; LMLSTP I, II and III; Mars Base 0; "Biohome"; Bio-Plex and ALSSIT; "LSSIF"; MELISSA; CEEF; The "Test module"; "Bios-1, -2, and -3; and Biosphere 2 Laboratory [our apologies to those not listed]. The future will undoubtedly include humanobiospheres of Lunar and Martian bases and perhaps extraterrestrial planetoids such as asteroids or comets.

4. ***Biospheres*** *are macrobiospheres that contain two or more biomes (major ecosystem units with distinctive life forms, e.g., rain forest, marsh, savannah, desert, tundra, steppe, etc.).* Presently, there are only two, the planet Earth (Biosphere 1) and Biosphere 2 Laboratory. Biosphere 1, of course, represents a unique closed system for humans and all the life systems that we presently know about.

2.3.4 The Term Biosphere

In the scientific literature, the term "*Biosphere*" was first coined in 1875 by an Austrian geologist, E. Suess (1875; 1909). The term was used to refer to the "inhabited" part of the Planet, *i.e.* to describe *the range limits to which life extends as a receptacle of life.* Later, Vladimir I. Vernadsky (1927; 1937) also used it from a geological standpoint to refer to the *totality of all life and organic material produced by life, i.e.* the life phenomenon or "life system", which included such things as the accumulated excreta (bioinert bodies) and metamorphic rocks (*e.g.*, coal).

Professor Vladimir I. Vernadsky, however, continued to define the Biosphere as the range or receptacle of life, *i.e.* as *the shell within which life, life-supporting material and reactions occur; and are supported by solar energy.* From a geometrical viewpoint then the Earth's Biosphere must follow the *Globe's shape or pattern* (Gitelson *et al.*, 2003). The creators of Biosphere 2, with its miniature biomes of the Earth's Biosphere, defined the proper name ***Biosphere*** *as a macrobiosphere with two or more biomes* (*i.e.* major ecosystem units with distinctive life forms, *e.g.*, tundra, steppe, desert, rainforest, *etc.*).

The roots of the term *biosphere,* however, are from the prefix *bio-*, (*bios = life*) and the suffix, *-sphere* (*sphaira = ball*); both terms are from the Greek. From the historical usage and per Webster (1994) the suffix *-sphere* clearly means *range* and when combined with the prefix *bio-*, the term *biosphere* should mean *the range limits to which life extends as per a receptacle.* This is a much more general definition of the term than as defined by the Biosphere 2 creators. In order to maintain continuity with both historic usages and the logical use of the term, we propose that ***Biosphere***, with a capital "***B***", is an appropriate proper noun for a closed system with two or more biomes, much as planet Earth is capitalized, whereas, ***biosphere***, with a small "***b***", is a more general term, *i.e. any receptacle or materially closed system containing life.*

A further description of the term has been provided by Professor Iosef I. Gitelson and colleagues (Gitelson *et al.*, 2003) wherein they surmise that "the" Biosphere (Biosphere 1) has the main properties of: *Integrity, Closure, Cyclicity, Steady state* and *Energy* dependence. When these properties are functional in the Biosphere, "everlasting" life is potentially assured under the conditions of the limited quantity of substrata of the Planet. In contrast, a ***biological sphere*** or ***biosphere*** is *a materially-closed life-containing entity* and is not limited by these constraints. As such, the concepts of *biosphere* and *Biosphere* have acquired definite scientific meanings, and therefore more detailed properties of such systems can be subjected to analysis by rigorous methods.

2.3.5 Noosphere

Vladimir I. Vernadsky used the term ***noosphere*** to describe *that sphere of life within the Biosphere that is controlled by human intellect with the goal of keeping the Biosphere in a condition suitable for human life*. From this point of view, it is important that humanity identify the characteristics that allow the understanding of the functionality and control of the Biosphere. The concept of "*criticality*" (see below) may be the most important aspect of this understanding.

Establishing the principles of criticality and other behaviors of closed systems can be comprehensively examined at different levels, such as with various kinds of microbiospheres (or in this case "***micronoospheres***" *i.e., systems controlled by human intellect for conditions suitable for specific life forms*) and human occupied ***macronoospheres*** where criticality is important for space exploration, colonization and planet Earth sustainability.

2.4 The Concept of Criticality

Criticality may be defined as *a relatively rapid decrease in the ratio of organic material to inorganic forms within a system*. The criticality slope of the decrease in organic matter over time is volume-time proportional, *i.e.* the relative time scale increases with increasing volume to generate the same criticality slope. The discussion of criticality slope is beyond the scope of this chapter.

In the construction of closed systems, the ratio of organic mass to inorganic mass is one of the first considerations. If an ecosystem's initial mass is 100% organic mass (massorg) and 0% inorganic mass, we might predict an immediate, rapid conversion of organic mass to inorganic mass (*e.g.*, a jar full of leaves and bacteria). Thus, in this case the *time to criticality* would be virtually instantaneous ($t \sim 0$, Figure 2.1). In the example, the closed container packed with wet leaves, would immediately begin to convert $(CH_2O)_n$ to CO_2 and H_2O via any free oxygen and / or anaerobic processes until limiting reactants are consumed.

In contrast, a closed ecosystem with a sufficiently low initial $mass_{org}$ might be more likely to maintain reactions between organic and inorganic forms such that there would be a balance. For a particular type of closed system, there would be an initial ***minimum*** $\boldsymbol{mass_{org}}$ ***(%) for criticality***, *i.e. a minimum amount of initial organics as a percent of the total mass below which criticality behaviour would not occur and the system would maintain an organic-inorganic equilibrium* (PE) (asymptote line in Figure 2.1). There must also be a maximum initial organic mass for surviving criticality, to establish a new balance between organic and inorganic masses.

For time to criticality as the initial percent of organics is reduced, the curve approaches an asymptote. The asymptote describes the maximum initial percent of organics, $\mathbf{P_E}$, where equilibrium will persist indefinitely. In Figure 2.1 $\mathbf{P_E} = 2\%$. The time to criticality curve follows the general formula (1):

$$\mathbf{P}_{io} = \frac{100 - \mathbf{P}_\mathrm{E}}{1 + t} + \mathbf{P}_\mathrm{E}. \tag{1}$$

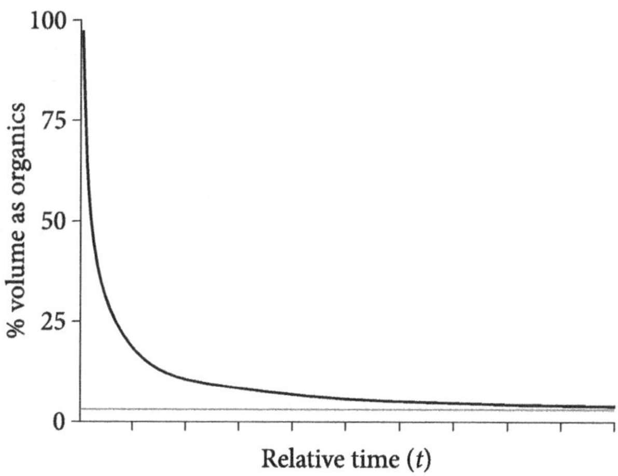

Fig. 2.1. Time to criticality. Percent volume as organics of total system on relative time to criticality. The maximum percent volume as organics, that will remain in equilibrium, $\mathbf{P_E}$, (light gray line) without reaching criticality, is arbitrarily given as 2% in the present example (adapted from Yensen and Biel, 2005)

Where: $\mathbf{P}_{io}$ is the initial percent volume as organics (living material); *t* is time to criticality; and the asymptotic value $\mathbf{P}_E$ is maximum percent for an equilibrium of organic material at which criticality will not occur (the $\mathbf{P}_E$ grey line in Figure 2.1 arbitrarily set at 2% to produce the graph). Skilled noospheric control of a system may allow high $\mathbf{P}_E$ values, important in the exploration of space where total volume and mass levels must be kept to a minimum.

The relative amounts and relationships of the phase states of matter (solid, liquid and gas) and their effects on criticality and ecosystem dynamics could be significant, but there is a paucity of information in the literature and the principles remain undeveloped. It may also be possible to design resilient systems which could go through criticality and rebound to a higher than initial $\boldsymbol{mass_{org}}$ level, or to a higher ***maximum post-criticality*** $\boldsymbol{mass_{org}}$.

Speciation models may be investigated using milder and / or extended forms of criticality conditions to select for two or more sub-populations with the end result of divergence to for two species. For instance, the one-celled green alga, *Chlorella sp.*, may be induced through selection to develop additional mating types that are not cross-fertile.

The creation of species under controlled conditions, without genetic engineering, could be an important new tool for evolutionary science. The ability modulates environmental factors and to predict speciation events in organisms could be very useful and would provide a better understanding of past evolution. More importantly, such an approach could be invaluable for the development of future human-guided nooevolution (*human-knowledge based evolution*) and for a more stable biospheric condition derived by intelligent design.

Thus, the concepts and lessons of closed-system criticality will also help humanity to generate a sustainable noosphere. And, perhaps this knowledge could increase the probability of noospheric ecosystems or parts thereof to survive the impact of an extraterrestrial object, which would otherwise collapse the Earth's ecosystems. In the meantime, it can help us to identify negative anthropogenic factors that could create criticality conditions and a rapid system change or global collapse of biogeochemical support system(s) of the Earth's Biosphere.

2.4.1 The Volume-Criticality Principle

Our understanding of the volume-criticality principle arose serendipitously from the mass production of one- and two-liter microbiospheres, called Ecospheres. In the commercial production tens of thousands of Ecospheres were shipped to various places around the world.

A relatively small percent of the total shipped Ecosphere production was returned for replacement, because the systems were guaranteed for replacement if there was a loss of shrimp. Still, this amounted to hundreds of "returns" in which one or more of the shrimp had died. We noticed that the one-liter Ecospheres would be returned in 1.5–2.0 months, while the two-liter Ecosphere would be returned in 2.0–3.0 months. Allowing for the time from "fabrication to shipment" and time from "collapse to return", the time to criticality was estimated at 1.0–1.5 months for the one-liter Ecosphere and 1.5–2.5 months for the two-liter Ecosphere. Some additional experience with other smaller and larger Ecospheres (with virtually identical ecosystems) and with other closed systems suggested that time to criticality was size dependent (Figure 2.2). This ***volume-criticality principle*** should be tested in other systems. We have, however, evaluated the volume-criticality principle with regard to microbiospheres, in general, with positive results. Capillary-tube microbiosphere systems evaluated under the microscope can reach criticality within 24–48 hours.

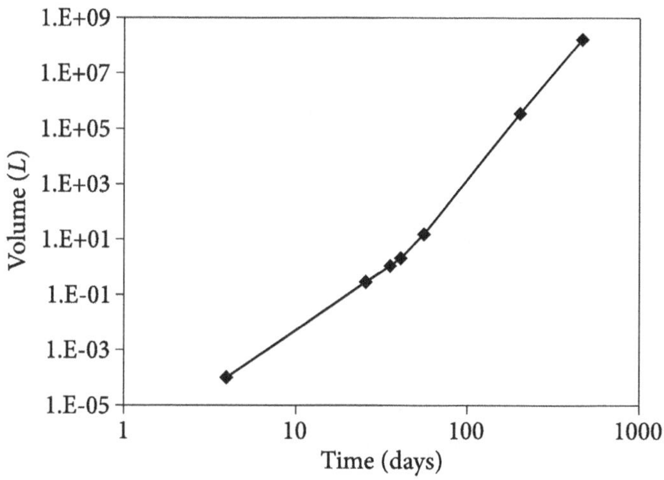

Fig. 2.2. Model System Time to Criticality. Log-log graph of ecosystem volume (L) on time (days) to criticality of various systems ranging from capillary systems to Biosphere 2 Laboratory (adapted from Yensen and Biel, 2005)

Note: Criticality times are subjectively determined, for instance, Bios-3 [6th point] was not a totally recycling system nor did it experience "criticality" *per se* (Gitelson *et al.*, 2003)

2.5 Microbiospheres: Descriptions and Discussion

2.5.1 The Ecosphere, a Synthetic Microbiosphere

Following the 1970's to 1980's work on the energetics of closed systems by Drs. C. Folsom and J.A. Hansen (1986) and in various semi-closed and closed microcosms by Drs. R. Beyers and H.T. Odum (1993), N. Yensen (1988), Drs. D. Harmony and L. Acker began constructing microbiospheres, called ***Ecosphere***® microbiospheres. The Ecospheres were initially one and two-liter systems, contained several brackish water shrimps, and were based on a NASA technology transfer recipe developed by Dr. J. Hansen. An initial attempt at mass production and marketing of the original Hansen system was disappointing due to frequent shrimp deaths. In addition, algal growths would cover the inner surface of the glass sphere making observation of the shrimp difficult and reduced the esthetic value and sale of the product.

To resolve these problems, and in contrast to the work of Dr. C. Folsom, Dr. J.A. Hansen and others, Dr. Nikolas P. Yensen discarded the concept of making the Ecospheres by "enclosing" a piece of a natural ecosystems. It was the paradigm that only such pieces of extant systems would have the requisite biota to cycle all the nutrients. Indeed, in a conference call between Hawaii, JPL-Pasadena and Tucson, we were summarily instructed that only closed-off pieces of natural ecosystems would survive. Fortunately, we eschewed the sage advice by these American "fathers of closed systems", and designed an artificial, shrimp-based ecosystem. This was done by identifying index components in successful Hansen-constructed systems. We specifically sought out and identified the key indicator organisms and microbes of the ecosystem.

Then, not unlike the contemporary work of Dr. F. Taub, we used ecological principles of trophic webs, food pyramids, and community selection and succession to invent an ecosystem that, at least for the shrimp, would avoid retrograde phylogenetic extinction. The development further involved the construction of several cleanrooms (filtered to 0.5 micron), the initiation of unialgal cultures, and the establishment of various microbial ecosystems. After a year of research, we finally developed our first successful synthetic ecosystem that could recycle nutrients and stay balanced while displaying a pre-programmed ecological succession.

Thus, the crystal-clear-walled, aesthetically pleasing Ecosphere was born (see Photo 2.2).

Ecosphere view Shrimps inside of Ecosphere

Photo 2.2. The crystal-clear-walled, aesthetically pleasing Ecosphere (photo images from N.P. Yensen's archive)

2.6 Bioboxes

Bioboxes are "bench-scale" closed systems, which may be made using a variety of materials, such as transparent and inverted "shoe" or "storage boxes". Clear-plastic storage boxes (2.5' × 2.5' × 2.5') are ideal Biobox systems. In the USA, they are available at the nationwide hardware and department stores for approximately $5. Greenhouse or polypropylene plastic may also stretch over a frame to make a simple Biobox. Bioboxes are excellent research tools and can serve for many educational activities.

Except for field respiration studies by ecophysiologists, the use of Biobox closed systems as a research tool has been underutilized. Biobox systems can be purchased "off-the-shelf" or constructed economically within a short period of time, allows the use of replicates and statistical analyses. "Off-the-shelf" nutrient solutions for initiating plant-based systems are also readily available. We have used Bioboxes in various research formats and have employed oxygen and carbon dioxide sensors. Thus, we have demonstrated an ease of Biobox system manipulation.

2.7 Experimental vs. Mathematical Models

We agree with Academician Iosif Gitelson (Gitelson *et al.*, 2003) that small closed ecosystems may serve as model systems on which to test paradigms and hypotheses. The relatively simple microbiospheres can have ecosystem complexities exceeding the computational level of the most advanced computers. The mathematical models therefore must be less complex than the complete system they are modeling. By letting the microbiosphere's ecosystems do the calculating, we can test simple models with statistically significant sample sizes of actual systems. This allows the identification of ***index parameters*** and ***index species*** for monitoring system behavior. These mathematical model indices, in turn, may be used to generate principles and concepts leading to predictability in various closed ecosystems, not the least of which is planet Earth. In this sense, mathematical modeling of closed systems can be the final outcome predictor of criticality events and the mathematician's role is essential, if not "critical". Only the experimentalists, however, can provide information on which to build proper theories and mathematical models.

Ironically, in the past, experimentalists, with a few notable exceptions, have limited their experimental designs to only one, two, or a few systems. In a new science, where a paucity of information exists, such studies may provide limited statistical significance, albeit providing anecdotal information and / or a gestalt for the requirements of certain systems.

2.7.1 Retrograde Phylogenetic Extinction

Dr. Bassett J. Maguire's (1980) experimental studies of macroscopic organisms in an array of one-liter microbiosphere biobottle systems resulted in a discovery of a basic closed system phenomenon. He discovered the principle of "***retrograde phylogenetic extinction***", wherein there is a tendency for larger and more-complex organisms to go extinct first and for the "simpler" organisms, such as cyanobacteria and bacteria to persist. While the exact order of retrograde extinction does not appear to have rigid rules, it does seem logical that organisms that have persisted for long geological periods would have more resilience than recently evolved organisms when confronted with the stress of enclosure in a microbiosphere. Also, systems entering criticality are more likely to encounter anaerobic conditions not unlike the low oxygen levels estimated for the early history of life on the Planet.

2.8 Humanospheres: Examples and Discussion

A description of all the various humanospheres conceived and / or trialed in part or parcel is well beyond the scope of this article. The following examples, Biotubes, Bios 1, 2 and 3, and Biosphere 2 Laboratory, were selected for inclusion partly because we have familiarity with them and partly because the work with Biotubes has not been published. The term for humans who inhabit a Biosphere (Biosphere 2) was ***biospherian***. We use the term ***bionaut*** to refer to a human inside of any biosphere (small "b") following the terminology for cosmonauts, astronauts, and aquanauts.

2.8.1 Biotubes

Biotubes are large (Photo 2.3), "walk-in-sized" closed systems made from clear polypropylene or greenhouse plastic. They offer an excellent educational activity and may, at some point, be developed as a research tool. Although no research *per se* has been conducted in the Biotubes, they do provide an excellent potential for short-term human research trials. At the present, they are an interesting and exciting educational aspect of closed ecosystems suitable for student experiential learning.

Photo 2.3. Biotube (photo images from N.P. Yensen's archive)

The development of the Biotube demonstrates that human-sized systems can be constructed, both economically and quickly compared to all other systems presently reported in the literature. So far, eight Biotube systems have been constructed for a cost of less than US$250 each. Students and professors design, purchase materials and construct each Biotube using materials readily available from hardware stores.

The Biotube effort is: first, to see if human macrobiosphere systems can be built with sufficiently low leak rate and second, to provide the experience of being inside a closed system for a brief period. The opportunity to design, construct, contemplate and attempt to understand these systems has also led to an appreciation of the complexity of the Earth's Biosphere. The over consumption of oxygen, air pollution, food production, waste treatment, and social psychology all come to the forefront of one's thoughts when one is in repose inside a Biotube. Humanity is faced with these difficulties and challenges, but these problems seem far away and not personal. In a Biotube they are all very personal and uncomfortable close. In this regard, an individual's contemplation inside a Biotube can make a profound impression on one's thinking.

Very stringent limits are set for occupancy of the Biotubes. If the oxygen drops from the normal atmospheric ~21% to 19%, all personnel inside the Biotube are required to evacuate. This oxygen level is not unusual in a class room of 30 students and is equivalent to the oxygen density at circa 2000 meters (~6500') elevation, but maintaining the level above 19% by using living plants in a "greenhouse" format presents a challenge. The use of mass balance, and direct sensor readouts, allows the estimation of the carbon dioxide level, which is not allowed to exceed 2%, a level well below that considered safe for bakers and brewers. The temperature and humidity are also monitored to assure a safe comfort levels in accord with national (USA) standards. And, perhaps "overcautiously", pulse-oximeters (to measure pulse rate and blood oxygen levels) may be attached to resting occupants inside a Biotube.

Two observers are always present when a bionaut is inside the system. Acuity information taken every half hour and verbal dialogue are reliable estimates of a bionaut's healthy mental condition. These considerations simulate checks that might be necessary for an extended occupancy. The ability to rapidly breach the 6 mil polypropylene plastic walls is present on both the inside and the outside (*e.g.*, a pair of scissors and / or a knife). The walls of the Biotubes are of polypropylene plastic, the same plastic that is safely used and approved for drinking water containers.

One Biotube system has produced oxygen using an aquatic system while all the others have used terrestrial systems of flats and / or potted plants. Nutrient solutions for the plant life are readily obtained from local garden shops.

Calculations show that inside the Biotubes there is adequate oxygen for several person-days. Still, the longest period has been only 30 hours, while most occupancy periods are even briefer (1–4 hours). Between trials the Biotubes are flushed with fresh air. The common bionaut complaint is that they did not get to stay inside long enough.

The Biotubes have been used strictly as an educational activity. No human experiments or studies have been conducted and to the best of our knowledge we are in compliance with the criteria for human experimentation.

2.8.2 *Shepelev, BIOS 1, 2, and 3*

In 1962 a Russian scientist, Professor Evgeni Y. Shepelev was the first person to inhabit a closed system. He used 45 liters of the algae *Chlorella* to produce oxygen in a closed steel container. And from 1964–1965 he conducted five experiments at the Institute of Biomedical Problems in Moscow (Gitelson *et al.*, 2003). Prof. Evgeni Y. Shepelev survived the malodorous experiences.

In Krasnoyarsk, Russia BIOS-1 was established as a test facility for space exploration-type closed systems. Between 1961 and 1965 research linked *Chlorella* systems with human habitat systems (Gitelson *et al.*, 2003:54). The Russians, ahead of the USA, had extensive experimentation in closed systems for habitation.

Academicians Iosif I. Gitelson, Leonid V. Kirensky and Ivan A. Terskov were among those who initiated the Krasnoyarsk work. They used 8 square meters of *Chlorella* per person to consume the carbon dioxide and cycle back oxygen. Food and water were not recycled in these trials, but came from storage, rather than being produced by the system.

The intent of this research was for short manned flights, with longer-term goals of extended space flights and planetary colonization. BIOS-1 was developed such that it could produce food, water and oxygen. The 1968 trials used wheat and a few other plants in BIOS-1. The plants and algae, as a separate system, regenerated the oxygen from respired CO_2 in sufficient quantity to support the human occupants (Gitelson *et al.*, 2003).

By 1972, the plant and human systems were combined and the system was renamed BIOS-2. BIOS-2 was then replaced by the grander scale BIOS-3 (Photo 2.4). It is estimated to cost a million dollars (USA). It contains crew quarters and plant cultivation areas.

Professor Genrikh M. Lisovsky next to the model of the BIOS-3
(photo image from the collection of Karl Biel)

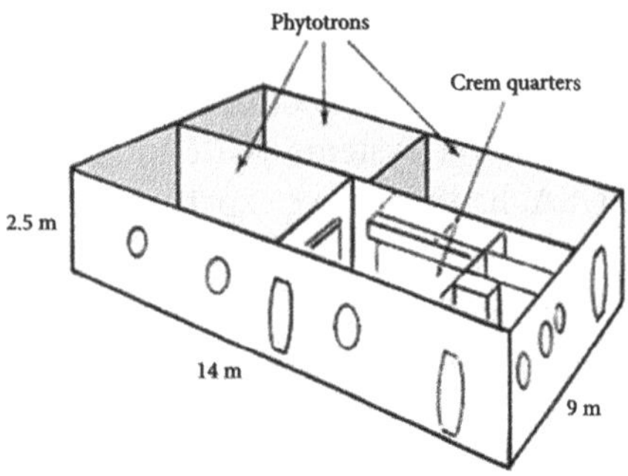

BIOS 3 Diagram
http://www.permanent.com

Wheat plants in BIOS 3
http://www.permanent.com

Photo 2.4. Author of dwarf wheat Professor Genrikh M. Lisovsky (top), BIOS-3 diagram (bottom left), and wheat plants in BIOS-3 (bottom right)

An early trial in 1976 showed that three men could survive inside BIOS-3 for five months with supporting plant life. They used a "Conveyor" approach with crops at three to seven different stages of growth, simultaneously. They planted foods to produce a balance of proteins, carbohydrates and vitamins and minerals. It was found that the percentage of edible biomass (*edibility coefficient*) was higher for salad foods than for grains or root plants (Salisbury, 2003). Meat was brought into the system prior to closure and human fecal waste was not recycled (Genrikh M. Lisovsky per. com.). The selection of participants was based on cooperative ability, knowledge and experience with experimental programs. Initially, participants were monitored several months before closure, continuously during closure, and for a month after emerging to obtain data on metabolism and cognitive skills. No significant detrimental effects were observed and subsequent monitoring became less stringent (Salisbury, 2003).

The Bios experiments established that humans and plants could function symbiotically in a small closed system. Potential crops were identified and gas recycling techniques were trialed. Perhaps most importantly, it stimulated scientific interest throughout the world.

Now the BIOSes are running in International Center of Closed Ecological Systems (Institute of Biophysics, Siberian Brach of the Russian Academy of Science, Krasnoyarsk) under the leadership of Academician Iosif I. Gitelson.

2.8.3 Biosphere 2 Laboratory

The Biosphere 2 Laboratory (B2L, Photo 2.5) was designed as an "air-tight-matter-closed" facility with an air leak rate of about 10% per year. Concepts which inspired the design and construction of Biosphere 2 Laboratory were adapted from the Russian geochemist Academician Vladimir I. Vernadsky. The Biosphere 2 Laboratory was designed and built as an Earth-like human-controlled noosphere, it could have controlled stewardship and could be operated so as to investigate Earth processes, such as, whole system behaviors, biogeochemistry, and microbial components. The biogeochemistry of the Biosphere 2 Laboratory system was controlled for temperature, humidity and atmospheric concentrations of carbon dioxide, oxygen, methane, and nitrous oxide. The dynamic range and evolution of atmospheric gas species were monitored with a large number and variety of sensors for operational and safety purposes.

The Biosphere 2 Laboratory, as a whole, was called a "***mesocosm***", *a complex mini-ecosystem under controlled environmental conditions* (Bil' and Osmond, 2003*a*, 2003*b*; Osmond, 2003, 2014). It is comprised of individual model ecosystems with differing climates, life forms and management strategies. These subsystems are also referred to as miniature "biomes" or individual mesocosms, including the desert, rainforest, savannah, thornscrub (ecotone), mangrove-marsh and the coral reef-ocean biomes. A human dominated intensive agriculture and dwelling mesocosm completed the interconnected system of Biosphere 2 Laboratory in its original form (Marino and Odum, 1999*a*, 1999*b*). The Biosphere 2 Laboratory is the world's largest controlled environment "apparatus" for experimental ecosystem research.

Photo 2.5. The view of Biosphere 2 Laboratory, Oracle, Arizona, USA (top: photo image and bottom: diagram)

2.8.4 *Closed System Missions*

Mission 1. Mission 1 ran for 2 years (1991–1993). This initial 2-year closure was intended to be the first of many such missions as the structure and equipment were constructed of materials so as to last up to 100 years. The first closed system experiments in Biosphere 2 Laboratory were thought to have "problems" (Marino and Odum, 1999*a*, 1999*b*) *which, in retrospect, were not problems at all, but great opportunities for insights in to the Earth's Biosphere and illustrated important atmosphere-life considerations.*

An opportunity for insight developed when the oxygen dropped to 14% and a million liters of oxygen needed to be added to the system a little over a year into the trial. While the carbon dioxide level increased it did not match the mass balance that would be predicted by the drop-in oxygen. The best hypothesis to date to explain this is that the underlying base of the Biosphere 2 Laboratory system, being made of concrete, continued to absorb carbon dioxide such that the plants were unable to convert it back to oxygen via photosynthesis. It is also possible, but difficult to document, that other soil processes could absorb carbon dioxide as there are many microbes that form calcareous structures. Planet Earth may be undergoing a similar loss in carbon dioxide that does not match the oxygen consumption of fossil fuels (see ***Oxygen Flux in Closed Systems***, below).

Mission 2. Mission 2 ran for 6 months (1994) with similar problems to mission 1.

2.8.5 Open System Missions

Interim Mission. From 1994 to 1995 the system was opened and the priorities were reset for studies of responses of plants and ecosystems to possible future global environmental change. During this period, research emphasis was placed on studying the processes within the biomes.

Columbia University Mission. From 1996 to 2003, under the auspices of Columbia University's Earth Institute, modifications and new designs were implemented, such as: the installation of a ventilation system for air flow from the outside to the inside, the modification of the wilderness area, the providing of control for the mean CO_2 content, and the reduction of nitrous oxide concentrations, and the separation of the agricultural biome from the wildness biomes. It became, in essence, the world's largest terrarium with carbon dioxide control.

From 2001 to 2003, Biosphere 2 Laboratory became a multi-user, international facility for experimental ecosystem science. During this time international collaboration (*e.g.*, Australia, China, England, Germany, Italy, India, Russia, Sweden, USA) produced a number of valuable studies, some of which have been published (see for example, Kudeyarov, Ponizovsky *et al.*, 2002; Demyanova *et al.*, 2003; Ponizovsky *et al.*, 2003; Osmond, Ananyev *et al.*, 2004; Rascher, Bobich *et al.*, 2004; Ananyev, Kolber *et al.*, 2005; Barron-Gafford, Martens *et al.*, 2005; Kolber, Klimov *et al.*, 2005; Van Haren, Handley *et al.*, 2005; Kudeyarov, Biel *et al.*, 2006; Barron-Gafford *et al.*, 2007; and others). Greater number of articles can be found in the list of "Literature Cited" of Professor Barry Osmond's review published in the "Annual Review of Plant Biology" (Osmond, 2014).

2.8.6 *The End of Biosphere 2 Laboratory or a New Era for Biosphere 2 Laboratory?*

In March of 2003 the USA Department of Energy considered funding the maintenance and operation of Biosphere 2 Laboratory as a National Scientific User Facility.

A negative assessment, however, was made by the USA Department of Energy's Biological and Environmental Research Advisory Committee: Pitelka L.F. - chair, Coleman J., Dale V.H., Fletcher M.M., Harriss R., Johnson D., Reynolds J.F., and Wofsy S.C. (see Pitelka *et al.*, 2003).

At the end of 2003 a new Columbia University administration returned the Biosphere 2 Laboratory to Mr. Ed Bass, the original visionary of the Biosphere 2 Laboratory, who had hoped that the system would teach us about the Planet Earth, how to take care of it and how we may be able to establish colonies outside of the Earth's Biosphere (E. Bass, per. com.).

The Biosphere 2 Laboratory is the only ecological closed system facility in the world. It alone can provide important answers to questions about the future of our Planet, as well as identifying criticality parameters necessary for space colonization.

But, for many years the Biosphere area has been considered for homes, resort, and commercial areas, and … It also has approval for educational and research uses as part of University of Arizona Campus (see *e.g.*, Explorer, June 13 and June 27, 2007, www.explorernews.com; Arizona Daily Star, June 27, 2007, contact reporter Eric Swedlund at: eswedlund@azstarnet.com, *etc.*).

And, fortunately, as mentioned recently Professor Barry Osmond (Osmond, 2014), former President of Biosphere 2 Laboratory: "…*After years of uncertainty, the magnificent "apparatus" is again in the hands of a cohort of scientists from many disciplines affiliated at the University of Arizona, Tucson* (Stokstad, 2011). *This broad local support base may yet see the potential of Biosphere 2 Laboratory more fully realized…*"

We also hope that the user(s) will have enough intelligence to use this unique facility wisely to answer these questions for benefit (and possible future salvation) of all humanity as *the Earth's Biosphere enters a period of criticality.*

Although Biosphere 2 Laboratory is obviously not a duplicate analog to the Earth, the two systems share many biological processes and interactions. Importantly, Biosphere 2 Laboratory with its smaller volume will reach criticality in a much shorter period and can there give predictability of some potential behaviors of Planet Earth. The ecosystems in Biosphere 2 Laboratory already resemble anthropogenic degraded or stressed natural habitats. Understanding the causes and modification necessary to improve these habitats offers a unique potential to understand restoration possibilities. Although simultaneous replicate experiments are difficult, temporal repetition has been performed to good success. The unique biota, which persists in Biosphere 2 Laboratory despite stress and unfavorable conditions, presents interesting answers to questions about closed and semi-closed systems. The growing awareness and acceptance of this approach is verified by the emergence of other large-scale closed or semi-closed experiments, such as CLIMEX (Dise and Jenkins, 1995), the Ecotron (Lawton, 1996) and NASA's human-rated test facility (Henninger *et al.*, 1996).

The capacities and parameters of the Biosphere 2 Laboratory facility are described in Appendix I.

2.9 The Earth (Biosphere 1) Description and Discussion

2.9.1 *Earth, a Sample Size of One*

Relatively few materially closed ecosystems have been constructed and fewer still have been studied in detail. One prominent, real-world, closed-ecosystem experiment, which unfortunately has a sample size of only one, is... Earth! Significantly, all humanity and known life depends on the outcome of this experiment. The importance of this experiment is faced with the daunting complexity with size of the Biosphere1 system.

2.9.2 *Biosphere 1 Properties*

The property of ***Integrity*** refers to the fact that all Biosphere life is part of a single system in that any life entity exchanges material with all other life entities. ***Closure*** considers that all such exchanges of the basic biogenic elements (especially carbon) are confined to the Biosphere. ***Cyclicity*** is defined as the property of having closed cycles of material exchanges, wherein different elements may vary in the degree of openness but do not exceed a fraction of a percent per cycle. Cyclic processes are also maintained in a ***Steady state*** varying about equilibrium constants of live biomass and essential biogenic elements over geological time. Such a steady state is ***Energy dependent*** upon a stable radiation flow through the system, most notably photosynthetically captured solar radiation, which drives the system through chemical energy transformation, ultimately dissipating as heat energy (Gitelson *et al.*, 2003).

Dr. Lovelock James E. (1972; 1990; 1991) used the term "***Gaia***" to refer to this self-regulating system of the Biosphere. The concept has been extended by others to the belief that Gaia is a super-organism and even to the extent that some believe Gaia is a sentient organism. The downside of this extensive non-scientific usage has been a confusion about the term and its proper usage. Regardless of the corruption of the term, the scientific community has come to realize that it is vitally important for humanity to comprehend the self-regulating mechanisms of biospheric cycling, closure and sustainability.

The Earth's physical Biosphere was formed some 5 billion years ago, and consists of a "delicate veil-like spherical surface" within which life, as part of a system, can exits. This thin "skin" surrounds the Planet at a thickness of about 1/1000 the Earth's diameter, extending 11 kilometers (~seven miles) up above sea level and 11 kilometers (~seven miles) below to the deepest sea floor in the Marianas trench. Some 90% of life occurs between 0.1 kilometers (~300 feet) below the surface to 3 kilometers (~10,000 feet) above sea level (Braun, 1990). As a whole, life is relatively sparse in the Biosphere, when compared to the patches normally inhabited by humans. The Biosphere life zone may be characterized elementally as a thin dispersed carbon layer "sandwiched" between a thick top slice of nitrogen-oxygen (the atmosphere, $N_2 + O_2$) on a thick bottom slice of silicon-oxygen (the crust, $Si + O_2$). The carbon layer of the "sandwich" is variably mixed with hydrogen-oxygen (water, $H_2 + O$) which acts as a "lubricant" to move the carbon around by energy derived from captured photon.

The Earth's Biosphere is not strictly a materially closed system because comets, meteorites, meteors, dust, gasses, cosmic rays, photons and neutrinos enter the system. Given that photons are only part of "matter" when they energize electrons and that there is debate as to whether neutrinos have mass (which typically depart as quickly as they arrive) we may still conclude that the Biosphere system is clearly *open to energy fluxes*. Not the least of which is incoming solar radiation reflected or absorbed and then re-radiated into space; or the Earth's nuclear core, whose energy eventually radiates into space. Notwithstanding Earth-extraterrestrial collision events and resultant debris dispersal, the Biosphere is now beginning to "radiate" a minute amount of anthropogenically propelled metal, carbon and silicon back into space, beginning with the Voyager space craft. The geological flux of material in and out of the Biosphere has had an insignificant effect on the Biosphere's properties. Alterations in the energy fluxes, however, can have significant effects on the life systems.

In recent decades, it has become evident that technological activity is upsetting global equilibriums. The alarmed world community has recently become more interested in the "flows of existence" of the Biosphere and the limits of its sustainability. Scientists conclude that after millions of years of relatively balanced flows, the Biosphere and its dynamics cannot be described without accounting for how human intellect-devised technological processes are causing serious interference and impact on the biospheric cycles.

The development of remote sensing from the space has made possible the accurate observations of many dynamic processes in the Biosphere. This, however, does not enable us to determine the factors that delimit the Biosphere's sustainability. The limits can be identified only by studying the Biosphere's response to extreme conditions and to challenge its very existence. Mathematical models, at first, appear to be the only practical experimental approach to study extreme perturbations. Yet models are barely, if at all, able to answer questions about the mechanisms responsible for the main quality of the Biosphere, such as the closure of the material cycles, or the comprehension of the laws providing the existence of this unique natural formation called the Biosphere. For instance, there is little agreement on the cause of the global temperature rise (Figure 2.3). It appears that global temperature is a result of many complex interactions. To help resolve some of these and other questions, experiments should be performed on real, physically existing ecosystems with varying degrees of closure.

The influence of anthropogenic factors may be positive or negative and some may counterbalance others. Because the Earth's Biosphere is massive compared to any other biosphere, the response times are much slower than for smaller systems and going to criticality could be a 300-year process (Figure 2.4) or with a cataclysm it could be very short. For instance, given a significant meteorite or comet impact upon the Earth, criticality may be virtually instantaneous. Biological and anthropogenic processes are predicted to require a few centuries for a critical collapse of the Earth's ecosystems. It is possible that the Biosphere is presently moving into a criticality state. In contrast, it may not be necessary for an extraterrestrial impact to generate a rapid collapse of the Earth's ecosystem, if negative anthropogenic factors reach a critical condition for rapid change, in a few decades the system may collapse its cycling global biogeochemistry, such as a methane eruption from the ocean floor. The answers to these probing critical questions need to be forthcoming.

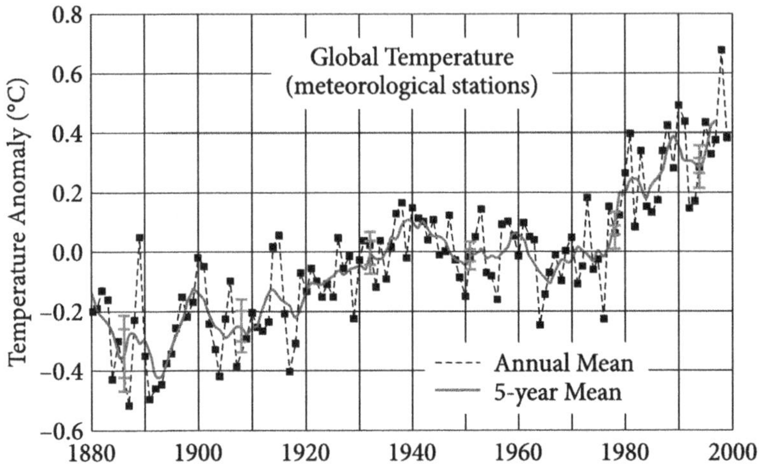

Fig. 2.3. Global temperatures, showing the well-known global increase of over a half degree from 1880 to 2000 (adapted from Yensen and Biel, 2005)

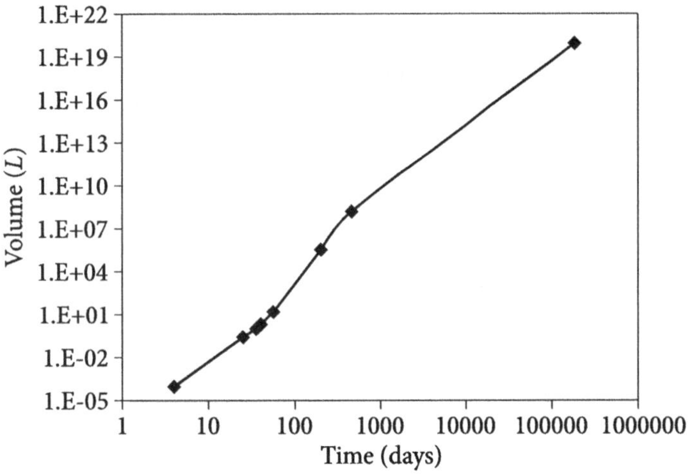

Fig. 2.4. Earth System Time to Criticality. Projected times are based on systems ranging from capillary systems to Biosphere 2 criticalities. The log-log graph is ecosystem volume (liters) on time (days) to criticality wherein the time of criticality may vary significantly depending upon degree of perturbation, ratio of biotic to non-biotic components and especially anthropogenic influence (adapted from Yensen and Biel, 2005)

It is hoped that, unlike the past Biospheric criticalities, humanity may manage the Biosphere through advanced noospheric technology into an era of sustainable stability.

2.10 Oxygen Flux in Closed Systems

Estimates of the free atmospheric oxygen in the Biosphere range from 1.5×10^{21} grams to 1.14×10^{24} grams (Vernadsky, 1960, 1998; Levine, 1985, respectively). The amount of estimated living material (in constant flux) on the Earth ranges from 1.84×10^{18} to 1.0×10^{21} grams (Gitelson *et al.*, 2003; Vernadsky, 1960; 1998, respectively).

We expect that the amount of free oxygen delimits the amount of aerobic living mass, *i.e.*, there must be a limit to the ratio of aerobic living mass to free oxygen. The precision of the current ratio of living mass to oxygen ($\boldsymbol{m_l} : \boldsymbol{O_2}$) on Earth is not well defined; it ranges from a maximum of one-to-one (1 : 1) to one to a half million parts oxygen mass (1 : 500,000). The imprecision of these ratios gives us numbers that look very unpromising as we diminish the amount of oxygen on the planet by burning fossil fuels. The sensitivity of closed systems to perturbation, however, increases as the ratio diminishes suggesting that accurately quantifying this ratio, and then monitoring it, is important to the well-being of the system. Knowing this ratio can also give us better information for designing more resilient human systems for space travel and planetary colonization.

The current oxygen consumption (principally by fossil fuel use in the United States and Europe) exceeds the oxygen production by all the plant life on our planet Earth by about ten times, although this number needs recalculation (Gitelson *et al.*, 2003). If the use of chemical fuels without oxygen in their composition is continued there will be a constant elimination of oxygen from the atmosphere. If the depletion of oxygen at the current rate expands to other rapidly industrializing countries, such as China and India, we may expect global problems with insufficient oxygen in as little as thirty to seventy years, particularly in cities such as Mexico City and Tokyo, where it is already possible to buy oxygen to breathe at street corners.

Concurrent to the human induced global drop in oxygen, like the Biosphere 2 Laboratory, we are observing an increase in carbon dioxide, but not to as high a level as we might expect from mass balance.

Thus, like Biosphere 2 Laboratory, we may expect that carbon dioxide is being sequestered somewhere on the Planet. The data supporting this global condition is not well known, but there are some studies that indicate that there is a doubling of the formation of calcium carbonate by organisms living in ocean (George Soto, per. com.). If the extra calcium carbonate on the sea floor explains the missing free oxygen from the global system, similar to Biosphere 2, then we are faced with a "permanent" atmospheric loss of oxygen and a proportional reduction of global biomass.

Moreover, from our standpoint, Biosphere 2 Laboratory was serendipitously a huge success in the initial two-year closed system mission, as it demonstrated this important principle related to the problem of atmospheric carbon dioxide sequestration into the calcium carbonate of the concrete and / or soil.

Many considered the missions a "failure" due to the consequent reduced oxygen to support the eight biospherians. Others considered it valuable information (Marino and Odum, 1999*a*, 1999*b*; Ponizovsky *et al.*, 2003).

We would like to point out that it has provided an excellent real-world model of a current critical problem facing the Earth, the oxygen-carbon dioxide imbalance.

This overlooked aspect elucidates new problems with respect to the Kyoto Agreement and carbon credits for the sequestering of carbon dioxide: firstly, it provides an incentive to remove oxygen from the atmosphere and secondly, the loss of cycling oxygen produced from the normal stimulation of plant growth by the increased carbon dioxide levels.

While carbon credits, due to the world's concern for global warming, appear justified, the long-term effects on the global oxygen supply and ratio of organic to inorganic carbon are potentially far more disastrous than a "failed" mission of Biosphere 2 Laboratory. If it were not for Biosphere 2 Laboratory (and its initial mission) we would have neither the tool nor the understanding that would allow us to demonstrate and test the global problem of the sequestering of atmospheric oxygen, via carbon dioxide, into calcium carbonate. It is imperative that we conduct these trials on small Biospheres. The continued use of oxygen-free hydrocarbons (petroleum and coal), instead of using of nuclear (see, for example, Koch, 2004), wind and other sources of energy (see also, Energy of Future, 2004), will exceed the free oxygen supply, thus the highly publicized fear of running out of oil and coal reserves becomes inconsequential. There is far more fossil fuel than there is oxygen to burn it. The sequestering of carbon dioxide only shortens the time to oxygen depletion.

The world is currently concentrating on the valid and important question of global warming and weather changes, which may significantly affect the short-term life style of humanity. We consider it more important to pay attention to the loss of oxygen and the potential to drastically reduce all life on the Planet for eons to come. The use of closed system research is critical to identifying the slopes and integrals of the biomass-oxygen curves over time. The use of systems like Biosphere 2 Laboratory, Bios systems and Biotubes systems could be instrumental in identifying criticalities and predicting the future of life on this Planet.

2.11 The Future of Closed System Work: Concepts and Strategies

The current situation with the geo-ecosystems of planet Earth is serious for the sustainability of humanity and for many of the other species presently co-existing on the Planet. If we consider on a relative scale that the length of time that life has been on this Planet is equivalent to 24 hours, we find that humanity has only been on the Planet in the last 30 seconds (Raven *et al.*, 1986) and that the industrial consumption of oxygen has been in only the last 0.2 milliseconds. And in another 0.2 milliseconds the bulk of the Earth's free oxygen could be consumed and the biomass of the Planet could plummet. The rate of change of the parameters of the global system is rapidly accelerating due to human activities, yet we do not yet know many of the basic principles that determine these changes or their effects, much less do we have the ability to predict the end result.

There are many speculations, but unfortunately, even though any one of them may be correct, we presently have very limited projections and no absolute knowledge of which mathematical projection is correct.

Biosphere 2 Laboratory in its initial mission achieved what no other test had done; it produced real, albeit tentative, data about how the Earth's Biosphere might react. The urgent challenge is to quickly conduct enough tests with Biosphere 2 Laboratory and / or similar systems to begin to make projections of the survivability of humanity and the global ecosystem.

2.11.1 *Education, Research and Consortium Concepts*

It should be the responsibility of governments to cooperate to conduct the tests and trials that could lead to the maintenance of planetary life. Public relations will be critical to disseminate information and continue support for facilities and research opportunities at Universities and Institutions.

Such a consortium should have: long-term sustaining members, research members and education partners. The sustaining members would contribute core funding in support of the operational costs and eventually afford cost-free access for research members. The affiliated research members, made up of various institutions, would prepare and coordinate selection of proposals and follow up on the results and dissemination of information. The educational partners would be of a larger group of partner schools that would generate both a knowledge based curriculum and a technically trained sector within the population to do the work necessary to preserve our distinctive Earth.

From this standpoint, the objective will be to do worldwide research utilizing top students from many countries (especially, Russia, Japan, Germany, Scandinavian countries, Czech Republic, UK, USA, Canada, Mexico, China, India, Australia, where nascent or full-fledged programs already exist).

A closed-system curriculum should be developed with a potential for bachelor, masters and Ph.D. degrees in closed systems.

Course work could include: *Introduction and History of Closed Systems; Principles and Natural History of Closed Systems; Thermodynamics of Closed Systems; Criticality in Closed Systems; Cycles in Closed Systems; Complexity of Closed Systems; Closed Eco-system Diversity* (including aquatic and marine biome comparisons in Biosphere systems); *Closed Systems to Study Principles of Natural Hygiene as a Salubrity and Maintenance of Ecologically Clean Environment; Global Closed Systems and Climate Change; Closed Systems to Study Environmental Problems* (pesticides, pollution, eco-terrorism, *etc.*); *Closed Systems in Space, Submarine, Mines, Arctic and Antarctic; CELSS-Closed Ecological Life Support System;* and *The Management of Closed Systems.*

2.11.2 Ecosystems for Space

Interest in using biological systems for life support outside of the Earth's atmosphere resulted from a Chinese invention, the rocket. It was the onset of rocket science (pioneered by the Russian *Konstantin Eduardovich Tsiolkovsky*, the American *Robert Hutchings Goddard* and the German *Hermann Oberth*) that, following the German V2 rockets of World War II, led to the "US-Russian 12-year space race." The "space race" began 4 October 1957 with the launching of the 83.6-kilogram Sputnik 1 and a month later Sputnik 2, which carried the first living creature to enter space, the dog Laika.

In 1958 NASA (National Aeronautics and Space Administration) was established. Four years later (April 1961) Yuri Gagarin became the first human into space. Twelve years later (20 July 1969) Neil Armstrong and Edwin ("Buzz") Aldrin stepped onto the moon and, ironically, NASA lost interest in closed ecological life support systems.

Although the US Air Force began work on creating human biological life support systems in the early 1950s, NASA carried out their work throughout the 1960s. In the USSR, experimental closed-life-support-systems work started in the 1960s (Institute of Biomedical Problems of the Russian Academy of Sciences, Russia, Moscow and the Institute of Biophysics, Siberian Branch of the Russian Academy of Sciences, Russia, Krasnoyarsk).

Today, there is a resurgence of attention for the closed systems in Europe (The European Space Agency) and Japan (Closed Ecology Experiment facilities - Mitsui, Hitachi, Kawasaki, and other companies). These and other facilities have, or should have, two common fundamental goals: 1) to model the Earth's Biosphere and 2) develop practical closed human-life-support systems for space exploration and colonization.

Space research has the virtue of being international in the sense that these objectives are important for humanity, albeit extremely costly to conduct. It should be noted that, in the field of space biology and medicine in particular, cooperation and data exchange between the USA and the USSR was regular and continuous from the early 1960s through the 1980s, in spite of the space race.

2.11.3 Closed System Challenges

- To show that the oxygen, carbon dioxide and volatile organic compound gases will balance with humans inside an ecological closed system such that the system can be shown to have long-term stability.
- To create enclosed working models of the Earth's Biosphere and its ecosystems and thus better understand the regularities and laws that control its life.
- To develop technologies for the solution of pollution problems in our urban areas and for developing high-yield sustainable agriculture. To create ground-based life support systems that provide a high quality of life in extreme conditions of the Earth's Biosphere, as at polar latitudes, deserts, mountains, under ocean, polluted cities, *etc.*
- To create Biospheres for human life support beyond the limits of Earth's Biosphere which are essential for permanent human presence in space.

2.12 General Conclusions

Cooperation in creating artificial closed ecological systems is a new issue in the realm of the natural sciences and has not been developed as intensively as it should have been. The problems, however, of creating closed ecosystems for use on Earth and for space applications really concerns humanity as a whole. For this reason and the new global geo-ecological realities facing the 21st century, the scientific communities dealing with these problems must be more far-sighted and integrated than in the past.

The concept of criticality is important in human development toward a sustainable noosphere, survivability of asteroid-comet impacts, and the identification of negative human impacts that could create conditions for a rapid system change toward a global collapse of any of the biogeochemical support systems for the Earth's Biosphere.

Abbreviations

ALS – Advanced Life-Support System
B2L – Biosphere 2 Laboratory
BIO-plex / ALSSIT – Advanced Life Support System Integrated Test Bed
BPC – Biomass Production Chamber
CEEF – Closed Ecology Experiment Facility
CELSS – Controlled Ecological Life Support Systems
LMLSTP – Lunar-Mars Life Support Test Project Phase I, II, and III
LSSIF – Life Support Systems Integration Facility
MELISSA – Micro-Ecological Life Support System Alternative

Literature Cited and Used

Ames Research Center Closed Life-Support Systems, NASA SP-134, Government Printing Off, Washington DC, 1967.

Ananyev G., Kolber Z., Klimov D., Falkowski P., Berry J., Rascher U., Martin R., Osmond B. Remote sensing of heterogeneity in photosynthetic efficiency, electron transport and dissipation of excess light in *Populus deltoides* stands under ambient and elevated CO2 concentrations, and in tropical forest canopy, using a new laser-induced fluorescence transient device, *Global Change Biol.*, 2005, Vol. 11, p. 1195–1206.

Anonymous. *Above and Beyond: The Encyclopedia of Aviation and Space Sciences, Life-Support Systems-Mercury,* New Horizons Publ., Chicago, USA, 1968, 8 p.

Anonymous. *Contractor Reports CR-543-545,* NASA, Washington DC [comprise several NASA publications in a single volume: Thermal Integration of Electric Power and Life Support Systems for Manned Space Stations; General Purpose Airborne Simulator-Conceptual Design Report; Experimental Study of Liquid Surface], 1966.

Arthurs K. *Terrariums and Miniature Gardens*, Lane Magazine and Book, Menlo Park, California, USA, 1975, 88 p.

Barron-Gafford G.A., Grieve K.A., Murthy R. Leaf- and stand-level responses of a forested mesocosm to independent manipulations of temperature and vapor pressure deficit, *New Phytol.*, 2007, Vol. 174, p. 614–625.

Barron-Gafford G., Martens D., Grieve K., Biel K., Kudeyarov V., cLain J.E.T., Lipson D., Murthy R. Growth of eastern cotton-woods (*Populus deltoides*) in elevated [CO2] stimulates stand-level respiration and rhizodeposition of carbohydrates, accelerates soil nutrient depletion, yet stimulates above- and belowground biomass production, *Global Change Biol.*, 2005, Vol. 11, p. 1220–1233.

Beyers R.J. *The Metabolism of Twelve Microecosystems*, Ph.D. Thesis, University of Texas, Austin, 1962*a*.

Beyers R.J. Relationship between temperature and the metabolism of experimental ecosystems, *Science*, 1962*b*, Vol. 136, p. 980–982.

Beyers R.J., Odum H.T. *Ecological Microcosms*, Springer-Verlag, New York, 1993, 557 p.

Bil' K.Y., Osmond C.B. Biosphere 2 laboratory: a multi-user, international facility for experimental ecosystem science, *New and Nontraditional Plants and Prospects of Their Utilization*, Proc. 5th Int. Symp., Moscow-Pushchino, Russia, 2003*a*, Vol. 3, p. 22–23.

Bil' K.Y., Osmond C.B. Biosphere 2 laboratory: a multiuser, international facility for experimental ecosystem science, *Emission and Sink of Greenhouse gases on the Northern Eurasia Territory*, Proc. 2nd Int. Conf., Pushchino, Russia, 2003*b*, p. 18–19.

Bjurstedt H. ed. *Basic Environmental Problems of Man in Space,* Proc. 1st Int. Symp., Paris, France, 29 October - 2 November 1962, Springer-Verlag, New York, 1965, 508 p.

Braun H. *The Phoenix Project: An Energy Transition to Renewable Resources,* Research Analysts, Scientific Publications and Productions, Phoenix, Arizona, 1990, 272 p.

Breder C.M., Howley T.H. The chemical control of closed circulating systems of sea water in aquaria for tropical marine fishes, *NY Zoological Society, Zoologica*, 1931, Vol. 9, No. 11, p. 403–442.

Broecker W.S., *GSA Today*, 1996, Vol. 6, No. 7, p. 1–7.

Demyanova E.G., Kudeyarov V.N., Bil' K.Y., Murthy R. Soil respiration and microbial biomass at different levels of CO_2 concentration, *Emis-*

sion and Sink of Greenhouse Gases on the Northern Eurasia Territory, Proc. 2nd Int. Conf., Pushchino, Russia, 2003, p. 43–44.

Dise N.B., Jenkins A. eds. *The CLIMES Project: Whole Catchment Manipulations of CO_2 and Temperature*, Climate Change Research Report 3/95, Norwegian Institute for Water Research, Oslo, Norway, 1995.

Folsome C., Hanson J.A. The emergence of materially-closed-ecosystem ecology, *Ecosystem Theory and Application*, Polunin N. ed., John Wiley, New York, 1986, p. 269–288.

Gentry L., Liptak K. *The Glass Ark: The Story of Biosphere 2*, Viking, New York, 1991.

Gitelson I.I., Lisovsky G.M., MacElroy R.D. *Manmade Closed Ecological Systems*, Taylor and Francis, London and New York, 2003, 402 p.

Golueke C.G., Oswald W.J. Role of plants in closed systems, *Annu. Rev. Plant Physiol.*, 1964, Vol. 15, p. 387–408.

Goto E. *et al.* eds. *Plant Production in Closed Ecosystems*, Proc. Int. Symp. Narita, Japan, August 26–29, 1996, Kluwer Academic Publ, Dordrecht, The Netherlands, 1996.

Henninger D.L., Tri T.O., Packham N.J.C. NASA's advanced life support systems human-rated test facility, *Advanced Space Res.*, 1996, Vol. 18, p. 223–232.

Hochachka P.W. *Living without Oxygen: Closed and Open Systems in Hypoxia Tolerance*, Harvard University Press, 1980, 192 p.

Khlebopros R., Okhonin V., Fet A. 2007 *Catastrophes in Nature and Society: Mathematical Modeling of Complex Systems*, Word Scientific Publishing Co, Singapore, 2007, 336 p. http://www.ebooks-share.net/catastrophes-in-nature-and-society-mathematical-modeling-of-complex-systems

Koch L.J. *Experimental Breeder Reactor-II (EBR-II). An Integrated Experimental Fast Reactor Nuclear Power Station*, Authorized by Argonne Natl. Lab., 2004. www.lpd.anl.gov/anlpubs/2004/03/49168.pdf

Kolber Z., Klimov D., Ananyev G., Rascher U., Berry J., Osmond B. 2005 Measuring photosynthetic parameters at a distance: laser induced fluorescence transients (LIFT) method for remote measurements of photosynthesis in terrestrial vegetation, *Photosynth. Res.*, 2005, Vol. 84, p. 121–129.

Kovrov B.G. Manmade microecosystems with closed material cycle as models of the biosphere, *Biophysics of the Cell Populations and Super-Organismic Systems*, Nauka, Novosibirsk, 1992, p. 62–70. [in Russian]

Kovrov B.G., Fishtein G.N. Microecosystems and experience of their using for studies of protists in a community of microscopical organisms, *J. General Biol.*, 1980, Vol. 46, No. 3, p. 336–344. [in Russian]

Kudeyarov V.N., Ponizovsky A.A., Bil' K.Y., Blagodatsky S.A., Semenov V.M., Kuznetsova T.V., Alekseev A.O., Kudeyarova A.Y., Marthy R. The soil in the intensive forestry biome at "Biosphere 2 Center" Columbia University (Arizona, the United States), *Eurasian Soil Sci.*, 2002, Vol. 34, p. S48–S56.

Kudeyarov V.N., Biel K.Y., Blagodatsky S.A., Semenov V.M., Demyanova E.G., Dorodnikov M.V. Fertilizing effect of the increasing CO_2 concentration in the atmosphere, *Eurasian Soil Sci.*, 2006, Vol. 39, No. 1, p. S6–S14.

Lawton J.H. The Ecotron facility at Silwood Park: the value of 'big bottle' experiments, *Ecology*, 1996, Vol. **77**, p. 665–670.

Levine J.S. *The Phytochemistry of Atmospheres - Earth, the Other Planets and Comets*, Academic Press, Orlando, 1985.

Lovelock J.E. Gaia as seen through the atmosphere, *Atmospheric Environ.*, 1972, Vol. 6, p. 579–580.

Lovelock J.E. *The Ages of Gaia*, Bantam Books, New York, 1990.

Lovelock J.E. *Healing Gaia, Practical Medicine for the Planet*, Harmony Books, New York, 1991, 192 p.

MacElroy R., Smernoff D., Rummel J.D. *Controlled Ecological Life Support System Design, Development and Use of a Ground-Based Plant Growth Module*, NASA CP-2479, 1987, 80 p.

Maguire B.J. Some patterns in post-closure ecosystem dynamics (failure), *Microcosms in Ecological Research*, Giesy J.P.Jr. ed., DOE Symposium Series 52, US Depart Energy, Washington DC, 1980, p. 319–332.

Marino B.D.V., Odum H.T. Biosphere 2: introduction and research progress, *Ecological Engineering*, 1999*a*, Vol. 13, p. 3–14.

Marino B.D.V., Odum H.T. eds., *Ecological Engineering Special Issue*, 2nd edn., Elsevier Science, 1999*b*, Vol. 13, No. 1–4, p. 1–374.

Mazur I.I. ed. *Energy of Future*, Pilot Issue, NPK INTELLEKT-C and ELIMA, December 2004, 193 p. www.magazine@rngs.co.ru [in Russian]

Michie J. History, *Closed Ecological Systems, an Introduction*, Yensen N.P. ed., Unpublished ms., 2004, 105 p.

Myers J. Study of a photosynthetic gas exchanger as a method of providing for the respiratory requirements of the human in a sealed cabin,

Publ. Air University School Aviation Medicine, United States Air Force, Randolph Air Force Base, Texas, 1958, p. 58–117.

Myers N. *The Gaia Atlas of Planet Management*, Pan Books, United Kingdom, 1985.

Nelson M., Soften G. eds. *Biological Life Support Systems*, Synergetic Press, 1988.

Odum H.T., Hoskin C.M. Metabolism of a stream microcosm, *Publ. Inst. Marine Sci.*, Texas, 1957, Vol. 4, p. 116–133.

Ortiz M., Vender G., Adami M., Leigh L.S., Biel K.Y., Yensen N.P. Earth-analog microbiospheres for evaluating biomass and other criticality conditions of global-system models, *Posters, 7th Int. Carbon Dioxide Conf.*, Boulder, Colorado, USA, 2005.

Osmond B.C. *Biosphere 2 Laboratory: A Multi-User, International Facility for Experimental Ecosystem Science*, Unpublished booklet, Biosphere 2 Center, Columbia University, New York, 2003, 13 p.

Osmond C.B. Our eclectic adventures in the slower eras of photosynthesis: from New England down under to Biosphere 2 and beyond, *Annu. Rev. Plant Biol.*, 2014, Vol. 65, p. 1-32.

Osmond B., Ananyev G., Berry J., Langdon C., Kolber Z., Lin G., Monson R., Nichol C., Rascher U., Schurr U., Smith S., Yakir D. Changing the way we think about global change research: Scaling up in experimental ecosystem science, *Global Change Biol.*, 2004, Vol. 10, p. 1–16.

Pitelka L.F. (chair), Coleman J., Dale V.H., Fletcher M.M., Harriss R., Johnson D., Reynolds J.F., Wofsy S.C. *An Evaluation of the Biosphere 2 Center as a National Scientific User Facility*, US Depart. Energy Biological and Environmental Research Advisory Committee, Washington DC, March 2003, p. 1–18.

Ponizovsky A.A., Kudeyarov V.N., Blagodatsky S.A., Alekseev A.O., Biel K.Y., Marthy R. Soil as a component of Biosphere 2, *Priroda (Nature)*, 2003, No. 7, p. 46–52. [in Russian]

Rascher U., Bobich E.G., Lin G.H., Walter A., Morris T., Naumann M., Nichol C.J., Pierce D., Bil K., Kudeyarov V., Berry J.A. Functional diversity of photosynthesis during drought in a model tropical rainforest - the contributions of leaf area, photosynthetic electron transport and stomatal conductance to reduction in net ecosystem carbon exchange, *Plant Cell Environ.*, 2004, Vol. 27, No. 10, p. 1239–1256.

Raven P.H., Evert R.F., Eichorn S.E. *Biology of Plants*, Worth Publishers Inc., 1986, 691 p.

Salisbury F.B. *Forward Manmade Closed Ecological Systems*, Taylor and Francis, London and New York, 2003, p. vii–x.

Spotte S.H. *Fish and Invertebrate Culture Water Management in Closed Systems*, 2nd edn., Wiley, 1979, xvi, 179 p.

Stokstad E. Cash advance, new approach aims to relaunch Biosphere 2, *Science*, 2011, Vol. 333, p. 146.

Suess E. *Die Entstehung der Alpen*, Wien, 1875. [cited by Gitelson *et al.*, 2003]

Suess E. *Das Antlitz der Erde*, Wien, Prague, Leipzig, 1909, Vol. 3. [cited by Gitelson *et al.*, 2003]

Taylor M.R. *Dark Life: Martian Nanobacteria, Rock-Eating Cave Bugs and Other Extreme Organisms of Inner Earth and Outer Space*, Simon and Schuster, 1999. [ISBN 0684841916]

Van Haren J.L.M., Handley L.L., Biel K.Y., Kudeyarov V.N., McLain J.E.T., Martens D.A., Colodner D.C. 2005 Drought-induced nitrous oxide flux dynamics in an enclosed tropical forest, *Global Change Biol.*, 2005, Vol. 11, p. 1247–1257.

Vernadsky V.I. *Geochemistry Essays*, Moscow–Leningrad, 1927. [in Russian]

Vernadsky V.I. On limits of the biosphere, *News Acad. Sci. USSR, Geology*, 1937, Vol. 1, p. 3–24. [in Russian]

Vernadsky V.I. *The Biosphere, (Selected Works)*, Publ. Acad. Sci. USSR, Moscow, 1960, 102 p. [in Russian]

Vernadsky V.I. *The Biosphere*, Springer-Verlag [annotated translation of original], New York, 1998, 192 p.

Warington R. Notice of observation on the adjustment of the relations between animal and vegetable kingdoms, *Quaternary L Chemical Society*, 1851, p. 3.

Webster's New Dictionary, Deluxe Books, Los Angeles, 1994, 437 p.

Yensen N.P. Ecosystems in glass, *Carolina Tips, Burlington, N Carolina*, 1988, Vol. 51, No. 4, p. 13–15.

Yensen N.P., Biel K.Y. The concept of criticality and some principles for sustainability in closed biological systems, biospheres and their classification, *The Complex Systems under Extreme Conditions*, Khlebopros R.G., Soukhovolsky V.G. eds., Krasnoyarsk, Russia, 2005, p. 134–158.

Appendix I

A Description of Biosphere 2 Laboratory
(Modified from Osmond, 2003; Bil' and Osmond, 2003a, 2003b)

Agroforestry mesocosm: recently used for cottonwood experiment to determine the sequestration of carbon under different carbon dioxide concentrations. ***Size***: west bay surface 553 m^2, volume 11,500 m^3; center bay surface 653 m^2, volume 12,400 m^3; east bay surface 553 m^2, volume 11,500 m^3. ***CO_2 injection***: controlled input. ***CO_2 suppression***: Push-pull fans extract high CO_2 air from inside and replace it with equal volumes of low CO_2 air from outside using variable speed controls. ***CO_2 monitoring***: Licor 6262 IRGA via a 16 port Valco rotary sample valve, usually twice daily. ***Temperature:*** Soil, leaf and atmospheric temperature are measured each bay with 4 instrument arrays of 3 air thermocouples and 3 soil thermocouples. ***Vapor Pressure Density:*** lowered using high pressure fogging nozzles with reverse osmosis treated water to achieve 1.2–3.5 kPa with temperatures of 32 °C. ***Irrigation:*** Conventional impact rotary sprinkler heads, mounted high in the space frame, delivered reverse osmotic water. The farm tank capacity was 36,400 liters. ***Water budget***: Water influx and efflux measured, including "rain", fog and the humidity at the in- and out-flow fans, condensation from the air handlers, and sub soil drainage. ***Light levels:*** Apogee quantum sensors in arrays of 12 per bays at 3 heights. ***Air handling:*** 3 air handlers per bay. Two 40 hp tower water coil for chilling and heating and one 60 hp with 3 conditioning coils: Air flow: 4,080 m^3; hot water coil: 1,303,992 btu / hour; chilled water coil: 3,875,916 btu / hour; tower water coil: 4,200,520 btu / hour (btu = British Thermal Units). ***Isolation:*** clear plastic curtains separated the 3 with air locks. Leak rates of 2-5% per hour between bays.

Tropical rainforest mesocosm: had control of and monitoring of: CO_2 concentration, soil and atmospheric temperature, atmospheric vapor pressure, irrigation, soil moisture, light level, air handling, closure. ***Size***: surface 1950 m^2, volume 27,000 m^3. ***CO_2 injection:*** controlled input. ***CO_2 suppression:*** During full summer growth, CO_2 concentrations suppressed to ±220 ppm during night respiration.

CO_2 monitoring: Licor 6262 IRGA via a 16 port Valco rotary sample valve, usually twice daily. ***Temperature:*** Soil, leaf and atmospheric temperature measured with 5 instrument arrays. ***Vapor Pressure Density:*** controlled as in the agroforestry mesocosm. ***Irrigation:*** As in the agroforestry mesocosm, but with a 57,400 liters tank. ***Water budget:*** As in the agroforestry mesocosm, but with condensation from the air handlers and floor accumulations from utility wash down. ***Light level:*** in arrays with high and low quantum levels per sensor. ***Air handling:*** by three with 40 hp and three with 60 hp with capacities similar to the agroforestry mesocosm; and air flows of 8,150 m^3 / min; hot water coil a surface of 24.3 m^2 moving 7,410 liters / min and transferring 11,627,748 btu / hour; tower water coil surface of 48.6 m^2 moving 4,089 liters / min transferring 8,145,864 btu / hour. ***Isolation:*** clear plastic curtains separate the system, but with air locks and a 61-cm diameter pressure-equilibrating tunnel to relieve pressure to / from the savannah. This tunnel, with ultrasonic flow transducers, allows the CO_2 and water balance to be adjusted.

Ocean mesocosm: The 2,650 m^3 coral reef-containing "ocean" is instrumented for light, dissolved oxygen, temperature, conductivity and CO_2 % and features: a two-stage heat exchanger to maintain 26.5 °C, a wave generator via vacuum lift, and manual chemical adjustment.

Savannah / thorn scrub mesocosm: The 1,250 m^2 surface area with a volume of 25,000 m^3 is not separated from the ocean and shares air handlers (four 60 hp units and three 40 hp units) with an air flow of 9,500 m^3 / min have control over air temperature, rain, and humidity, but no CO_2 control. The hot water coil with 32.4 m^2 surface conducts 9,880 liters / min moving 5,215,968 btu / hour; *chilled water coil* – 32.4 m^2, 43,920 liters / min, 15,503,664 btu / hour; *tower water coil* surface of 56.7 m^2, moving 46,570 liters / min transferring 9,375,920 btu / hour.

Desert mesocosm: The surface area of 1400 m^2 has a volume of 22,000 m^3. Atmospheric temperature, rain, and humidity are controlled, but not CO_2. Three 60 hp air handlers move 4,077 m^3 of air / min; with air heated on surface of 24.3 m^2 coil moving 7,410 liters / min transferring 3,911,976 btu / hour; air cooled on surface of – 32.4 m^2 coil moving 32,940 liters / min of water exchanging 11,627,748 btu / hour; a 56.7 m^2 tower water coil moving 17,040 liters / min exchanging 3,690,168 btu / hour.

Data Acquisition and Control: There are two separate Ethernet networks within the Biosphere 2 Laboratory from which data is automatically collected on a schedule. External users can view and retrieve any research data.

The Power and Water Energy Center: electricity is supplied off the local grid or with two redundant backup generators (a natural gas 1,500 kW prime mover and a diesel 1,500 kW emergency generator). During the 10 dry months, the tower water loop of 18,925 liters / min provides a fraction of the cooling. The rest is by one of two redundant 600 tons water chillers (4 °C) which provide up to 18,925 liters / min. A 302,800-liter eutectic tank is a third back up. The two natural gas boilers deliver 1,500 liters / min of hot water at 85 °C.

3
Accelerated Method for Measuring and Predicting Plants' Stress Tolerance

Karl Y. Biel[1,2,*], **John N. Nishio**[3,*]

[1]*Institute of Basic Biological Problems, Russian Academy of Sciences, Pushchino, Moscow Region 142290, Russia*
[2]*Biosphere Systems International Foundation, Tucson, Arizona 85755, USA*
[3]*Biocompatible Plant Research Institute, Chico, California 95973, USA*

What is essential is invisible to the eye, –
The Little Prince (by Antoine De Saint-Exupéry)

Abstract

A method for measuring stress-tolerance and the metabolic interactions between mitochondria and chloroplasts in leaf slices is described. Oxygen evolution of the leaf slices under stress treatments, such as high light, high temperature, and / or anoxia was monitored. After the stress treatments, different rates of photosynthetic oxygen evolution and rates of recovery from the stress were used as measurements of stress tolerance. We illustrate application of the method with many higher plants and unicellular algae. The method is useful to examine the interactions among photosynthesis, photorespiration, and respiration as well as to estimate photoprotective function of exogenous oxidants, such as nitrate anion and oxaloacetic acid. The approach can be used for investigation of photosynthetic metabolism and the cooperation among organelles. Utilizing the method to follow changes in such metabolic activities as protein synthesis, protein conformation states, enzymes activity, carbon

* *Corresponding authors*: karlbiel@hotmail.com; nishio@biocompatible.org

metabolism, and gene expression at different points during the treatments will be educational.

Keywords: Oxygen electrode, Leaf slices and unicellular algae, Anoxia, High light and temperature, Photosynthesis, Respiration, Stress-resistance of plants

3.1 Introduction

In the present chapter, we describe a simple method using the oxygen electrode that allows examination of the response of respiration and photosynthesis in leaf slices or algae to anoxia and high light under different temperatures. The method is useful to examine the interactions among photosynthesis, photorespiration, and respiration. The method provides a quantifiable assessment of stress tolerance that also permits us to examine fundamental biochemically and genetically related responses involved in stress tolerance and the cooperation among organelles. Additionally, we demonstrated a role for compounds, such as nitrate anion (NO_3^{2-}) and oxaloacetic acid (OAA), as protective agents against photoinhibition, and we examined the role of dark adaptation in the activation of photosynthesis and NO_3-dependent oxygen evolution. A physiological and ecological role of a dark period (night) in stress tolerance is presented. Utilizing the method to follow changes in such metabolic activities as protein synthesis, protein conformation states, enzymes activity, carbon metabolism, and gene expression at different points during the treatments will be educational. A practical example of screening for drought tolerance is presented in the Appendix II.

3.2 Background

Plant abiotic stresses, due to drought, high light, nutrient deficiency, heat, cold, or combinations thereof, are unavoidable conditions for most plants. The loss of productivity to abiotic factors is significantly higher than that lost to biotic factors (Boyer, 1982). Measurements of total productivity are reliable indicators of stress. Chlorophyll fluorescence measurements in conjunction with gas exchange or oxygen evolution; measurements of metabolites, such as betaine or polyols; and activity measurements of enzymes, such as superoxide dismutase and catalase, have been used as indicators of stress tolerance. It is also possible to examine metabolic pools by pulse-chase methods, and to examine specific enzymes related to stress tolerance and photosynthesis. Such measurements, alone or in combinations, have been very useful in assessing whether stress has occurred or not, and they have allowed us to examine certain fundamental aspects of how or why certain plants are more tolerant of environmental stress. As more "stress" genes are identified, measurements of their expression and examination of the functional proteins will be extremely useful.

Part of the reason there is little information about how different organelles and parts of the leaf are involved in stress tolerance is technological. There are really no practical systems that readily allow examination of interactions between anabolic and catabolic reactions in relation to stress tolerance in physiological intact systems. Furthermore, our understanding of carbon metabolism within leaves is far from complete.

We have developed a method using a simple oxygen electrode for measuring stress tolerance that also can be used to investigate the interactions between mitochondria and chloroplast, *in vivo*. Many fundamental aspects of plant metabolism can be investigated using the approach. Because the chapter in some part discusses interactions between the respiratory and photosynthetic organelles and photosynthesis under anaerobic conditions, the subjects are briefly reviewed below.

3.2.1 Interaction between Anabolism and Catabolism

Mitochondria have two well-known and important functions. Firstly, mitochondria are considered the "power houses" of the cell, because they are the site of the tricarboxylic acid (TCA) cycle, which is responsible for most of the NADH formed during the oxidation of reduced carbon, and electron transport. During electron transport, the NADH (and $FADH_2$) produced by the TCA cycle are oxidized to drive the formation of the $\Delta\mu_{H}{}^{+}$, which is used to produce ATP. The ATP, NADH, and $FADH_2$ are used for a host of energy consuming anabolic reactions such as nitrogen and sulfur reduction, amination of keto-acids for protein synthesis, lipid synthesis, nucleic acid synthesis, and so forth. Secondly, intermediates of the Krebs cycle provide carbon backbones for anaplerotic reactions. How the mitochondria might help the chloroplasts perform their function is an intriguing question.

In plants the chloroplasts are the "power houses", and the mitochondria are the "substations", passing power from the chloroplasts to the growing plant. From a biological perspective, even mitochondria in heterotrophs ought to be called substations, because they transfer energy from plants and chemoautotrophs into usable energy forms. Chloroplasts are the site of photosynthesis that involves light energy capture, photosynthetic electron transport, and ultimate conversion of light energy to "functional" chemical energy, ATP, Fd_{red}, and NADPH.

We will focus on one of the main downstream events relating to utilization of the ATP and NADPH, carbon fixation.

In the late 1930's, Ruben's pioneering work using ^{11}C to investigate carbon metabolism in plants (Ruben *et al.*, 1939; Ruben and Kamen,

1941; Ruben, 1943) eventually led to the elucidation of the primary paths of carbon metabolism in C_3 plants in the USA (Chapter 11; see also Calvin and Benson, 1948; Benson and Calvin, 1950*a*, 1950*b*; Bassham, Benson *et al.*, 1954; Bassham and Calvin, 1957; Benson, 2002) and in the USSR (Doman, 1952; Nichiporovich, 1953, 1955; Zalensky *et al.*, 1955; Doman *et al.*, 1958).

Until the 1950's, it was generally accepted that carbon was "fixed" into carbohydrates during photosynthesis, and plant growth was due to respiratory metabolism of the photosynthetic products. It was assumed that there was little interaction between photosynthesis and respiration. The radiotracer studies revealed common biochemical pathways for carbon reduction and carbohydrate oxidation (Benson and Calvin, 1950b). It was apparent that there were complex interactions between photosynthesis and respiration (McLachlan and Porter, 1959), and it was also thought that the two processes were separated spatially and temporally (Zalensky, 1961).

Professors Andrew A. Benson and Melvin Calvin examined the possibility that glycolytic reactions were inhibited by light, and they showed that pyruvate oxidation occurs in the light (Benson and Calvin, 1950*b*). The results were later confirmed by others (Hirt *et al.*, 1971; Karpilov *et al.*, 1978). Additionally, ultrastructural studies showed that there were distinct differences in the structure of chloroplasts and mitochondria from the palisade mesophyll (PM) and spongy mesophyll (SM) (Chugunova *et al.*, 1976; Karpilova *et al.*, 1982).

Evidence for interactions between chloroplasts, mitochondria, and the cytosol as "a continuum with respect to phosphate bond energy" (McLachlan and Porter, 1959) was greatly enhanced, when it was shown that chloroplasts have specific permeases to transport triosephosphates and organic acids, such as malic acid and OAA (Heber, 1974). Such compounds not only provided carbon skeletons but also carried reductive power and ATP to the cytosol (Simonis and Urbach, 1973; Heber, 1974) for processes other than carbon fixation (Raven, 1976*a*, 1976*b*); see review (Semikhatova and Zalensky, 1982). The transported compounds could also be utilized by mitochondria, or even the reverse could occur; that is, OAA and malic acid could be transported to the chloroplast from the cytosol (Magomedov, 1982; Nasyrov, 1982; Bil' and Nishio, 2000*a*).

One approach to understanding the role of oxygenic respiration in metabolism is to conduct experiments in the absence of oxygen. Tobacco leaves placed in anoxic conditions survived longer when exposed to light and dark cycles compared to just dark alone (McLachlan and Porter, 1959). The simple interpretation is that in tobacco, photosynthetic oxygen evolution was possible under anoxic conditions, and the products of photosynthesis were somehow utilized under anoxic conditions.

Drs. R. L. Travis and R. C. Huffaker (1970) examined the role of respiratory "energy" on nitrate synthetase. They showed that under aerobic conditions, light (photosynthetic "energy") was required to maintain nitrate synthetase activity and protein synthesis (polyribosome accumulation) (Travis and Huffaker, 1970). Thus, it was clear that photosynthesis could supply energy for growth and maintenance and that photosynthesis and respiration must interact in some way to provide for normal growth of plants (Raven, 1976*a*, 1976*b*; Zalensky, 1977; Zalensky *et al.*, 1980; Semikhatova and Zalensky, 1982).

While simple in principle, the interactions between photosynthesis and respiration are quite complex (Semikhatova and Zalensky, 1982). In particular, it was shown that the activity of the tricarboxylic acid cycle is not completely inhibited in the light, because some photosynthetic products are influenced and metabolized by respiratory pathways (Heber *et al.*, 1964; Ried, 1970; Raven, 1972; Chapman and Graham, 1974*a*, 1974*b*; Chapman and Osmond, 1974; Karpilov *et al.*, 1978; Ivanova *et al.*, 1993; Pärnik and Keerberg, 1995; Pärnik, Voronin *et al.*, 2002). Moreover, ^{14}C label organic acids of the TCA cycle can enter the Benson-Bassham-Calvin cycle, not only by recarboxylation of CO_2, but directly through the transfer of carbon skeletons (Chapman and Graham, 1974*a*, 1974*b*; Chapman and Osmond, 1974). Oxy-acids, produced by the tricarboxylic acid cycle, can be transported to the chloroplast, where they can participate in transamination reactions and the synthesis of amino acids (Larsson, 1979). The amount of interaction between photosynthesis and respiration can change significantly during the development of plants (Raven, 1972).

3.2.2 Cooperation between Photosynthesis and Respiration under Stress

The interaction between chloroplasts and mitochondria under stressful conditions is not well understood. Probably, Professor Igor A. Tarchevsky and Dr. Tatyana M. Konyukhova (see review Tarchevsky, 1982) were the first to show, *in vitro*, that under extreme conditions, mitochondria can provide chloroplasts not only CO_2 from respiration, but also ATP. Low rates of photoautotrophic carbon fixation are possible in mixotrophically grown photosystem-I-minus or photosystem-II-minus mutants of *Chlamydomonas*, which supports the notion that respiratory products (ATP and NADH) can contribute to photosynthetic carbon fixation (Hertz *et al.*, 1985). Further confirmation of the contribution of mitochondria to photosynthetic metabolism (Krömer *et al.*, 1988, 1993; Krömer and Heldt, 1991; Gardeström and Lernmark, 1995; Hurry, Tobiæson *et al.*, 1995) and in the protection of photosynthesis against photoinhibition (Saradadevi and Raghavendra, 1992) has been more recently published. The just mentioned experiments were done mainly with mutants and protoplasts. Intercellular relations are lacking in protoplasts, and photosynthetic metabolism is altered, because normal sink / source relations are changed. In the mutants, pleiotropic effects may alter the overall metabolism of the cells. Consequently, regulatory mechanisms may be compromised.

A recent review concludes that while we know that TCA cycle activity and respiratory activity occur in the light at reduced rates compared to the dark, the specific contribution, and even the type of activity, remains to be elucidated (Krömer, 1995). However, see the work of Professor E. A. Chapman with colleagues, which shows specific carbon fluxes between the tricarboxylic acid cycle and chloroplasts in the light (Chapman and Graham, 1974*a*, 1974*b*; Chapman and Osmond, 1974).

Anatomical characteristics are also important. Leaves are heterogeneous organs, and the distribution of metabolism across leaves (Nishio *et al.*, 1993; Sun and Nishio, 2001; Sun *et al.*, 1996*a*, 1996*b*; Biel *et al.*, 2003, 2008, 2010; Fomina and Biel, 2014; see also Chapter 7) should be considered. A recent model illustrated our hypothesis that abiotic forces impact the distribution of metabolism (Soukhovolsky *et al.*, 2002; Biel *et al.*, 2010; Fomina and Biel, 2014) across leaves. The growth environment significantly alters leaf anatomy (Clements, 1905; Thoday, 1931), and the role of such changes and how they affect the interactions between photosynthesis and respiration are not known. The method described here allows us to begin to address such issues.

3.3 How Is Stress Tolerance Measured?

The stress test is conducted by monitoring the oxygen status of plant material. Oxygen evolution of leaf slices and algae is measured polarographycally with a Rank electrode in a temperature-controlled, water-jacketed cuvette containing 40 mM HEPES, pH = 7.0 or 7.6, 300 mM sorbitol, and 10 mM $NaHCO_3$. After steady state dark respiration is established, light saturation curves are measured for each species and set of plants grown under the same conditions. The time required to conduct the light saturation curve is 15–20 minutes. Once conditions are established for each set of plants, the initial light-saturated photosynthetic rate can be measured on subsequent samples.

For oxygen evolution measurement of leaf slices (cross-sections), slices approximately 300–400 μm thick were prepared. The leaf thickness was adjusted for each species, as different species had different tendencies to remain intact while spinning in the oxygen electrode, which is likely due to cell size and growth conditions. A light saturation curve was conducted on all of the dark-adapted samples.

Figure 3.1 illustrates a model oxygen evolution trace during the oxygen evolution stress assay (OESA). All portions of the curve, such as R2, the post illumination respiratory "burst", are not always discernible in all experimental cases. Post illumination respiration (PIR) was discernible in isolated tissue. The PIR rate was not always observed in the whole leaves and the *Chlorella vulgaris* utilized. Often R4 is the same as R1. The temperature is varied depending on the species, but 37 °C is used most commonly. After the dark respiration rate (R1) is established, the steady state oxygen evolution rate (P1) is measured under saturating light (1350–2200 μmol quanta / $m^2 \cdot s$ (quanta measured as PAR–future reference to photon flux will be in shortened units of μmol / $m^2 \cdot s$).

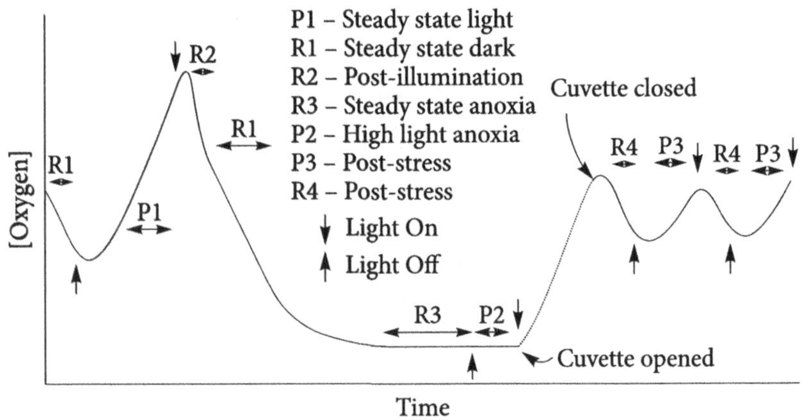

Fig. 3.1. Model trace of oxygen evolution stress assay
Note: All other traces in all other figures are real traces (not models). Oxygen evolution of leaf slices and / or algae is measured polarographycally with a Rank electrode in cuvette containing 40 mM HEPES, pH = 7.0 or 7.6, and 10 mM $NaHCO_3$, temperature – 37–42 °C

Then the light is turned off, and in the dark, the leaf slices deplete the oxygen in the cuvette. The "anoxic" conditions, thus attained (R3), are maintained for ~30 min. Saturating light is then turned on for 2–7 min (P2). After the light is turned off, the chamber is opened (in the dark) to allow oxygen back into the chamber. After several min, the cuvette is closed, and rates of respiration (R4) and light dependent oxygen evolution (P3) are again measured for several cycles.

The photosynthetic stress resistance index is calculated as % of P1, [P3 rate / P1 rate] × 100, where P1 is the initial photosynthetic rate, and P3 is the rate after the photoinhibitory treatment. When respiration rates change during the treatment another formula to calculate the photosynthetic stress resistance index (PSRI) is utilized, - [P3 rate –R4 rate / P1 rate - R1 rate] × 100.

The amount of leaf tissue to buffer volume used for each sample must be adjusted to reach anoxia in a reasonable amount of time (which is dependent on the rate of respiration). In some cases, self-shading is possible, and this is evident in the light saturation curves. If light saturation is not reached, the samples should not be utilized.

During development of the assay, we examined the effect of temperature (25, 30, 35, and 40 °C (as high as 42 °C)) on the stress assay. Initial experiments were conducted at 42 °C to insure an adequate photoinhibitory response. The degree of stress can be adjusted for different species by controlling the temperature, time in light, and time under anaerobic conditions. Most of our experiments are now conducted at 37 °C, which is the normal upper limit for many C_3 plants, and similar results were obtained by regulating the time of anoxia and length of illumination.

3.3.1 Testing Possible Artifacts of the Stress Test

It was difficult to fully explain the desired decrease in oxygen evolution after the anoxic treatment, high temperature, and high light modeled in Figure 3.1. Therefore, we tested several possibilities to better understand the factors that may have contributed to the decrease in photosynthesis.

Figure 3.2 illustrates that *Beta vulgaris* leaf slices at 42 °C were photoinhibited by the light treatments under low oxygen concentration, while there was little effect on respiration. The photoinhibitory effects were observed only when anoxia, light, and temperature were appropriately maintained.

It is possible that the decreased rates of light-driven oxygen evolution (Figure 3.2) were due to anoxia alone, so additional experiments to test the possibility had to be conducted. Experiments shown in Figure 3.3 were done under the same conditions as in Figure 3.2. In panel of Figure 3.3A cycles of light and dark were given over the same length of time as in Figures 3.2, 3.3B, and 3.3C. The leaf slices were not damaged by the time in the cuvette at 42 °C, so the decrease in photosynthesis was not due to time in the cuvette.

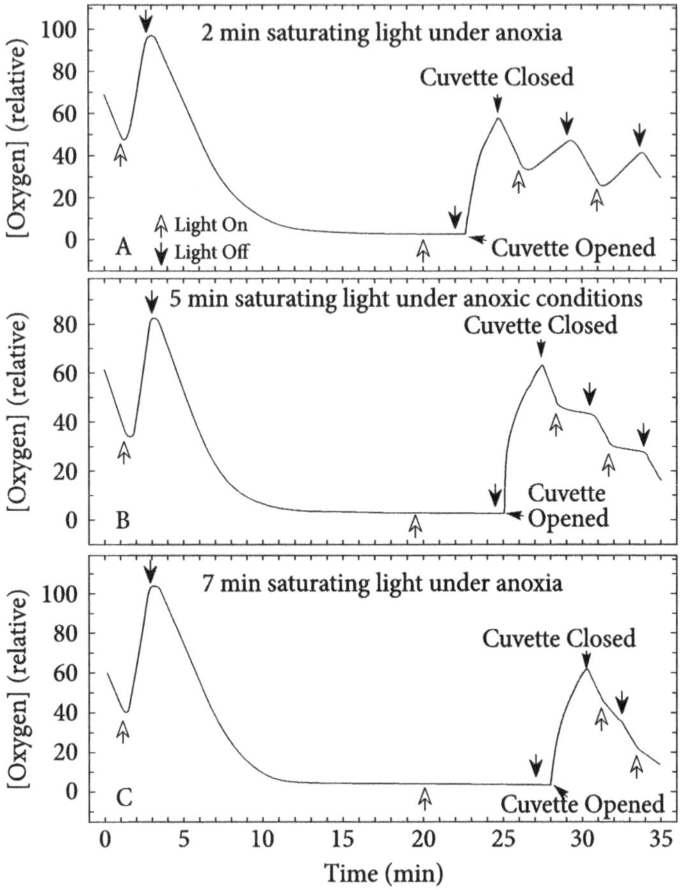

Fig. 3.2. Effect of anoxia and high light on photosynthetic HCO_3^--dependent oxygen evolution of sugar beet (*Beta vulgaris*) leaf slices vs. anoxia time. A – 2 min saturating light under anoxia; B – 5 min saturating light under anoxic conditions; C – 7 min saturating light under anoxia

In panel of Figure 3.3B the leaf slices were put in the dark under aerobic conditions, which allowed respiration to occur unhindered. After the initial photosynthetic time period, the leaf slices were allowed to respire in the dark for 2-3 minutes. Then the chamber was opened and allowed to equilibrate with the air. After about 18 min, the cuvette was closed and respiration monitored. The saturating light was then turned on, and oxygen evolution was nearly identical to the baseline rate.

Panel of Figure 3.3C shows the effect of dark treatment at the depleted oxygen concentration. The experiment was conducted as in Figure 3.2, except no light was given under anoxic conditions. After oxygen was allowed back into the chamber, the rate of oxygen evolution was slightly lower than the baseline rate, but significantly greater than the slices given a light treatment under low oxygen concentration (Figure 3.2).

The results of Figure 3.3B and 3.3C panels, taken together show that the decreased rate of oxygen evolution in Figure 3.2 is not due to anoxia alone. The experiments repeated with other C_3 species showed similar results (not shown).

Figure 3.3 shows that the photoinhibitory decrease in oxygen evolution illustrated in Figure 3.2 is not due to artifacts, such as time in cuvette, high temperature, and / or anoxia.

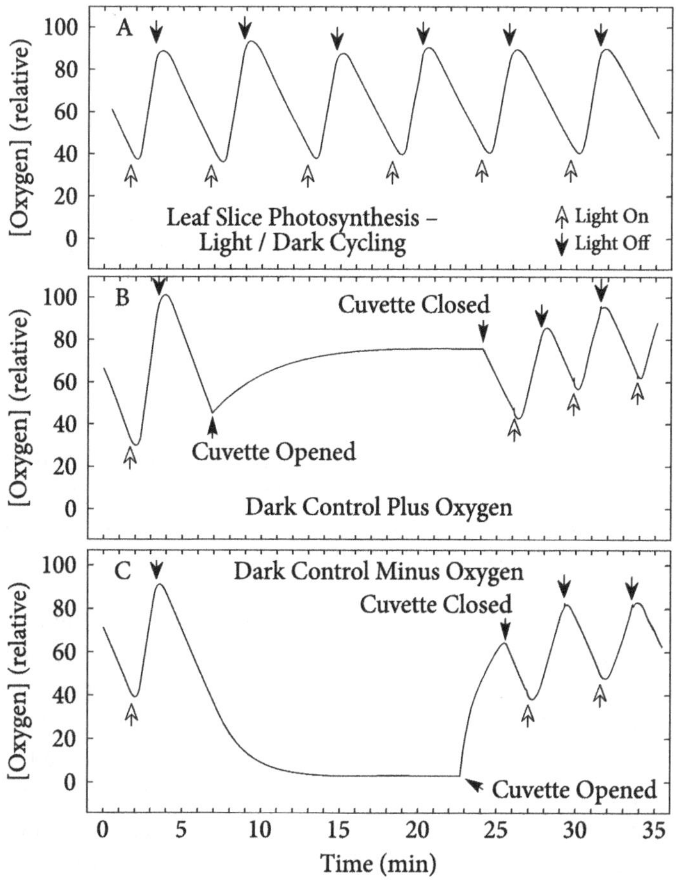

Fig. 3.3. Anoxia and time in cuvette do not impact photosynthetic HCO_3^--dependent oxygen evolution

Note: Sugar beet (*Beta vulgaris*) leaf slices photosynthesis is stable 35 minutes. Anoxia has minimal impact on oxygen evolution: A – light / dark cycling; B – dark control plus oxygen; C – dark control minus oxygen

3.3.2 *Effect of Temperature and Chemical Additions on the Oxygen Evolution Stress Assay*

Figure 3.4 shows the effect of temperature on sugar beet HCO_3^--dependent oxygen evolution. The initial rate of oxygen evolution at 25 °C was lower than at 30, 35, and 40 °C (left portion of Figure 3.4). Unlike HCO_3^--dependent oxygen evolution, respiration did not saturate at 30 °C, but increased with temperature up to 40 °C (Figure 3.4).

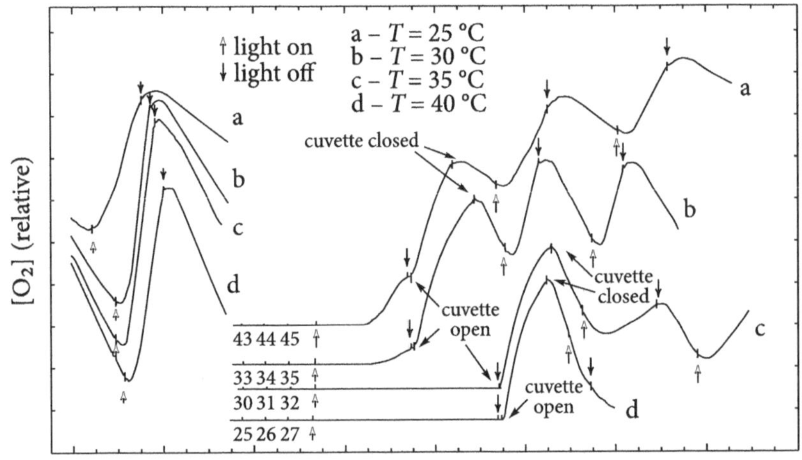

Fig. 3.4. Effect of temperature on photosynthetic HCO_3^--dependent oxygen evolution of sugar beet (*Beta vulgaris*) leaf slices under anoxia. Shown on the left are R1, P1, and R2. The break in the traces is because the majority of the extended time under anoxia has been deleted from the graphs

Higher temperatures decreased or inhibited photosynthetic electron transport under anoxia. On the right of Figure 3.4 are the oxygen evolution traces after different lengths of anoxia. At 25 and 30 °C red beet (*Beta vulgaris*) leaf slices, under anoxic conditions, exhibit HCO_3^--dependent oxygen evolution after 2-3 min. The greater rate occurred at 25 °C, even though the leaf slices at 25 °C were under anoxic conditions 10 min longer than those measured at 30 °C. At 35 °C and after 33 min of anoxia the leaf slices did not exhibit light dependent oxygen evolution, even after 8 min of illumination (length of illumination also causes stress (Figure 3.2). There is no HCO_3^--dependent oxygen evolution at 40 °C and after 28 min of anoxia.

Lower photosynthetic stress resistance index values were measured under higher temperatures. When the leaf slices were treated for shorter times at the higher temperature of 35 °C (Figure 3.4, trace *c*, right side of panel), oxygen evolution was significantly decreased from the original rate (Figure 3.4, trace *c*, left side of panel), and at 40 °C, oxygen evolution was completely inhibited by the treatment (Figure 3.4, trace *d*, right side of panel). The apparent decrease in the rate of oxygen uptake at the end of trace (*d*) is not due to photosynthesis but is due to a lack of oxygen (see Figures 3.2 and 3.3). Thus, elevated temperatures, more than anoxia, are responsible for the lowered photosynthetic stress resistance index. Extended anoxia will eventually lead to responses similar to that with the elevated temperature, however.

The results shown in Figure 3.4 suggested that at the higher temperatures an oxidant is depleted or the ability to accept electrons is somehow inhibited by the high temperature treatment. To test the possibility, we added an exogenous electron acceptor, sodium nitrate ($NaNO_3$).

Figure 3.5 show that exogenous $NaNO_3$ allows rapid oxygen evolution under anoxic conditions after a short activation period. The assay exhibited the same results, whether $NaNO_3$ was added more than 30 min before or only 2 min before the oxygen evolution measurements under anoxia were conducted (data not shown). Thus, nitrate anion rapidly penetrated the leaf slices, and during the 30-min incubation period was not reduced significantly to alter the results. Unlike the other experiments (Figures 3.2, 3.3 and 3.4), the light was left on during the period that oxygen was allowed back into the cuvette (cuvette open). This was done to minimize production of metabolites in the dark, after the light period.

The control leaf slices, to which no exogenous compounds were added, exhibit photoinhibition, as the leaf slices showed net oxygen uptake after the photoinhibitory stress (Figure 3.5). After about 3 min of oxygen uptake, a low oxygen evolution rate was observed. Similar results were obtained when serine was added to the leaf slices (Figure 3.5). In contrast, under anoxic conditions in the presence of $NaNO_3$, oxygen evolution is comparable to the pretreatment rate. Exogenous ATP or glycine, added to the solution in which the leaf slices were being monitored, increases photoinhibitory tolerance and the PSRI compared to the controls (Figure 3.5). Unlike the $NaNO_3$ treatment, ATP and glycine do not stimulate photosynthetic electron transport under anoxic conditions, but in contrast to the control, oxygen evolution after the stress treatment was immediately positive. Additions of glycine or oxaloacetic acid (see Figure 3.14B) caused results similar to that of ATP additions, except there is a very brief period of oxygen uptake after the cuvette is closed.

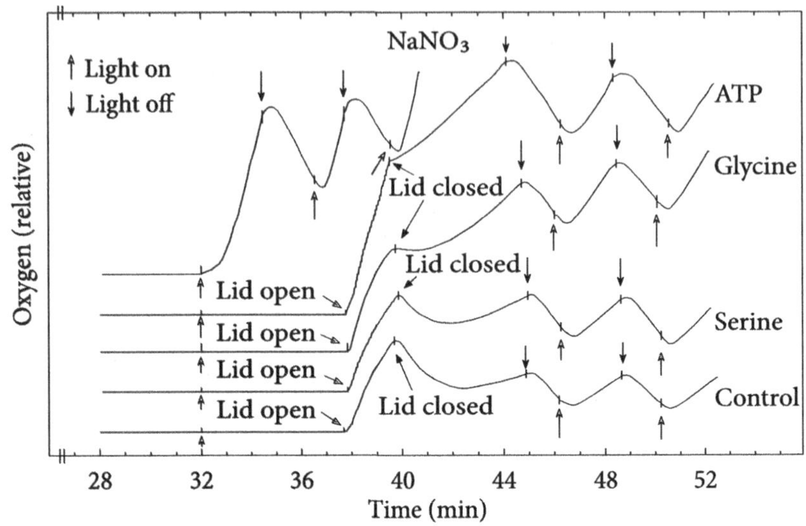

Fig. 3.5. Substrates were added to sugar beet (*Beta vulgaris*) leaf slices into the cuvette after light curves were measured (in the dark). $NaNO_3$ allowed oxygen evolution under anoxic conditions, and no photoinhibition was observed. While ATP did not allow oxygen evolution under anoxic conditions, oxygen evolution immediately occurred when oxygen was resupplied. Glycine decarboxylation yields NADH that can be a source of ATP, and glycine protected like ATP did. Serine, which is formed during photorespiration, and is "recycled" back to RuBP (and requires ATP) did not protect and exhibited traces similar to the control, to which no exogenous substrates were added

Taken together, the results shown in Figures 3.4 and 3.5 suggested the possibility that some compound, possibly from the mitochondria, was being supplied to the chloroplasts that protected the leaf slices from photoinhibition.

Taken together, the results shown in Figures 3.4 and 3.5 suggested the possibility that some compound, possibly from the mitochondria, was being supplied to the chloroplasts that protected the leaf slices from photoinhibition.

To test the possibility that dark respiration might play a role, the temperature dependence of oxygen evolution was first measured under conditions of light / dark cycling followed by constant dark. As the temperature is elevated, the 5-min period becomes an adequate time period to deplete all the oxygen in the cuvette. At 40 °C the rate of respiration is still increasing (Figure 3.6A). There is an abrupt inhibition of respiration at 45 °C due to the anoxic condition that is rapidly reached at 45 °C in the dark. Oxygen evolution following the 45 °C treatment is also inhibited. If a mitochondrial metabolite were required for activation of photosynthesis, we hypothesized that maintaining oxygen in the cuvette at 45 °C would maintain photosynthesis.

Figure 3.6B shows that leaf slices given a light / dark cycling during the 5-min period, required to raise the temperature, decreased the damaging impact of high temperature on light dependent oxygen evolution and respiration, even at 50 °C (net photosynthesis would be positive). It is likely that prolonged treatment at 50 °C would eventually kill the leaf slices, as in Figure 3.6A.

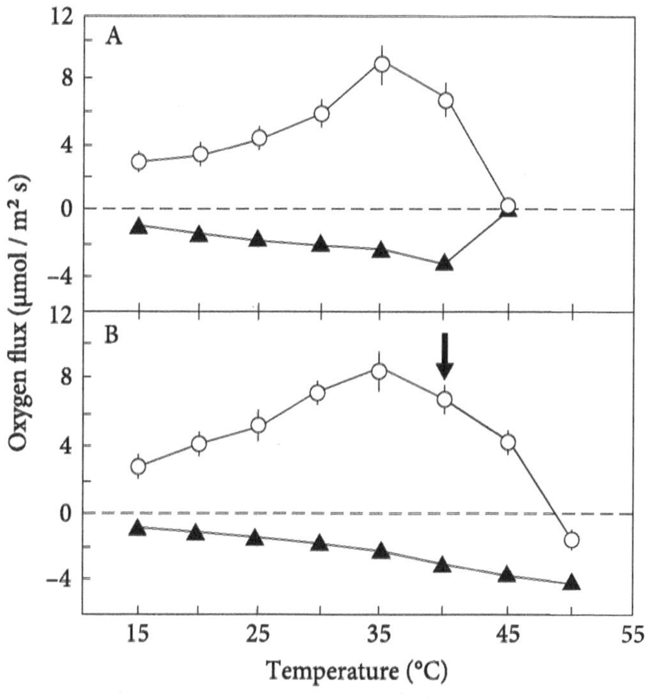

Fig. 3.6. Oxygen flux of red beet (*Beta vulgaris*) leaf slices vs. of temperature (A and B). Rates of photosynthetic oxygen evolution (○) and respiration (▲) were measured during two cycles of light and dark; then the temperature was raised over a 5-min period in the dark

Note: Measurements were done polarographycally with leaf slices in 40 mM HEPES-KOH buffer, pH 7.0, plus 10 mM $NaHCO_3$. Excitation light was 1350 μmol / $m^2 \cdot s$. Plants were grown in the soil in a growth-chamber at a constant 23 °C and a 16 hours light period of 500 μmol / $m^2 \cdot s$. Age of plants was 75–78 days. (B) Arrow indicates a cycle of dark / light was provided during the 5-min period when the temperature was raised from 40 to 45 °C and from 45 to 50 °C. The cycle of light / dark prevented death at 45 °C. Rate of raising temperature in cuvette was 1.0–1.5 °C / min, $n = 3$

3.4 Practical Applications

Figure 3.7 clearly illustrates that the oxygen evolution stress assay can distinguish between different metabolic states in leaves, and that it can be used to screen for improved stress tolerance from foliar treatments. In the experiments shown, photosynthesis is maintained in plants treated to improve drought tolerance. One day after treatment, the plants exhibited much higher rates of oxygen evolution after the stress test than did the controls (the treated plants had higher PSRI's). The experiment in panel of Figure 3.7D was conducted 4 days after the plants were treated.

The control traces in Figure 3.7 show how different plants respond to the stress test. Plants of different species and ages responded differently to the stress assay. This is an important consideration in using the oxygen evolution stress assay in applied situations. For example, the least difference is shown for cabbage; however, the control cabbage exhibited the greatest increase in stress tolerance in the untreated controls, compared to the other species. Species and varietal differences in stress tolerance and responses to the treatments must be taken into consideration when using the oxygen evolution stress assay.

Direct comparison of the photosynthetic stress resistance index from different species or the same species of different ages (leaf age) should be done with caution; however, if the purpose of the experiment is to compare species responses or the PSRI of different aged leaves or plants, then the OESA can provide important information.

The oxygen evolution stress assay is useful for screening varieties. In all cases shown in Figure 3.7, cycles of light and dark after the photoinhibitory treatment increased the post-stress rate of oxygen evolution, but even in soybean complete recovery did not occur. Plants exhibiting improved drought tolerance also had high photosynthetic stress resistance indexes (not shown).

In a series of experiments, we demonstrated that the green alga, *Chlorella vulgaris*, was very resistant to anoxia. Anoxic treatments varied from 12 minutes to 3 hours. As in C_3 higher plants in the presence of $NaNO_3$ (Figure 3.4), the alga did not require the cuvette to be opened to allow oxygen back into the chamber, to exhibit competent electron transport (not shown). Thus, a photosynthetic stress resistance index as described (see: Additional Materials and Methods in Appendix I and Figure 3.1) cannot be calculated. Calculations of the photosynthetic stress resistance index, when oxygen evolution occurs directly from the anoxic conditions will be referred to as PSRI*.

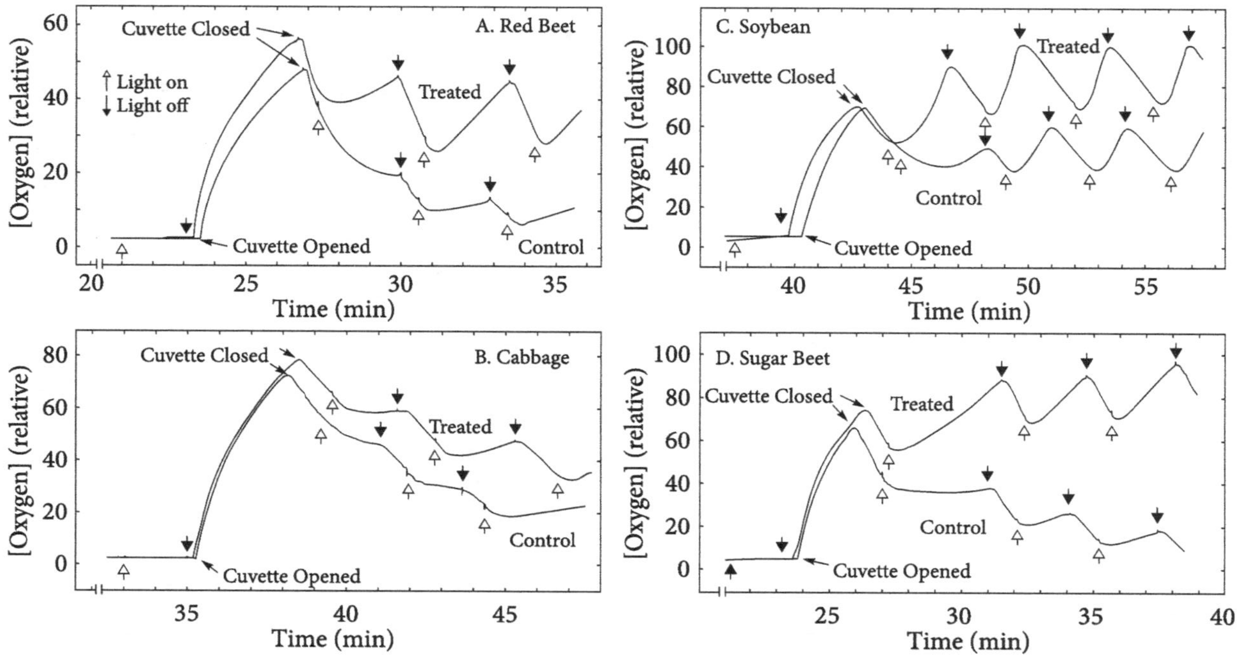

Fig. 3.7. Oxygen evolution in leaf slices of red beet (A) (*Beta vulgaris*), (B) cabbage (*Brassica oleracea*), (C) soybean (*Glycine max*), and (D) sugar beet (*Beta vulgaris*) after anoxia and saturating light treatment

Note: The traces show how different plants respond to the stress test. There are species and varietal differences in stress tolerance and responses to the treatments. The upper traces in each panel illustrate improved photosynthetic stress resistance indexes after foliar treatments to improve drought tolerance

Whereas in the higher plants, the rate of respiration decreased as the oxygen concentration approached zero, in *C. vulgaris,* the rate of respiration was linear until it reached a minimum near zero oxygen concentration.

3.4.1 Whole Leaf Physiological Responses

Conditions of growth affect the degree of stress tolerance within species and varieties. For example, in one experiment cabbage (*Brassica oleracea*), red beet (*Beta vulgaris*), and radish (*Raphanus sativus*) responded differently to drought stress (Figure 3.8). Cabbage exhibited an elevated photosynthetic stress resistance index compared to the initial well-watered PSRI value (Figure 3.8A). On the other hand, red beet and radish had the lower photosynthetic stress resistance index values after drought treatment (Figure 3.8B and 3.8C). While the PSRI for red beet remained low during the two weeks experiment, radish exhibited an increased the photosynthetic stress resistance index values after two weeks of drought treatment.

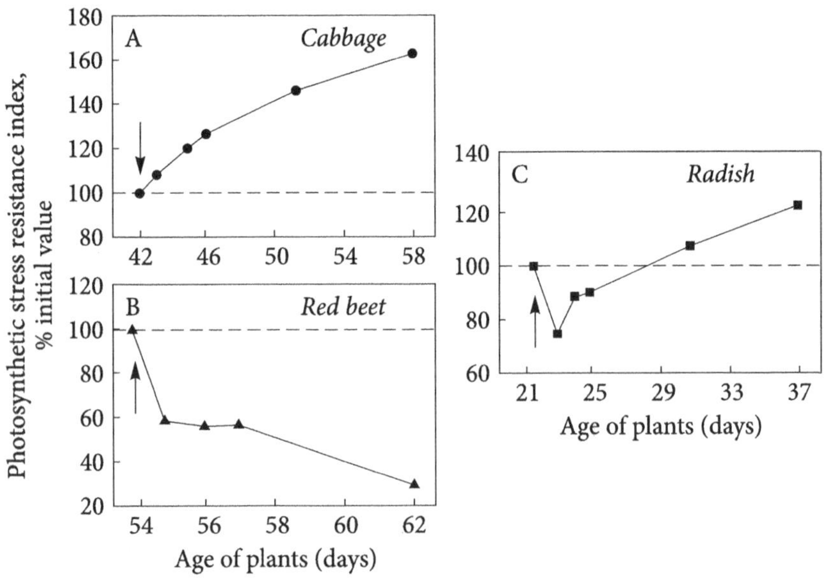

Fig. 3.8. Influence of soil drought stresses on the tolerance of photosynthesis of (A) cabbage (*Brassica oleracea*), (B) red beet (*Beta vulgaris*), and (C) radish (*Raphanus sativus*) plants after irradiation of leaf slices with saturating light during anoxia

Note: The soil humidity was 90% at the start of the experiment. In the middle of the subsequent days plants exhibited drought stress. Plants were watered one time per day at 16 hours. Measurements were done between 12 and 14 hours polarographycally with leaf slices at 37 °C. Arrow shows beginning of soil drought

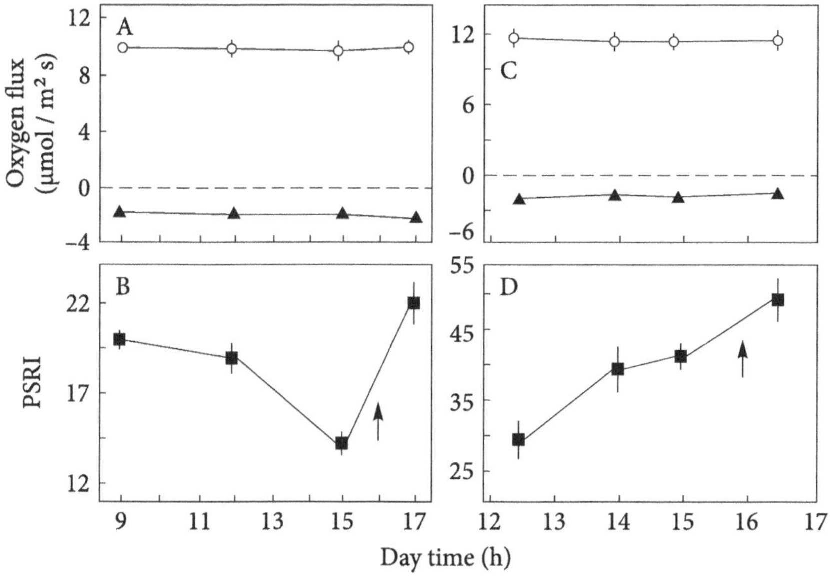

Fig. 3.9. Oxygen evolution stress assay used to examine stress tolerance throughout the day. During the OESA, saturating light was supplied during anoxia. A and C: Oxygen flux; B and D: PSRI. A and B: red beet (*Beta vulgaris*); C and D: radish (*Raphanus sativus*), $n = 2$

Note: For both species relative humidity of the soil was 70-80%. Arrows show time of watering the plants. Each point in graphs is the average of 5 discs from same age leaves from 5–6 different plants. Red beet (*Beta vulgaris*) was soil grown–constant 23 °C and light period of 16 hours (500 μmol / m^2 · s). Oxygen flux measured with leaf slices at 42 °C in 40 mM HEPES-KOH buffer, pH 7.0 and 10 mM $NaHCO_3$, with excitation light of 1350 μmol / m^2 · s; age of plants – 82–83 days. Radish (*Raphanus sativus*) was soil grown–constant 30 °C and 16 hours light period (350–400 μmol / m^2 · s). Oxygen evolution measured with leaf slices at 37 °C in 40 mM HEPES-KOH buffer, pH 7.61 and 10 mM $NaHCO_3$ with excitation light of 1350 μmol / m^2 · s; age of plants – 22–23 days

The photosynthetic stress resistance index value changed during the day, whereas the oxygen evolution rate remained relatively constant (Figure 3.9). Red beet, which exhibited a low photosynthetic stress resistance index under drought stress (Figure 9B), was also exhibited a low PSRI at midday. In contrast, the photosynthetic stress resistance index of radish increased during the afternoon (Figure 3.9D).

The photosynthetic stress resistance index changed dramatically compared to several other physiological parameters assayed (Figure 3.10; see also Kosobryukhov *et al.*, 2004). The PSRI after three days was about 50% higher in the drought-treated sugar beets (*Beta vulgaris*) (Figure 3.10A).

One week after treatment, the photosynthetic stress resistance index was almost double that of the control, and about double the initial PSRI. On the other hand, oxygen evolution of leaf slices (Figure 3.10B) during the two weeks drought treatment (Figure 3.10D) exhibited little differences. The rate of leaf growth decreased during the onset of the drought treatment, but after 2 weeks, there was little difference in leaf area between the well-watered and drought-treated plants (Figure 3.10C).

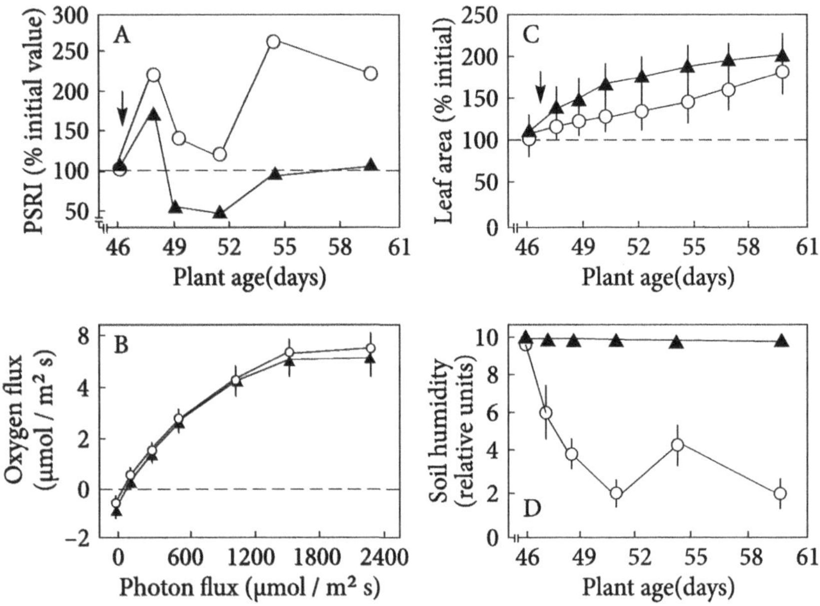

Fig. 3.10. Influence of drought stress on sugar beet (*Beta vulgaris*)
Note: A – Photosynthetic stress resistance index of leaf slices; B – light saturation curve of oxygen evolution; C – leaf area; D – relative soil humidity during experiment; (▲) – control; (○) – drought treatment. Arrow shows start of drought. Age of plants – 43 days, $n = 3 \div 6$

A distinct advantage of the new oxygen evolution stress assay method is that a direct effect is being monitored. There is no need to separate down regulatory mechanisms, occurring at the light harvesting level, from actual declines in performance. However, if one wants to investigate the role of down-regulation in the stress tolerance that is possible.

To further test the physiological application of the oxygen evolution stress test, we examined the response of *Tradescantia fluminensis*, a plant known to be tolerant of many environmental stresses, including anoxia (Fomina *et al.*, 1993). Therefore, we examined the response of *T. fluminensis* to the oxygen evolution stress test at 42 °C, rather than 37 °C, which was utilized for *Spinacia oleracea* (Figure 3.11). As predicted *T. fluminensis* exhibits high rates of oxygen evolution under anoxic conditions. Additionally, it exhibits a high photosynthetic stress resistance index of 73%, whereas control *Spinacia oleracea*, which was measured at a temperature 5 °C lower than *T. fluminensis*, had a PSRI of only 21% (Figures 3.11 and 3.12); however, the light treatment was lower.

Since the photosynthetic stress resistance index varied throughout the day (see Figure 3.9), we also examined the effect of nitrate anion on oxygen evolution under anoxia throughout the day. Oxygen evolution, under conditions that inhibited controls, did not vary throughout the day in sugar beet (*Beta vulgaris*) plants, when exogenous $NaNO_3$ was added (Figure 3.13). The percentage of photosynthesis under anoxic conditions remaining (the PSRI) compared to the initial rate was 40.0 ± 4.3, 36.8 ± 5.5, 38.8 ± 3.2, and 41.0 ± 3.5 at 9:00, 11:00, 14:00, and 16:00 (hours), respectively.

However, the induction rate of light-dependent oxygen evolution in the presence of $NaNO_3$ under anoxic conditions was dependent on the time of day. The induction time at 9:00 hours was about 2.5 minutes, whereas at 16:00 hours, the induction period was less than 30 seconds (Figure 3.13A).

To test whether the decrease in induction time of NO_3-dependent oxygen evolution under anoxic conditions was due to light, a circadian cycle or both, we measured the rate of photosynthesis under anoxic conditions that completely inhibited photosynthesis in the controls (see Figure 3.14A).

A 17:00 hours dark period was sufficient to alter the light induction period, as samples measured after 17:00 hours dark (9:00 h), 19:00 hours dark (11:00 h), and 23:00 hours dark (15:00 h) all had induction periods of about 6–7 minutes (Figure 3.13B). However, after 29 hours (9:00 h) and 31 hours (11:00 h) in dark the induction period was about 4 min and after 36:00 hours (15:00 h) the induction period was 10 minutes (data not shown). Although the samples exhibited different activation periods, the percent photosynthesis remaining was similar at the different time points (see above).

Thus, whether the difference in induction times is due to light activation or is due to a circadian rhythm needs to be further examined. In the samples where there was a long induction period, there was a rapid transient in oxygen evolution (Figure 3.13), that was observed only in the presence of $NaNO_3$.

Since inverting leaves causes light in the palisade mesophyll to decrease significantly (Sun and Nishio, 2001), and the PSRI in the palisade mesophyll was significantly increased after leaf inversion (Figures 3.11B and 3.12), we predicted that prolonging the dark period in plants would increase the photosynthetic stress resistance index.

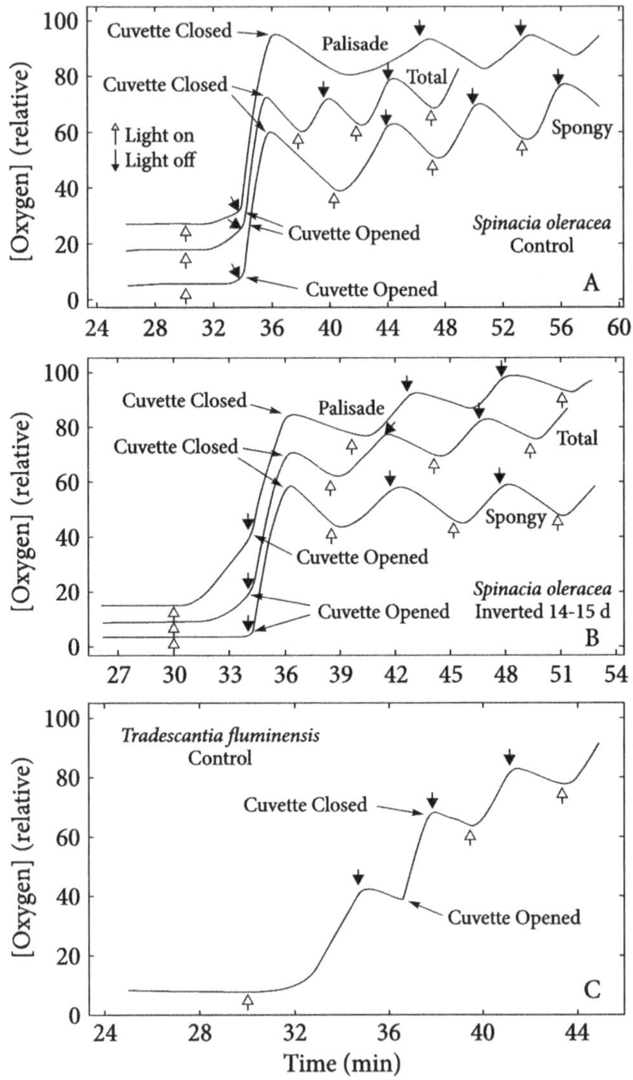

Fig. 3.11. Traces from oxygen evolution stress assay of tissue layers of *Spinacia oleracea* leaf slices and whole leaf slices of *Tradescantia fluminensis.* A – *S. oleracea* control; B – *S. oleracea* flipped leaf – 14–15 days; C – *T. fluminensis* control – total leaf

Note: For *S. oleracea* the stress temperature was 37 °C; for *T. fluminensis* – 42 °C. *S. oleracea* received 2200 μmol / m^2 · s for 4 min and *T. fluminensis* received 1300 μmol / m^2 · s for 5 min after the leaf slices had been in anoxic conditions for 30 min. *T. fluminensis* is known to be tolerant of anoxic stress (Fomina *et al.*, 1993), and is illustrated as an example

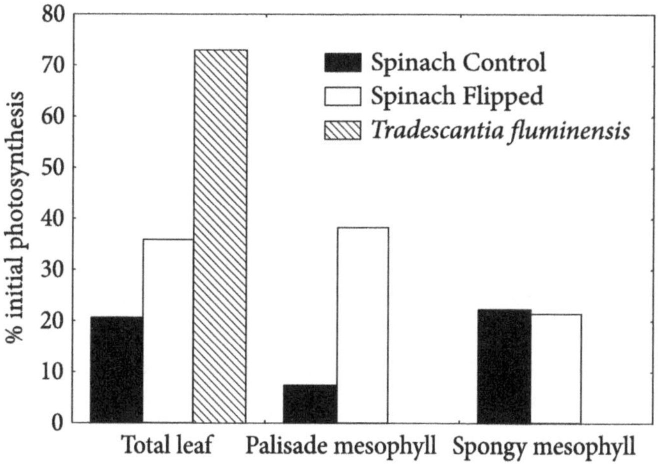

Fig. 3.12. Photosynthetic stress resistance index in total leaf slices, and palisade mesophyll and spongy mesophyll of *Spinacia oleracea* and in total leaf slices of *Tradescantia fluminensis* (from Figure 3.11)

Note: *Tradescantia* is known to be anoxia tolerant, and is shown as an example. As mentioned in text direct comparisons of species must be done with caution. The *Tradescantia* experiment was conducted at a higher temperature, but a lower light flux than for the spinach experiment

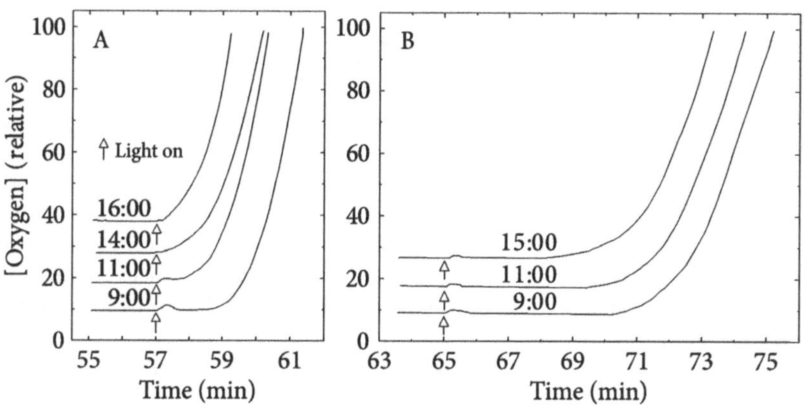

Fig. 3.13. Influence of $NaNO_3$ on oxygen evolution stress assay trace of sugar beet (*Beta vulgaris*) leaf slices at different times of the day. A: control; B: after 34–35 hours dark treatment

Note: Plants were dark-adapted for various lengths of time in a temperature and humidity controlled growth chamber. After the tissue plugs were removed from the plants [light (<5 μmol / m^2 · s) from the room was allowed into the chamber], the leaf plugs were floated on 40 mM HEPES buffer, pH = 7.6, containing 300 mM sorbitol for 20 minutes under photon flux 100 μmol / m^2 · s (see also Figure 14)

Additionally, a dark treatment might result in increased photosynthesis and total plant productivity under stressful conditions because shade grown *Ulva fenestrata* when placed in high light exhibits higher photosynthesis and productivity than high light grown *U. fenestrata* (Titlyanov *et al.*, 1990).

Figure 3.14A illustrates that a 55-hour dark treatment significantly increased the photosynthetic stress resistance index in *Beta vulgaris* plants. The PSRI of the dark treated leaf slices was 43%, whereas, the control leaves (normal dark-light cycle) exhibited little to no oxygen evolution after the 60-min stress treatment. The controls and dark-adapted leaves exhibited the same light saturation points, but the slope of the dark-adapted leaves was slightly higher, though not statistically significantly (data not shown).

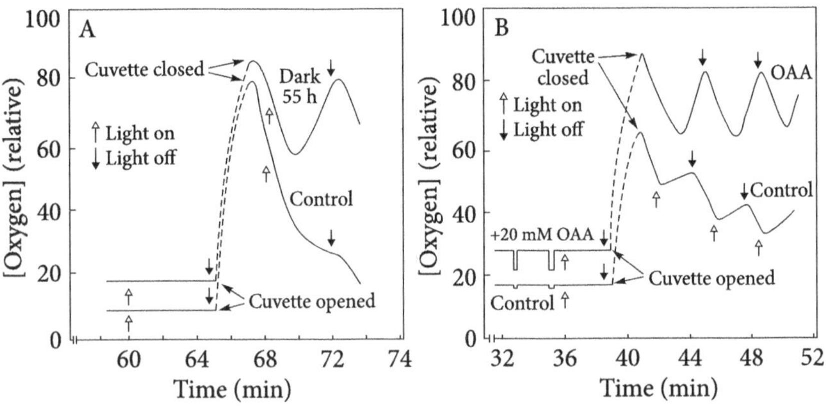

Fig. 3.14. Oxygen evolution stress assay traces of improved stress tolerance caused by physical and chemical means. A: leaf slices of sugar beet (*Beta vulgaris*) after 55 hours dark treatment of plants; B: sugar beet leaf slices treated with 20 mM of oxaloacetate in dark during respiration followed by anoxia (treatment in oxygen electrode cuvette)

Note: Age of plants for B – 50 days old; similar result was obtained with age of plants 90–106 days old

The experiment was conducted on similar plants after 59 hours of dark with the same result. Two weeks after the plants were removed from the 55 hours dark treatment the six largest controls out of nine were compared to the six plants that received the 55 hours dark treatment. The dark treated plants were 20% larger ($p = 0.110$) in total (roots and shoots) dry weight compared to the control plants. Plants were harvested 74 days from sowing.

3.4.2 *Effect of Dark and Sodium Nitrate on the Photosynthetic Stress Resistance Index and Photosynthesis in Leaf Slices under Anoxic Conditions*

Since the upper portion of sun leaves are known to be more stress tolerant than the abaxial portion of the leaf (Sun *et al.*, 1996*a*, 1996*b*), we hypothesized that the palisade mesophyll would have a higher photosynthetic stress resistance index than the spongy mesophyll. Surprisingly, the oxygen evolution stress assay on fresh paradermal *Spinacia oleracea* leaf sections showed that the photosynthetic stress resistance index was lower in the palisade mesophyll than in the spongy mesophyll. However, Figure 3.11A illustrates that in *Spinacia oleracea* leaves, oxygen evolution occurred under anoxic conditions, as Figure 3.4 showed for sugar beet (*Beta vulgaris*) leaves at 25 and 30 °C. The palisade mesophyll was responsible for the majority of the oxygen evolution under anoxic conditions (Figure 3.11A).

Figure 3.5 illustrated that compounds such as $NaNO_3$, glycine, and ATP protected leaves from photoinhibition during the oxygen stress test. However, $NaNO_3$ was the only compound tested that allowed oxygen evolution under anoxic conditions. Since the palisade mesophyll was responsible for oxygen evolution under anoxic conditions, and the palisade mesophyll had a lower PSRI, we then hypothesized that the ability to evolve oxygen under anoxic conditions was partly responsible for the known stress tolerance of the palisade mesophyll (Sun *et al.*, 1996*a*, 1996*b*).

Additionally, we hypothesized that the protection must be due in part to some sort of oxidant, such as nitrate anion. If such a capacity did not exist, then the palisade mesophyll would be more susceptible to photoinhibition than it is. Thus, we predicted that oxygen evolution under anoxic conditions would increase in the spongy mesophyll of inverted leaves because we thought that the level of the unknown oxidant was related to light. Again, we had hypothesized incorrectly, as the opposite occurred. The photosynthetic stress resistance index was only 7% in the palisade mesophyll of control *Spinacia oleracea*, but it dramatically increased to 38% two weeks after leaves were inverted (Figures 3.11A,B and 3.12), whereas, the photosynthetic stress resistance index of about 20% in the spongy mesophyll was little changed by leaf inversion. In whole leaves the photosynthetic stress resistance index almost doubled to 36% after 2 weeks. In control leaves the photosynthetic stress resistance index of the total leaf was similar to that of the spongy mesophyll, whereas in inverted leaves it was similar to that of the palisade mesophyll (Figures 3.11A,B and 3.12).

3.4.3 *Post-Illumination Respiration*

Table 3.1 shows that in the palisade mesophyll of control *Spinacia oleracea* leaves, the post-illumination respiration (PIR) rate (R2, see Figure 3.1 for details) was 46% higher than the steady state dark respiration rate (R1), whereas in the spongy mesophyll the R2 and R1 were the same. Steady state dark respiration was not affected by leaf inversion. Flipping leaves did not change the R2 / R1 ratio in the palisade mesophyll. In the spongy mesophyll of inverted leaves, however, the R2 increased 44%, so the R2 / R1 ratio was 1.36, compared to 0.95 in control leaves.

Table 3.1. Post-illumination respiration (PIR) in the palisade mesophyll and spongy mesophyll from control and inverted *Spinacia oleracea* leaves

Variant	Tissue type	PIR	Steady-state respiration	Ratio PIR / Steady-state respiration
		μmol O_2 / $m^2 \cdot s$		
Control	Palisade	0.60 ± 0.05	0.41 ± 0.03	1.46
	Spongy	0.55 ± 0.12	0.58 ± 0.07	0.95
Inverted	Palisade	0.65 ± 0.10	0.43 ± 0.12	1.51
	Spongy	0.79 ± 0.05	0.58 ± 0.05	1.36

3.5 Discussion

Oxygen evolution stress assay. We have described a new method to quantify one aspect of photosynthetic stress tolerance in plants. The new oxygen evolution stress assay method is useful for screening plants, as there are clear differences in the capacity for various plants to withstand the stress treatment applied. The primary effect of our treatment is photoinhibition.

The exact site of damage during our assays has not been examined. In many plants photosystem II is the main site of damage under anoxic conditions, however, the nature of our treatment will allow examination of many biochemical parameters at different, known physiological states under varying treatments and conditions.

The oxygen evolution stress assay is also useful for screening plants. Standard conditions appropriate for "controls" for each species can be utilized for comparing the photosynthetic stress resistance index of different varieties, including recombinants; growth conditions, such as drought, high or low temperature, nutrient stress, pH stress and so forth;

and treatments, such as foliar sprays (see *e.g.*, Figure 3.7) or challenge with pathogens.

One blind experiment comparing 3 varieties of *Zea mays* with known stress tolerance characteristics (based on years of field work!), was confirmed in one month by the oxygen evolution stress assay (see Figure 3.A in Appendix II).

We have the ability to measure pulse amplitude modulated fluorescence and CO_2 gas exchange, but we chose to develop a method that allows more in-depth analysis. Whereas, screening for chlorophyll fluorescence characteristics is valuable, little information about the actual biochemical or physiological effects can be gained from such measurements. It may be possible to use chlorophyll fluorescence in conjunction with our method or as a supplement, because decreased oxygen evolution should cause measurable changes in chlorophyll fluorescence. CO_2 gas exchange is also a valuable diagnostic tool (Voskresenskaya *et al.*, 1974), however, adding exogenous compounds and investigating their impact on gas exchange is more difficult than with the oxygen electrode.

Photoinhibition and low oxygen. In our experiments oxygen was limiting, but CO_2 was present in saturating amounts. The final oxygen concentration and ratios of other gases in the oxygen electrode cuvette were established by the metabolism of the leaf sections. Plant shoots would rarely, if ever, experience anoxia under elevated temperatures, so the anaerobic conditions utilized for our experiments were non-physiological. However, such a physiologically derived condition is pertinent because the different responses in plants could be correlated to stress tolerance.

It is possible to rapidly develop anoxic conditions by degassing the buffer or by bubbling nitrogen gas (N_2) through it, which can reduce the time to reach anoxia by ten minutes or more, depending on the sample (Fomina *et al.*, 1993). Conditions established by such methods would not represent physiologically pertinent gas ratios in the cuvette. Thus, further investigations to understand the biochemical nature of the stress tolerance response should be conducted under experimental conditions that do not utilize nitrogen gas, argon gas, or vacuum, for example.

Furthermore, the leaf slices during the onset of anoxia must utilize available substrates for respiration. If the leaf slices were immediately placed under anoxia, the carbon balance would be different from that attained "naturally" by the plant. In plants or cells that exhibit low rates of respiration, artificially changing the oxygen may be useful. Sodium dithionite can also be used to temporarily adjust the oxygen concentration.

The capacity to utilize substrate and respire under low oxygen tension varies among species. For example, as the lower oxygen concentration is reached, the respiration rate decreased (Figures 3.2, 3.3, and 3.4). In *Chlorella vulgaris* the biphasic nature of the respiration rate as it reached the lowest oxygen concentration was not observed. Instead, the respiration rate was constant until it reached the minimum oxygen concentration (data not shown). The constant rate of respiration (linear utilization of oxygen) by *Chlorella vulgaris* suggests that such organism may be useful for space travel. Breeding or genetically manipulating higher plants for such capacity could also be beneficial. *Arabidopsis* had a much more linear rate near the anoxic conditions (Gómez, Bil' *et al.*, 2003) than did the other C_3 species (not shown).

The variation in the respiration rates during the onset of anoxia suggests different substrates for respiration and / or K_M's for various respiratory enzymes exist in the various species, and there may be substrate differences at different time points along the respiratory curves. For example, the initial rapid respiration after illumination decreases as the rapidly utilizable photosynthate is metabolized, and the following rates of respiration could be due to hydrolysis of starch, that possibly is followed by β-oxidation of lipids or catabolism of proteins. Each process could be occurring at various rates simultaneously, but may be affected by the oxygen concentration differentially.

The presence of $NaNO_3$ did not appear to affect the shape of the oxygen concentration curve during onset of anoxia, but nitrate anion, possibly due to NO_3-dependent respiration, somehow improved photoinhibitory stress tolerance.

Therefore, to study the various phenomena at the biochemical level, use of nitrogen gas and the like to attain anoxia is not recommended. The extended time for the onset of anoxia, during the natural onset of anoxia by the leaf slices, can be used for incubating the samples with exogenous compounds, and the effect of the exogenous compounds on respiration can be monitored. If a compound requires a longer period of time for uptake, then the cuvette can be opened and closed in the dark during the incubation period, to allow longer periods of respiration before the anoxic condition is established. The increased period of oxidative respiration could also be done in the absence of exogenous compounds to examine how depletion of substrates impacts oxygen evolution under anoxia and the photosynthetic stress resistance index. The opening and closing of the cuvette can be done at different ranges of oxygen concentrations $[O_2]$ to test the effect of $[O_2]$ and the $[O_2]$ / $[CO_2]$ ratio on various aspects of respiration, such as substrate utilization.

Some plants exhibited a rapid rate of respiration immediately following the light period (Table 3.1). Increased respiration following illumination was first observed at the end of the 19th century (Borodin, 1881; Paladin, 1893), and later in the first quarter of the 20th century (Warburg and Negelein, 1922), and it was shown in higher plants (Spoehr and McGee, 1923; van der Paauw, 1932) (*cf.* Rabinovitch, 1945). More recently the finding has been confirmed in *Chlamydomonas* (Bunt and Heeb, 1971) and many marine plants (Downton *et al.*, 1976).

The very early data and that of others (Azcón-Bieto and Osmond, 1983; Raghavendra *et al.*, 1994) suggest that products of photosynthesis are responsible for the increased respiration. In the presence of host factor dinoflagellates exhibit an accelerated post-illumination respiration, and the original steady state dark rate is eliminated, indicating that the added substrate accelerated respiration (Gates *et al.*, 1999).

We interpret our data similarly, that is, when there is an increased post-illumination respiration, it indicates that recently produced photosynthate is being used for respiration; and such utilization was occurring in the light. The related CO_2 release has also been reported (Atkin *et al.*, 1998).

Under less stressful conditions during the oxygen evolution stress assay than that used in Figures 3.2, 3.3, and 3.4, the photosynthetic organisms evolved oxygen under anoxic conditions (Figure 3.11). We have now established that in whole leaf samples, but not in isolated tissue layers, a higher capacity to evolve oxygen under anoxia correlates well with a higher photosynthetic stress resistance index (data not shown). Additionally, the response under anoxia and light was identified as a key experimental point for further investigating the mechanism of tolerance to photoinhibitory stress in plants. There appears to be an important link between photosynthesis and respiration in the stress tolerance observed.

Why is there photoinhibition under light and anoxia? The oxygen evolution stress assay allows to test the role of secreted compounds in the observed stress tolerance. For example, *Chlorella vulgaris*, which secretes glycolate during photosynthesis, exhibited a high photosynthetic stress resistance index. It is simple to test the role of the secreted glycolate by rinsing the cells after the initial 15 min of light, when light saturation curves were measured. If the ability to photosynthesize under anoxia or the photosynthetic stress resistance index is reduced by such treatment, then we could assume that some secreted compound was responsible. Simple labeling experiments could then follow to identify the substrates. Additionally, by rinsing and then adding exogenous compounds specific roles for various substrates could be established without labeling experiments. However, by adding labeled substrates and chasing, one could see how the compounds are metabolized. Of course, similar experiments could be conducted on the leaf slices. Apoplastically transported products would be secreted into the buffer, and they also could be rinsed away, identified by labeling, and / or added back to the cuvette solution. Clearly, such additions and subtractions could be conducted at various time points, where different experimental conditions have been established during the assay. In our experiments with $NaNO_3$, reduction of nitrate anion by secreted reductant in the apoplast or buffer is a possibility, but the many possibilities were not tested.

Biochemical responses. As mentioned above, the oxygen evolution stress assay can be used for many purposes. *Glycine max* grown under high temperature and adequate water, exhibited very high stress tolerance in our tests conducted at 42 °C (Figure 3.7). On the other hand, *Brassica oleracea* and two *Beta vulgaris* varieties, red beet and sugar beet, did not withstand the stress treatment as well (Figure 3.7). The species differences observed could be due to plant age and the age of the leaves measured. We also have successfully used the test on *Raphanus sativus*, *Spinacia oleracea*, *Zea mays* (see Figure 3.A in Appendix II), and a green organism, *zoochlorellae*, that is symbiotic with the freshwater sponge, *Spongilla lacustris* (Nishio *et al.*, 2000), *Solanum melongena* and *Chlorella vulgaris* (not shown). In making comparisons within varieties, it is important that similar age plants and leaves be used for the direct comparisons. The age and concentration of tissue, algae, or photosynthetic bacterial cultures should also be similar.

The oxygen evolution stress assay conducted at 42 °C on sugar beet suggested that this plant was not stressed tolerant compared to soybean (Figure 3.7). However, when sugar beet was grown under drought conditions, the photosynthetic stress resistance index was increased (Figure 3.10). When the OESA was conducted at 42 °C, the time under anoxia could be shortened compared to experiments conducted at 37 °C. It is possible that some "protective" metabolite is more rapidly consumed or destroyed at 42 °C compared to at 37 °C. The elevated temperature could also increase the onset of some sort of structural damage to the photosynthetic apparatus. It is likely that structural damage is not the primary effect, as cycling light and dark after the anoxic photoinhibitory treatment resulted in recovery of photosynthesis. Repair of D_1 protein, for example, could not occur in that length of time.

Additionally, Figure 3.3A showed that when light and dark were cycled during the time course of an experiment, there was little change in the rates of photosynthesis and respiration compared to the initial rate. The "protective" metabolite(s) could be enhanced in stress tolerant plants compared to wild types or controls.

With the oxygen evolution stress assay, it was possible to examine biochemical interactions between respiration and photosynthesis in a relatively intact system. It was also possible to test protective metabolites by applying various molecules exogenously. The data strongly support the notion (Osmond and Björkman, 1972; Osmond, Badger *et al.*, 1997) that photorespiration directly protects chloroplasts from stress damage, and that mitochondria can play an important role in the light activation of photosynthesis.

ATP, a major product of respiratory electron transport was applied in the dark. Somehow, ATP protected the leaf slices from high light stress under anoxia because when the light was turned on the leaf slices rapidly activated oxygen evolution in samples with exogenous ATP, compared to controls. Glycine and oxaloacetate also protected the plants from photoinhibition (Figures 3.5 and 3.14B). Glycine decarboxylation forms NADH that can be used by the respiratory electron transport chain to form ATP. Glycine is a key component of C_2 photorespiration. On the other hand, serine, which also is an important metabolite of C_2 photorespiration, did not protect the leaf slices from photoinhibition. The serine formed during photorespiration is recycled back to the 3-PGA pool, and recycling back to RuBP requires ATP and NADPH. Thus, as predicted and unlike glycine and ATP, serine had no effect on the stress tolerance of plants.

The upper portion of leaves are more stress tolerant compared to the bottom of leaves (Sun *et al.*, 1996*a*, 1996*b*; Biel, Fomina *et al.*, 2008, 2010; Fomina and Biel, 2014). We hypothesized that the palisade mesophyll of *Spinacia oleracea* leaves would exhibit a higher photosynthetic stress resistance index than the spongy mesophyll, but the hypothesis was not supported by the results (Figure 3.12). The result may be partly explained by the experiment that showed that ATP exogenously applied to the leaf slices resulted in a higher the photosynthetic stress resistance index compared to controls (Figure 3.5). The palisade mesophyll has mitochondria that are more dependent on chloroplasts for ATP. In the dark, they can produce ATP, but not as much as the spongy mesophyll can, since the spongy mesophyll had two times higher levels of *Cyt-c*-oxidase activity, than the palisade mesophyll (Bil', Fomina *et al.*, 2003; Biel, Fomina *et al.*, 2008, 2010). Consequently, the photosynthetic stress resistance index of the palisade mesophyll was lower than that of the spongy mesophyll, probably, because the ATP from photosynthesis was depleted during the dark treatment time. If ATP is a key protector against photoinhibitory stress in the light, the palisade mesophyll would have sufficient and more ATP than the spongy mesophyll. Additionally, other photosynthetic compounds may be involved in the stress tolerance.

Thus, in control plants, the spongy mesophyll had a higher photosynthetic stress resistance index because it has a greater capacity to make ATP in the dark compared to the palisade mesophyll. While oxygen is required for normal mitochondrial electron transport the data suggest that the spongy mesophyll may have a greater oxidative respiration capacity. Interestingly, fumarase activity, indicative of tricarboxylic acid cycle activity, was more than 5-times higher in the spongy mesophyll (Bil', Fomina *et al.*, 2003; Biel, Fomina *et al.*, 2008, 2010). Additionally, the results imply that the ATP pool in the spongy mesophyll is not consumed during the dark treatment, whereas it is consumed in the palisade mesophyll.

Likely ATP is only one component of an array of mechanisms involved in maintaining the photosynthetic stress resistance index. One possibility is that there is an oxidant pool in leaves that leads to higher stress tolerance. For example, exogenously applied $NaNO_3$ apparently acts an oxidant during the oxygen evolution stress assay and results in an elevated photosynthetic stress resistance index compared to controls. Therefore, an oxidant such as nitrate anion at the top of the leaf could protect the palisade mesophyll from photodestruction.

To test the effect, we measured stress tolerance in the palisade mesophyll and spongy mesophyll after leaves were inverted which decreased the light intensity on the palisade mesophyll. After leaf inversion, the palisade mesophyll exhibited oxygen evolution under anoxic conditions, unlike the control (Figure 3.11). Thus, it is very likely that an oxidant accumulated and protected the cells from photoinhibition in the palisade mesophyll of inverted leaves. The identity of such an oxidant(s) is not known at the present, but it appears to be important in protecting the leaf from photoinhibition.

Possibly nitrate reduction takes place mainly in the palisade mesophyll, and when the leaves are inverted, such reduction is greatly decreased, so nitrate accumulates. Sulfate reduction or any other number of reducing reactions could also play a similar role. Besides inorganic oxidants, organic oxidants may play a role. Cytochromes, ferredoxin, plastoquinone, ascorbate, α-tocopherol, glutathione, betaine, formate, oxalate, *etc.* are other possible oxidants – probably the upper part of the leaf has a larger plastoquinone pool, for example.

While such speculation about the importance of oxidants in stress tolerance is reasonable, we conducted another experiment, which strongly supports our hypothesis.

Plants that were placed in the dark for extended times (34–55 hours), exhibited very high photosynthetic stress resistance indexes compared to plants that were maintained on the normal day / night cycle, and measured at least 5–7 hours after sunrise (Figure 3.14A). In the experiments, the conditions were purposefully set so the control would have a negative photosynthetic stress resistance index (strong photoinhibition). We have not measured the photosynthetic stress resistance index in plants at sunrise or the beginning of the photoperiod, nor in the middle of the dark period. But it is possible that oxaloacetic acid can accumulate during dark period of time in some desert plants (Bil', Lyubimov *et al.*, 1982) and some marine higher plants (Pärnik, Bil' *et al.*, 2002).

The palisade mesophyll is "armed" with many defenses to high light and oxygenic stress, including biochemical pathways, such as superoxide dismutase, catalase, peroxidase, the alanine effect, Karpilov pathway, and C_1 / C_2 photorespiration (Sun *et al.*, 1996*a*, 1996*b*; Bil', Fomina *et al.*, 2003; Biel and Yensen, 2005; Biel, Fomina *et al.*, 2008, 2010; Fomina and Biel, 2014; see also Chapter 7).

It is well known that photorespiration can protect plants from photodamage. There are, however, other metabolic pathways, that can also contribute to such protection. In addition to catalase, superoxide dismutase, and peroxidase, oxidation of oxaloacetic acid to glycolate scavenges active oxygen species and forms glycolate (see Figure 3.13 in Chapter 3.2), as an extra source of substrate for C_1 photorespiration. The C_1 photorespiration utilizes redox potential and decreases the amount of nitrogen cycling required for C_2 photorespiration. These activities are likely more active in the palisade mesophyll (Biel, Fomina *et al.*, 2010).

The spongy mesophyll is more involved in synthetic reactions (anaplerotic). The tricarboxylic acid cycle and normal mitochondrial electron transport capacity is greatly increased in the spongy mesophyll. The mitochondria of the palisade mesophyll are more involved in the reactions mentioned above, such as C_1 / C_2 photorespiration.

While the palisade mesophyll is responsible for the midday depression (see Chapter 7; see also Nishio, Bil' *et al.*, 2000; Biel, Fomina *et al.*, 2010), we clearly showed that it is really armed for stress tolerance. If the ability to scavenge active oxygen were to be decreased, the palisade mesophyll would be greatly damaged. Thus, the upper palisade mesophyll, in many ways, acts as a shield to protect the lower part of the leaf. Of course, the upper palisade mesophyll is also capable of photosynthesis.

After leaf inversion, the palisade mesophyll exhibited oxygen evolution under anoxic conditions, unlike the control. Thus, it is very likely that an oxidant accumulated and protected the cells from photoinhibition in the palisade mesophyll of inverted leaves. The identity of such an oxidant(s) is not known. One possibility is that under anoxic conditions, phosphoenolpyruvate is used to reduce CO_2 to OAA. Oxaloacetic acid can then be reduced to malic acid, which oxidizes NADH formed during glycolysis. Essentially this represents another form of fermentation.

Another possibility is that nitrate reduction takes place mainly in the palisade mesophyll, and when the leaves are inverted, such reduction is greatly decreased, so nitrate accumulates.

The idea that mitochondria and chloroplasts may be functionally different in the spongy mesophyll and palisade mesophyll is not new, because the ultrastructure of the two organelles is different in the two tissues. However, biochemical data supporting the structural information is lacking.

Many fundamental questions remain to be answered about metabolic specialization across leaves that can help us understand the role of oxidative and reductive carbon metabolism in stress tolerance. Are there functionally different types of chloroplasts and / or mitochondria within leaves? What aspects of the palisade mesophyll make it more resistant to stress? Is there a direct role for carbon metabolism in stress resistance? Does the palisade mesophyll act as a light shield (down-regulation)? What is the role of antioxidant system scavenging? What is the nature of the interaction between palisade mesophyll and spongy mesophyll with regard to controlling carbon metabolic activity? What biochemical parameters in the two cell types might be "fixed", and what role does gene expression play in the flexibility of carbon metabolism across leaves?

When we made oxygen evolution stress assays on plants in the presence of oxaloacetate (Figure 3.14B), as a Krebs cycle intermediate, and presumed cytoplasmic constituent, the photosynthetic stress resistance index of plants was greatly enhanced. Oxaloacetate, like nitrate, could act as an oxidant. While nitrate allowed high levels of oxygen evolution under anoxia (Figure 3.5), oxaloacetate did not (Figure 3.14B).

Thus, the oxygen evolution stress assay method also allows one to add radioisotopes or other labeled products to examine directly the effect of such additions at different stages of metabolism (*e.g.*, dark respiration, anoxia in light or dark, or post illumination). Regarding down-regulation, the recovery period of recycling may be due to D_1 protein repair. Such a possibility could be addressed using the method described. Additionally, extended periods of dark increased the photosynthetic stress resistance index, which further implies a direct connection between stress tolerance and mitochondrial performance.

How is carbon metabolism related to stress tolerance? Using the oxygen electrode, we showed that the treated plants are more tolerant to high temperature, anoxia, and high light. Possibly plants that are tolerant to stress have an increased ability to utilize or dissipate light energy or reductive energy, and there is less oxidative damage. This may be possible for the following reasons:

- Activation of alternative CO_2 fixation pathways including the C_4 pathway, direct fixation of CO_2 to formate, and the reductive tricarboxylic acid cycle would provide additional sinks for photosynthesis.
- Cycling of nitrogen may be decreased because serine is utilized for synthetic pathways, rather than as a source of recycled carbon for the RuBP pool. Additionally, direct formulation of tetrahydrofolate eliminates one deamination of glycine.
- Increased photorespiratory utilization for synthetic pathways also provides a "sink" for electrons, but the energy is not "wasted".
- Induction of greater disconnection of the light harvesting complexes from the reaction centers or other photoprotective mechanisms.

3.6 Perspectives for Application of Method

The above possibilities along with increased protective enzymes, that detoxify active oxygen species, would result in healthier plants. Further research in plant productivity and stress tolerance can benefit from using the oxygen evolution stress assay method. One example is given below.

A three-component approach to assessing stress tolerance in plants is possible. Two complementary approaches utilizing the oxygen electrode can be employed to gain insight into the possibility that altered metabolism and ATP status maintain photosynthesis under stress conditions. The third component of the holistic approach to understanding carbon metabolism is to examine reductants and protective enzymes such as ascorbate, superoxide dismutase, glutathione reductase, catalase, and peroxidase in plants with higher CO_2 fixation rates. The two oxygen electrode approaches are summarized below.

1. Oxygen evolution experiments can be used to investigate the role of specific metabolites in the improved stress tolerance.

Figures 3.5 and 3.14B show that metabolites directly applied to leaf slices increased photoinhibitory tolerance. It is important to further investigate the exciting results by testing other metabolites of photorespiration, such as glycolate and glyoxylate. Also, some intermediates of the tricarboxylic acid cycle, such as malate, fumarate, and succinate can be tested. Glycolate and glyoxylate will be a source of glycine, if there is sufficient nitrogen. Particular Krebs cycle metabolites can provide any number of possible sources of ATP from NADH oxidation in mitochondria. Additionally, malate can be utilized by both chloroplasts and mitochondria. Since nitrate allowed photosynthetic electron transport to be initiated under anoxic conditions, it is possible to test other endogenous oxidants such as sulfate, trivalent iron, and phosphate.

We are uncertain about the consequences of adding artificial electron acceptors, such as ρ-benzoquinone and 2,6-dichlorophenolindophenol to intact leaf slices. In isolated chloroplasts ρ-benzoquinone and 2,6-dichlorophenolindophenol readily cross membranes and accept electrons from photosystem II. It may be possible to show that nitrate is acting as an oxidant by testing ρ-benzoquinone or 2,6-dichlorophenolindophenol, as we did nitrate. The rate of oxygen evolution initiated under anaerobiosis should be similar.

2. Oxygen evolution experiments can be applied to testing alternative fixation pathways.

There appears to be a difference in energy status between controls and stress-treated plants, which may be related to alternate carbon fixation pathways. Leaf slices from plants treated to improve productivity and stress tolerance are more rapidly light activated after a stress treatment in the cuvette (Figure 3.7). It is possible to monitor ^{14}C-fixation in leaf slices at different time points during the stress treatment as illustrated in Figure 3.1. Listed below are the different points for labeling and some expected results in plants with $^{14}CO_2$ in the light (P1 and P2), as well as in the dark (D1, D2, D3):

P1 – *Steady state photosynthesis* in air was higher in the treated plants. Unlike in air, the rate of oxygen evolution of leaf slices was similar at the start of the oxygen evolution experiments in treated and control samples. We predict that ^{14}C-fixation will be higher in the treated plants, as there is a possibility that carbon fixation will be elevated in the treated plants due to increased phosphoenolpyruvate-carboxylase activity or some other form of carboxylation, as suggested by ^{14}C-labeling of leaves experiments (not shown);

D1 – *Post-illumination dark CO_2 fixation.* Measuring ^{14}C-fixation immediately following the light period can test our hypothesis that the fixation products will be altered in the treated plants because of light independent carboxylation. The dark fixation may be higher in the treated plants. Under the conditions of our experiment, in which saturating bicarbonate is present, the post illumination CO_2 burst will not affect measurements;

D2 – *Steady state dark CO_2 fixation.* Steady state dark fixation measurements will test the hypothesis that there will be higher dark CO_2 fixation in the treated plants due to phosphoenolpyruvate carboxylase and possibly other dark carboxylation reactions;

D3 – *Steady state anoxic dark CO_2 fixation.* Dark CO_2 fixation measurements under anoxic conditions will test the hypothesis that the treated plants will produce different fermentation products, such as formate, which can be reduced by reducing power formed during glycolysis;

P2 – *Steady state anoxic light fixation.* In many plants oxygen evolution does not occur under aerobic conditions, in part because Rubisco activity is inhibited. Additionally, the electron transport system becomes reduced in the dark due to endogenous reductants. We hypothesize that light CO_2 fixation under anoxic condition will be elevated in the treated plants due to higher rates of the reductive carboxylic acid cycle and possibly direct reduction of CO_2 to formate by ferredoxin.

The simple method developed for comparing levels of stress tolerance in intact leaf slices using oxygen evolution shows great promise. As we have demonstrated, there are many possible applications of the oxygen evolution stress assay. With the method it is possible to add different metabolites and compounds to observe the effect of the compounds. Uptake of the compounds was not measured, but obvious effects were measured. ATP, glycine, sodium nitrate, and oxaloacetic acid protected leaf slices from photoinhibition.

The data strongly support the idea that respiration and photorespiration directly protect chloroplasts from stress damage. Plants that showed increased drought tolerance also exhibited higher stress tolerance using the oxygen evolution stress test. The method provides a useful way to examine interactions among various organelles, enzymes, and biochemical pathways.

Acknowledgments

We thank Dr. Irina R. Fomina (Institute of Basic Biological Problems of the Russian Academy of Sciences, Pushchino, Moscow Region 142290, Russia) and Dr. Tiit R. Pärnik (Institute of Agricultural and Environmental Sciences, Estonian University of Life Sciences, Tartu, Estonia, deceased) for productive discussion and Mr. Vitaly K. Biel (Tucson, Arizona 85755, USA) for participation in some experiments.

Abbreviations

Fd_{red} – reduced ferredoxin
OAA – oxaloacetic acid
OESA – oxygen evolution stress assay
3-PGA – 3-phospoglyceric acid
PIR – post-illumination respiration
PM – palisade mesophyll
PSRI – photosynthetic stress resistance index
Rubisco – ribulose-1,5-bisphosphate carboxylase / oxygenase
RuBP – ribulose-1,5-bisphosphate
SM – spongy mesophyll

References

Atkin O.K., Evans J.R., Siebke K. Relationship between the inhibition of leaf respiration by light and enhancement of leaf dark respiration following light treatment, *Aust. J. Plant Physiol.*, 1998, Vol. 25, p. 437–443.

Azcón-Bieto J., Osmond C.B. Relationship between photosynthesis and respiration. The effect of carbohydrate status on the rate of CO_2 production by respiration in darkened and illuminated wheat leaves, *Plant Physiol.*, 1983, Vol. 71, p. 574–581.

Bassham J.A., Benson A.A., Kay L.D., Harris A.Z., Wilson A.T., Calvin M. The path of carbon in photosynthesis. XXI. The cyclic regeneration of carbon dioxide acceptor, *J. Am. Chem. Soc.*, 1954, Vol. 76, p. 1760–1770.

Bassham J.A., Calvin M. *The Path of Carbon in Photosynthesis*, Prentice-Hall, Inc., Englewood Cliffs, New Jersey, 1957, 104 p.

Benson A.A. Following the path of carbon in photosynthesis: a personal story, *Photosynth. Res.*, 2002, Vol. 73, p. 29–49.

Benson A.A., Calvin M. Carbon dioxide fixation by green plants, *Annu. Rev. Plant Physiol.*, 1950*a*, Vol. 1, p. 25–42.

Benson A.A., Calvin M. The path of carbon in photosynthesis. VII. Respiration and photosynthesis, *J. Exp. Bot.*, 1950*b*, Vol. 1, p. 63–68.

Biel K.Y., Yensen N.P. 2005 The hypothesis of halosynthesis, photoprotection, soil remediation via salt-conduction, and potential medical benefits, *Complex Systems under Extreme Conditions*, Khlebopros R.G., Soukhovolsky V.G. eds., Krasnoyarsk, Russia, 2005, p. 86–115.

Biel K.Y., Fomina I.R., Nazarova G.N., Soukhovolsky V.G., Khlebopros R.G., Nishio J.N. Untangling metabolic and spatial interactions of stress tolerance in plants. 1. Patterns of carbon metabolism within leaves, *Protoplasma*, 2010, Vol. 245, p. 49–73.

Biel K.Y., Fomina I.R., Nishio J.N., Soukhovolsky V.G. Optimization of photosynthesis within leaves of C_3 and C_4 plants. Example of complex system self-regulation when energy source is also stress inductor, *System Identification and Control Problems,* Proc VII Int Conf SICPRO'08 (ISBN 978-5-91450-002-0), Moscow, 2008, p. 566–593.

Bil' K.Y., Fomina I.R., Nazarova G.N., Nishio J.N. Light dependent organization of carbon metabolism within spinach leaf, *Physiol. Biochem. Cult. Plants*, 2003, Vol. 35, No. 5, p. 392–402.

Bil' K.Y., Lyubimov V.Y., Demidova R.N., Gedemov T. Assimilation of CO_2 by plants of the family *Chenopodiaceae* with three types of autotrophic tissues in the leaves, *Soviet Plant Physiol.*, 1982, Vol. 28, No. 6, p. 808–815.

Bil' K.Y., Nishio J.N. Measurements of stress tolerance and the interaction between photosynthesis and respiration, *in vivo*, *Abstracts in 9th Western Regional Photosynth. Conf.*, January 6–9, 2000, Asilomar Conf Center, Pacific Grove, California, 2000*a*, p. 18.

Bil' K.Y., Nishio J.N. Preliminary analysis of the utility of a novel stress resistance assay on three Garst lines of corn, a C_4 plant, *Report for Garst*, Experiment Conducted 23/10/2000–28/11/2000, 2000*b*, p. 1–4.

Boyer J. Plant productivity and environment, *Science*, 1982, Vol. 218, p. 443–448.

Bunt J.S., Heeb M.S. Consumption of O_2 in the light by *Chlorella pyrenoidosa* and *Chlamydomonas* reinhardtii, *Biochim. Biophys. Acta*, 1971, Vol. 226, p. 354–359.

Calvin M., Benson A.A. The path of carbon in photosynthesis, *Science*, 1948, Vol. 107, p. 476–480.

Chapman E.A., Graham D. The effect of light on the tricarboxylic acid cycle in green leaves. 1. Relative rates of the cycle in the dark and the light, *Plant Physiol.*, 1974*a*, Vol. 53, No. 6, p. 879–885.

Chapman E.A., Graham D. The effect of light on the tricarboxylic acid cycle in green leaves. II. Intermediary metabolism and the location of control points, *Plant Physiol.*, 1974*b*, Vol. 53, No. 6, p. 886–892.

Chapman E.A., Osmond C.B. The effect of light on the tricarboxylic acid cycle in green leaves. III. Comparison between some C_3 and C_4 plants, *Plant Physiol.*, 1974, 2000*b*, Vol. 53, No. 6, p. 893–898.

Chugunova N.G., Bil' K.Y., Chermnykh L.N. The structure and photosynthesis of cucumber leaves at different temperatures in the root zone, *Soviet Plant Physiol.*, 1976, Vol. 22, p. 584–591.

Clements E.S. The relation of leaf structure to physical factors, *Trans. Am. Micr. Soc.*, 1905, Vol. 26, p. 19–102.

Doman N.G. A question about the nature of rapidly labeled photosynthetic products, *Proc. Acad. Sci. USSR*, 1952, Vol. 84, p. 1017–1019.

Doman N.G., Khadzhi-Murat L.N., Demina S.E. Unities and peculiarities of carbon assimilation pathway(s) in different plant species, *Proc. Acad. Sci. USSR*, 1958, Vol. 122, No. 1, p. 111–113.

Downton J.S., Bishop D.G., Larkum W.D., Osmond C.B. Oxygen inhibition of photosynthetic oxygen evolution in marine plants, *Aust. J. Plant Physiol.*, 1976, Vol. 3, p. 73–79.

Fomina I.R., Biel K.Y. Photosynthetic carbon metabolism: Strategy of adaptation, *Contemporary Problems of Photosynthesis*, Allakhverdiev S.I., Rubin A.B., Shuvalov V.A., eds., Institute of Computer Sciences, Moscow–Izhevsk, 2014, Vol. 2, p. 415–483. [in Russian]

Fomina I.R., Bil' K.Y., Zakrzhevskii D.A., Balakhnina T.I. The photosynthetic carbon metabolism in *Tradescantia fluminensis* leaves preincubated without light in an argon or hydrogen atmosphere, *Soviet Plant Physiol.*, 1993, Vol. 40, No. 1, p. 21–27.

Gardeström P., Lernmark U. The contribution of mitochondria to energic metabolism in photosynthetic cells, *J. Bioenerg. Biomembr.*, 1995, Vol. 27, p. 415–421.

Gates R.G., Hoegh-Guldberg O., McFall-Ngai M.J., Bil' K.Y., Muscatine L. Free amino acids exhibit anthozoan "host factor" activity: They induce the release of photosynthate from symbiotic dinoflagellates *in vivo*, *Proc. Nat. Acad. Sci. USA*, 1995, Vol. 92, p. 7430–7434.

Gómez S.M., Bil' K.Y., Aguilera R., Nishio J.N., Faull K.F., Whitelegge J.P. Transit peptide cleavage sites of integral thylakoid membrane proteins, *Mol. Cell Proteomics*, 2003, Vol. 2, No. 8, p. 22–25.

Heber U. Metabolite exchange between chloroplasts and cytoplasm, *Annu. Rev. Plant Physiol.*, 1974, Vol. 25, p. 393–421.

Heber U., Santarius K., Urbach W., Ullrich W. Photosynthese und Phosphathaushalt. Intrazellulärer Transport von ^{14}C- und ^{32}P-markierten Intermediärprodukten zwischen den Chloroplasten und dem Cytoplasma und seine Folgen für die Regulation des Stoffwechsels, *Z. Naturforsch.*, 1964, Vol. 19*b*, p. 576–587.

Hertz S.M., Biel K.Y., Krupenko A.N., Ladygin V.G. Certain peculiarities of photosynthetic metabolism of carbon in *Chlamydomonas* mutants with inactive photosystem I or / and photosystem II, *Physiol. Biochem. Cult. Plants*, 1985, Vol. 17, No. 3, p. 249–253. [in Russian]

Hirt G., Tanner W., Kandler O. Effect of light on the rate of glycolysis in *Scenedesmus obliquus*, *Plant Physiol.*, 1971. Vol. 47, p. 841–843.

Hurry V., Tobiæson M., Krömer S., Gardeström P., Öquist G. Mitochondria contribute to increased photosynthetic capacity of leaves of winter rye (*Secale cereale* L.) following cold-hardening, *Plant Cell Environ.*, 1995, Vol. 18, p. 69–76.

Ivanova H., Keerberg O., Pärnik T. Influence of oxygen concentration on the rates of carbon fluxes in the biochemical system of CO_2 assimilation, *Proc. Estonian Acad. Sci. Chem.*, 1993, Vol. 42, No. 4, p. 185–197.

Karpilov Y.S., Novitskaya I.L., Belobrodskaya L.K., Biel K.Y., Maslov A.I., Kuzmin A.N., Karpilova I.F., Kerimov S.X., Popova E.I., Petrukhin Y.A., Hertz S.M. Glycolytic reactions in autotrophic cells: Their role in photosynthetic metabolism and photorespiration, *Mechanisms of Photorespiration and Its Specificity in Different Types of Plants*, Karpilov Y.S., Romanova A.K. eds., NCBI, Pushchino, Russia, 1978, p. 90–187. [in Russian]

Karpilova I.F., Chugunova N.G., Bil' K.Y., Chermnykh L.N. Ontogenetic changes of chloroplast ultrastructure, photosynthates, and photosynthate outflow from the leaves in cucumber plants under conditions of reduced night temperature, *Soviet Plant Physiol.*, 1982, Vol. 29, p. 96–104.

Kosobryukhov A.A., Bil' K.Y., Nishio J.N. Sugar beet photosynthesis under conditions of increasing water deficiency in soil and protective effects of a low-molecular-weight alcohol, *Appl. Biochem. Microbiol.*, 2004, Vol. 40, No. 6, p. 581–587.

Krömer S. Respiration during photosynthesis, *Annu. Rev. Plant Physiol. Plant Mol. Biol.*, 1995, Vol. 46, p. 45–70.

Krömer S., Heldt H.W. On the role of mitochondrial oxidative phosphorylation in photosynthesis metabolism as studied by the effect of oligomycin on photosynthesis in protoplasts and leaves of barley (*Hordeum vulgare*), *Plant Physiol.*, 1991, Vol. 95, p. 1270–1276.

Krömer S., Malmberg G., Gardeström P. Mitochondrial contribution to photosynthetic metabolism, *Plant Physiol.*, 1993, Vol. 102, p. 947–955.

Krömer S., Stitt M., Heldt H.W. Mitochondrial oxidative phosphorylation participating in photosynthetic metabolism of a leaf cell, *FEBS Lett.*, 1988, Vol. 226, p. 352–356.

Larsson C. $^{14}CO_2$ fixation and compartmentation of carbon metabolism in a recombined chloroplast-cytoplasm system, *Physiol. Plant.*, 1979, Vol. 46, p. 221–226.

McLachlan G.A., Porter H.K. Replacement of oxidation by light as the energy source for glucose metabolism in tobacco leaf, *Proc. Roy. Soc. London*, 1959, ser. B 150, p. 460–473.

Magomedov I.M. *Physiology and Biochemistry of Photosynthesis in Plants Exhibiting Primary Carboxylation of Phosphoenolpyruvate (Dr. Sci Dissertation)*, Leningrad, Leningrad State University, 1982, 418 p. [in Russian]

Nasyrov Y.S. Genetical regulation of formation and activity of photosynthetic apparatus, *Physiology of Photosynthesis*, Nichiporovich A.A. ed., Nauka, Moscow, 1982, p.146–164. [in Russian]

Nichiporovich A.A. Photosynthetic products and the physiological function of the photosynthetic apparatus of plants, *Works Inst. Plant Physiol. Acad. Sci. USSR*, 1953, Vol. 8, p. 3–41. [in Russian]

Nichiporovich A.A. *Light and Carbon Nutrition of Plants - Photosynthesis*, Publ. Acad. Sci. USSR, Moscow, 1955, 284 p. [in Russian]

Nishio J.N., Bil' K.Y., Fomina I.R., Nazarova G.N. Light dependent organization of photosynthetic metabolism within spinach leaves, *Abstracts in 9th Western Regional Photosynth. Conf.*, January 6–9, 2000, Asilomar Conf. Center, Pacific Grove, California, 2000, p. 72.

Nishio J.N., Bil' K.Y., Gómez S.M. Photosynthetic characterization and pigment analysis of symbiotic organisms from a freshwater sponge living in Swastika Lake, *Abstracts in 9th Western Regional Photosynth Conf.*, January 6–9, 2000, Asilomar Conf. Center, Pacific Grove, California, 2000, p. 73.

Nishio J.N., Sun J., Vogelmann T.C. Carbon fixation gradients across spinach leaves do not follow internal light gradients, *Plant Cell*, 1993, Vol. 5, p. 953–961.

Osmond B., Badger M., Maxwell K., Björkman O., Leegood R. Too many photons: photorespiration, photoinhibition and photooxidation, *Trends Plant Sci.*, 1997, Vol. 2, No. 4, p. 119–121.

Osmond C.B., Björkman O. Simultaneous measurements of oxygen effect on net photosynthesis and glycolate metabolism in C_3 and C_4 species of *Atriplex*, *Carnegie Inst. Wash. Yearbook*, 1972, Vol. 71, p. 141–148.

Pärnik T., Keerberg O. Decarboxylation of primary and end products of photosynthesis at different oxygen concentrations, *J. Exp. Bot.*, 1995, Vol. 46, p. 1439–1447.

Pärnik T.R., Voronin P.Y., Ivanova H.N., Keerberg O.F. Respiratory CO_2 fluxes in photosynthesizing leaves of C_3 species varying in rates of starch synthesis, *Russ. J. Plant Physiol.*, 2002, Vol. 49, No. 6, p. 729-735.

Pärnik T.R., Bil' K.Y., Kolmakov P.V., Titlyanov E.A. Photosynthesis of the sea grass *Thalassodendron ciliatum*: leaf morphology and carbon metabolism, *Photosynthetica*, 1992, Vol. 26, No. 2, p. 213–223.

Rabinovitch E. *Photosynthesis and Related Processes*, Wiley (Interscience), New York, 1945, Vol 1, 599 p.

Raghavendra A.S., Padmasree K., Saradadevi K. Interdependence of photosynthesis and respiration in plant cells – interactions between chloroplasts and mitochondria, *Plant Sci.*, 1994, Vol. 97, p. 1–14.

Raven J.A. Endogenous inorganic carbon sources in plant photosynthesis. 1. Occurrence of the dark respiratory pathways in illuminated green cells, *New Phytol.*, 1972, Vol. 71, p. 227–247.

Raven J.A. Division of labour between chloroplast and cytoplasm, *The Intact Chloroplast*, Barber J. ed., Elsevier, Amsterdam, 1976*a*, p. 403–443.

Raven J.A. The quantitative role of "dark" respiratory processes in heterotrophic and phototrophic plant growth, *Ann. Bot.*, 1976*b*, Vol. 40, p. 587–602.

Ried A. Energetic aspects of the interaction between photosynthesis and respiration, *Prediction and Measurement of Photosynthetic Productivity*, Setlic I. ed., Centre Agricultural Publishing and Documentation, Wageningen, 1970, p. 231–246.

Ruben S. Photosynthesis and phosphorylation, *J. Am. Chem. Soc.*, 1943, Vol. 65, p. 279–282.

Ruben S., Kamen M.D. Long-lived radioactive carbon: C^{14}, *Phys. Rev.*, 1941, Vol. 59, p. 349–354.

Ruben S., Kamen M.D., Hassid W.Z., DeVault D.C. Photosynthesis with radio-carbon, *Science*, 1939, Vol. 90, p. 570–571.

Saradadevi K., Raghavendra A.S. Dark respiration protects photosynthesis against photoinhibition in mesophyll protoplast of pea (*Pisum sativum*), *Plant Physiol.*, 1992, Vol. 99, p. 1232–1237.

Semikhatova O.A., Zalensky O.V. Coupling of photosynthetic and respiratory processes, *Physiology of Photosynthesis*, Nichiporovich A.A. ed., Nauka, Moscow, 1982, p.130–145. [in Russian]

Simonis W., Urbach W. Photophosphorylation *in vivo*, *Annu. Rev. Plant Physiol.*, 1973, Vol. 24, p. 89–114.

Soukhovolsky V.G., Fomina I.R., Bil' K.Y., Nishio J.N., Khlebopros R.G. An optimization model of the photosynthetic leaf. The model of optimal photosynthetic CO_2 fixation within leaves of mesophytic C_3 plants, *Dokl. Biol. Sci.*, 2002, Vol. 382, p. 28–30.

Skupchenko V.B. The vibrating microtome of the soft tissues. The new scientific methods, *The New Scientific Methods*, Kalinin E.P. ed., Komi Branch of the Acad. Sci. USSR, Syktyvkar, 1979, Vol. 2, p. 1–56. [in Russian]

Sun J., Nishio J.N. Why abaxial illumination limits photosynthetic carbon fixation in spinach leaves, *Plant Cell Physiol.*, 2001, Vol. 42, p. 1–8.

Sun J., Nishio J.N., Vogelmann T.C. ^{35}S-Methionine incorporates differentially into polypeptides across leaves of spinach (*Spinacia oleracea*), *Plant Cell. Physiol.*, 1996*a*, Vol. 37, p. 996–1006.

Sun J., Nishio J.N., Vogelmann T.C. High-light effects on CO_2 fixation gradients across leaves, *Plant Cell. Environ.*, 1996*b*, Vol. 19, p. 1261–1271.

Tarchevsky I.A. Mechanisms of drought effects on photosynthetic absorption of CO_2, *Physiology of Photosynthesis*, Nichiporovich A.A. ed., Nauka, Moscow, 1982, p. 118–129. [in Russian]

Thoday D. The significance of the reduction in the size of leaves, *J. Ecol.* 1931, Vol. 19, p. 297–303.

Titlyanov E.A., Leletkin V.A., Kolmakov P.V., Lapshina A.A., Bil' K.Y. Effect of light intensity on the productive potential and its realization in thalli of the marine green algae *Ulva fenestrata* P. et R., *Soviet Plant Physiol.*, 1990, Vol. 37, No. 3, p. 366–374.

Travis R.L., Huffaker R.C. Light-induced development of polyribosomes and the induction of nitrate reductase in corn leaves, *Plant Physiol.*, 1970, Vol. 46, p. 800–805.

Voskresenskaya N.P., Polyakov M.A., Karpushkin L.T. Effect of oxygen concentration on CO_2 gas exchange of bean and maize plants, *Soviet Plant Physiol.*, 1974, Vol. 21, No. 3, p. 455–461.

Zalensky O.V. Investigation of the relationship between photosynthesis and respiration in leaves of plants, *News Acad. Sci. USSR*, 1961, Biol. Sect. 2, p. 202–212. [in Russian]

Zalensky O.V. *The Study of Ecology-Physiological Aspect of Photosynthesis*, Reported in XXXVII Timiryazev's Reading, 03 June 1976, Inst. Plant Physiol., Russian Acad. Sci., Nauka, Leningrad, 1977, 57 p. [in Russian]

Zalensky O.V., Semikhatova O.A., Voznesenski V.L. *The Application of ^{14}C-Methods for Research of Photosynthesis*, Publ. USSR Acad. Sci, Moscow-Leningrad, 1955, 90 p. [in Russian]

Zalensky O.V., Zubkova E.K., Mamushina N.S., Fillipova L.A. 1980 On the relationships between photosynthesis and respiration in the assimilating cell. 1. A study in the glycolysis and Krebs cycle in the light and dark conditions in *Chlorella* cells, *Soviet J. Bot.*, Vol. 65, No. 9, p. 1141–1149. [in Russian]

Appendix I

Additional Materials and Methods

Plant Growth Conditions. Crop plants, red beet (*Beta vulgaris* L., subsp. *esculenta* Guerke), sugar beet (*Beta vulgaris* L. cv. Monohikari, Seedex, Longmont, Colorado); cabbage (*Brassica oleracea* var. Capitata), soybean (*Glycine max,* cv. Corsoy), and radish (*Raphanus sativus* L., var. sativus) were grown in 15 × 15 × 15 cm pots in Scotts Professional Potting Soil, containing complete nutrient pellets (Sierra 17-6-12 Plus Minors, Grace Sierra, Milpitas, California) or Scotts Osmocote (18-6-12) fertilizer pellets under standard conditions in a growth chamber. The relative humidity was 70%, the daytime temperature was 26 ± 2 °C and the night time temperature was 20 ± 2 °C. Light of 500–600 quanta (400–700 nm) / $m^2 \cdot s$ was provided during the 16 hours photoperiod. Five- to six-week-old plants were used for all experiments. Spinach plants (*Spinacia oleracea* cv. hybrid 424, Harris Moran Seed Company, Modesto, CA) were grown as above, except the temperature of the growth chamber was 23 ± 2 °C during the day and 17 ± 2 °C at night, and the photoperiod was 16 hours. Alternatively, plants were grown in the glasshouse with supplemental light provided by 1,000 watts metal-halide lamps and 150-watt incandescent bulbs for 12 hours per day. The air temperature near plant level was 28–31 °C during the day and 20–26 °C at night. The age of the plants from sow date is presented with the figures.

Tradescantia fluminensis were grown in an adjacent glasshouse, where the temperature was 23 °C during the day and 16 °C at night. Plants were watered daily with measured amounts of reverse osmosis purified water.

Drought Treatment. Plants were given a measured amount of water that decreased the relative soil humidity to about half of the control plants. The soil humidity was measured with a simple conductivity meter (Rapitest, distributed by Luster Leaf Products, Inc., Wood-stock, Illinois). Soil humidity was measured at least 3 times per day at 4 different locations in each 15 × 15 × 15 cm pot.

Leaf Area. Leaf area was measured during development of leaves over a two-week period. Leaves were traced every two to three days. A planimeter was used to measure the leaf area.

Fresh Paradermal Sections. Paradermal leaf sections were made with a lab-built vibratome (Skupchenko, 1979; Bil' *et al.*, 2000, 2003; Biel, Fomina *et al.*, 2008, 2010). 80-120 μm sections were used for the assays.

Appendix II

Below is a short report illustrating the functionality of the oxygen evolution stress assay. Garst provided us 4 varieties to test for drought tolerance. One variety was eliminated from testing due to poor growth. The other three were ranked for drought tolerance based on the oxygen evolution stress assay, and our results were in accord with 3–5 years of field trials conducted by Garst. Our tests were completed within 30 days. Younger plants or possibly sprouts could be used, which would speed up screening even more.

Preliminary Analysis of the Utility of a Novel Stress Resistance Assay on Three Garst Lines of Zea mays, a C_4 Plant

Abiotic stress is the biggest factor in decreasing crop productivity. Thus, improving stress tolerance in plants is an important step in maintaining and improving crop productivity worldwide. A method for assaying stress tolerance in plants and predicting performance under stress could greatly decrease the time for screening and selection. Great monetary savings could result from such an assay. As shown in the main body of the chapter, we have developed an assay that works well with C_3 crops. In the following experiment we conducted a practical test of the stress assay on a C_4 crop, corn. Four unknown corn varieties were supplied by Garst. One variety did not grow well under our growth conditions, so the assay was conducted on only three varieties (Var #1, Var #2, Var #3, and Var #4).

Material: Seeds were germinated in distilled water in Petri dishes at room temperature (25 °C). Var #4 showed germination after 24 hours, whereas the others were much slower (2 days more for equivalent germination). Germination in all varieties was uneven.

Soil: From 30-40 seed, 20-25 germlings were selected for transplanting into 10.2 cm diameter pots containing Scotts Professional Potting Mix plus Scotts Osmocote (18-6-12) fertilizer pellets. Plants were watered twice daily between 9:00–10:00 and 16:00–17:00 (hours).

Growth Conditions: Plants were grown in the glasshouse at the University of Wyoming, Botany Department under supplemental light 16 hours supplied by 1,000 watts metal-halide lamps and 150-watt incandescent bulbs. The temperature of the glasshouse was 28 °C during the day and 23 °C during the night. Pots were rotated 8-9 times during the 3 weeks of growth in the greenhouse.

Soil Humidity: maintained near 80-85% in controls was measured with a simple conductivity meter (Rapitest, distributed by Luster Leaf Products, Inc., Woodstock, Illinois). For the drought treatment plants were watered once daily between 16:00–17:00 hours. For measurements, plants from 6 pots (of the 20–25 pots) with soil humidity of approximately 50% were utilized for the drought samples (Figure 3.A).

HCO_3-Dependent O_2 Evolution: Leaf slices 400–500 μm in width were cut from 6 leaf plugs (0.64 cm^2) from 6 different plants taken from the older third of the fifth leaf. Bicarbonate-dependent oxygen evolution was measured polarographycally with a Rank O_2 electrode in 2 ml of 40 mM HEPES-KOH buffer, pH 7.61, 300 mM sorbitol, and 10 mM $NaHCO_3$, at 37 °C. Oxygen evolution (light curves) was measured at photon fluxes of 0, 120, 220, 450, 940, 1400, and 2200 $\mu mol / m^2 \cdot s$. Different light levels were produced by placing neutral density filters between the light source (Philips 500 W C2X / DAB) and the water-jacketed sample cuvette. The rate of oxygen evolution was expressed per m^2 of leaf surface area (cellular content lost due to cutting of leaf was not subtracted). Photosynthetic stress resistance index (PSRI) was measured as described above.

Results

Growth Characteristics: Of the four different lines supplied, Var #3 exhibited the lowest rate of germination and growth. Var #3 was very sensitive to humidity, and it appeared to have tip rot. Although there is adequate circulation in the glasshouse, Var #3 never really recovered from the high humidity during germination. One week after transplant, Var #1 also exhibited some tip rot, but the plants recovered (some tips were removed). However, Var #3 never recovered, and growth was very slow. The plants did not have five leaves when the others had six or more. Var #3 was eliminated from the study.

Visually, Var #4 appeared most healthy and was the most prolific initially. After 1.5-2 weeks, Var #1 and Var #2 appeared healthy also, but they were smaller than Var #4 at that point. Var #2 was the next largest size, and Var #1 was the shortest, but the stalks were thicker and the leaves were wider than the other two varieties.

Additionally, Var #1 and Var #2 exhibited less uniformity in size compared to Var #4. The experiment was completed November 21, 2000, but the plants have been maintained in the glasshouse. Until December 08, 2000, the plants maintained their relative height differences with Var #4 being the tallest and Var #1 the shortest. Var #1 still had wider leaves and thicker stalks.

Oxygen Evolution: All varieties in well-watered conditions exhibited low rates of photosynthesis (2–4 $\mu mol\ O_2 / m^2 \cdot s$) and respiration (–0.6 to –0.8 $\mu mol\ O_2 / m^2 \cdot s$) at 37 °C (Figure 3.A). One day after the start of the drought treatment, photosynthesis and respiration at 37 °C increased (a well-watered control was not measured, because that would be 3 extra measurements / day). The rate of photosynthesis and respiration continued to increase throughout the 7 days drought treatment. Var #1 maintained the highest rate of photosynthesis and respiration during the experiment. At the end of the experiment, all varieties exhibited the same rate of photosynthesis. Var #1 exhibited a respiration rate that was 40% higher than the other two varieties at the end of the experiment.

Gross photosynthesis is shown in Figure 3.A. Although Var #4 was the fastest growing variety, it exhibited the lowest rates of oxygen evolution at 37 °C. One possibility is that they have a more efficient metabolism for endosperm utilization, so they developed greater leaf are more quickly. Their rapid net growth compared to the other varieties may be due to a lower maintenance respiration rate. Photosynthesis per plant may have been higher in Var #4.

Photosynthetic Stress Resistance Index (PSRI): Under well-watered conditions the 3 varieties exhibited low PSRI's. However, the PSRI of Var #1 was 9.8-times higher than in Var #2, and 3.5-times higher than in Var #4. The range of the photosynthetic stress resistance index was from 1.6–15.5%.

The drought treatment linearly increased the PSRI in all lines (Figure 3.A). The rates of increase were different among the varieties. On the first day after the start of the drought treatment the PSRI of Var #2 was 66% of Var #1. By the third day the PSRI of Var #2 reached a plateau at about 80% of the Var #1. One day after the start of the drought treatment, the photosynthetic stress resistance index of Var #4 remained constant. By the third day, however, the PSRI of Var #4 was similar to that of Var #1.

At no point during the experiment did the photosynthetic stress resistance index of either Var #2 or Var #4 exceed that of Var #1.

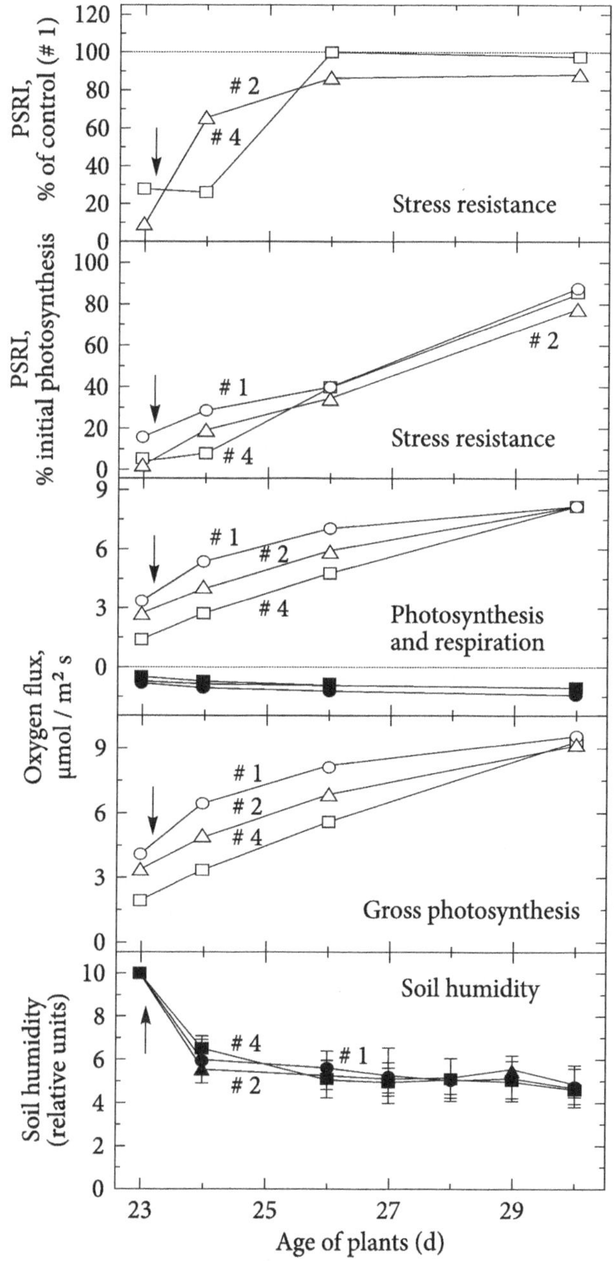

Fig. 3.A. Effects of soil drought on three Garst's *Zea mays* lines (Bil' and Nishio, 2000*b*)

Note: Measurement of photosynthetic stress resistance index (PSRI) has been done under 37 °C. Arrow shows beginning of the soil drought

General Conclusion

The germination and growth characteristics of the four Garst *Zea mays* lines were variable. Overall Var #4 exhibited the greatest growth, but Var #1 exhibited the most robust growth anatomy, as estimated by visual examination. Var #3 was very sensitive to humidity and apparently developed tip rot.

Under drought treatment all of the varieties exhibited increased photosynthesis, respiration, and stress tolerance (higher PSRI's). However, the sensitivity to the drought stress was variable.

Var #1 exhibited the most rapid response to drought, and Var #1 had the highest PSRI at the start of the experiment. Var #4 exhibited the most sluggish response. Although the photosynthetic stress resistance index of Var #2 was low under well-watered conditions, it exhibited a rapid increase during the onset of drought (but not as rapidly as Var #1).

In rapidly changing conditions, such as normal field conditions, the period of the time for induction of the stress response can be significant.

Without knowing the physiological traits that Garst has associated with the three varieties tested, the assay showed clear differences. Our interpretation of the results could be changed based on known traits. For example, we know that a low PSRI does not necessarily mean that the plants are sensitive to photoinhibition.

Suggestions

To repeat the experiment to see if the plants have a memory for stress response and to measure how long the photosynthetic stress resistance index remains elevated under well-watered conditions. For example, if Var #4 can survive the induction period of stress, but has a longer "memory", then it might be considered a good candidate for stress tolerance.

To conduct the assays efficiently, a bank of oxygen electrodes with computer controls is required. Many experiments should be done for a complete understanding of the results and to develop a routine assay for selecting stress tolerant *Zea mays*. For example, the assay could be used productively to screen recombinants and to examine carbon metabolism, protein synthesis (gene expression), and interactions between photosynthesis and respiration in *Zea mays*.

HYPOTHESES

4
The Hypotheses of Halosynthesis, Photoprotection, Soil Remediation via Salt-Conduction, and Potential Medical Benefits

Karl Y. Biel[1,2,*], **Nicholas P. Yensen**[3]

[1]*Institute of Basic Biological Problems, Russian Academy of Sciences, Pushchino, Moscow Region 142290, Russia*
[2]*Biosphere Systems International Foundation, Tucson, Arizona 85755, USA*
[3]*NyPa International, Tucson, Arizona 85705, USA; Biosphere Systems International, Tucson, Arizona 85705, USA*

Science is something that cannot be,
but that what can be — is technology -
Academician Peter Kapitsa

Abstract
The present chapter proposes that all terrestrial higher plants may be divided into three general categories based on their salt management: 1) Excluders, 2) Accumulators and 3) Conductors. Excluders exclude salt at the root level and are typically glycophytes, but there are a few euhalophytes as notable exceptions. Many miohalophytes and euhalophytes are salt accumulators, while virtually all crinohalophytes are salt conductors. Salt conductor plants move salt from the soil through the roots and plant tissues to salt glands on the plant surfaces. We describe the hypothetical mechanisms and pathways of moving salt through the plant and predict that some plants with the ability to do this may acquire energetic benefits

* *Corresponding authors*: karlbiel@hotmail.com

through a process described herein. This process we call ***Halosynthesis. Halosynthesis is any salt-mediated mechanism, by which exogenous energy is stored / utilized by organisms under appropriate conditions***. We predict in the present case, that the appropriate conditions for halosynthesis may be met in salt glands and possibly other structures. Halosynthesis utilizes environmental energy, in the face of otherwise stressful conditions, to optimize growth and other metabolic functions. The process of halosynthesis may have great potential to benefit conductor plants due to evapotranspirational energy and / or through the accumulation of a salt-mediated and / or salt-dependent, photon-electron, surface charge. The first and best-known mechanism of energy extraction process is photosynthesis while the second discovered environmental energy source is that of thiobacteria which can extract heat energy from undersea hot-water vents to fix and reduce carbon. Conductor halophyte plants have been less well studied in this respect and this chapter specifically examines the potential of the genus *Distichlis* as a good example of the quintessential halosynthesizing conductor plant. Photo-halosynthesis mechanisms may "energize" or stimulate certain processes, including cancer fighting processes, in the human body via photon impingement upon cations in epidermal papillae, protrusions, pilous structures, and / or saline eye surfaces. We speculate that UV light impinging on metallic salts in or on the face, shoulders and extremities may also produce beneficial effects and may involve a limited amount of red light which is known to protect against ulcers, gastritis, pathogens, and potentially other diseases.

Keywords: *Distichlis*, Halosynthesis, Photoprotection, Soil remediation, Salt-conduction

4.1 Introduction

The estimated 1–10 billion hectares of salinized farm soils worldwide are an untapped resource opportunity for new halophilic (salt-loving) crops. Much of this "waste" cropland occurs in North America, South America, the Middle East, Central Asia, Northern Africa and Australia. To meet the potential demand for new halophilic crops, as water resources become scarce, many salt-tolerant plants and euhalophytes (plants that grow better under high salinities) are being studied by the emerging science of halophytology. Various new halo-technologies, such as the serial biological concentration of salts and biodrainage, incorporate halophytes for both practical (crops) and aesthetic (ornamental) reasons.

One of the first patents for a halophyte crop was issued in 1985 (Yensen, 1985). But the development of additional better-suited crops has been slow. We have just begun to develop new "crops" that can be used to eliminate toxic compounds and elements (*e.g.*, Se, Pb, Cr, Cd, Zn, and U), petroleum products, asphalt, radioactive nuclides via phyto-remediating and bio-remediating halophytes. There is even work being done with vacuolar-salt-accumulating algae that can convert seawater to "brackish" water by sequestering salt within its tissues (H. Leith, per. com.). And in new developments, ornamental halophytes could be irrigated with brackish or seawater, thus freeing up large amounts of fresh water for municipalities. Over 67 genera and hundreds of species are considered to have potential for development as cultivars (Yensen, 2002). Of the estimated 10,000 salt-tolerant species, there maybe exist as many as 250 potential staple-crop halophytes ... plus ornamentals, dune stabilizers, pharmaceuticals and environment-improving halophytes.

The development of these needed crops is a major task and will be a significant milestone in the advancement of agriculture. The modicum of "embryonic halophyte crops", developed to date, herald from classical breeding of wild halophytes with minimal input from biotechnology, tissue culture, and / or plant screening techniques.

The incipient era of halotechnology must develop and carry forward a knowledge base. For this to happen, the support level will need to be at least on par with that of the US Salinity Laboratory by the Honourable George E. Brown and the early investigators: H.E. Hayward, L.A. Richards, F.M. Easton, A.D. Ayres, L. Bernstein, L.V. Wilcox, Mass, G.J. Hoffman and numerous other visionaries, who in 1928 established and / or carried out the early investigations in Riverside, California under the auspices of the USDA. Their efforts led the world in developing a knowledge base to understand how to manage "low salinity" and how it affects freshwater crops and soils. The US Salinity Laboratory still leads the world in how salt affects our fresh water crops. The new International Center for Biosaline Agriculture in Dubai, United Arab Emirates is now the world's pre-eminent center for high-salinity crop research.

Today, with the growing millions of acres of saline agricultural worldwide and the potential to export halophyte technology, we are entering a new era of halophytology. Some of the first observations of the increased productivity with increased salinity on a cereal grain on high salinity were noted in the genus *Distichlis* (Yensen *et al.*, 1988). Fortunately, with regard to *Distichlis*, there is now a growing body of information on ecology, physiology, morphology, anatomy (including salt gland structure), *etc.* Still, the basic understanding of how "*productivity can increase (reaching levels similar to or greater than freshwater crops) with an increase in salinity*" still remains unexplained (Yensen, 2002).

Photohalosynthesis in salt-conducting plants may provide some answers to this conundrum and an opportunity to increase food production while remediating salt-affected soils. As a broad concept, it may provide plant and animal protection effects, as well as other medical benefits.

4.2 The Haloconductor Theory

4.2.1 The Remediation of Saline Soils

Worldwide, saline soils reduce productivity by trillions of dollars annually. Thus, the remediation of salt-affected soils while growing productive salt-tolerant crops could be of great benefit ... particularly in arid regions. The remediation of salt-affected soils by mechanical, chemical, and / or leaching, although effective, may be difficult, expensive and cause additional problems. Biological remediation may have the advantage of permanently restoring the land to its natural productive capacity. Certain plants may gradually remove salts from soil in beneficial ways.

The terms glycophyte, oligohalophyte, miohalophyte, and euhalophyte describe groups of plants with increasing tolerances to salt. Little attention, however, has been paid to classifying plants according to their ability to remove salt from the soil. In Figure 4.1 it may be noted that we predict that both the glycophyte *Triticum* and the euhalophyte, *Salicornia*, do not effectively remove much salt from the soil. This is because *Triticum* excludes salt from entering its tissues, while *Salicornia* has limited removal ability once its tissues are full of salt.

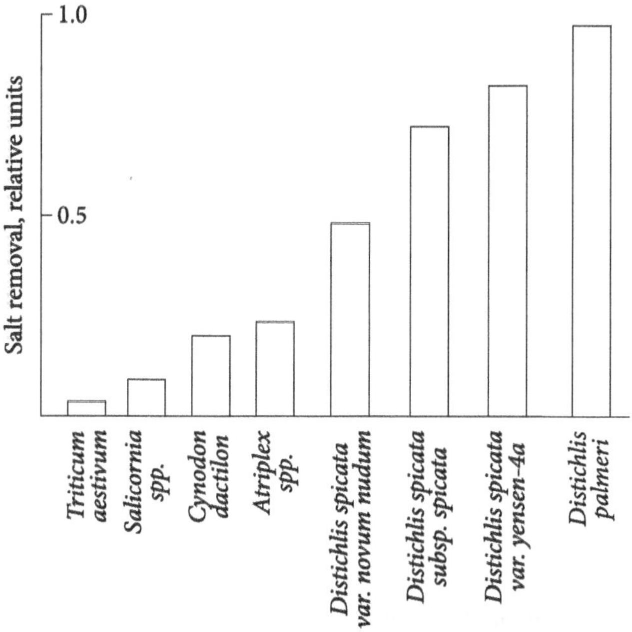

Fig. 4.1. Efficiency of soil-salt removal by different plants (from low to high removal ability): *Triticum aestivum* (wheat), *Salicornia* spp. (pickle weed), *Cynodon dactylon* (bermuda grass), *Atriplex* spp. (salt bushes), *Distichlis* spp. [4 NyPa lines] (salt grass) (adapted from Biel and Yensen, 2005*a*, 2005*b*)

In the past, it has been considered uneconomical to remediate saline soils using salt-accumulating halophytes to extract the salt from the soil for three reasons:

1. More salt is added to the soil via the water required to grow the halophyte than is extracted by the harvest of salt-laden plant material,
2. It is uneconomical to mechanically harvest and remove the salt with the otherwise useless plant material, and
3. The plant material is generally inedible and / or of little value due to the heavy load of salt in the plant tissue with 25–54% of dry weight as salt (Ungar, 1991).

The present concept considers salt-excreting halophytes that can be used in a method to potentially mobilize up to many tons of salt per hectare per annum with a certain percentage of this salt becoming airborne. This airborne salt may move to areas leached of salts. It has been long known that many crinohalophytes excrete salt onto their surfaces. The concept of using this ability to move salt from the soil through the plant, onto the plant surface, and potentially into the air, is a new concept and has not been described in detail or quantified in the literature. We term this process ***haloconduction***.

The application of this process for the remediation of soils promises to be an environmentally friendly economical approach for the remediation of salt-contaminated soils (Biel and Yensen, 2006*a*, 2006*b*). This process may be applied using the haloconducting principle with a number of species from various families and genera, such as, but not limited to: *Distichlis, Aeluropus, Spartina, Limonium, Avicennia,* and *Atriplex* (to a greater or lesser degree via bladder cells). Further, salt accumulators, such as *Climacoptera crassa* through senescence of leaves of accumulated salt may disperse salt only if the leaves become windblown or are otherwise are dispersed.

The haloconducting approach and process may function in other plants with different genetic morphologies and physiological adaptations. The process may be applied to "salt-gland-bearing" species, which range from being mildly salt-tolerant to highly euhalophytic.

4.2.2 New Approach for Soil Remediation via Salt Conducting Plants

A new classification system is necessary to understand salt balance and to remediate salinized lands. In the past, because some salt-accumulating halophytes may have up to 50% of their dry weight as salt, it is often suggested by neophytes and laypeople that, "harvesting halophytes would remove salt from the soils". The typical response from halophyte and irrigation experts, however, has been that, "*the water required to grow halophytes carries an influx of more salt than is removed by the harvested halophytes*". We believe, however, that a re-evaluation of the different field situations, in conjunction with a new way to classify plants according to how much salt they can permanently remove from soil, may significantly alter this paradigm.

A paradigm shift occurs when we realize that not all field situations require salt-water irrigation. Many areas of primary (nature caused) and secondary (human caused) salinization have influxes of water that are relatively low in salinity, such as in parts of Australia, western North America, northeastern China, northern Chile, Kara-Kum and Kizil-Kum regions of Asia and other areas salinized by salt-bearing "*produce*" water of petroleum and gas wells. In these situations, where the salinization has occurred over many years from relatively-fresh-water sources, or in the case of oil and gas wells where a one-time event has created a hypersaline condition, it may be possible to remove salt from the soil using halophytes (see for example, Figure 4.2).

Fig. 4.2. Salt excretion from the soil by halophytic grass *Distichlis* spp. (photo image from the collection of Dr. Elisabeth Poscher)

In the past, there have been many ways to classify plants regarding salinity and the various mechanisms that the plants employ to live in saline environments (see review Yensen, 1995). However, none of these classification systems have specifically considered, with respect to all plants, how a plant removes salt from the soil, how much salt is removed and where the removed salt goes.

We have therefore divided all plants into three groups with respect to their ability to remediate salty soils (Figure 4.3).

- **Excluders**, the first category of plants, exclude salts from entering their tissues at the roots. Most terrestrial plants fall into this category although many have limited abilities to exclude salts and when the salt loads increase they die from excess penetration of salt into the plant tissues. Some salt-excluding plants, however, are euhalophytes the red mangrove, *Rhizophora* spp.). These plants, in general, have limited or no ability to remediate saline soils. And, because they selectively absorb water, excluding salts at the root level, they tend to accumulate soil salts excessive of their soil-salt tolerance. This typically results in the demise of the plant with "plant-induced" salinization of the soil (Figure 4.3-*1*).
- **Accumulators**, the second category, uptake salts and accumulate the salts in the vacuoles of the plant cells. These may be glycophytes, salt-tolerant or euhalophyte plants, with or without succulence. In general, these plants have some exceptions such salt-laden plants often do not have

very valuable products, which typically do not cover the cost of removal. Also, they may become too salty to use as forage for grazing livestock. These plants remove a limited amount of salt from the soil, *i.e.* 50% of their dry weight, when their tissues become salt "saturated" (Figure 4.3-*2*).

- **Conductor** plants, the third category (Figure 4.3-*3*), absorb salts from the soil, transport the salt through the plant to the shoot surface and, with the aid of wind, disperse the salt away from the site. These plants potentially have great ability to remove salt from the soil and widely distribute the salt via wind dispersal. It may be typical to have very wide dispersal of aeolian salt via air currents, for instance, it is well known that 80% of all inland salt has arrived from the oceans via air currents carrying microscopic salt particles.

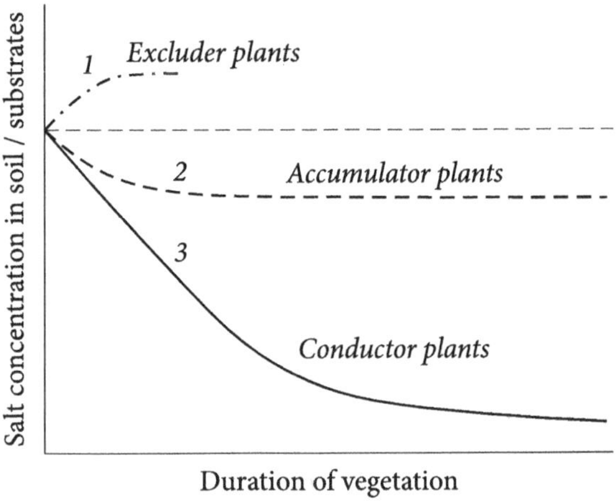

Fig. 4.3. Soil salt concentration over time with respect to (1) Excluder, (2) Accumulator, and (3) Conductor plants (adapted from Biel and Yensen, 2005*a*, 2005*b*)

While this classification is phylogenetically artificial (and is not necessarily related to salt tolerance) it does allow the classification of plants with regard to their ability to remediate saline soils. Further, the application of the principals of this classification system may have additional economic advantages (such as for non-salt-laden forage).

Distichlis, Spartina, Aeluropus and other conductor plants may serve as a biological tool to remove salts from the soil by placing the salts onto the shoot surfaces without accumulation of salts in the tissues. Moreover, as just noted, a certain proportion of the excreted salt located on the surface may be released into the air and carried aloft with air currents, thus permanently eliminating the salt from the remediation area (see, for example, Biel and Yensen, 2006a).

Putative advantages of salt conducting plants are that the salt is removed from the soil and released without accumulation in the plants. Our speculation is that in conductor plants the salts are essentially excluded from the cells. We suggest that the salts initially transported from the roots symplastically (xylem) and then apoplastically (and / or interstitially) in the vacuoles of collecting cells (when present as such in *Distichlis*) so that nearly all cell volumes become vacuolar-apoplastic continua (Semenova *et al.*, 2010), of which the solutions flow to the salt glands (or to the papillae in some species). Driving force, giving ordered motion to salt solution from the vacuole of the collecting cell through the basal cell of the salt gland to cap cell, arises from the impulses of a mechanical compression–expansion of plasma membrane that penetrates the basal cell in the form of extracellular channels (Semenova *et al.*, 2010). Finally, the salt is released externally on the surface of cap cells (and / or papillae) and possibly in some cases (as of yet undocumented) as a "hollow" salt crystal filament. Outlined a system of salt conducting is largely speculative, since is not fully confirmed by experimental data, even the works related to the structure and function of salt glands are contradictory (see ref. in Semenova *et al.*, 2010). This speculation, though simplified, is important because of the potential value to plants and the environment to uptake, transport, and eliminate salts from the immediate vicinity, as will be discussed later herein.

4.2.3 *Advantages of Conductor Plants for Soil Remediation*

Figure 4.4 illustrates the advantage of conductor plants when compared to salt accumulator halophytes. Salt accumulating plants, although with great ability to tolerate high salinity, reach an absorption plateau, which limits their ability for continued uptake of additional salt. The maximum tolerance and initial rate of salt absorption varies between the many species of salt-accumulating plants, but in the long term they are highly restricted with regard to developing improved lines with greater ability to remove salt from the soil (2b, dotted line of Accumulator plants in Figure 4.4).

All salt conductors (even if limited in total flow of salt) have a continuous flow of salt from the soil to the shoot surface and potentially to the air. From this standpoint, it is important to understand the mechanisms and pathways of salt movement. It is likewise important to identify and / or develop plants with high capacity for an increased salt flow, cell-protection mechanisms, and potential for aeolian salt dispersal from the cap cell surface (3b, dotted line of Conductor plants in Figure 4.4).

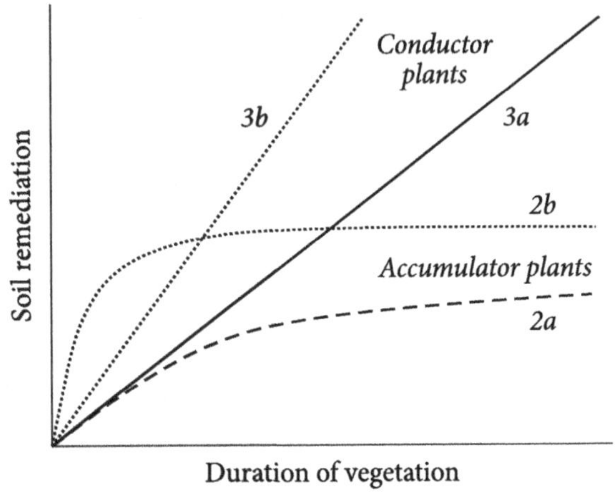

Fig. 4.4. Graph of generalized soil-salt remediation ability of existing and developed salt-accumulating plants (2a, 2b) versus conductor plants (3a, 3b). The 2a dashed line represents a wild-type accumulator plant. The 2b dotted line represents the theoretical limit of salt removal via accumulator plants through selection and breeding. The 3a solid line represents typical wild-type salt-conducting plants. The 3b dotted line represents the improved salt-conductor plants and their greater potential for soil remediation (adapted from Biel and Yensen, 2005*a*, 2005*b*)

4.2.4 Productivity Considerations

Regarding productivity, higher plants are often classified into three types according to their sensitivity to salinity. For instance, *glycophytes* have productivity inversely related to salinity. The second, *miohalophytes* (salt-tolerant plants) maintain productivity with increasing salinity to some threshold, above which they have a linear decrease in productivity. The third, *euhalophytes*, have a convex curvilinear increase in productivity with salinity up unto some maximum level and then have a curvilinear decreased productivity with increased salinity.

Figure 4.5 considers the theoretical relative productivity limits for Excluder, Accumulator and Conductor plants.

Our new classification of plants according to how the plants manage the salts considers that most *glycophytes* are excluders, but not exclusively so. With regard to the *miohalophytes* and *euhalophytes* there can be major paradigm shifts with regard to whether a plant's ability to grow in highly saline substrates affects its ability to remove salts from the system. Conceptually, some *miohalophytes* may be able to conduct significant quantities of salt, over many years, while in contrast some *euhalophytes*, which tolerate very high salinities, may remove little or no salt from the system.

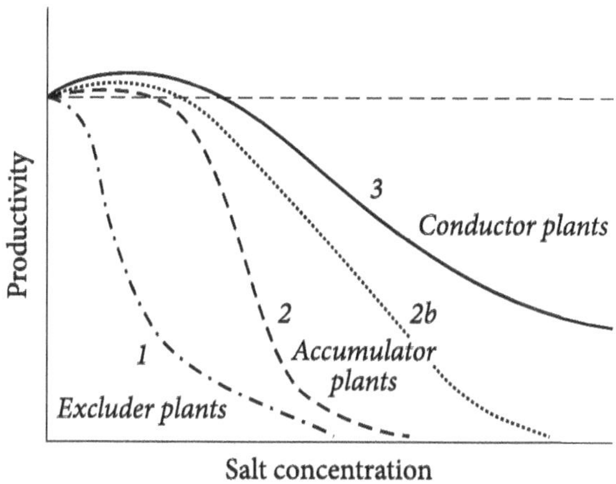

Fig. 4.5. Theoretical productivity of three types of plants with increasing salinity. Salt Excluder plants (1) typically perish in high salinity; Salt Accumulator plants (2, 2b) and Salt Conductor plants (3) often have the ability to survive in high salinity. The 2b shows the potential of some salt-accumulating plants to have high productivity at high salinity and have productivities similar to Conductor plants (3) even though they may accumulate up to 50% of dry weight as salt with insignificant release to the plant surface (adapted from Biel and Yensen, 2005*a*, 2005*b*)

Salt excluder plants, such as beans (*Phaseolus* spp.), wheat (*Triticum* spp.), maize (*Zea* spp.), have a very rapid productivity drop rate with increasing salinity. Salt accumulator plants, such as salt bush (*Atriplex* spp.), pickle weed (*Salicornia* spp.), and seep weed (*Suaeda* spp.), are in general much more salt tolerant, often due to their tissues and metabolic capacity functioning well under high salinity. Because salt-accumulator plants have evolved to tolerate high levels of tissue salt content, many are able to survive in remarkably high salinities, but still succumb to some threshold salt level above which their productivity drops precipitously even though they can survive the hypersaline conditions. From a practical consideration, their accumulation of high levels of salts (some of which may be very toxic to animals) can make their use uneconomical and in some cases very negatively so.

In comparison to excluders and accumulators, salt-conductor plants, such as saltgrasses (*Distichlis* spp.), chord grass (*Spartina* spp.), black mangroves (*Avicennia* spp.), *etc.* have the advantage of maintaining low-salt levels in the shoot tissues (thus being palatable for animals) and some can grow under very high salinities to use this underutilized resource. Photosynthetic and "non"-photosynthetic (halosynthesis) energy of conductor plants gives the additional advantage of being able to function with either basal-cells mitochondrial energy and / or energy derived from salt glands (see below) to create a high flow rate from root to the shoot surface and maintain low volumes of salt solution within the plant.

4.2.5 *Intriguing Productivity Curves in a Clonal Conductor Plant*

Already for almost three decades we have conducted salinity trials with a fast-growing clone of *Distichlis spicata* var. yensen-4a (Yensen, 1985). Extensive experience with this euryhalic variety, which shows high productivities over a broad range of environmental conditions (Figure 4.6), demonstrated its advantages as a salt-conducting plant for soil-salt remediation work.

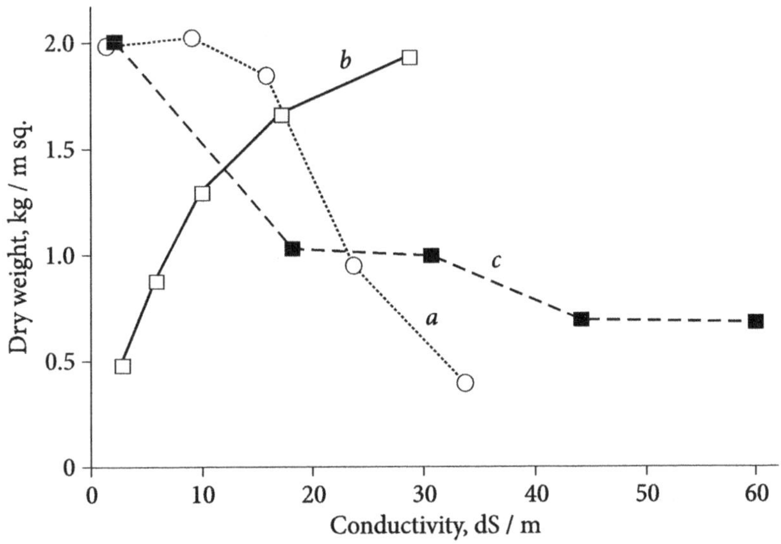

Fig. 4.6. Differing productivity-on-salinity behavior of clonal *Distichlis spicata* var. yensen-4a showing: *a* – *threshold behavior* with high-light, warm, humid conditions (Arizona; -o-); *b* – *euhalophyte behavior* with high-light, hot, arid, sulfate-nutrient conditions (California; -□-); *c* – *glycophyte-salt-tolerant behavior* with low-light, high-humidity, cool conditions (Western Australia; -■-) (Shannon, Greive, Barrett-Lenard, Perfumo, and Sergant, per. com.) (adapted from Biel and Yensen, 2005*a*, 2005*b*)

Experimental work with this variety to better understand its full potential range of productivity under increasing salinity resulted in well-documented growth curves over salinity. The curves, under varying light, temperature and humidity conditions, however, showed differing productivity-on-salinity behaviors: *i.e.* similar to a *glycophyte*, a salt-tolerant *miohalophyte*, and a *euhalophyte* (Figure 4.6). Theses growth curves on clonal material are so remarkably different that our disbelief prompted an investigation into the possible causes.

Consider the first trial in Tucson, Arizona (a two-month, open-air, high-light, hot-summer, humid / arid, rainy-season, modified-Hoagland's-nutrient solution study in the Sonoran Desert) a modest increase in productivity was found followed by a threshold point, and thereafter a steep linear drop in productivity (Figure 4.6). This curve is typical of a threshold salt-tolerance type *halophyte*. This was initially not a cause for concern.

A second trial (a one-month, closed-greenhouse, low-light, cool-winter, humid, modified-Hoagland's-nutrient solution study conducted in Perth by the Western Australia Agriculture Department) found an initial steep drop in productivity followed by a leveling out which extended into very high salinities (Figure 4.6). This is a curve more commonly associated with *glycophytic* C_3 plants that can tolerate high salinity by shutting down metabolic process to survive the higher salinities. Two other trials produced similar results under cool, low light, humid greenhouse conditions in Arkansas (Martin, 2003) and in humid, low-light growth chambers in Melbourne (Sergant, per. com.). These results suggested that the first trial may have had some possible unexplained anomalies.

A fifth trial (a two-year, arid, open air, high-light, hot to cool, high-sulfate, drainage-water-analogue, nutrient-solution study conducted by the United States Department of Agriculture, Salinity Laboratory, in Riverside, California) found a dramatic increase in productivity with increasing salt levels up to salinities approaching ocean water (Figure 4.6). This growth curve is analogous to the productivities anecdotally observed for this variety in arid field situations around the world (N. Yensen, per. ops.). This is a curve more commonly associated with obligate euhalophytic plants that optimize productivity in high salinities. This trial, conducted by highly reputable scientists (K. Grieve and M. Shannon) over a two-year period, was in stark contrast to other trials, but conformed closely with our experience to the plant under field conditions.

The fact that material of the same clonal plant variety can have such varying productivities over increasing salinity appears astounding and first suggested that we must be dealing with experimental error. Such a phenomenon has not been described in the literature. If true, it might call into question all the published salinity trials conducted in greenhouses, growth chambers or at different seasons.

Because these trials used different parameters for the productivity measurements, the absolute productivities may not be directly comparable. Still, the absolute productivities aside, the slopes of the curves clearly show how seriously the environmental conditions dramatically affected the productivity behavior of conductor plants with respect to salinity.

These data infer that *Distichlis* may make efficient use of the available energy, particularly under arid, salty conditions. The mechanisms that allow salt grasses to grow more rapidly in saline conditions are not understood.

We do know that most salt grasses have bi-cellular salt glands on the shoot surface. There is an established correlation between salt-gland density and productivity under increasing-salt conditions (K. Marcum, per. com.). It is also known that either higher temperatures or light levels can improve growth rates with increasing salinity (Kemp and Cunningham, 1981).

Our preliminary result with clonal *Distichlis spicata* trials under different environmental conditions produced very different growth curves and implies that an unknown mechanism is affected by environmental factors, salt-gland density, structure and physiological functions.

4.3 The Halosynthesis Hypothesis

Science advances through the process of observation followed by the developing and testing of hypotheses. Theories and laws are subject to change and revision, to wit, Newtonian physics changed dramatically by Arthur Eddington's measurement of light passing near the sun, heralding in the relativistic interpretation. Apropos to the present discussion, Einstein's Nobel-winning, corpuscular-photon, photoelectric effect changed our concept of light and paved the way for Feynmann's Nobel-winning electrochromodynamic.

The following hypothesis is not presented to presume such revolutionary changes to the paradigms of halophyte growth, but does advance hypotheses developed to satisfy an accumulation of contradictory observations regarding the growth of halophytes under stressful environments, and the discrepancy between observation and standard theory of halophytism. We would appreciate the reader's indulgence in considering that the following is presented to stimulate thought and discussion, and not necessarily to adequately describe the truth.

4.3.1 Concept Description and Terminology

The halosynthesis hypothesis is pure speculation, and to date is only supported by circumstantial evidence that may, or may not, potentially be explained through other known mechanisms. It is, however, an interesting hypothesis that could produce valuable results irrespective of its success as a correct hypothesis and may potentially lead to discoveries of new phenomenon related to the conversion of light energy without plant pigments and the protection of plants against destructive UV light (*e.g.*, UV-C, $\lambda_{220\text{-}280\,nm}$).

The potential destruction of organic compounds by certain light frequencies is analogous to how the frequency of a microwave oven quickly destroys flavonoid compounds in broccoli and other vegetables, (whereas steamed vegetables, when cooked to the same degree, lose few of their flavonoid compounds) (Vallejo *et al*, 2003). Plant structures and chemicals are known to have important protective functions against strong (γ and UV) radiation (Chirkova, 2002).

Likewise, excreted salts and / or other compounds of "halosynthesis-related" structures, such as epidermis cells, salt glands, trichomes, bladder cells, idioblasts, may have important protective functions, including protection from certain destructive frequencies for light-sensitive metabolites and proteins. The lengthening of destructive short light-wave frequencies via Compton scattering (Compton, 1922; 1923) is not likely to contribute significantly to the protection *per se*, however, other wave-lengthening modalities are not ruled out.

Halosynthesis may utilize electromagnetic light energy to protect and / or benefit metabolic life functions. Certain aspects of this vision may potentially be applied to organisms other than plants and particularly in arid or non-electrical-conductive environments. Because the concept of halosynthesis (and associated concepts) is new and not yet tested, it is a field without extensive terminology. We have therefore taken some care in defining this term:

Halosynthesis is any salt-mediated (non-classical photosynthetic) mechanism (e.g. photon quanta or evapotranspiration related) by which energy from the environment is stored and / or utilized by organisms.

In the present case, described below, halosynthesis principally considers salt glands of plants and other relevant structures and organs. In salt-gland-bearing conductor *crinohalophytes* the process of halosynthesis may potentially benefit plants via one or more of the following processes: mechanical and osmotic pressure, evapotranspirational-hydraulic energy, photoelectric effect, Gaussian surface-charge distribution and photon / electron chain generation.

4.3.2 Hydraulic Considerations and Salt Gradient from Soil to Shoot Surface

A generalized model of water and salt conduction from the soil to the salt glands, based upon the literature, considers that salt and water penetrate the root through the cells apoplastically, symplastically and / or trans-membranally across temporary and permanent porous membranes (transvacuolar) (Newman, 1976; Zholkevich *et al.*, 1989) to enter the xylem where the vascular tissue carries it upward. It is common knowledge that the vascular system of plants carries fluids from the roots to the shoots (xylem) and back to the roots (phloem). The mechanism of upward water flow in the xylem may be by passive (capillary and evapotranspirationally), active ("peristaltic" and hydraulic pumping) and / or by both (see Zholkevich *et al.*, 1989). It is assumed that in conductor plants the salt may be transported as salt water and via the same mechanisms within the xylem. Conductor-type plants may increase the hydrostatic pressure in the xylem by a peristaltic mechanism (per the "few-minute-contraction" form as opposed to the "diurnal" form) similar to that of *Zea mays* and *Helianthus annuus.*

From the xylem we suggest that the principal path of transport is apoplastically around the bundle sheath cells and then around the mesophyll cells to reach the epidermis, which may act as "salt collectors" and therefore as photoelectric collectors (see ***Photoelectric effect*** below).

Within the epidermal cells the salt and water may move apoplastically, symplastically and / or transmembranally to the basal cell of the salt glands. In the literature, there is considerable information regarding different membrane transporters for various salts in the roots, stems and leaf tissues. Because of the complexity of all the possible pathways for the various salts and fluids, we have created a simplified hydraulic model of fluid and salt transport in order to better understand the overall role of the salt glands with respect to the system as a whole.

A mechanistic theoretical model (Figure 4.7) of a conductor plants may further help to explain the relationship between sap pressure, salt conductance, water flow and volume. We may identify three hypothetical conductor plants: A. plants with an average number of salt glands, B. plants with below average number of salt glands and C. plants with above average number of salt glands. If we consider that the total conductive volume is equal for A, B, and C, we find that in:

A. The conduction flow rate is constant throughout the shoot resulting in average pressure in root (Figure 4.7A).

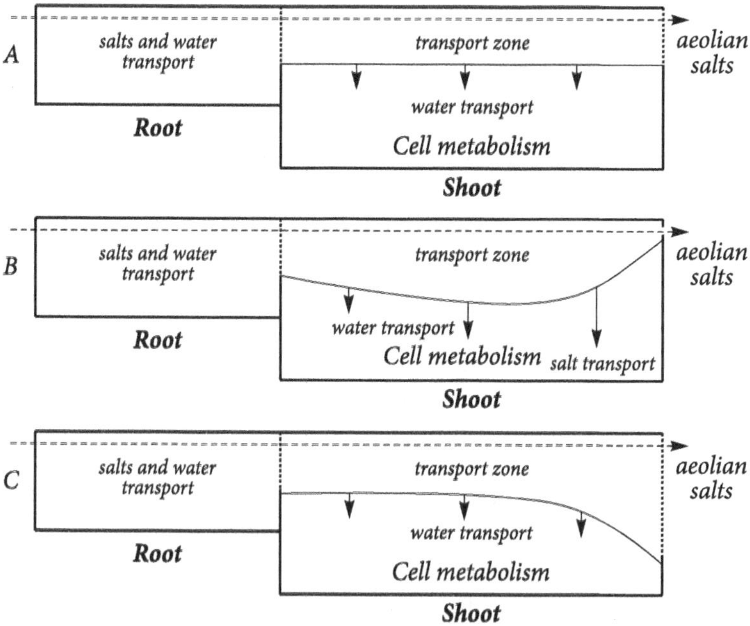

Fig. 4.7. Diagrammatic representation of theoretical conductor plants: A. Conductive rate of salt and water remain constant throughout the shoot; B. The conductive rate decreases at the salt gland potentially increasing salt entrance to the metabolic area; C. Increased number of salt glands causing increased flow rate of salt and water (adapted from Biel and Yensen, 2005*a*, 2005*b*)

B. The conduction flow rate may be reduced throughout the shoot, resulting in an increased pressure in root and shoot, potentially resulting in reduced water and salt absorption in the roots and increased salt movement into the metabolic area (Figure 4.7B).

C. The conduction flow rate may be increased throughout the shoot, resulting in a decreased pressure in root and shoot and thereby a potential for an inverse salt pump (Figure 4.7C).

From this simple diagram we can predict that plants with a greater number of salt glands may be better for both the goals of soil remediation and productivity due to reduced influx of salt into the metabolic area. The overall result is an increase in salt removal and remediation from the soil (3b in Figure 4.4).

In a conductor plant the salinity increases faster in a slower moving volume of water given the same stomatal conductance [note: conductor plants may also have a high or low volume with the same flow rate]. The flow rate may be limiting at the root uptake, entrance to the root-shoot conduction system, within the conducting zone, or at the salt glands. The salt-gland limitation may be dependent upon the number of glands and / or gland efficiency.

The conductance of any given volume through a conduction zone may be limited by a restriction at the entrance (Figure 4.7B), at the exit (Figure 4.7C) or at some place(s) within the conduction zone (not figured).

Both the relative volume of liquid entering or departing the conducting system may vary (as might the relative flow rates and associated energetic cost). In accord with Figure 4.7B a high transport-zone volume with restricted salt excretion may result in a relatively higher salt concentration in the transport zone. This in turn may either increase the salt level in the metabolic area or increase the cost to protect metabolism from salt intrusion. Such a cost may take several forms ranging from salt pumps and osmotica to a putative ammonia-ion protector (see below). A higher rate of salt-water departure from the transport zone (Figure 4.7C) may reduce the hydraulic pressure in both the transport zone. A reduced hydraulic pressure in the transport zone may result in a reduced pressure in the roots, which may facilitate the uptake of water from the soil. Also, the reduced hydraulic pressure in the shoot-transport zone may help to lower the cost for protection of the metabolic area and reduce the salt concentration within the shoot. Reduced shoot salt content has the potentially economic advantage of providing low-salt-content shoot tissue for animal forage.

The differential excretion of salt and the transpiration of water from the salt glands may affect the relative concentrations of salts and energetic costs in the above considerations. Likewise, an increased conductive volume may provide a larger potential volume available to the shoot-cell cytoplasm at a given salinity and flow rate. In order to provide an equivalent amount of water to shoot-cell cytoplasm under a reduced conductive volume, either a more rapid transport of salt water will be required or a higher energetic cost to move water against a greater osmotic gradient.

Some species and varieties of salt-conducting plants may have large underground root biomass, and significantly, may have an increasing salt-concentration gradient from soil to salt gland. This concentration gradient is determined by the existing soil water salt concentration versus the 100% solid salt crystals on the shoot surface. It should be clear that the salinity gradient will move water toward the increasing salt concentration, *i.e.* from root to shoot surface. It is likewise expected from this salt gradient that the diffusion of salt would be to move salt in the opposite direction, *i.e.* from shoot surface to the root, against the water movement.

At least two factors can prevent or reduce reverse salt movement from shoot surface to roots: active fluid transport and unidirectional ports. The active fluid transport mechanism and pathways are mentioned and referenced (see Zholkevich *et al.*, 1989) above. If the salt concentration in the immediate transport pathway between the roots and the epidermis and / or salt glands is isotonic the salt will move passively with the flow of the fluid (Figure 4.7A). Active transport in the roots and in the salt glands should also work synergistically at both ends of the transport zone to prevent the backflow of salt ions. The efficiency of preventing backflow could benefit from an increased number of salt glands at the shoot surface (Figure 4.7C). Unidirectional ports for salts may function at the root and salt gland, salt-concentration- gradient interfaces.

4.3.3 Salt Glands and Evapotranspirational Halosynthesis

In the genus *Distichlis*, and other *crinohalophyte* grasses, their bi-cellular salt glands have subdermal collector basal cells that take up internal salts which move via symplastic transport to a hypodermal, excreting cap cells. The basal cell–cap cell interface is noted for having a high density of mitochondria closely associated with the area of plasmodesmata. Chloroplasts, however, are not known from either the salt gland cells or adjacent cells (see Liphschitz and Waisel, 1982; Semenova *et al.*, 2010).

It is generally assumed that the transport of salt from the basal collector cells to the excreting cells requires energy generated by the basal cell mitochondria. But the literature does not discuss how or what carbon substrate for Krebs cycle might reach the basal cells. Because the best potential source of a carbon substrate for the salt gland mitochondria is limited to only the mesophyll cells, we can predict that the products available to go to the salt glands are limited to the organic acids which are involved in the C_4 pump of the C_4 photosynthesis. In this case, the capacity of the C_4 pump and overall photosynthesis should be decreased. Therefore, we suggest that this transport of photosynthetic products to the basal cells during illumination of the plants may exist, but that it may be only temporary like a "starter motor" or that it is semi-catalytic to the overall process. Idealistically, salt glands should utilize and / or absorb solar energy from a source independent of mesophyll photosynthesis, if such a source existed.

We hypothesize that there is another energy source that may be tapped during, or before, the excretion process. For instance, in the excretory cap cell the evapotranspiration process itself may generate an energy gain; *i.e.*, transpiration may result in an internal negative hydraulic pressure, resulting in useful potential energy at the plasmalemma between the basal collector cell and the excretory cap cell. Another possibility may be a result of the photoelectric effect.

4.3.4 *The Photoelectric Effect*

Due to the photoelectric effect, photons impinging on sodium ions (and other metals or compounds) will expel energized electrons. With sodium ions photons of sufficiently energetic light waves (λ = 220–550 nm) will expel electrons. These wavelengths are also potentially destructive to certain chemicals and pathways. Sodium in the epidermis will protect the underlying tissues from damaging UV-B and UV-C light (see Figure 4.12), by reflection, by converting the high energy waves into expelled electrons and by Compton scattering (Compton, 1922; 1923) wherein the resulting wavelengths will be longer, less energetic and may be accepted by the pigment-protein light harvesting complexes of chloroplasts and / or other photoreceptors, such as phytochromes. At the same time, the light can create energetic electrons (up to 3000 mV potentials) which are critical to the photohalosynthesis processes (see below).

4.3.5 *Epidermal Electro-Halosynthesis*

Sodium (and presumably other photoactive metals) is located in the shoot epidermis over virtually the entire light gathering surface (Koyro, 2003) potentially generating photoelectrically energized electrons (see above). These electrons, unable to move to the surface due to the non-conductive properties of the cuticular lipids, will move from cell to cell through the conductive cellular saline solution (see "hydraulic considerations" above) to the excretory cells and / or papillae. Upon reaching the surface they will assume a Gaussian[1] electrostatic distribution. In this electron collection process, the current from the epidermal cells to the excretory cells may potentially be converting directly to stored energy when passing across the tonoplast and plasmalemma membranes (see below).

4.3.6 *Salt-Gland Electro-Halosynthesis*

We suggest that the electrons produced by the photoelectric effect will move to the surface at the excretory cells, *i.e.* the salt glands and / or papillae. Because the lower and lateral surfaces of cap cells are typically cov-

---

[1] Type of distribution might be different; it does not necessarily have to be Gaussian. Regardless of the type of distribution, the qualitative characteristics of the phenomenon are the same.

ered with an insulating cuticle / lipid matrix, the electrons are thus predicted to move on the cap cells to the non-cuticular distal surfaces (Figure 4.8) and to any externally excreted salt crystals.

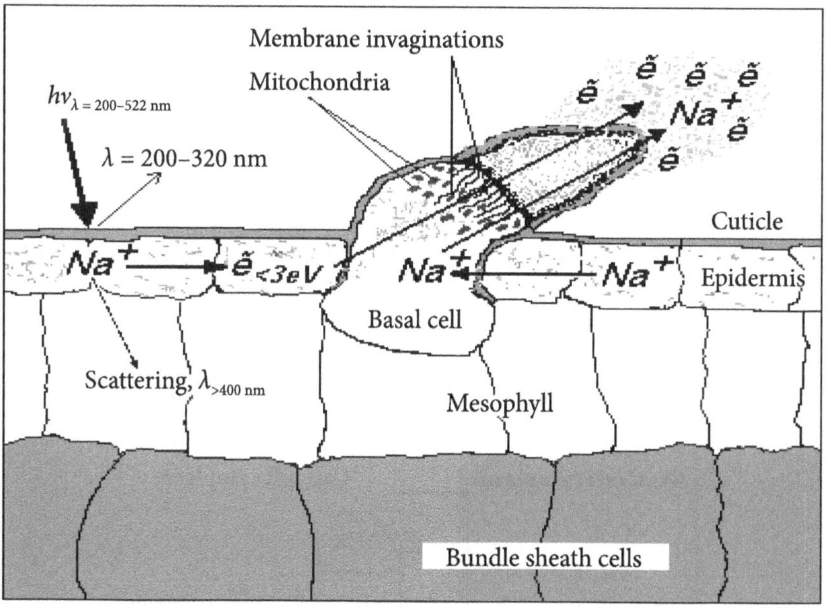

Fig. 4.8. Diagrammatic representation of generalized salt gland and some of the hypothesized photo-halosynthesis and inverse sodium pump using ATP synthase molecules to generate ATP (adapted from Biel and Yensen, 2005*a*, 2005*b*)

This electrostatic negative charge distribution would ultimately be distributed in accord with Gauss' Law ($\boldsymbol{\Phi} = \boldsymbol{E} \times \boldsymbol{A} \times \boldsymbol{cos\ \theta}$, where $\boldsymbol{\Phi}$ is the electric flux through the surface, $\boldsymbol{E}$ is the electric field normal [perpendicular] to the surface, $\boldsymbol{A}$ is the area, and $\boldsymbol{\theta}$ is the angle of the electric flux to the surface). Thus, a negative electrostatic charge would be formed on the distal surfaces of the cap cells and papillae and would be more-or-less "equally" distributed on said "hemispherical" cap cells and papillae.

With a relatively strong electrostatic negative electron charge (albeit relatively low voltage) gathered from the surrounding leaf epidermis which will be concentrated at the tip of the salt glands (and / or at the tip of an attached crystal), and an electrostatic force may be exerted on the Na^+ ions (and other positive ions). This force may be calculated based upon the external voltage potential and the distance between the charged sodium ions within the surface and the distal surface of the salt gland / salt whisker where the negative electrostatic charge may reside, *i.e.* $F = (k_e \times C_{s\bar{e}} \times C_{Na+}) / d^2$; where F is the force in newtons (1 N = 1 kg × 1 m / 1 s²), $k_e = 8.987 \times 10^9$ N m² C⁻², $C_{s\bar{e}}$ is the surface electron charge in coulombs (1 C = 1 A × 1 s), C_{Na+} is the ion charge in coulombs, d is the distance between charges in meters (Serway and Jewett, 2003). Simple calculations presuppose very strong forces at the small distances involved. Dimroth *et al.*, (2000) indicate that as little as > 50 mV potential is sufficient to drive the rotational ATP synthase structure. According to our calculations, the photoelectric effect on sodium may produce free electrons with acceleration energies of up to 3000 mV.

Various forms of ATP synthase have been described since its remarkable structure and functions was first identified and for which Paul Boyer, John Walker and Jens Skou were awarded a Nobel Prize (see Boyer, 1993, 1997; Abrahams *et al.*, 1994; Skou and Esmann, 1992). This would allow production of ATP on the "plant-side" when the sodium or other cations move to the "excretory-side" of the membrane, thus keeping the useful ATP within the plant tissues (Figure 4.9).

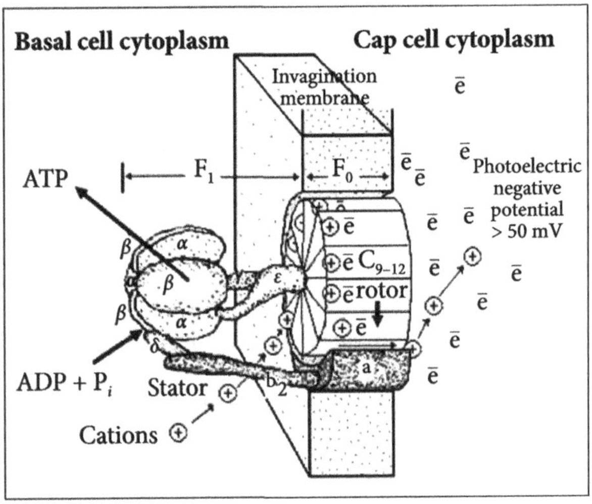

Fig. 4.9. Diagram of proposed "inverse sodium / cation pump". By arranging the ATP synthase F_1 unit (α, β, γ, ε, δ and b_2 stator) on the "plant-side" of the plasma membrane, the production of ATP will occur within the epidermal tissue. An external electrostatic charge > 50 mV is sufficient to drive the rotor for cation transport to the surface of papillae, cap cells, bladder cells or other charged structures **Note**: the photoelectric effect on sodium ions should produce electrons with up to 3000 mV. The rotational torque of the rotor drives ATP of the β subunit from where the ADP + P_i have been "stator-attached" (modified from Boyer, 1993, 1997; Skou and Esmann, 1992; Abrahams *et al.*, 1994; Dimroth *et al.*, 2000; Biel and Yensen, 2005*a*; Yensen and Biel, 2006; Fomina and Biel, 2013, 2014)

Chlorine and other anions, due to their size and behavior are predicted to move toward the surface of a solution (Garrett, 2004), but may get an additional electrostatic boost from external cations during salt-crystal formation. As the salt is excreted, as in the case of *Distichlis* spp. and other salt-gland-bearing species, the whiskers and crystals formed may facilitate or inhibit the electrostatic force depending on charge, direction of the whisker, crystal shape, humidity and other environmental factors. In appropriate conditions, the salt whiskers (typically 2 μm to > 500 μm long) have been observed (Figure 4.10) to exceed several millimeters as fine crystalline "hairs". It is speculated that, on occasion, the whiskers may be "tubular" analogous to the geological formation of hollow "soda straw" stalactites found in limestone caves wherein the salt water passes from the salt gland (and potentially thorough the salt-tube whisker) to the electrostatic charged tip. The salts would follow an osmotic and salt gradient maintained by the evaporation of water at the tip of the salt crystal where the highest salt concentration is expected to exist. We predict that in dry air the evapotranspiration and electrostatic effects may be more pronounced, in part due to dry air's greater electrostatic capacitance and in part due to a greater evapotranspiration potential. Halosynthetically generated ATP could help to explain the results of the different growth curves of the clonal *Distichlis* observed (see Figure 4.6) under varying environmental conditions.

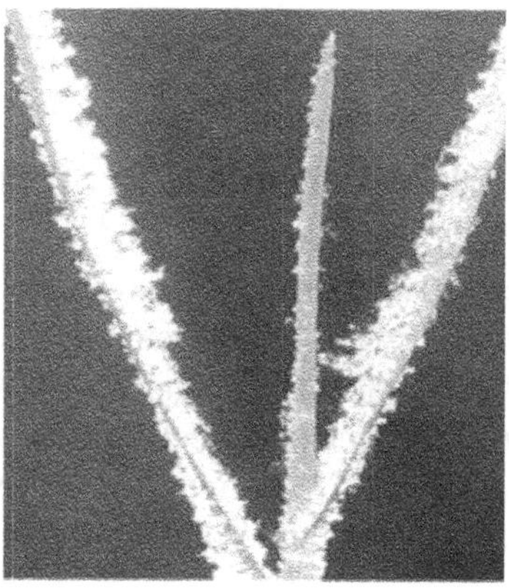

Fig. 4.10. *Distichlis spicata* with microscopic salt crystal filaments on the leaf surface, grown in virtually 100% still air (carefully protected from air currents), to demonstrate the capacity of the plant to excrete salt microcrystals, which are easily dislodged into the air with the slightest air movement (adapted from Semenova *et al.*, 2010)

Note: talking or slight vibration of the stem can put virtually all the salt filaments into air where will drift up or down with the slightest air currents

Under appropriate conditions, then, the electrostatic charge, and potentially in combination with hydrostatic pressure across a plasma membrane, may generate usefully energy. As the salt passes across the tonoplast, plasmalemma invaginations, and other membrane ports, in response to a > 50 mV electrical potential and / or equivalent hydrostatic potential, the energy may be converted to a biochemical energy storage product.

As mentioned above (Liphschitz and Waisel, 1982; Semenova *et al.*, 2010) the centrifugal zone of the basal cells has a high density of mitochondria located near and surrounding in the area of the invaginations of the plasmalemma and the plasmodesmata between the basal cell and the cap cell.

In conclusion, we predict the photoelectrically produced electrons from metallic salt cations in the epidermis will be intercepted by the basal cells before moving to the cap cells, papillae, bladder cells, and / or other surfaces, and in the process producing ATP. At that time, it is interesting to note that the electrons can reach the mitochondria located along the path to the cap cells. Some of these electrons may have the potential to reduce mitochondrial NAD^+ to NADH and / or convert directly into $\Delta\mu_{Na}{}^+$ and $\Delta\mu_H{}^+$; and therefore, indirectly into ATP (Figure 4.11). By these hypothetical schemes we present the possibilities of the conversion of photoelectric energy into biochemical storage products without classical photosynthesis.

4.4 Physico- and Bio-Chemical Protection Synergism

We predict that all terrestrial plants may have similar protective mechanisms to defend themselves against high-energy photons. The epidermis cells are the primary protection layer against UV light. The protection may be provided by regular epidermal cells, trichomes, salt glands, cuticle, and other types of cells such as hypoderma, hypodermal cells with calcium oxalate crystals (hypodermal idioblasts) and other salt containing cells or structures.

The physical protection mechanisms include, at least: reflection, Compton scattering, conversion and absorption. And, we consider the absorption and conversion of UV light energy by metallic salts as perhaps the most important mechanism for the protection of plants against negative UV energy ... and possibly for the protection of animal skin (Biel and Yensen, 2005*a*, 2005*b*; also see below).

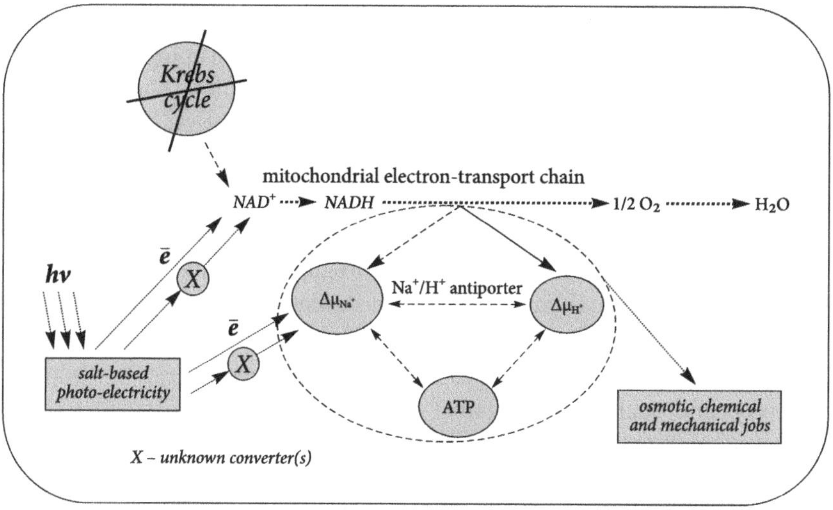

Fig. 4.11. Hypothetical scheme of photo-halosynthesis as a non-photosynthetic mechanism by which solar energy is utilized and / or stored by halophyte glands through the benefit saline conditions (modified from Biel and Yensen, 2005*a*; Yensen and Biel, 2006; Fomina and Biel, 2013, 2014)
Note: Slashed Krebs cycle means the termination of its supply by organic substrates. The interconversion of $\Delta\mu_{Na}{}^{+}$, $\Delta\mu_{H}{}^{+}$, and ATP is well known in the literature (*e.g.*, Skulachev, 1997; Semikhatova and Chirkova, 2001)

We thus assume that this may be a universal mechanism for many plants and animals, not only for protection, but also in the collection of photon energy (via conversion to electrons, Compton-generated lower-energy photons, surface charge and production of $\Delta\mu_{H}{}^{+}$) in the photo-halosynthetic process.

In leaves, due to the sodium and other metallic salts, which may be sequestered in (or near) the epidermis, we find that the plants may receive protection against some of the destructive light (Figure 4.12). We would also predict that these metallic ions may serve to refocus and alter this potentially destructive UV light via the Compton effect resulting in longer, less energetic wavelengths, and more energized electrons.

Further the anatomical structure of the focusing apparatus should be such that they may take advantage of the reflective angle of the emissions in accord with $\boldsymbol{\lambda_f - \lambda i = \Delta\lambda = h\,(1 - \cos\theta)\,/\,m_e \times c}$ (where $\boldsymbol{\lambda_f}$ is the resultant wavelength, $\boldsymbol{\lambda i}$ is the incident wavelength, $\boldsymbol{h}$ is Planck's constant = $6.626 \cdot 10^{-34}$ J s, $\boldsymbol{\theta}$ is the angle of divergence from the incident photon vector, $\boldsymbol{m_e}$ is the electron rest mass and $\boldsymbol{c}$ is the speed of light in a vacuum) (Compton, 1922, 1923).

We propose these considerations because, as mentioned above, sodium and other metallic salts are primarily located in the epidermis (Koyro, 2003) and do three important things:

a) photo-electrically generate energized electrons via the halosynthesis processes (see above),

b) through Compton scattering may lengthen the wavelengths of destructive high-energy light to lower-energy non-destructive light more appropriate for photosynthesis and photoregulation reactions, and, in the case of salt on animal surfaces, to non-damaging potentially beneficial wavelengths to the animal body, and

c) via the preceding, protect the proteins, membranes, biochemical pathways and nuclear acids against destruction (Figure 4.12).

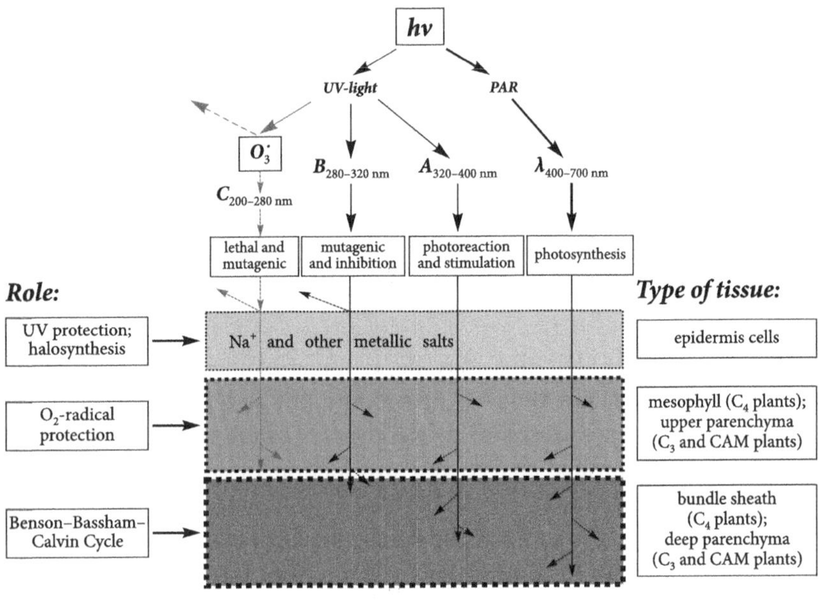

Fig. 4.12. Hypothetical scheme of UV and excessive light resistance of photosynthetic organs of terrestrial plants (modified from Bil', Fomina *et al.*, 2003; Biel and Yensen, 2005*a*; Yensen and Biel, 2006; Fomina and Biel, 2013, 2014)

From this point of view, it is easy to understand why C_4 plants should be less sensitive to UV destruction compared to mesophytic C_3 plants, as noted in the literature (Muzafarov *et al.*, 1995), but not explained. Further, halophytes should have increased resistant to UV light when their epidermal tissues contain metallic salts. We would even expect this concept to apply to many non-halophytes which have epidermal metallic salts. Also, some desert species (for example, *Haloxylon aphyllum*, *Suae-*

da acuminata) have a vacuolized hypodermic layer and other species (for example, *Salsola dendroides*) may have calcium-oxalate crystals in this hypodermal layer of cells which essentially are hypodermic idioblasts layers (Biel, 1988). These hypodermal layers presumably function similar to the halo-protection mechanism of the epidermis cells.

There is a steep gradient of decreasing light levels from the top to the bottom of the leaves (Cui *et al.*, 1991; Vogelmann, 1993). Professor Thomas Vogelmann (1993) found that the epidermal-cell-focusing of light may result in beneficial light conditions for the cells and chloroplasts deep within a leaf. At the same time, this may potentially result in destructive levels of light deep within the leaf and / or activate the production of oxygen radicals (in the electron transport chain of chloroplasts) that are likewise destructive (Bil' *et al.*, 2003).

4.4.1 Biochemical Protection against Oxygen Radicals

While photo-halosynthesis may utilize some of the incident light energy, there are many wavelengths that may penetrate further into the plant tissues irrespective of (and even due to) the halide elements, other metals, and other photoprotective compounds. Because these interceptive protections against destructive photons are not always sufficient, plants need "back-up" biochemical mechanisms to counter the potentially excessive light and effects of oxygen radicals formed in the chloroplasts. Some elements of this protection may include pigments (e.g. xanthophyllic cycles), enzymes (e.g. superoxide dismutase, catalase, peroxidase), photorespiration pathway and some miscellaneous compounds (e.g. ascorbic acid, oxaloacetic acid, *etc.*).

From our previous experience we have found that the upper part of mesophyll in the leaf of C_3 plants acts as a shield against excessive light for the deeper mesophyll. In particular, we have found that Rubisco activity near the surface of spinach (*Spinacia oleracea*) leaves is minimal, but catalase activity is maximal. After flipping the leaf to an inverted position, patterns of the activities of both enzymes also inverted (Chapter 7; see also Soukhovolsky, Fomina *et al.*, 2002; Bil', Fomina *et al.*, 2003; Biel, Fomina *et al.*, 2008, 2010). It is logically assumed then that the low level of Rubisco near the upper surface of the leaf may be due to its destruction by active oxygen radicals. Similar situations may exist in other and perhaps all terrestrial plant species. It would be clear then why Rubisco in C_4 plants is located in the bundle sheath cells. We speculate that the mesophyll of C_4 plants not only fixes atmospheric CO_2 and transports carbon via C_4-acids into the bundle sheath cells, but also functions as a protector of the Benson–Bassham–Calvin cycle and other important processes against the negative influence of oxygen radicals and excessive light (Figure 4.13).

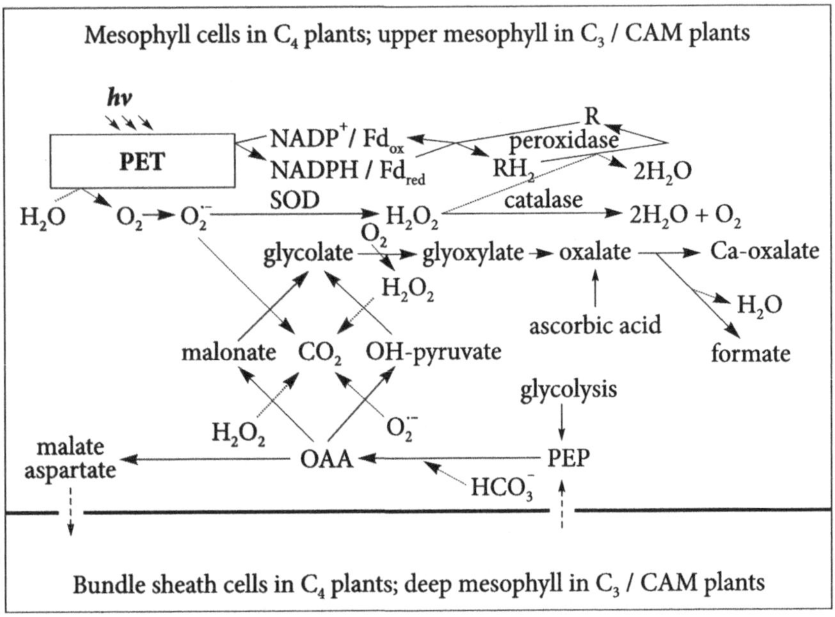

Fig. 4.13. Interrelationships among various metabolic activities in mesophyll of C_4 plants and in upper mesophyll of C_3 and CAM plants related to stress tolerance of photosynthetic organs under high light illumination (modified from Lyubimov *et al.*, 1978; Bil', Fomina *et al.*, 2003; Bil' *et al.*, 2003; Yensen and Biel, 2006; Biel, Fomina *et al.*, 2008, 2010; Fomina and Biel, 2013, 2014; Professor Thomas Vogelmann, per. com.)

We further speculate that this protection uses some portion of oxaloacetic acid, the first product of PEP-carboxylative reaction. It has long been shown (Lyubimov *et al.*, 1978) that, in the mesophyll cells of C_4 plants, up to 7% of the photosynthetically-formed oxaloacetic acid is oxidized by hydrogen peroxide and oxygen anion radicals to form glycolate, the first product in the photorespiration pathway. We predict that the percent of oxaloacetic acid necessary to neutralize the oxygen radicals is not constant and may change with respect to various protective enzyme activities and with the light environment within the leaf.

Thus, we have identified two critical protective mechanisms, one against UV light and the other against active oxygen radicals. ***Photo-haloprotection*** against UV light occurs in the epidermal cells and / or in some other tissues, *e.g.* hypoderma, idioblasts, trichomes, scales, *etc.* ***Biochemical protection*** processes protect against active oxygen radicals resulting from excessive light in the mesophyll of C_4 plants and in the upper mesophyll of C_3 and CAM plants. These protective mechanisms allow optimization of the energy structure and function of the plant under varying light conditions.

4.5 A Case Study, *Distichlis*

Because conductor plants have been less well studied than excluders and accumulators, the following specifically examines what may be considered a quintessential halosynthesizing conductor plant, specifically plants of the genus *Distichlis*.

4.5.1 Ecophysiology

We have selected the genus *Distichlis*, a C_4 grass, as a plant that has increased productivity in saline environments and has the ability to conduct salts from soil to air. Preliminary studies indicate that *Distichlis* lines have varying capacities to conduct salt from the soil to the leaf surface for wind dispersal, halodispersion, a mechanism not well known or described.

Salt grasses, such as *Distichlis*, may serve as productive crops under high salinities while removing salt from the upper soil horizons. How much salt is moved through the plant, and under what conditions, is not known, although estimates range from 2–150 tons / hectare / year. Anecdotal observations of salt whiskers and airborne salt dust suggest that, some percent of this salt becomes airborne. Aerially dispersed salt, to surrounding higher-elevation regions of salt-leached soils, may have the potential of stimulating productivity. It has been shown that many species, and in particular those having the C_4 photosynthesis, require sodium as a micronutrient (Brownell and Crossland, 1972).

4.5.2 Taxonomy and Geographic Distribution

World-wide there are nine species in the genus *Distichlis* (*Poaceae*). The genus is placed by various authors in different subfamilies: *Festucoideae* (Hitchcock, 1950; Gould, 1951) or *Chloridoideae* (Texas A&M University herbarium; Marcum and Murdoch, 1994). Similarly, it has been placed in different tribes by various authors: *Festuceae* (Gould, 1951; Hitchcock,

1950) or *Aeluropodeae* (Texas A&M University herbarium). Seven species occur in the Americas, one in Australia, and one in Sudan. The Australian species *Distichlis distichophylla*, however, is considered by Beetle (1955) a synonym of *Distichlis spicata* and *Distichlis sudanensis* (known from a single specimen) is most likely *Aeluropus massauensis*. Note, N. Yensen in agreement with Beetle (1943), treats *Distichlis stricta* as a subspecies of *Distichlis spicata* (NP Yensen per. obs.). Bermuda grass, *Cynodon dactylon*, is closely related and, like *Distichlis*, is in the subfamily *Festucoideae* (Hitchcock, 1950) or the subfamily *Chloridoideae* (Marcum and Murdoch, 1994), and like *Distichlis* it is in the tribe *Festuceae* (Hitchcock, 1950) or, unlike *Distichlis*, in the tribe *Chlorideae* (Avdulov, 1931 in Karpilov, 1970; Texas A&M University herbarium).

4.5.3 Root-Soil Restructuring Capacity

Species of *Distichlis (Poaceae)* have strong penetrating rhizomes (see rhizocanicular percolation effect below) with the ability to open heavy-clay, deflocculated soils (Yensen, 2002; Leake, 2003). Further, they have the ability to grow aerobically within anaerobic sites due to the aerenchyma air passageways of the rhizomes and the roots (Leake, 2003). The oxygenation of black anaerobic mud is visible due to the brown iron oxide coloration that occurs around the roots. The rhizocanicular percolation effect of the rhizomes and roots physically opens up sealed soils such that water and gasses may penetrate more rapidly. The roots have air passageways (aerenchyma) and a ± biannual nature, which leaves the soil with holes full of organic matter. The organic material of the roots and rhizomes promotes biological activity on soil particle surfaces causing soil flocculation and complete restructuring of saline soils (Leake, 2003) to a depth of up to one meter or more in as little as three years (N. Yensen, per. obs.).

In *Distichlis* spp. it is known that the rhizomes typically have a limited lifetime (annual to biannual) and when the internode regions atrophy the nodes produce new rhizomes. Over a number of years, the opening of the soil, via new rhizomes and roots, results in channels of organic matter and can produce a dramatic increase in soil percolation rate, according to field observations (Yensen, 2002). Even with "cement-like" deflocculated soils the rhizocanicular effect facilitates percolation. For instance, in Mexico, Australia and USA (California) heavy-clay, deflocculated soils planted with *Distichlis* spp., with water at 10–20 dS / m salinity, had increased percolation and salt leaching to the extent that the soils became amenable to fresh-water plants and earth worms (Yensen, per. obs.).

Other genera with rhizocanicular ability are: *Aeluropus* spp., *Agropyron* spp., *Agrostis* spp., *Arundo* spp., *Cynodon* spp., *Eragrostis* spp., *Jouvea* spp., *Monanthochloe* spp., *Odyssea* spp., *Pennisetum* spp., *Phragmites* spp., *Spartina* spp., and *Sporobolus* spp. (Yensen, per. obs.; Heuperman *et al.*, 2002). But, perhaps the rhizocanicular effect phenomenon is better understood with *Distichlis* from which it was first described (Yensen and Bedell, 1995; Yensen *et al.*, 1997; Leake, 2003).

4.5.4 Salt Tolerance

Some populations of *Distichlis* are euhalophytes and have the ability to grow better in salty conditions than under fresh water conditions, while others do not have this ability, but are still salt tolerant (Yensen, 2001). And, as noted above, within a single clonal line the environmental conditions can switch the plants from euhalophytic behavior to a "salt-tolerant" behavior, *i.e.* with reduced growth on increasing salinity (Yensen unpubl.). It is not known how much of the salt tolerance of "salt-loving" varieties are due to whole plant structures and strategies and how much is due to the plant cell structures and metabolism. That is, it becomes more or less clear now that the salt tolerance of *Distichlis* is due to its ability to keep salt out of the metabolic zones (Figure 4.14), and may be, in partly, cells are able to carry out total metabolism in the presence of salt ions.

4.5.5 Photosynthesis

Distichlis is a C_4 plant with typical "Kranz" leaf anatomy (Liphschitz and Waisel, 1982; Semenova *et al.*, 2010). Close to 50 years ago C_4 species were classified on the basis of their first product, *i.e.* as a malic and / or aspartic type of C_4 plants (Karpilov, 1969; 1970; Black, 1973; *etc.*), and, on basis of decarboxylating enzymes of C_4 acids, enzymes as NADP-ME (ME – malic enzyme), NAD-ME and PEP-carboxykinase in C_4 species (Edwards *et al.*, 1971; Hatch *et al.*, 1975; Edwards and Walker, 1983; *etc.*). It has also been found that the malate / aspartate ratio in the first photosynthetic products is not stable in C_4 plants and changes during leaf ontogenesis (Biel and Fomina, 1985; Biel, 1993), or may change due to environmental factors (Biel, 1988; 1993; Biel *et al.*, 1990). It is important that, for example, even between different lines of *Zea mays* there are some hybrids which have "aspartic C_4-photosynthesis", and, as a result of that, these lines produce more protein in total biomass than the "malic C_4-photosynthesis" lines (Biel, 1988; Biel *et al.*, 1990).

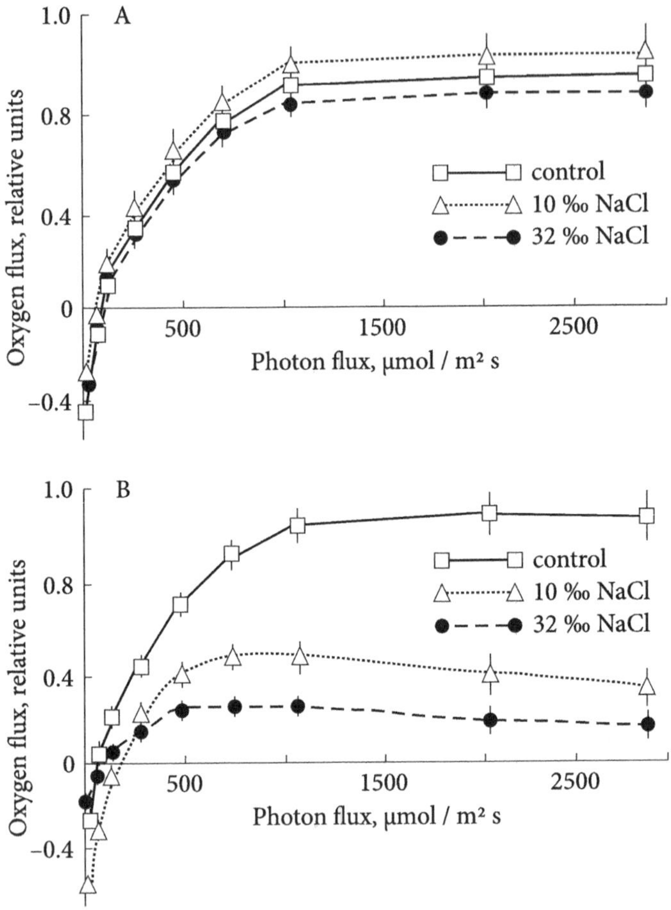

Fig. 4.14. Effect of different sodium chloride concentration on oxygen flux of whole-shoot-leaf blades (A) and leaf-blade slices (B) of saltgrass *Distichlis spicata*; □ - control; Δ - 10‰ NaCl; • - 32‰ NaCl, $n = 3$ (modified from Biel and Fomina, 2018)

Note: A. Shoots of *D. spicata* were exposed at room temperature in 0, 10, and 32‰ of NaCI during 20 hours. Then cross sections have been made and the oxygen flux in dark and under different light intensity was measured. The experiment shows that there is no difference of oxygen flux between control and salt treatment of shoots. B. Leaf cross sections of *D. spicata* were exposed for 1 hour in 0, 10, and 32‰ of NaCI, and then oxygen flux was measured. The photosynthesis was dramatically decreased (compare to control) after salt treatment of leaf slices

A review of the literature (Liphschitz and Waisel, 1982; Semenova *et al.*, 2010) did not identify the first photosynthetic products for *Distichlis*. But, from our previous observations (Bil' and Gedemov,

1980; Bil', Lyubimov *et al.*, 1981; Lyubimov *et al.*, 1987; Biel, 1988; 1993) with some Kara-Kum desert plants, growing in saline soil, we hypothesize that, under high salinity, the C_4-photosynthesis of *Distichlis* will produce more aspartic acid than malic acid. And further, that protein production will increase over carbohydrate production. We believe that this phenomenon can correlate with the quality of the *Distichlis* grain and with total productivity of this genus and other C_4 desert plants.

4.5.6 Ammonia Nutrition as a Protector against Salinity

In certain marine animals' ammonia is used as osmotic protection against high salinity (Tam *et al.*, 2003). It is also known that NH_3 is toxic to photosynthesis and other metabolic functions in plants. This toxicity can be eliminated if ammonia is maintained away from metabolic processes. For example, ammonia may be maintained in small quantities near the plasmalemma where its positive charge may serve to protect from Na^+ intrusion into the metabolic area and to maintain the sodium ion within the apoplastic area. We therefore speculate that in *Distichlis* (and possibly other *halophytic* plants) ammonia may function as an electrostatic protector against toxic cations and a "binder" of anions. If verified, this would be a new approach to salt protection in plants.

Our preliminary data show that ammonia nutrition can change the rate of CO_2 photoassimilation and the quality of photosynthetic products and plant productivity compared to nitrate nutrition in C_4 species (Biel and Fomina 1985; Bil' *et al.*, 1987; Shabnova *et al.*, 1997). That is, under ammonium nutrition a high aspartate / malate ratio is maintained. This is due to: 1) the stimulation of aspartate aminotransferase, 2) the inactivation of the enzymes that form and decarboxylate malate, 3) decreasing the C_4 pump activity, with the result of shifting the leaf metabolism to amino acid and protein synthesis (Biel and Fomina, 1985; Bil' *et al.*, 1987; Biel, 1988; 1993).

Under nitrate nutrition, however, the aspartate / malate ratio decreases (Biel and Fomina, 1985; Bil' *et al.*, 1985). Thus, ammonium nutrition has the overall effect of increasing the total nitrogen and protein content over that of nitrate nutrition in malic C_4 species such as *Z. mays* (Bil *et al.*, 1987). Importantly, under ammonia nutrition the ammonium is evenly distributed throughout the plant parts. Under nitrate nutrition, however, virtually all the nitrate was concentrated in the stem (Shabnova, Fomina *et al.*, 1993; Shabnova *et al.*, 1997). Different genotypes of *Z. mays* present varying responses to ammonia nutrition. Some responses are great (*e.g.*, early-ripening Sterling maize variety) and some responses are less (KVC-701 maize variety) (Biel *et al.*, 1990).

Distichlis is taxonomically closely related to *Z. mays* and, likewise, has shown to have varying responses to ammonia vs. nitrate nutrition (Yensen, unpubl. field trials). We expect that the different *Distichlis* varieties, with their different salt tolerances will have different protein / carbohydrate ratios and productivities under ammonia vs. nitrate nutrition.

4.5.7 Soil Salt Removal and Benefits to Changes in Soil Properties

Successful remediation of salt-affected soils can be described as establishment of self-sustaining vegetation and the absence of migration of salts into objectionable locations (Yensen *et al.*, 1999). It is further desired that the self-sustaining vegetation have beneficial value. Halophyte remediation trials in the Smackover Creek area of southwestern Arkansas have demonstrated that it is possible to take deflocculated, sealed soils with a pH ranging from 3.5 to 5.0, that have been heavily impacted with oil and brine to the extent that no plants have grown *in situ* for over 50 years and completely remediate the soil within a 3–5-year period using rhizocaniculating saltgrasses (Yensen *et al.*, 1999). The trial involved 83 warm seasons and 33 cool season species, with less than 4–5 species showing remediation capabilities.

Certain rhizocaniculating saltgrass species, such as *Distichlis* spp., have penerhizomes allowing penetration through hard soils and extensive, deep roots with aerenchyma tissue, which contains spaces in the roots to permit gas exchange within anaerobic sites (Figure 4.15).

The rhizomes are biannual and leave the soil full of channels of organic matter, which allows bacteria to re-aggregate the clay particles (D. Beck, per. com.). The presence of organic matter will maintain aggregation of soil particles (I. Labron, per. com.).

In conclusion, the genus *Distichlis* serves well both as a model for investigating photo-halosynthesis and as a practical forage cultivar for salt conduction, dispersion and soil remediation. The model of photo-halosynthesis as applied to *Distichlis* may have much broader applications and can be introduced genetically into other important agriculture plants. Moreover, the process may exist not only in terrestrial plants but in animals as well. Likewise, some of the concepts may be applicable for humans.

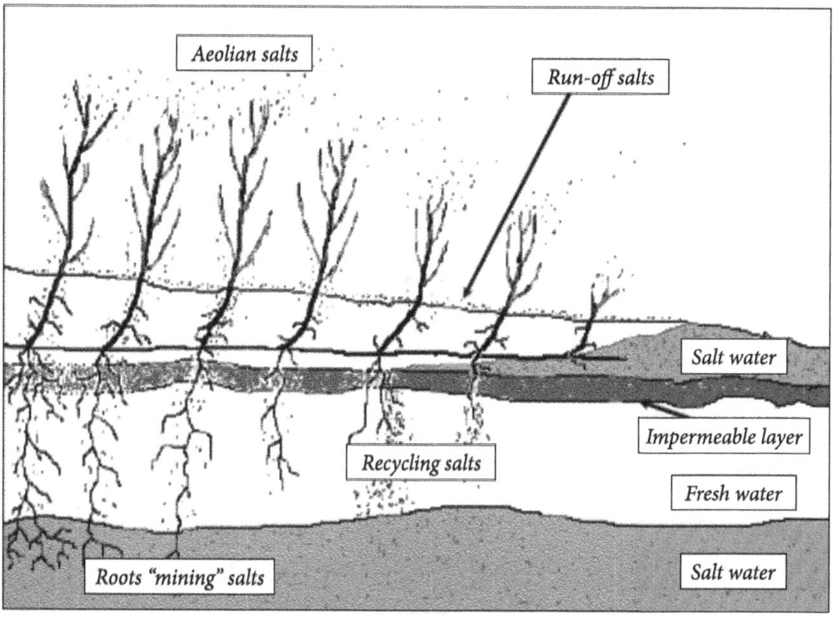

Fig. 4.15. Diagram of a generalized rhizocaniculating *Distichlis* plant. Sharp penerhizomes open heavy clay soils and even asphaltic soils in search of salt water. As the deeper aerenchymatous roots penetrate into anaerobic and impermeable aquacludes they facilitate the draining of parched hypersaline water tables to be later recycled back to the plant. The roots may extend over 2 meters deep in active search of the deeper, heavier, hypersaline waters. The salt is then conducted to the leaf-blade surface where wind may disperse microscopic aeolian salt "whiskers" to the air. Some salt washes or falls back to the soil surface and is either washed away or, more commonly reabsorbed *in situ* by the vegetated surface (modified from Biel and Yensen, 2005*a*; Yensen and Biel, 2006)

4.6 Potential Medical Benefit of Photo-Halosynthesis

The human body has not always been covered with clothes, nor have the eyes been protected with sun glasses against the negative influence of UV light. From an evolutionary point of view, mammals have a special covering (hair) which protects the body from such negative radiation (although hair most likely evolved primarily for body temperature maintenance). Conversely, it has been known for a long time that some parts of solar spectrum of light helps to increase resistance to certain environmental factors, including destruction of pathogens, and, as such, can even help to cure illnesses.

According to our photo-halosynthesis hypotheses "negative" UV light may have positive benefits for the human body (Figure 4.16). We speculate that metallic salt ions, principally sodium, may be protective, energize the epidermis, and stimulate some beneficial biochemical pathways.

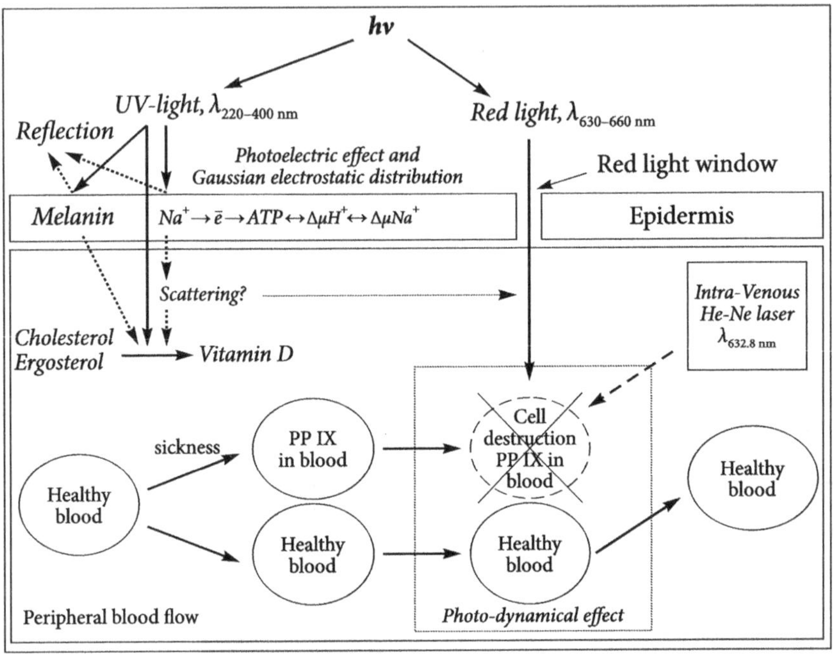

Fig. 4.16. Influence of solar radiation on human blood. Sodium and other metallic ions can protect the skin and body against negative UV light. Some UV light is reflected, some is Compton scattered to longer wavelengths, and some will photo-electrically energize electrons, which can reduce mitochondrial NAD^+ / FAD^+ to NADH / FADH and / or convert directly into $\Delta\mu_{Na}{}^+$ and $\Delta\mu_H{}^+$, and indirectly into ATP in basal lay of skin. Concurrently, red light (630-660 nm) enters through skin's "light window" and is absorbed by protoporphyrin IX (PP IX), common to "sick" blood cells, and photo-dynamically destroys the sick cells (modified from Prokopev, 2004; Biel and Yensen, 2005*a*)
Note: hydrogen peroxide (**H_2O_2**) forming through excitation of PP IX by red light, in part, can be source of atomic oxygen (**'O'**), which is one of the highest antioxidant. In tissues, atomic oxygen will eliminate local hypoxia and destroy pathogenic microflora: viruses, bacteria, fungus, *etc.* (Neumivakin, 2007)

Contrary to the cancer risk associated with UV light, UV light may be converted into beneficial light forms and / or may energize surface tissue. When incident light strikes the epidermis, protrusions, pilous structures, and / or the saline eye surfaces, we speculate that some beneficial red light is produced by Compton scattering of UV light from metallic salts located in or on the face, shoulders and extremities. This Compton-scatter-produced red light and the natural red light of wavelengths (λ) of 630–660 nm can penetrate up to 5–10 mm deep into the body and into the peripheral circulatory system. This light has been shown to have health benefits. For instance, it is known that the monochromatic light from He-Ne laser ($\lambda_{632.8\ nm}$, see Figure 4.16) has a positive influence against ulcers and gastritis in the stomach and duodenum (Udut, Karpov *et al.*, 1990). Prokopev (2004) conducted esophageal-inserted red-light-emitting-catheter experiments with 420 ulcer patients and found that 83% (against control) had positive results. The benefits included a decrease of pain and a reduced recovery period of 16 days from the typical 24-day period.

Simplified, external, broad-spectrum red-light methods, which included all wavelengths between 635 nm and 665 nm administered simultaneously, provided positive results similar to intravenous treatment (Prokopev, 2004).

Prokopev's (2004) experiments show that the solar spectrum has an effect like monochromatic and broad-spectrum red light (Figure 4.16). His calculations of photon functional distribution suggest that the maximum quantity of photons from solar energy may be found at 632.7 nm, which is close to the He-Ne laser of 632.8 nm. After taking into account the solar photon diffraction and scattering in the upper layers of the skin, he showed that red light penetrates to 2 to 5+ mm deep into the peripheral blood circulation system. The absorbed dose of energy by the blood at the wavelength of 600–660 nm throughout the exposed portions of the extremities in the northern temperate zone for a period of 3–4 hours is comparable to or greater than the laser and / or broad-spectrum red-light exposure treatment.

Further studies (Prokopev, 2004) with intravenous blood treatment by photo-illumination produced data that show that the spectral characteristics of whole blood and its components have, as the first acceptor of photons, porphyrin and porphyrin-containing structures. Photon absorption by these acceptors causes significant photooxidative reactions with oxygen molecules and oxygen-active particles. This induces a chain of oxygen-radical reactions in unsaturated fatty acids of the membranes, which in turn causes the destruction of pathologically changed cells, increased antioxidant activity, the proliferation of cells and bone marrow tissues; and, indirectly changed the immunological and hormonal status. These phenomena stimulated the growth of reticulocytes, erythrocytes, hemoglobin, and osmotic resistance. The antioxidant activity of erythrocytes resulted in an increased oxygenation in the blood.

Also, there was an increase of fibrinolytic activity with decreased aggregation of erythrocytes, a decrease of hemocrit, and vasodilatation. All of which, resulted in decreased blood viscosity and improved circulation including increased microcirculation and improved oxygenation of tissues. Oxygenated tissues had increased growth, energetic and synthetic processes and a greater proliferation of blood cells and tissues.

Thus, the treatment of blood by He-Ne laser and / or natural red light caused an increase in non-specific pathogen resistance and stimulated "anti-stress mechanisms" on the molecular, cell, organ and organism levels. And, as a result of this, the body was able to mobilize bodily functions and important chains against pathogenic etiologies (Prokopev, 2004).

From this it follows that the present popular point of view about negative role of solar radiation for our body may not be entirely correct. Solar energy can be very useful for body chemistry, human health, and psychology, if used properly. There is presently a reduced ozone protection resulting in increased penetration of negative photons, particularly the UV-B and UV-C light. It appears, however, that nature may have dealt with this problem in the past and has, at least, some remedial solutions.

As noted above, plants may protect themselves from negative and excessive light with metallic ions, elements, enzymes systems and organic molecules. Similarly, this raises the question of animal protection against destructive light, "Why are metallic ions located in / on the epidermis?" and, "Can they protect the body against UV light?" We suggest, "Yes, it's possible ... but this should be tested." We predict, following the above described plant model, that Na^+, Ca^{++}, Mg^{++} and other metals located in or on the upper part of the skin can reflect and / or decrease energy of UV-B and UV-C according to Compton scattering; or be converted by the photoelectric effect to generate electrons with up to 3000 mV of energy. These electrons should disperse according to a Gaussian charge distribution on the surface protrusions of the skin papillae formed by the basal layer. If this flow of electrons passes through the mitochondria located in the basal layer of the papillae, they can reduce mitochondrial NAD^+ to NADH and / or convert directly into $\Delta\mu_{Na}{}^+$ and $\Delta\mu_H{}^+$, and indirectly into ATP (see Figures 4.11 and 4.16). This energy located in the surface of the skin might be used for many purposes including regeneration of DNA structure and / or to fight mutagenesis.

The photo-halosynthetic effects may function, for example, in the protection of eye ball via tear fluids and other eye components containing metallic ions, in this case, specifically sodium. From this and the above, it is clear that one of the prophylactic aims for human health is the maintenance of the body's salt balance, which diminishes with the age.

We consider that limited sunbathing has the potential to be very positive for total health, particularly during the times of day with a greater component of red light, such as mornings and evenings. From this standpoint, it would be interesting to conduct a skin cancer survey of coastal populations or other groups commonly with "salt-covered skin". We suggest if the above phenomenon is verified, that sunbathing should be carried out in the presence of salt water and / or upon perspiring. And ironically, that "fresh-water sunbathers" might be well advised to rinse off in salty water or use saline (specifically with Na^{+} or Ca^{++}) suntan lotion. In this case part of the negative UV light will be reflected, part will be Compton scattered to longer useful wave-lengths, and some light will serve to photo-electrically energize electrons.

4.7 Predictions and Potential Tests of Hypotheses

4.7.1 *Salt Conduction*

The convergent evolution of salt tolerance has allowed many species to persist, and in some cases, thrive, in saline environments. The unique ability of certain of these plants to act as conductors of salts through their vascular system to the shoot surface has given them significant advantages over other plants which can only survive by excluding the salt, or over those plants which accumulate salt in vacuoles until an accumulation limit of salt is reached. Although conductor plants range from minimal to maximal ability to pass salts from roots to the shoot surface, we predict that:

- Conductor lines are the best candidates for soil remediation, forage and other products,
- Conductor lines have the ability to conduct "unlimited" amounts to salt from the soil to their shoot surfaces and into the air and they do not accumulate salt in their tissues,
- The potential presence of ammonia-nutrition-protection in the "root → transport → salt-gland" system protects against the toxic effects of salts on the cytoplasm and metabolism of cells.

The predictions of this concept will determine:

- the salt uptake from soils,
- the quantification of the potential amount of salt removed from soil,
- the salt movement through the plants, excretion and the aeolian dispersal,

- the specific environmental conditions (including protective nitrogen nutrition) for optimization of soil remediation, salt conductance and plant productivity,
- the degree of soil re-flocculation, permeability, humic content, and fertility,
- the percentage of salt that becomes airborne, re-enters the soil, or is washed away, regarding specific weather and surface conditions.

At this time, there are excellent lines of conductor grasses for the development and realization in evaluating the salt-conductor plant strategy. In Figure 4.17 we illustrate a one example of high productive conductor grass *D. spicata*, which has a very strong resistance to extremely adverse conditions. In particular, cv. NyPa Forage has recently been shown, having an advantage in constructed wetlands receiving aquaculture effluent that is high in both nutrients and salt levels (Lymbery *et al.*, 2013).

Fig. 4.17. Salt-conductor grass *Distichlis spicata* var. yensen-4a planted in the Tulare Lake Drainage District, California, USA (top – view of the field; bottom – a greater increase) (photo images from the collection of Karl Biel)

The testing of the above concepts using specific salt-excreting halophytes like that, as well as some of those presented in Figure 4.1, can verify the above predictions.

4.7.2 Halodispersion

We predict that under specific conditions (Biel and Yensen, 2006*a*, 2006*b*) over 50% of the excreted salt may become airborne. In the case of certain lines of the saltgrass, *Distichlis*, we estimated that 5–50 tons / ha / year could be dispersed. For example, reduction of dominant salt ions (Na^+, Cl^-, and SO_4^{2-}) found in commonly compromised agricultural soils (Figure 4.18). Trial was in an established stand of a conductor plant and was compared to adjacent bare soil. Irrigation with drainage water was once or twice a month. The ion content is from the end of irrigation cycle and shows a minimum of 4.4 ± 0.6 metric tons / hectare / month of NaCI removed by *D. spicata* (Biel and Fomina, 2018).

The concept of halodispersion is new and estimates of salt uptake and aerial dispersal is presently not available for any plant. Further application (Biel and Yensen, 2006*a*, 2006*b*) of this principal and associated knowledge may contribute to the remediation of soils in many parts of the world.

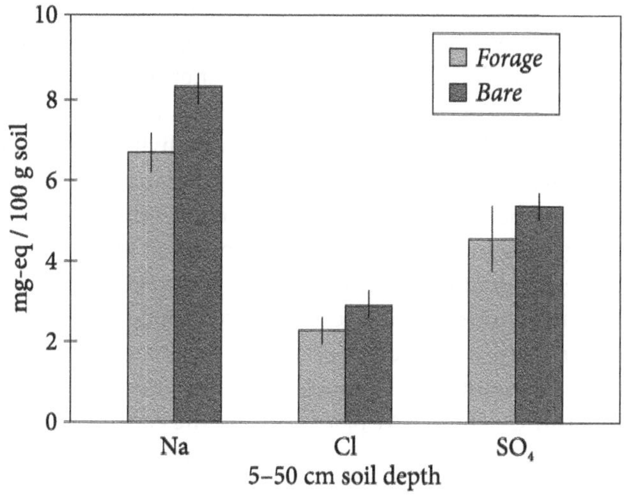

Fig. 4.18. The ion content of the *Distichlis spicata* "Forage" trial and adjacent "Bare" soil. Calculation shows a minimum of 4.4 ± 0.6 metric tons / hectare / month of NaCI removed by *D. spicata*

Note: The soil samples were from the tile-drained experimental plot check 7, Tulare Lake Drainage District, Corcoran, California, 08 September 2005, $n = 3$

4.7.3 *Metabolism*

Work based upon these hypotheses may answer some fundamental questions about the effect of high salinity on metabolism and mechanisms important to salt grass lines and / or other crop plants. The realized knowledge of these predictions may facilitate the selection of other commercial forages for practical application, *e.g.* role of the "ammonia protection" and optimization of protein for dairy vs. carbohydrate for meat production with minimal accumulated salts. Future cooperation with geneticists could create additional information useful for the development of more salt-tolerant glycophyte crops, such as maize, wheat, barley, oats, rye, rice, *etc.* and may provide information regarding nutritional and anti-nutritional characteristics of plants, seeds, and storage products in salty environments.

The study of the conductor phenomenon has led to predictions of basic physiological concepts and principals that were heretofore unknown. The ability of "salt conducting plants" with "salt membrane protection" to handle salt without negative effect on the growth and production of the plants can allow the dual benefits of remediating soils while achieving the production of photosynthetic products assisted by "halosynthesis".

4.7.4 *Protection*

From the above descriptions, we can extrapolate that similar protection strategies may be utilized by other plants. For example, high altitude plants should resist high levels of UV light energy, and we predict that such plants use sodium and / or other elements and compounds in, or near, the epidermis for protection of the mesophyll against destruction. Likewise, the intertidal organisms, such as cyanobacteria that occupy the spray zone, may be able to utilize photo-halosynthesis effects to their benefit, as might the human eye which is also awash in salt water. By contrast, submerged aquatic plants would not be expected to use or need similar protective strategies due to the near complete absorption of UV light in the first few centimeters of water. Protection against excessive light may also be manifested by biochemical processes located in the upper mesophyll in C_3 and CAM plants, in the mesophyll of C_4 plants, resulting in optimal function of the plant under varying light conditions.

4.7.5 *Halosynthesis*

If our hypothesis of halosynthesis is proven correct it would represent the new mechanism by which life can access energy from sunlight and from the environment. The first and the best-known mechanism is photosynthesis while the second-best known mechanism is the discovery of an environmental energy source via thiobacteria which can extract heat energy from undersea hot-water vents to fix and reduce carbon.

4.8 General Conclusions

Haloconduction, halodispersion, halosynthesis and UV light protection may be universal concepts and mechanisms that occur in many plants and animals. They involve photo-electrically energized electrons, Compton-generated, lower-energy photons, Gaussian surface charge distribution, inverse-cations-ATP pumps, mitochondrial electron transport chain, and metabolic protections. In photo-halosynthesis these processes serve for protection and the collection of photon energy, and in the mitochondria according to the equation: $\boldsymbol{hv + Na^+ => \Delta\mu_{Na}{}^+ / \Delta\mu_H{}^+ \Leftrightarrow ATP}$. The physical laws and function of the mechanisms may operate similarly in salt glands, papillae and other surface structures of organisms, suggesting that the concept of photo-halosynthesis may have broad and extensive applications.

These new concepts for plants and animals and their predictions should be of value to researchers, doctors, farmers and politicians, and could provide a myriad of potential opportunities for humanity in a world of global salinization, medical and anthropogenic problems.

Acknowledgments

Many thanks to corresponding member of the Russian Academy of Sciences, Professor Valery N. Kudeyarov and Dr. Iraida I. Skripnichenko (Institute of Physical and Chemical Soil-Science Problems, Russian Academy of Sciences, Pushchino, Moscow Region 142290, Russia) for kind help with soil samples analysis. Also, special grateful to Dr. Irina Fomina (Basic Biological Institute, Russian Academy of Sciences, Pushchino Moscow Region 142290, Russia) for productive discussion and useful advice.

References

Abrahams J.P., Leslie A.G., Lütter R., Walker J.E. Structure at 2.8 Å resolution of F1-ATPase from bovine heart mitochondria, *Nature*, 1994, Vol. 370, No. 6491, p. 621–628.

Beetle A.A. North American variations of *Distichlis spicata, Bull. Torr. Bot. Club*, 1943, Vol. 70, No. 6, p. 638–650.

Beetle A.A. The grass genus *Distichlis*, *Revista Argentina de Agronomía*, 1955, Vol. 22, No. 2, p. 86–94.

Biel K.Y. *Phototrophic Carbon Metabolism in Plants with Differential Organization of Photosynthetic Apparatus*, Doctor Biol. Sci. Thesis, Pushchino, 1988, 543 p. [in Russian]

Biel K.Y. *Ecology of Photosynthesis*, Nauka, Moscow, 1993, 221 p. [in Russian]

Biel K.Y., Fomina I.R. The effects of sodium chloride concentration on photosynthesis, respiration and pigment content in systems of whole-shoot-leaf blades versus leaf-blade slices in saltgrass *Distichlis spicata* (*Poaceae*), 2018. [in preparation]

Biel K.Y., Fomina I.R., Nishio J.N., Soukhovolsky V.G. Optimization of photosynthesis within leaves of C_3 and C_4 plants. Example of complex system self-regulation when energy source is also stress inductor, *System Identification and Control Problems,* Proc. VII Int. Conf. SICPRO '08 (ISBN 978-5-91450-002-0), Moscow, 2008, p. 566–593.

Biel K.Y., Fomina I.R., Nazarova G.N., Soukhovolsky V.G., Khlebopros R.G., Nishio J.N. Untangling metabolic and spatial interactions of stress tolerance in plants. 1. Patterns of carbon metabolism within leaves, *Protoplasma*, 2010, Vol. 245, p. 49–73.

Biel K.Y., Yensen N.P. The hypothesis of halosynthesis, photoprotection, soil remediation via salt-conduction and potential medical benefits, *The Complex Systems under Extreme Conditions*, Khlebopros R.G. and Soukhovolsky V.G. eds., Krasnoyarsk, Russia, 2005*a*, p. 86–115.

Biel K.Y., Yensen N.P. The hypothesis of photo-halosynthesis and its application for some agricultural and medical problems, *Functional Foods for Cardiovascular Diseases. Functional Food Can Help Reduce the Risks of Cardiovascular Diseases*, Martirosyan D.M. ed., Functional Foods Center D & A, Inc., Richardson, Texas, USA, 2005*b*, p. 215–227.

Biel K.Y., Yensen N.P. *Salt Removal by Conductor Plants,* Patent Pending, Submitted to the United States Patent and Trademark Office, 2006*a*, File #8416521, 26 p.

Biel K.Y., Yensen N.P. ICBA associates propose revolutionary new concept for halophytes: photohalosynthesis, *Biosalinity News (ICBA Newsletter)*, 2006*b*, Vol. 7, No. 1, p. 6–7.

Bil' K.Y., Fomina I.R. Chlorophyll-protein complexes from green algae and higher plants. On the role of metabolic regulation of the photosystem II complex formation in bundle sheath cell chloroplasts of C_4 plants, *Biochemistry-USSR*, 1985, Vol. 50, No. 2, p. 219–224.

Bil' K.Y., Fomina I.R., Nazarova G.N., Nishio J.N. Light dependent organization of carbon metabolism within spinach leaves, *Physiol. Biochem. Cult. Plants*, 2003, Vol. 35, No. 5, p. 392–402.

Bil' K.Y., Fomina I.R., Nishio J.N. Stress tolerance and light dependent organization of carbon metabolism within spinach leaves, *New and Nontraditional Plants and Prospects of Their Utilization*, Proc. V Int. Symp., Publ. Russian University Int. Friendship, Moscow–Pushchino, Russia, 2003, Vol. 3, p. 24–27.

Bil' K.Y., Fomina I.R., Shabnova N.I. Content of nitrogen compounds in maize shoots, structural and functional features of photosynthetic apparatus under ammonium and nitrate nutrition, *Photosynthetica*, 1987, Vol. 21, No. 4, p. 525–534.

Bil' K.Y., Fomina I.R., Shabnova N.I., Terekhova I.V., Tsenova E.N. The influence of ammonium nitrogen nutrition and conditions of illumination on activity of enzymes of photosynthetic carbon metabolism in corn of different genotypes, *Biochemistry-USSR*, 1990, Vol. 54, No. 7, p. 920–925.

Bil' K.Y., Fomina I.R., Tsenova E.N. Effects of nitrogen nutrition on photosynthetic enzymes activities, type of photosynthates, and photosystem II activity in maize leaves, *Photosynthetica*, 1985, Vol. 19, No. 2, p. 216–220.

Bil' K.Y., Gedemov T. The Structural and functional properties of assimilation apparatus in *Suaeda arcuata* Bunge (*Chenopodiaceae)* related to cooperative photosynthesis, *Proc. Acad. Sci. USSR*, 1980, Vol. 250, No. 4, p. 9–13.

Bil' K.Y., Lyubimov V.Y., Demidova R.N., Gedemov T. Assimilation of CO_2 by plants of the family *Chenopodiaceae* with three types of autotrophic tissues in the leaves, *Soviet Plant Physiol.*, 1981, Vol. 28, No. 6, p. 808–815.

Black C.C.Jr. Photosynthetic carbon fixation in relation to net CO_2 uptake, *Annu. Rev. Plant Physiol.*, 1973, Vol. 24, p. 253–286.

Boyer P.D. The binding change mechanism for ATP synthase – some probabilities and possibilities, *Biochim. Biophys. Acta*, 1993, Vol. 1140, No. 3, p. 215–250.

Boyer P.D. The ATP synthase – a splendid molecular machine, *Annu. Rev. Biochem.*, 1997, Vol. 66, p. 717–749.

Brownell P.F., Crossland C.J. The requirement for sodium as a micronutrient by species having the C_4 dicarboxylic photosynthetic pathway, *Plant Physiol.*, 1972, Vol. 49, No. 5, p. 794–797.

Chirkova T.V. *Physiological Bases of Plant Resistance*, Publ. St. Petersburg University, St. Petersburg, 2002, 240 p. [in Russian]

Compton A.H. Total reflection of X-rays from glass and silver, *Phys. Rev.*, 1922, Vol. 20, p. 84.

Compton A.H. A quantum theory of the scattering of x-rays by light elements, *Phys. Rev.*, 1923, Vol. 21, No. 5, p. 483–502.

Cui M., Vogelmann T.C., Smith W.K. Chlorophyll and light gradients in sun and shade leaves of *Spinacia oleracea, Plant Cell Environ.*, 1991, Vol. 14, p. 493–500.

Dimroth P., Kaim G., Matthey U. Critical role of the membrane potential for ATP synthesis by F_1F_0 ATP-synthases, *J Exp Biol.*, 2000, Vol. 203, p. 51–59.

Edwards G.E., Kanai R., Black C.C. Phosphoenolpyruvate carboxykinase in bows of certain plants which fix CO_2 by the C_4–dicarboxylic acid cycle of photosynthesis, *Biochem. Biophys. Res. Commun.*, 1971, Vol. 45, No. 2, p. 278–285.

Edwards G.E., Walker D. *C_3, C_4: Mechanisms, and Cellular and Environmental Regulation, of Photosynthesis*, University of California Press, Berkeley–Los Angeles, 1983, 542 p.

Fomina I.R., Biel K.Y. Photosynthetic carbon metabolism: Strategy of adaptation, *Photosynthesis: Open Questions and What We Know Today,* Allakhverdiev S.I., Rubin A.B., Shuvalov V.A. eds., Institute of Computer Science, Izhevsk–Moscow, 2013, Chapter 20, p. 707–774. [in Russian]

Fomina I.R., Biel K.Y. 2014 Photosynthetic carbon metabolism: Strategy of adaptation, *Contemporary Problems of Photosynthesis,* Allakhverdiev S.I., Rubin A.B., Shuvalov V.A. eds., Institute of Computer Science, Izhevsk–Moscow, 2014, Vol. 2, Chapter 29, p. 415–483.

Garrett B.C. Ions at the air / water interface, *Science*, 2004, Vol. 303, p. 1146–1147.

Gould F.W. *Grasses of Southwestern United States*, University of Arizona Press, Tucson, USA, 1951, 352 p.

Hatch M.D., Kagava T., Craig S. Subdivision of C_4 pathway species based on different C_4-acid decarboxylating systems and ultrastructural features, *Aust. J. Plant Physiol.*, 1975, Vol. 2, p. 110–128.

Heuperman A.F., Kapoor A.S., Denecke H.W. *Biodrainage: Principals, Experiences and Applications*, FAO, Rome, 2002, 79 p.

Hitchcock A.S. *Manual of the Grasses of the United States*, USDA Publ. 20, Washington DC, 1950, 1051 p.

Karpilov Y.S. *Specificity of the Structure and Functions of the Photosynthetic Apparatus in Some Tropical Plant Species, Proc. Moldavian Inst. for Research in Irrigation Farming and Vegetable Growing,* Kishinev, 1969, Vol. 11, Special Issue, 35 p. [in Russian]

Karpilov Y.S. (ed.) *Cooperative Photosynthesis of Xerophytes, Proc. Moldavian Inst. for Research in Irrigation Farming and Vegetable Growing,* Kishinev, 1970, Vol. 11, No. 3, 67 p. [in Russian]

Kemp P.R., Cunningham G.L. Light, temperature and salinity effects on growth, leaf anatomy, and photosynthesis of *Distichlis spicata* (L.) Greene, 1981, *Am. J. Bot.*, Vol. 68, No. 4, 507–516.

Koyro H.-W. Study of potential cash crop halophytes by a quick check system: Determination of the threshold of salinity tolerance and the ecophysiological demands, *Cash Crop Halophytes: Resent Studies. 10 Years after the Al Ain Meeting,* Lieth H., Motchenko M. eds., Kluwer Acad. Publ., Dordrecht–Boston–London, 2003, p. 5–17.

Leake J.E. The role of NyPa *Distichlis* spp. cultivars in altering groundwater and soil conditions - a work in progress, *Salinity Seminar,* Hydrological Society of South Australia, Australian Geomechanics Society, South Australian Chapter, and Intern. Association of Hydrogeologists, Australian National Chapter, North Adelaide, 2003, 1–8.

Liphschitz N., Waisel Y. Adaptation of plants to saline environments: salt excretion and glandular structure, *Tasks for Vegetation Science, Contribution to the Ecology of Halophytes,* Sen D.N., RajPurohit K.S. eds., Junk Publ., The Hague, 1982, Vol. 2, p. 197–214.

Lymbery A.J., Kay G.D., Doupé R.G., Partridge G.J., Norman H.C. The potential of a salt-tolerant plant (*Distichlis spicata* cv. NyPa Forage) to treat effluent from inland saline aquaculture and provide livestock feed on salt-affected farmland, *Sci. Total Environ.*, 2013, Vol. 445–446, p. 192–201.

Lyubimov V.Y., Atakhanov B.O., Bil' K.Y. Adaptation of photosynthetic apparatus to extreme conditions in plants of the Kara-Kum desert, *Soviet Plant Physiol.*, 1987, Vol. 33, No. 5, p. 680–686.

Lyubimov V.Y., Biel K.Y., Karpilov Y.S. The peculiarities of the glycolate formation in assimilation tissues of C_4 plants, *Plant Physiology,* Materials 6th Natl. Conf., Publ. Bulg. Acad. Sci., Sofia, 1978, Vol. 4, Part 1, p. 41–56.

Martin R. *The Salinity Tolerance of Six Bermudagrasses (Cynodon dactylon (L.) Pers.), One Dropseed (Sporobolus virginicus (L.) Kunth) and One Saltgrass (Distichlis spicata var. yensen-4A) for Phytoremedia-*

tion of Salt-Affected Soils in Oil, Gas and Brine Fields, M. S. Thesis, Louisiana Tech University, Ruston, 2003, 66 p.

Marcum K.B., Murdoch C.L. Salinity tolerance mechanisms of six C_4 turfgrasses, *J. Am. Soc. Hort. Sci.*, 1994, Vol. 119, No. 4, 779–784.

Muzafarov E.N., Kreslavski V.D., Nazarova G.N. *Light- and Hormone-Dependent Regulation of Photosynthesis and Plants Growth,* NCBI Press, Pushchino, 1995, 140 p. [in Russian]

Neumivakin I.P. *Hydrogen Peroxide. On Guard of Health*, Dilya Press, Moscow–St. Petersburg, 189 p. [in Russian]

Newman E.I. Water movement through root systems, Phil. Trans. R. Soc., Lond., 1976, Vol. 273, No. 927, p. 463–478.

Prokopev V.E. *Biophysical Mechanisms of Influence of Low-Intensity Laser Light on Biological Tissues and Optical Methods for Diagnosis Their Condition*, Doctor Sci. Thesis, Tomsk, 2004, 284 p. [in Russian]

Semenova G.A., Fomina I.R., Biel K.Y. Structural features of the salt glands of the leaf of *Distichlis spicata* "Yensen 4a" (Poaceae), *Protoplasma*, 2010, Vol. 240, p. 75–82.

Semikhatova O.A., Chirkova T.V. *Physiology of Plants Respiration*, Publ. St. Petersburg University, St. Petersburg, 200, 220 p. [in Russian]

Serway R.A., Jewett J.W. *Physics for Scientists and Engineers with Modern Physics*, 6th edn., Brooks / Cole, Pacific Grove, California, 2003, 1680 p.

Shabnova N.I., Fomina I.R., Biel K.Y. The productive process and qualitative features of overhead mass of corn plants with regard to a source of nitrogen nutrition, *J. Agr. Biol.*, 1997, Vol. 1, p. 57–63. [in Russian]

Shabnova N.I., Fomina I.R., Biel K.Y., Stakhova L.N. The amino acid composition of the maize green mass in connection with the conditions of nitrogen nutrition, *Physiol. Biochem. Cult. Plants*, 1993, Vol. 25, No. 4, p. 346–352. [in Russian]

Skou J.C., Esmann M. The Na, K-ATPase, *J. Bioenerg. Biomembranes*, 1992, Vol. 24, p. 249–261.

Skulachev V.P. Laws of bioenergetics, *Soros Educational J.*, 1997, Vol. 1, p. 9–14. [in Russian]

Soukhovolsky V.G., Fomina I.R., Bil' K.Y., Nishio J.N., Khlebopros R.G. An optimization model of the photosynthetic leaf: The model of optimal photosynthetic CO_2 fixation within leaves of mesophytic C_3 plants, *Dokl. Biol. Sci.*, 2002, Vol. 382, p. 28–30.

Tam W.L., Wong W.P., Loong A.M., Hiong K.C., Chew S.F., Ballantyne J.S., Ip Y.K. The osmotic response of the Asian freshwater stingray

(*Himantura signifier*) to increased salinity: a comparison with marine (*Taeniura lymma*) and Amazonian freshwater (*Potamotrygon motoro*) stingrays, *J. Exp. Biol.*, 2003, Vol. 206, No. 17, p. 2931–2940.

Udut V.V., Karpov A.B., Prokopev V.E., Naumov C.A., Siryanov B.N. About mechanisms of intravenous treatment of blood by He-Ne laser light, *Bull. TNZ Acad. Medical Sci.* USSR, 1990, Vol. 2, p. 65–78. [in Russian]

Ungar I. *Ecophysiology of Vascular Plants*, CRC Press, Boca Raton, Florida, 1991, 209 p.

Vallejo F., Tomás-Barberán F.A., Garcia-Viguera C. Phenolic compounds content in edible parts of broccoli inflorescences after domestic cooking, *J. Sci. Food Agr.*, 2003, Vol. 83, No. 14, p. 1511–1516.

Vogelmann T.C. Plant tissue optics, *Annu. Rev. Plant Physiol. Plant Mol. Biol.*, 1993, **44**: 233–251.

Yensen N.P. Saline agriculture and new halophyte crops, *Proc. 61st Annual Meeting of the Southwestern and Rocky Mountain Division, AAAS*, Tucson, Arizona, 1985, p. 19–23.

Yensen N.P. International Symposium on high salinity tolerant plants summary of papers presented, *Biology of Salt Tolerant Plants*, Khan M.A., Unger I.A. eds, Ohio University, Athens, 1995, p. 1–12.

Yensen N.P. 2001 *Halófitas del Golfo de California y sus Usos (Halophytes of the Gulf of California and their uses)*, Traducción Julio Rodriguez-Casas, Universidad de Sonora, Rosales y Blvd. Luis Encinas, Edificio Museo y Biblioteca, Hermisillo, Sonora, Mexico, 2001, 296 p.

Yensen N.P. New developments in the world of saline agriculture, *Prospects for Saline Agriculture*, Proc. Int. Seminar (April 2000), Pakistan Acad. Sci., Pakistan Agricult. Res. Council and COMSTECH, Islamabad, 2002, p. 10–12.

Yensen N.P., Bedell J.L. The use of saline water for forage production in coastal ecosystems, *Salt-Affected Lagoon Ecosystems*, Abstr. Int. Symp., Valencia, Spain, 1995, 221–222.

Yensen N.P., Bedell J.L., Yensen S.B. The use of saline water for forage production in coastal ecosystems, *Salt-Affected Lagoon Ecosystems*, Batlle-Sales J. ed., Materials Int. Symp., Universitat de Valéncia and Int. Soil Sci. Soc., Spain, 1997, p. 297–307.

Yensen N.P., Biel K.Y. Soil remediation via salt-conduction and the hypothesis of halosynthesis and photoprotection, *Ecophysiology of High Salinity Tolerant Plants*; *Tasks for Vegetation Science 40*, Khan M.A., Weber D.J. eds., Springer, The Netherlands, 2006, p. 313–342.

Yensen N.P., Hinchman R.R., Negri M.C., Mollock G.N., Settle T., Keiffer C.S., Carty D.J., Rodgers B., Martin R., Erickson R. Using halophytes to manage oilfield saltwater: disposal by irrigation / evaporation and remediation of spills, *Environmental Issues and Solutions in Petroleum Exploration, Production and Refining*, Proc. 6th Int. Petroleum Environ. Conf., Sublette K.L. ed., Houston, 1999, p. 16–18.

Yensen N.P., Yensen S., Weber C.W. A review of *Distichlis* spp. for production and nutritional values, *Arid Lands: Today and Tomorrow*, Proc. Int. Res. and Development Conf. (October 1985), Whitehead E.E., Hutchinson C.F., Timmermann B.N., Varady R.B. eds., West View, Boulder, Colorado; Belhaven, London, 1988, p. 809–822.

Zholkevich V.N., Gusev N.A., Kaplya A.V., Pakhomova G.E., Pilshchikova N.V., Samuilov F.D., Slavnii P.C., Shmatko I.G. *Water Exchange in Plants*, Nauka, Moscow, 1989, 256 p. [in Russian]

5
Protective Role of Silicon in Living Organisms

Vladimir V. Matichenkov[1,*], **Irina R. Fomina**[1,2,*], **Karl Y. Biel**[1,2,*]

[1]*Institute of Basic Biological Problems, Russian Academy of Sciences, Pushchino, Moscow Region 142290, Russia*
[2]*Biosphere Systems International Foundation, Tucson, Arizona 85755, USA*

None of organisms could exist without silicon, –
Academician Vladimir I. Vernadsky

Abstract
Review about the role of silicon in the soil-plant-animal system is presented, and the participation of silicon in stress tolerance promotion of organisms is discussed. It is hypothesized that the human body, as well as animals, plants and microorganisms have a unique protective mechanism which includes, in particular, mobile silicon compounds. According to our hypothesis, the function of silicon constituent can provide the additional synthesis of stress-protective molecules under genetic control but without "physical" participation of the genetic apparatus. The assumption is based on the indirect experiments, and on a basis of subordination of two constituencies: a) response of genetic apparatus on stress, insuring synthesis of stress-protective compounds, such as antioxidant enzymes, stress-proteins, glutathione, phenols, and other antioxidants, *etc.*, and b) on additional non-enzymatic formation of the same protective compounds on the matrixes of polysilicic acids. The active forms of silicon are considered as a matrix-depot for formation of compounds, which are assistance in any organisms to maintain the positive homeostasis under stress conditions.

Keywords: Silicon, Monosilicic acid, Polysilicic acids

* *Corresponding authors:* vvmatichenkov@rambler.ru; irafomi@rambler.ru; karlbiel@hotmail.com

5.1 Introduction

This chapter provides a review of the role of silicon in the soil-plant-animal system and discusses the role of silicon in improving stress tolerance of organisms. We hypothesized that the human body, as well as animals, plants and microorganisms have a unique protective mechanism which includes, in particular, mobile silicon compounds. Active forms of silicon are considered as matrix-depot for the formation of compounds that the body needs to maintain a positive homeostasis under stress.

5.1.1 Agriculture

Protection of agricultural plants against diseases and insect attacks by silicon compounds was started many centuries, or, possibly, several thousand years ago because even agriculturists of the Roman Empire and Ancient China were familiar with unique capacities of silicon remedies and successfully applied them. Today, the list of scientific publications related to application of the active forms of silicon for plant protection against various stressors (nematodes, funguses, negative climatic conditions, *etc.*) counts more than 4000 references (see ref. in Voronkov *et al.*, 1978; Savant *et al.*, 1997; Matichenkov *et al.*, 2001*a*; 2001*b*; 2013; Ma and Takahashi, 2002; Matichenkov and Bocharnikova, 2004*a*; 2004*b*; Bocharnikova *et al.*, 2011; and others).

Under stress conditions, the active forms of silicon have also positive influence on the health of animals, fishes and birds (Carlisle, 1997; Jurkic *et al.*, 2013; Price *et al.*, 2013). For example, small dietary supplement by Diatomaceous Earth, zeolite, silicon water, or other forms of active silicon harshly provides disease tolerance of the domestic animals, and decreases their death-rate (Carlisle, 1972; 1984; D'yakov *et al.*, 1990; Carlisle *et al.*, 1977). At the same time, the presence of active silicon supplements in food does not increase the level of silicon concentration in the blood of robust animals (Iler, 1979; Mayras *et al.*, 1980).

A long time ago, it was observed that the wild animals aspire not only to take, as food diet, various clays, Diatomaceous Earth, and zeolite, but also to take up their residences close to wedging out high-silica soft minerals. The animals used these minerals as nutrition.

In the mid-seventies of the 20th century, the Soviet Union's and American scientists proved the importance of silicon for improving the natural development of domestic and agricultural animals and birds; at least 6% of their fodder should be amorphous silicon dioxide (Carlisle, 1972; Voronkov *et al.*, 1978; D'yakov *et al.*, 1990; Carlisle *et al.*, 1977). It was also demonstrated that silicon promotes disease and stress tolerance of domestic animals (Carlisle, 1984; Bgatov *et al.*, 1987; Anokhin *et al.*, 1997; Vladimirov *et al.*, 1998), disease-tolerance of fishes (Exley *et al.*, 1997; Camilleri *et al.*, 2003), and increases the rate of cancroids growth (Pellenard, 1969).

5.1.2 *Medicine*

Prior to the beginning of the 20th century, amorphous silicic dioxide, silica, was the most commonly used pharmaceutical remedy in the World. It is well known, that silicon imbalance in the human body is accompanied with general metabolic disease and induces anemia, osteomalacia, shedding of hair, joint illness, and other diseases (Carlisle *et al.*, 1977; Voronkov *et al.*, 1978). Silicon compounds can prevent atherosclerosis development; and such of organism problems as headaches, cardiovascular and skin diseases, sicknesses of joints and osseous tissues have been treated successfully by silicon medications (Voronkov *et al.*, 1978; Birchall and Chapell, 1988; Kaufmann, 1995*a*; 1995*b*; Jugdaohsingh *et al.*, 2002; Perez-Grandos and Vaquero, 2002; Sripanyakorn *et al.*, 2004; McNaughton *et al.*, 2005; Jurkic *et al.*, 2013; Price *et al.*, 2013).

It was shown that concentration of silicon in blood increased several times around wound (Voronkov *et al.*, 1978; Mayras *et al.*, 1980), and the bandage sodden by silicic acids hastened wound repair (Kuzin *et al.*, 2000; Brown, 2002). It is also commonly known that silicon concentration in blood of pregnant women and in infants is increased (Voronkov *et al.*, 1978; Kaufmann, 1995*a*; 1995*b*; Van Dyck *et al.*, 2000).

In 1878, Professor Louis Pasteur predicted that "*therapeutic action of silica has a great future*" (see Voronkov *et al.*, 1978). Really, modern medicine has again returned to the use of silicon medications, which are not only completely safe for humans, but also very effective for the treatment of a wide range of diseases.

5.1.3 *Microorganisms and Plants*

Silicon-rich compounds have positive effect on growth and development of microorganisms (Al-Wajeeh, 1999; Wainwright, 2005). In particular, silicon is one of the basic nutrient elements for phytoplankton, diatoms and cyanobacteria (Siegel and Siegel, 1973; Cha *et al.*, 1999; Brummer, 2003; Martin-Jezequel and Lopez, 2003; Velikova *et al.*, 2005; Tuner *et al.*, 2006). At the same time, silicon is also a very important nutrient for organisms, which do not have so much silicon as diatoms and cyanobacteria have. For example, the cultures of microorganisms are developing in the glass Petri dishes better than in plastic ones (pers. obs. and cons. with microbiologists).

The literature data testify that silicon has positive effect on growing of a number of microorganism species (Mohanty *et al.*, 1990; Yoshon, 1990; Wainwright *et al.*, 1997; Soomro, 2000). Furthermore, if the active silicon was applied to the grounds or soils contaminated with oil or oil derivatives, populations of soil microorganisms increased (Bocharnikova *et al.*, 1999).

Positive influence of silicon on plant growth and development had been known since Professor Justius von Leibigh published the work (Leibigh, 1840) about mineral nutrition of plants. Today, the positive effects of silicon nutrition are estimated on various soils for the following cultivated plants: rice (*Oryza sativa* L.), sugar-cane (*Saccharum officinalrum* L.), barley (*Hordeum vulgare* L.), wheat (*Triticum aestivum* L.), oats (*Avena sativa* L.), rye (*Secale cereale* L.), sorghum (*Sorghum vulgare* L.), corn (*Zea mays* L.), sunflower (*Helianthus annuus* L.), beans (*Vicia faba* L.), soy (*Glycine max* L.), trefoil (*Trifolium pretense* L.), lucerne (*Medicago sativa* L.), millet (*Panicum miliaceum* L.), tomato (*Lycopersicum esculentum* L.), cucumber (*Cucumis sativus* L.), marrow (*Cucurbito pero* L.), lettuce (*Lactuca sativa* L.), tobacco-plant (*Nicotiana tabacum* L.), sugar-beet (*Beta vulgaris* L.), lemon tree (*Citrus* × *limon* (L.) Burm. f.), tangerine-tree (*Citrus reticulate* L.), vine (*Vitis vinifera* L.), apple-tree (*Malus silvestris* L.), melon (*Cucumis melo* L.), and others (D'yakov *et al.*, 1990; Savant *et al.*, 1997; Epstein, 1999; Matichenkov *et al.*, 2001*a*; Ma and Takahashi, 2002; Matichenkov and Bocharnikova, 2004*b*).

Summarizing of the data about influences of silicon fertilizers on plant productivity and soil fertility the following effects were determined:

1. Silicon enrichments provide the protective functions of plants on the mechanical, physiological, and biochemical levels (Aleshin *et al.*, 1987; Savant *et al.*, 1997; Ma and Takahashi, 2002; Belanger, 2005).
2. Agricultural activity results in deficiency of active silicon forms in plants and drives to degradation soil processes (Matichenkov and Bocharnikova, 1994; Bocharnikova *et al.*, 1995; Matichenkov *et al.*, 1999*a*).
3. Silicon fertilizers and silicon-rich soil amendments increase and retain phosphorus in accessible forms and as a result it facilitates absorption of phosphorus by plants (Matichenkov and Ammosova, 1994; Matichenkov and Bocharnikova, 2001; Matichenkov, 2008).
4. Silicon nutrition could be applied as a lime material in acid soils (D'yakov *et al.*, 1990; Savant *et al.*, 1997; Ma and Takahashi, 2002).

5. Silicon-rich soil amendments are able to optimize the physical properties of soils (Emadian and Newton, 1989; Matichenkov and Ammosova, 1994; Matichenkov and Bocharnikova, 2001).

The known facts show the positive influence of silicon on organisms throughout a certain general mechanism which, likely, connects with self-protective function of living systems. It is also logical to presume that biochemical basis of this mechanism is uniform for all organisms.

5.2 Forms of Silicon

Silicon is the second, after oxygen, element in the Earth's crust, and its concentration is increased with decrease of the lithosphere layer depth, and achieves the maximal value in the surface of soil (Reimers, 1990; Bashkin, 2005). In soil, the main part of silicon is presented as a silicon dioxide and various forms of aluminosilicates (Sokolova, 1985). Quartz, crystalline variety of silica, is characterized by stability to eolation (Orlov, 1985; Olier, 1990) and together with other macrocrystalline silicates [feldspar, plagioclase, pyroxenes] and secondary or clay silicon-containing minerals [kaolinite, vermiculite, smectite, and others] forms the skeleton of soil (Orlov, 1985).

Stability of silicon minerals to eolation is reflected in the classification of soil elements based on their mobility, where the silicon is considered as inert element (Polynov, 1952). However, in the same classification, silicon is also attributed to the group of mobile elements. Actually, in soil solutions are permanently presented the mono- and polysilicic acids and organo-silicon compounds, which have a high chemical and biological activity (Matichenkov, Calvert *et al.*, 2001; Matichenkov and Bocharnikova, 2004*a*; 2004*b*; Matichenkov, 2008). Thus, the soil silicon consists of two main groups - inert elements and biogeochemically active compounds.

Classification of silicon compounds in the soil, presented in Figure 5.1, is focused on the biogeochemical active forms of silicon. Traditionally (Sokolova, 1985), the silicon compounds in the solid phases of soil are classified on the bases of mineralogical structure, and are a priori considered as inert. But silicon compounds of the liquid phases are considered as active because they have numerous physico-chemical and biological properties. Undoubtedly, these active silicon forms are presented in the tissues of living organisms.

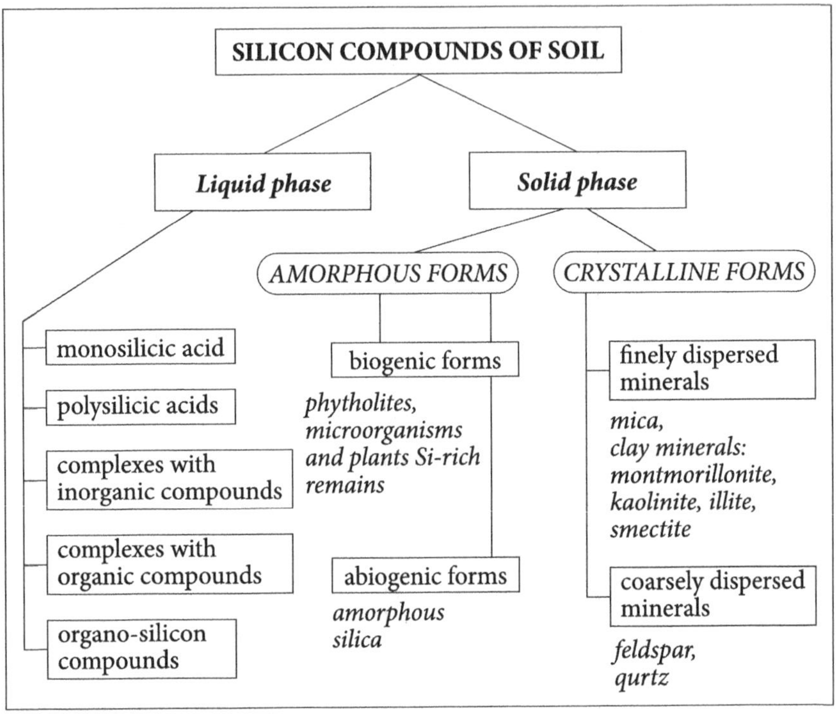

Fig. 5.1. Forms of silicon compounds in soil (modified from Matichenkov *et al.*, 1999*a*; Biel *et al.*, 2008)

Monosilicic acid is a weak inorganic acid with a slight buffering capacity at pH ≈ 7.0. This acid can react with ions of aluminum, iron, calcium, manganese, magnesium and heavy metals (Bocharnikova *et al.*, 1995; 1999; Matichenkov and Bocharnikova, 2001). Monosilicic acid is also able to replace phosphate anions from the calcium, iron, and aluminum phosphates (Matichenkov and Ammosova, 1994; Matichenkov, 2008); this process results in facilitating phosphate uptake by plants and soil microorganisms (Yoshida, 1975; Trevors, 1997).

Polysilicic acids are compounds consisting of two and more silicon atoms linked by siloxane bonds in a line [-Si-O-Si-] with hydroxyl groups bond with silicon atoms (Iler, 1979).

There are two groups of polysilicic acids in Nature: 1) oligomers of silicic acid contain from two to several hundred of silicon atoms per one molecule. They also were named as low-molecular weight polysilicic acids. It was supposed that in the diluted solutions (in most of natural waters, soil waters, lakes, rivers, and ocean waters) these compounds are not stable, and are locating as intermediates between monosilicic acid and high-molecular weight polysilicic acids (Iler, 1979; Dietzel, 2000; 2002); 2) high-molecular weight polysilicic acids are containing till several thousands of silicon atoms per one molecule (Davies, 1964; Varshal *et al.*, 1980; Stumm and Morgan, 1996). These acids are chemically inert, but, anyway, can influence the physical properties of soils (Yatsynin, 1989; Matichenkov *et al.*, 1994).

Chemical inertness of polysilicic acids is specified by the ability of their molecules "to twist" and by this way to compensate the negative charges, which are formed during dissociation of hydroxyl groups (Iler, 1979). Polysilicic acids are also able to form silicon bridges between soil particles (Yatsynin, 1994). These bridges, due to permanent changing of soil humidity, undergo dehydration with formation of silicon dioxide.

The processes of periodical drying-wetting of soils are formatting the siliceous films on a surface of practically all soil minerals. And exactly this film is the main source of monosilicic acid, when moisture comes upon the soil. Formation of polysilicic acids occurs as the monosilicic acid's molecules condensation (Iler, 1979). Though this process is poorly investigated, however, it is known, that in open water-bases the process can go, if the concentration of monosilicic acid is above 100 mg Si / liter (Iler, 1979). But usually, in the soil solutions the concentration of monosilicic acid is not higher than 1–50 mg Si / liter (Matichenkov, 2008). Theoretically, in such diluted solutions polysilicic acids should be depolymerized. Nevertheless, that does not happen (Varshal *et al.*, 1980; Alvarez and Sparks, 1985; Matichenkov and Snyder, 1996; Matichenkov *et al.*, 1997; Matichenkov, 2008).

It is logical to suppose that formation of polysilicic acids and their stable existence are due to location of anti-ions in the diffusion layer surrounding mineral surface. And only in this layer it is possible to keep permanently high concentration of monosilicic acid, which is formed during dissolving of siliceous films with following condensation to polysilicic acids (Frolov, 2004).

High molecular-weight polysilicic acids are also able to form the gels with stable structure, and the colloids containing nucleus with silica microcrystals inside it (Iler, 1979).

The ratio between mono- and polysilicic acids is usually stable in various soils or water-mineral systems (Matichenkov and Snyder, 1996; Matichenkov *et al.*, 1997; Matichenkov and Bocharnikova, 2001). However, this ratio could be varied by coming in or coming out of soluble silicon compounds, as well as by changing in mineralogical composition or in conditions for secondary minerals formation. Similar processes take place on the surface of soil microorganisms (Matichenkov *et al.*, 2001*b*).

The discussion about presence of organo-silicon compounds and soluble complexes with non-organic and organic ligands in soil solutions has been expounded in some publications as well (Fotiev, 1971; Matichenkov and Snyder, 1996).

5.3 Silicon Cycle in Soil–Plant System

Biological cycle of silicon on our Planet is most intensive in terrestrial ecosystems where plants uptake silicon in the range from 20 to 7000 kg Si / he / year (Matichenkov and Bocharnikova, 1994). Silicon is the 4th element in plant biomass, after oxygen, carbon, and hydrogen (Kovda, 1956; Perelman, 1975; Bazilevich, 1993).

As noted above, the plants and soil microorganisms are able to uptake only monomers of silicic acid and its anions (Yoshida, 1975; Ma, 2003). In higher plants, this process takes place through the roots and leaves. But, the silicon is distributing within the plant irregularly, according to needs of organism (see below).

Absorbed molecules of monosilicic acid may be polymerized and by this way participate in the formation of organo-silicon compounds (Figure 5.2). In their turn, polysilicic acids in plants are able to be dehydrated with forming of *phytholites* – amorphous silicon dioxide with complicate configuration. The formed phytholites are located within plant cells and in the intercellular space (Dobrovolsky *et al.*, 1988; Gol'eva, 2004).

It was shown that the size, structure, and quantity of phytholites depend not only on the presence of silicon forms available in soil, but also on the soil humidity, temperature, and on availability of other nutrient elements (Gol'eva, 2001; Hodson *et al.*, 2005). It should be noted that formation and rate of growth of phytholites in plants are controlled by specific proteins (Harrison, 1996; Perry and Keeling-Tucker, 2000). After plants death or their pulling-down, the biogenic silicon-forms (polysilicic acids, phytholites, organo-silicon compounds, *etc.*) are transported into soil where are subjected to processes of dilution and / or decomposition. The final product of the silicon cycle process is monosilicic acid, which is again involving into biological cycle and influences on a number of physical-chemical and biological soil properties (Matichenkov and Bocharnikova, 1994; Bashkin, 2005; Bocharnikova and Matichenkov, 2012).

It was distinguished *eluvial* and *accumulative* types of silicon cycles, in which biologically active silicon forms are either accumulated in soil or permanently are moving out from the soil (Bocharnikova and Matichenkov, 2012). Both processes are initiated by the anthropogenic and natural factors (Matichenkov and Ammosova, 1994; Matichenkov *et al.*, 1994; Bocharnikova and Matichenkov, 2012). It should be also noted the presence in Nature the phenomenon of *silicatization.* Under this process in the soil or in the ground takes place the replacement of carbon atoms (from the remains of plants and animals) by silicon atoms (Milnes and Twidale, 1983; Cid *et al.*, 1988). The opportunity of such replacing is conditioned by the high concentration of monosilicic acid in soil solution.

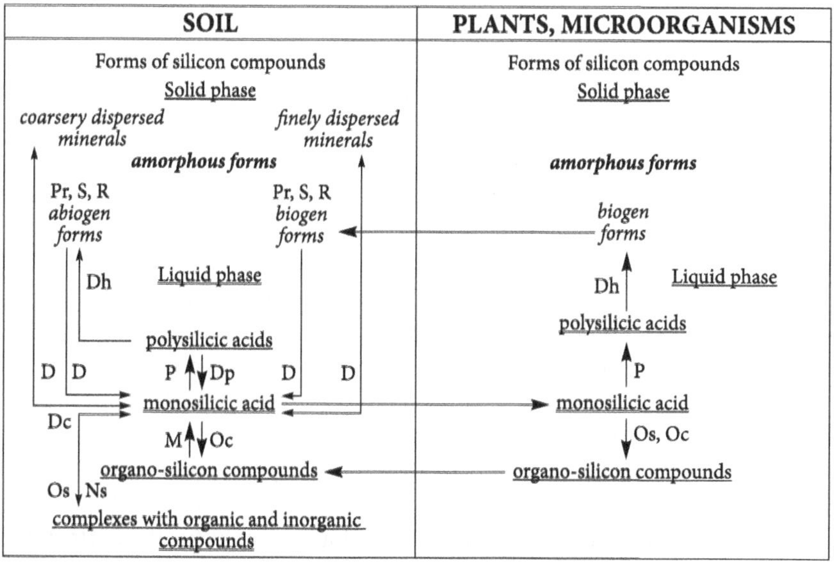

Fig. 5.2. Silicon cycle in system soil–plant–microorganisms (modified from Biel *et al.*, 2008)

Note: D – dissolving, Dh – dehydratation, Dp – depolymerization, Dc – decomposition of complexes, M – mineralization of organo-silicon compounds, Ns – formation complexes with inorganic compounds, Oc – formation of organo-silicon compounds, Os – formation complexes with organic compounds, P – polymerization, Pr – precipitation, R – replacement of inorganic anions, S – salt formation

5.4 Silicon and Flora

The intensity and type of silicon cycles in native *soil-plant* systems are connected not only with presence of active silicon forms in the parent material (Olier, 1990; Bashkin, 2005), landscapes, and secondary minerals (Kovda, 1956; 1985), but also with the type of plant association (Bazilevich, 1993; Matichenkov and Bocharnikova, 1994). For example, the plant associations with predominating of scanty silicon species such as pine forest and meadow non-crop grasses are formed on the active-silicon depleted soils (Lanning and Eleuterius, 1981; Bocharnikova and Matichenkov, 2012). On the other hand, the intensive silicon cycle embracing a lot of active silicon forms in soil, promotes formation of the crop associations or mixed wood (Bazilevich, 1993; Matichenkov and Bocharnikova, 1994). Accumulation and moving out of the soil silicon compounds determine succession of plant communities (Bocharnikova and Matichenkov, 2012). This determination causes the increase of stress sensitivity of silicon-accumulative plants at silicon-deficient conditions. And, on the other hand, the sufficient amount of active silicon forms grants for *silicophilic* plants more chances in competitive activity with silicon-non- accumulative species.

The main part of agricultural lands of the Planet has deficiency in active silicon forms (Matichenkov and Bocharnikova, 1994; Matichenkov *et al.*, 1999*a*; 1999*b*; Matichenkov, Calvert *et al.*, 2001; Matichenkov and Bocharnikova, 2004*a*; 2004*b*; Snyder *et al.*, 2006) due to principle of agricultural activity - removing nutrient compounds from the soil. Usually, deficiency of macroelements such as phosphorus (P), nitrogen (N), potassium (K), and calcium (Ca) is partially restored by mineral and organic fertilizers, but the lack of silicon is not compensated. It was estimated that movement of silicon from the ground is ranged between 30–7000 kg Si / ha / year. As a result, the upper soil horizon has a deficiency of active silicon; and, consequently, the trophic chain: *agricultural plants–farm animals–people* feels a serious deficiency of silicon.

5.4.1 Localization of Silicon in Plants

Distribution of silicon between plant organs is not equal and may vary from 0.001% in the pulp of fruit to 10–15% in the epidermal tissues (Voronkov *et al.*, 1978). According to the total silicon content, the plants are subdivided into two big groups: *silicophiles* [silicon-accumulative plants] and *non-silicophiles* [silicon-non-accumulative plants]. *Silicophiles* are the species, which have 1% or more percent of silicon per total dry mass (see, *e.g.*, Figure 5.3), when *non-silicophiles* have less than 1% of silicon per dry mass (Epstein, 1999; Ma and Takahashi, 2002). Usually, the main amount of silicon is located in needles, husks, bark, and some other organs (Ma and Takahashi, 2002; Snyder *et al.*, 2006). The high content of silicon was also shown in root caps (Hodson, 1986; Ma *et al.*, 2001), where monosilicic acid is polymerized into polysilicic acids, which further are crystallized into microcrystals of amorphous silica.

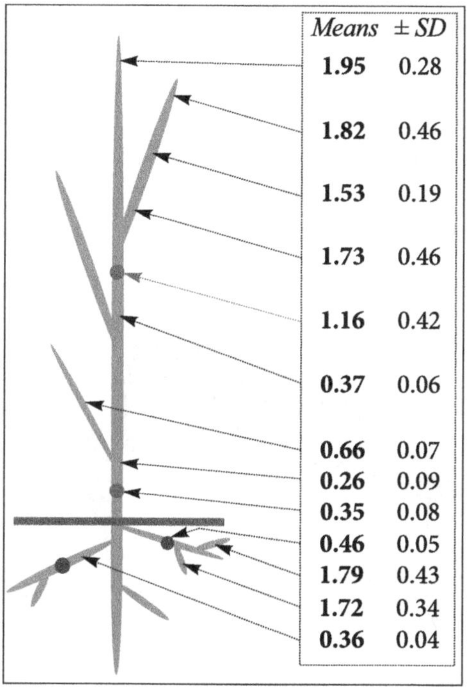

Fig. 5.3. Total silicon content in different parts of *Distichlis spicata* (n = 8), % from dry weight (Biel, Matichenkov *et al.*, 2018)
Note: Measurement of total silicon in *Distichlis spicata* was done according to (Eliot and Snyder, 1991)

It is possible that microcrystals of silica play a role of "mechanical drill" in the root system. Indirectly, it was confirmed by the fact that under deficiency of silicon nutrition the formation of the secondary and tertiary plant's roots is sharply decreased (Matichenkov *et al.*, 1999*b*; Matichenkov *et al.*, 2001*a*; Matichenkov and Bocharnikova, 2004*b*).

Absorbed molecules of monosilicic acid are accumulated in the epidermal tissues (Yoshida, 1975; Hodson and Sangster, 1989) and form the silicon-cellulose envelope where silicon is bonded with pectin and calcium (Waterkeyn *et al.*, 1982). As a result, the double cuticular layer (Figure 5.4) protecting and mechanically strengthening the plants is formed (Yoshida, 1975; Ma, 2003). This layer also prevents dehydration of the leaves (Emadian and Newton, 1989).

Perhaps higher plants have a special mechanism for selective uptake of monosilicic acid from the soil solution. Indirectly, this idea was confirmed in some experiments. For example, it was shown (Ma and Takahashi, 2002) that *Oryza sativa* L. and *Lycopersicum esculentum* Mill.

plants can reduce the concentration of monosilicic acid in the nutrient solution from 50 to 5 mg Si / liter if they grow hydroponically with the presence of silicon compounds. In our experiment, *Hordeum vulgare* L. absorbed monosilicic acid from soil solution (2–20 mg Si / liter) and collected this compound in roots up to 500–520 mg Si / liter of plant sap (Matichenkov *et al.*, 2005). Therefore, plants are really able to absorb monosilicic acid from the diluted solutions.

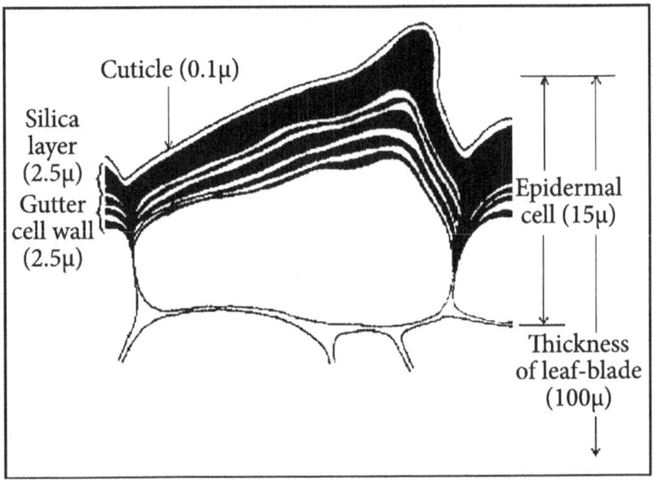

Fig. 5.4. Scheme of epidermal cell of the leaf of *Oryza sativa* (modified from Yoshida, 1975)

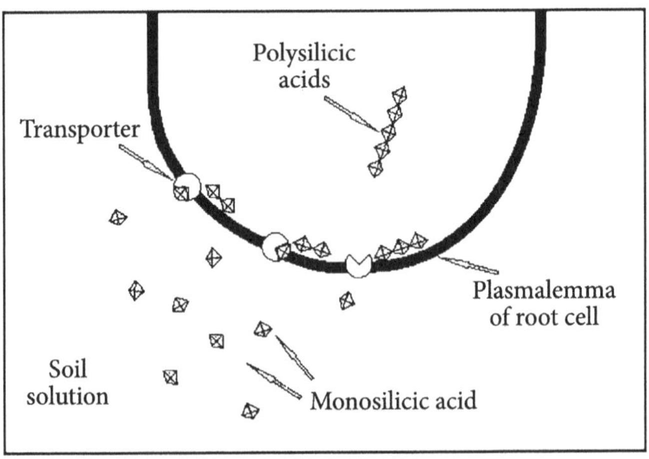

Fig. 5.5 Hypothetical scheme of silicon uptake by plants: molecules of monosilicic acid penetrate into the cells of root filaments and are immediately polymerized into polysilicic acids (modified from Biel *et al.*, 2008)

Table 5.1. Content of monosilicic and polysilicic acids in different parts of *Hordeum vulgare* L.

Part of the plant	Monosilicic acid		Polysilicic acids		Total silicon content, mg Si / kg dry mass
	mg Si / liter plant sap	% from total Si content	mg Si / liter plant sap	% from total Si content	
Leaf	100–400	1.0–5.0	2700–5000	30–55	9000–12000
Node of stem	130–200	1.6–2.5	3800–5000	40–62	8000–9000
Interstitial space	350–400	4.3–5.0	3500–5700	40–70	8000–9500
Root	500–520	3.0–4.0	5000–8000	30–53	15000–19000

Note: Plant material [fresh mass of 0.1–0.2 g] was ground with a mortar and pestle in distilled water at room temperature and centrifuged for 15 min at 6000 g [Centrifuge T-30, Janetzky, Germany] to precipitate colloids and solid particles. Concentration of monosilicic acid was immediately determined in the supernatant samples by the method developed by Drs. Mallen and Raily (see Iler, 1979). This method allows eliminating the effect of phosphate anions on the monosilicic acid measuring. Our modification allows to measure monosilicic and polysilicic acids separately. For this remained supernatant was treated as described earlier (Matichenkov *et al.*, 1997) to depolymerize polysilicic acids, and concentration of monosilicic acid was determined again as noted above. The difference between the second and the first measurements shows amount of polysilicic acids in tested solution. Concentrations of silicon compounds are expressed per liter plant sap; sap volume per kg fresh mass of the leaves was measured especially. Total silicon content was analyzed in the dried plant material according to Elliot and Snyder (1991). Each measurement was made in 8–12 independent replications.

Since the roots contain both monosilicic acid and polysilicic acids (Table 5.1), but the plants are not capable of absorbing polysilicic acids from solutions (Epstein, 1999; Ma and Takahashi, 2002), it is logical to assume that the polymerization of monosilicic acid molecules proceeds inside the root cells; and this process begins immediately after the absorption of monosilicic acid by the roots (Figure 5.5).

5.4.2 *Forms of Silicon in Plants*

Traditionally it was considered till now that about 90% of total silicon is presented in plants as amorphous silicon dioxide – phytholites and epidermal films (Savant *et al.*, 1997; Epstein, 1999; Ma and Takahashi, 2002).

This point of view was a subject of much controversy and was based on the "chemical rules" and chemical properties of silica (Iler, 1979). According to silicon chemistry, the occurrence of monosilicic acid and polysilicic acids of high concentrations is impossible in water or other solutions. In particular, it was known that at pH < 8 the amount of monosilicic acid in natural solutions is usually ranged between 0.1–100 mg Si / liter. Further, increase of the monosilicic acid concentration initiates its polymerization or its precipitation in a form of amorphous silica (Iler, 1979).

However, it was supposed that such conditions are not character for the intracellular and intercellular spaces of terrestrial plants (Medvedev, 2004), and therefore practically all polysilicic acids in plants should coagulate with forming of microcrystalline lattice of silicon dioxide (Iler, 1979; Yatsinin, 1994).

Because of that, despite the absence of actual data about real concentration of soluble silicon in plants, in plant physiology has been "created" the point of view that plants tissues have a very low concentration of mono- and polysilicic acids.

And only relatively recently, Japanese scientists demonstrated the existents of mono- and polysilicic acids at concentrations of 100–600 and 2000–3000 mg Si / liter, respectively, in the sap of *Oryza sativa* (Ma, 1990; Ma and Takahashi, 2002). The measurements presented in Table 5.1 (see also Matichenkov *et al.*, 2008) are consistent with these data. Really, 30–70% of the total silicon compounds in different organs of *Hordeum vulgare* represented polysilicic acids. Monosilicic acid in this experiment was about 1–5% of the total silicon content.

Thus, up to 75% of silicon in plant occurs in the form of water-soluble mono- and polysilicic acids. Undoubtedly, high concentration of these compounds is stipulating for important role of silicon in plant vital functions.

5.4.3 Silicon and Water Storage in Plants

Important property of silicon is the ability to accumulate and to store the water within organism throughout forming polysilicic acids and their gels. The model experiment showed that one atom of silicon is able to hold up to 146 molecules of water (Table 5.2). By this way about 20–30% of total silicon in organism may be involved into the process of maintaining internal water reserve. For example, under optimal silicon nutrition, the medium-latitude culture *Hordeum vulgare*, usually containing around 1.2–1.4% of silicon per dry mass, is able to store only with silicon assistance from 6 to 37 g water per 100 g fresh mass of individual plant.

Table 5.2. Content of water molecules and silicon atoms in different gels formed from silicic acid

Gel number	Si, % from total gel mass	H_2O, % from total gel mass	H_2O / Si, molecules / atom
1	1.28	97.9	146
2	1.60	97.7	119
3	2.40	97.1	82
4	3.20	96.6	63
5	4.80	95.3	40

Note: Standard silicic acid [16% silicon, product of TerraTech LLC] at 24 °C and atmospheric pressure of 765 mm mercury was mixed with water and mineral acids [HCI, H_2SO_4, and others] in different proportions until steady-state gels were formed. Final amounts of water molecules and silicon atoms in gels were calculated based on initial amounts of the compounds, which participated in the gel(s) formation, $n = 10$

Capacity of silicon to hold water is very important for tropical and arid plants, which are usually *silicophiles* (Ahmad *et al.*, 1992; V. Matichenkov pers. obs., 1986–2015). From this standpoint, it makes sense to assume, that *silicophiles* have a high tolerance to drought, in part, because during drought season they can use some "extra" water, which was collected in plants with silicon "assistance" during humid season. At least, some of our data are consistent with that.

In particular, the special experiment with woody tropical species *Hura crepitans*, *Hibiscus elatus*, *Ceiba pentandra*, and *Clitoria racemosa* maintained in the prolonged drought created in tropical mesocosm of Biosphere 2 Center Columbia University (Oracle, Arizona, USA) has argued for such a belief. The analyses showed that water potential (Rascher *et al.*, 2004) and total water content in leaves of these trees, even in the middle of the day(s) (Figure 5.6) were not decreased of the essence, when the rainy season was changed by the long-term season of the air and soil drought.

5.5 Silicon and Plants' Resistance to Extreme Environments

The number of publications, related to how active silicon compounds influence on plant growth and development, suggest the existence of the total mechanism, by which silicon improves plant stress-tolerance

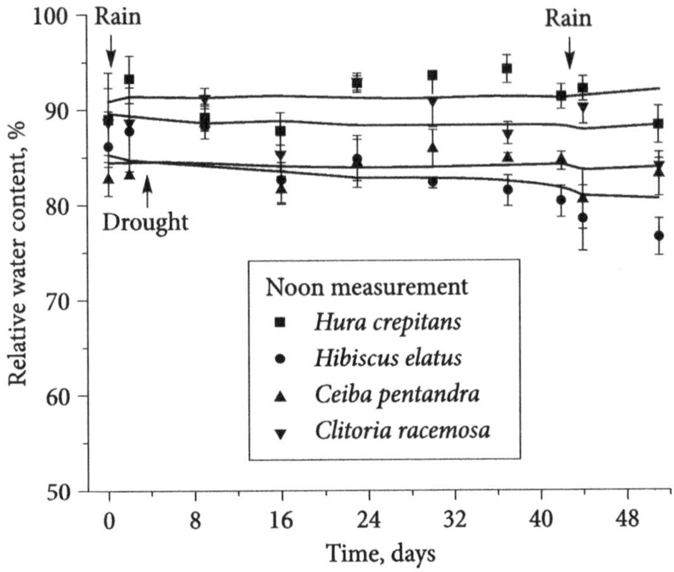

Fig. 5.6. Relative water content in the middle of the days in the leaves of tropical woody cultures *Hura crepitans, Hibiscus elatus, Ceiba pentandra,* and *Clitoria racemosa* during the long-term drought, maintained in tropical mesocosm of Biosphere 2 Center Columbia University (Oracle, Arizona, USA). Data points represent the mean ± SE, $n = 3 \div 4$ (modified from Biel *et al.*, 2008)

(Cherif *et al.*, 1994; Savant, *et al.*, 1997; Ma and Takahashi, 2002; Belanger, 2005). This mechanism seems to be so universal, that even only, for example, optimization of silicon nutrition, it will induce plant resistance against biogenic [infections, mycosis, insect injurious, *etc.*] and abiogenous [low or high temperature, salt stress, heavy metals, drought, *etc.*] factors (Bocharnikova *et al.*, 1999; Savant *et al.*, 1997; Matichenkov and Bocharnikova, 2004*a*; 2004*b*; Belanger, 2005; Snyder *et al.*, 2006).

There are several suppositions about protective roles of silicon in plants. In particular, such as thickening of epidermis, increasing chemical stability of DNA, RNA, and molecules of chlorophyll(s), functional activation of organelles, optimization of transport and redistribution of compounds within the plant, and others (Voronkov *et al.*, 1978; Aleshin, 1982; Aleshin *et al.*, 1987; Savant *et al.*, 1997; Ma and Takahashi, 2002; Matichenkov and Kosobryukhov, 2004; Matichenkov, Bocharnikova *et al.*, 2008) can support the involving of silicon into numerous adaptive responses of the plants.

But we speculate that all of these "adaptations" are the consequence of functioning of uniform protective mechanism, part of which is connected with the active silicon forms. Such opinion is indirectly supported by the fact that stress-tolerance of plants is increasing with rising silicon content within the tissues (Matichenkov *et al.*, 1999*b*; 2000). Moreover, in an experiment with *Distichlis spicata* growing at high salinity, the greatest amount of silicon is collected in the organs, which are more "suffering" from the salt stress (Matichenkov, Bocharnikova *et al.*, 2008). By other words, living organisms have a special ability to coordinate absorption and rapid redistribution of silicon into the parts of the body, exposed to stress (Biel, Matichenkov *et al.*, 2018).

5.6 Silicon as Matrix for Organic Compounds Synthesis

In the middle of the 20th century, Soviet scientists (Strelko *et al.*, 1963; Vysotsky *et al.*, 1967) had discovered and determined the possibility of catalytic synthesis of the organic compounds into the silica gel matrix. In this work, the removal of the synthesized molecules from silica gel matrix and conservation of the matrix allowed to realize non-limiting low-temperature synthesis of initial molecules from more primitive organic "row material".

On the basis of that Professor Vladimir V. Strelko and co-workers (1963) even advanced an idea that "*...the reactions of amino acids polycondensation, concentrated by their adsorption on the silica surface, could lead to formation of pre-proteins and might play an important role in origin of life on the Earth*".

In the modern organic chemistry, the polycondensation technology is developed in greater detail. But the main principle is the same: if certain organic molecules present in solution during the formation of silica gel, its surface as a matrix retains "imprints" of those template molecules. Then, if template molecules are removed, their "prints" on the surface of the silica gel become the matrix or the center of replication for synthesis of template molecules.

It is important to note, that the silica gel matrixes are playing the role of catalyst in a number of organic synthesis reactions. Some of them are: cyclization, rearrangements, reduction and / or oxidation, condensation, hydratation and / or dehydratation, formylation, and others (Banerjee *et al.*, 2001).

At present time, many chemical and pharmaceutical companies are using this approach for synthesis of complex molecules.

5.6.1 *Hypothesis on Silicon Participation in Protection of Living Organisms under Stress Conditions*

5.6.1.1 *Premises of hypothesis*

The possibility of biochemical compounds synthesis into the gels of polysilicic acids within living cell was not considered until now. The general reason for that, as mentioned above, was conception, that in living organisms the polysilicic acids are existing in very low concentrations, or even do not exist totally.

However, some of our new data cast doubt on that conclusion. Especially convincing are the results related to location and amount of mono- and polysilicic acids in the shoots of *Distichlis spicata*, — one of the most silicophilic species (Biel *et al.*, 2008; Matichenkov, 2008; Matichenkov *et al.*, 2008; Biel, Matichenkov *et al.*, 2018). In particular, the dynamics of washing out of two water-soluble silicon forms, monosilicic acid and polysilicic acids, from the leaves and stems of *Distichlis spicata* plants, which were grown with and without the application of plant available silicon, is presented in Figures 5.7 and 5.8. Even when the substrate is depleted of the active silicon, plants still contain certain amounts of monosilicic acid and polysilicic acids in the apoplast of leaves and stems. Further, during the experiment, careful grinding plant samples (see note in Figure 5.7) made it possible to extract much greater amount of monosilicic acid (Figures 5.7A and 5.8A) and polysilicic acids (Figure 5.7B and 5.8B) from symplastic area compared with apoplast. The biggest difference between the apoplastic and symplastic fractions was observed for polysilicic acids in the stem (Figure 5.8B). Application of plant-available silicon led to doubling the monosilicic acid content in the stem apoplast (Figure 5.8A) and more than two-fold increase in the polysilicic acids content in the leaf symplast (Figure 5.7B).

Thus, this test allows concluding that significant amount of mono- and polysilicic acids are located in the apoplast and symplast of the *Distichlis spicata* leaves and stems.

The accumulated soluble forms of silicon can be redistributed between plant organs under stress (Biel *et al.*, 2008; Matichenkov, 2008). But here it is important to note, that shoots of plants are able to accumulate the water-soluble silicon forms in amount, sufficiently enough for silicon creative job in organism.

The additional pre-suppositions for organic compounds synthesis on gels of polysilicic acids in living cell are following:

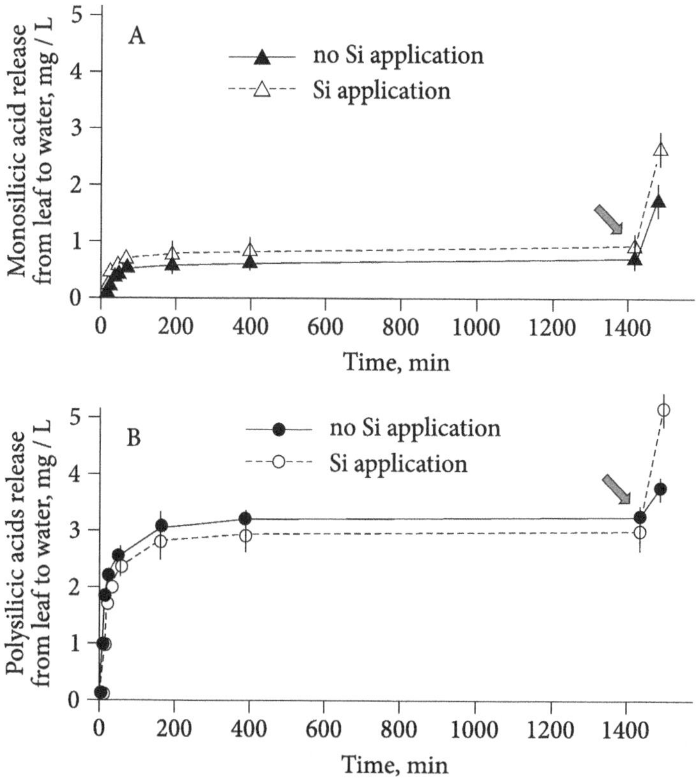

Fig. 5.7. Dynamics of washing out of mono- (A) and polysilicic (B) acids from the leaves of *Distichlis spicata* grown without silicon fertilizers (▲, •) and with silicon fertilization (Δ, O), added to substrate 12 days before measurements (modified from Biel *et al.*, 2008)

Note: Plants were grown in plastic pots (20 × 50 × 20 cm) with sand at temperatures of 22–24 °C / 17–19 °C [day / night], and natural illumination. Soil salinity was simulated by irrigation of water with 10 g / liter of NaCI. Plants were watered by regular water containing no more than 0.10–0.15 mg Si / liter. Plant-available silicon (MiracleGro: www.miracle-gro.com) was added into ground 1–2 times per month. Fresh plant material (leaves here) were cut into fragments 2.0–2.5 cm in length (the total weight of the samples was the same), put into flasks containing distilled water at a ratio of 1 / 100 by weight, and shaken. The apoplast content of the silicic acids was measured in a series of 1- to 5-ml aliquots sampled 0, 1, 5, 15, 30, 60, 180, 360, and 1440 minutes later. The samples of plant material obtained after filtering the remaining solutions were grinding and mixed with the filtrates again. After that, the suspensions were shaken for 60 min, and the amounts of mono- and polysilicic acids were measured once more. Measurements were made as described in the Note to Table 5.1. Arrows show the moment of grinding of the leaves and further returning leaf homogenates into initial water-soluble fractions. Data points represent the mean ± SE, $n = 4$

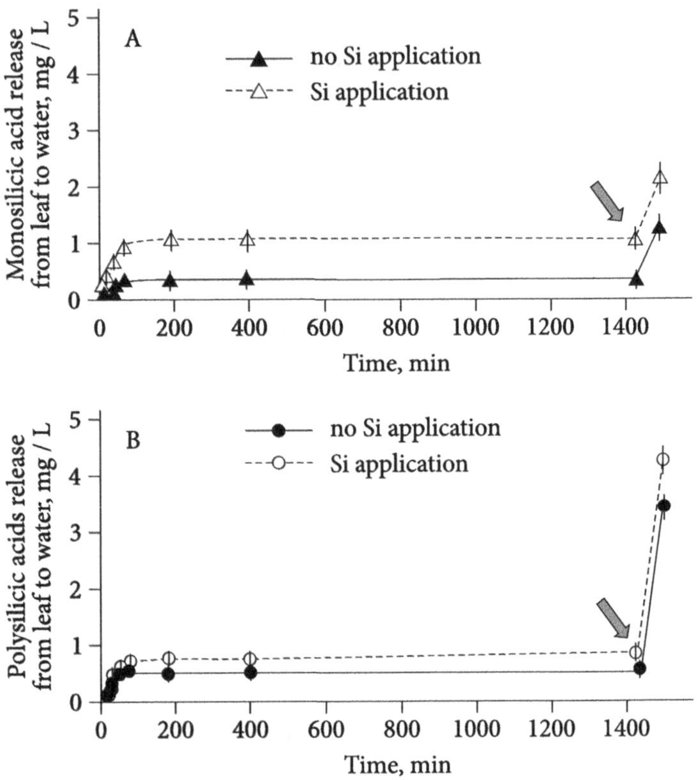

Fig. 5.8. Dynamics of washing out of mono- (A) and polysilicic (B) acids from the stems of *Distichlis spicata* grown without silicon fertilizers (▲, •) and with silicon fertilization (Δ, O), added to substrate 12 days before measurements (modified from Biel *et al.*, 2008)

Note: Experiment was carried out as described above (see Note to Figure 5.7). Arrows show the moment of grinding of the stems and further returning stem homogenates into initial water-soluble fractions. Data points represent the mean ± SE, $n = 4$

- Active silicon forms coming into living systems increase their stress-tolerance (Cherif *et al.*, 1994; Savant, 1997; Epstein, 1999; Ma and Takahashi, 2002; Matichenkov and Bocharnikova, 2004*a*; 2004*b*; Belanger, 2005);
- It has been found a specific protein (transporter), which regulates absorption of silicon from environment, and its redistribution within the plants (Ma *et al.*, 2006);
- Plants absorb silicon from the environment only in the form of monosilicic acid (Yoshida, 1975; Ma and Takahashi, 2002);

- Depending on plant species and type of tissue, the total silicon content in plants varies, on average, from 0.2 to 1.5% of dry mass (Epstein, 1999; Ma and Takahashi, 2002);
- In plants, the silicon exists in the forms of mono- and polysilicic acids, amorphous silica [phytholites], and organo-silicon compounds. The percentage of mono- and polysilicic acids may be in the range of about 1–5% (if not more), and 30–70% of the total silicon content, respectively (see Table 5.1);
- Water-soluble silicon compounds in plants purposefully redistributed directly into the parts and / or organs, which are exposed to stress (Matichenkov *et al.*, 2000; Matichenkov and Bocharnikova, 2004*a*; 2004*b*; Matichenkov, 2008);
- Polysilicic acids gels are able to form matrixes, imprinting in themselves the form of organic "molecule-formers". These matrixes can provide low-temperature synthesis of new "molecule-formers" from simpler precursor molecules (Strelko *et al.*, 1963; Vysotsky *et al.*, 1967; Banerjee *et al.*, 2001).

5.6.1.2 Hypothesis

During endeavor (ontogenesis) of total realization of genetic information, the organisms (Biel *et al.*, 1990; Biel and Fomina, 2005) usually feel deficiency of nutrient elements or energetic resources, or are affected by negative climatic factors, diseases, and other stressful pressures. At such conditions, the living systems are spending substantial part of energy for maintenance, adaptation and defense against internal and external problems. As a result of that, the plants [and other organisms as well] are usually slowed in development, and do not realize interior potential. But, the situation may be improved by supplying organism with essential nutrition and by creating optimal environment and internal conditions.

However, in real life this way is not practicable, especially in a large-scale agriculture. Improvement could be achieved, if the responsibility for the part of repairing, metabolic and protective functions would be shifted from genetic apparatus to the other system(s).

Using the plants as an example, we'll consider one of such system, although it is most likely that this mechanism is universal and functioning in all organisms, which are living now on our Planet.

Silicon uptake – Molecules of monosilicic acid located in soil penetrate across the root's plasmalemma [cell "sluice"] inside of the cells and are immediately condensed into active chains of polysilicic acids (Figure 5.5). Then the silicon compounds are distributed within all plant.

Silicon distribution – Some part of silicon compounds goes to epidermal layer, root caps, cell walls, and other tissues and organs for synthesis of phytholites and other silicon-containing structures. The second part of silicon is used to form polysilicic acids gels – the basis of further low-temperature synthesis of organic compounds. The third part of absorbed silicon is preserved and stored at the normal conditions as polysilicic acids or their gels with non-active surface within the cells and in the intercellular space.

Synthesis of organic compounds on the polysilicic acid gel matrixes at optimal conditions – Silicon compounds located inside of the cells are shaped into polysilicic acid gels. At certain moment, the gel is activated. Namely, the template molecules (*e.g.*, proteins, simple or complicated protective compounds, and so on) are imprinted thereon, thus creating a polysilicic acid gel matrix for the replication of the pattern molecules (Figure 5.9). After printing and moving out replicating substance, activated polysilicic acids gel as a "gel-plate" becomes to be able to carry out the catalytic synthesis of copies of the template molecules from simpler building blocks and components that are located in cytoplasm.

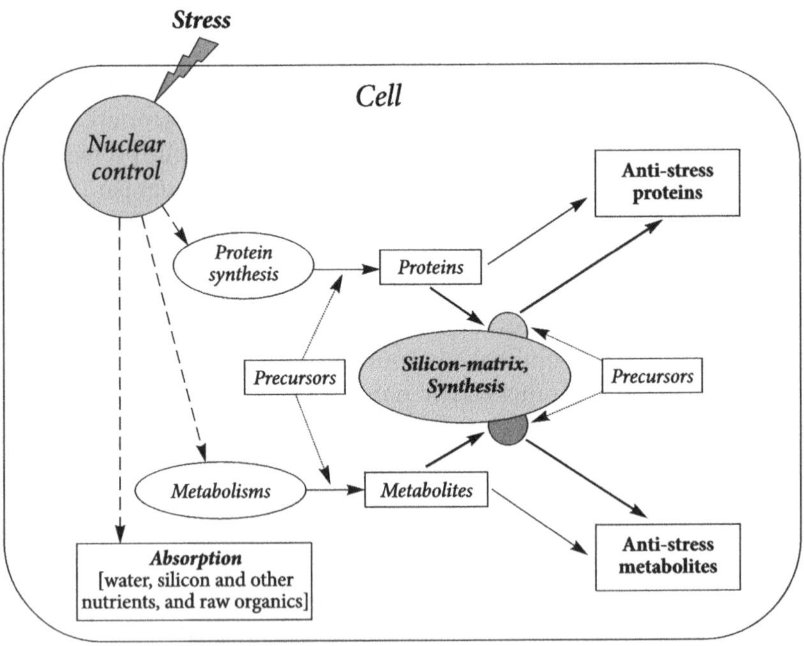

Fig. 5.9. Hypothetical scheme of the universal auxiliary stress-protective mechanism of living systems with participation of the movable silicon compounds (modified from Biel *et al.*, 2008)

Silicon-dependent synthesis of protective compounds during stress – After external irritation, the signal system is switching on the mechanism of stress identification. Simultaneously, organism additionally uptakes the silicon from the environment and transports it to stressed area. Information about the stress that is transmitted to a signal system, induces an adequate response in the genome, provides a synthesis of the necessary protective compounds such as stress-induced proteins, antioxidant enzymes, antioxidants and other substances (Figure 5.9). Synthesized stress protective molecules not only play their protective roles, but are also used as templates for synthesis of additional copies into the polysilicic acid gel matrixes.

When the organism is under extreme stress or is exposed to various stressors simultaneously, the rate and quantity of synthesizing stress-response-material(s) may be not enough to solve the problem because some other vitally-important functions should be done in organism at the same time. As a result of escalating energy deficiency and informational resources, the synthesis of "routine" compound, essential for cells is slowed down or even ceased.

We suppose that in living cell some of the protective compounds can activate polysilicic acid gel by printing their replicas on the gel surface as template molecules. Then, the template molecules are transported to stressed-zones. At the end, the polysilicic acid gel matrix begins to clone (to print) the same molecules from the "at hand material" of cytoplasm. By this way the polysilicic acid gel matrixes give to organism the possibility (on a background of the stress) to release informational-commanding resources and part of energy for cell functioning in former "before-stress" regime.

Thus, additional synthesis of protective compounds can carry out into polysilicic acids gel matrixes without direct participation of the genetic apparatus.

5.7 New Technologies

High pollution level of the environment and other negative consequences of human civilization bring up a question about the manner and conditions of humanity's survival. Application of synthetic drugs and pesticides initiates a number of negative consequences, which now are sensed by everybody. For example, antibiotics are not only doing to death of pathogenic microorganisms, but also to destroying the customer's immune system. Fungicides kill in soil and on the plants surface, together with pathogenic fungus, the useful organisms. Absolutely unwarrantable to apply in agriculture insecticides: with trying to protect the plants against insects, people provide penetration of the toxins inside of animals and, as a consequence – to human body. And, it is absolutely unallowable [in modern stage of genetic engineering development] to introduce genetically modified organisms into agriculture, food industry, pharmacology, and other fields because such organisms are the harbor for a number of unstudied and frequently negative consequences for humanity and Nature.

The way out from this situation, in some cases, could be the use of silicon compounds, although we are far away from the conclusion that silicon is "panacea from all troubles". Competent application of silicon compounds may help to solve some large-scale problems, and to modify whole series of economically-important technologies.

Below some of the really solvable actual problems are presented:

- Planetary problem of fresh water deficiency [the most of which is used for irrigation] can, in part, be solved by corrective application of silicon fertilizers, because the "activated" silicon facilitates decrease of irrigation water amount on 30–50% without diminution of quantitative and qualitative crops characteristics (Matichenkov and Kosobryukhov, 2004; Snyder *et al.*, 2006);
- The application of silicon compounds, which contribute to the resistance of plants to infection (and are often more effective than fungicides and insecticides) will reduce the use of pesticides in agriculture (Datnoff *et al.*, 1991; Savant *et al.*, 1997);
- The application of silicon fertilizers promotes plant tolerance to high / low temperatures, saline intoxication, and *etc.* and, as a result permit to get guaranteed yields enriched with silicon compounds under extreme environment (Matichenkov and Kosobryukhov, 2004; Snyder *et al.*, 2006);
- Consumption of the foodstuffs with high level of silicon compounds makes people and animals more resistant to diseases and unfavorable environment. And besides, the overdose of silicon compounds is not possible in principle, because the extra dose is transforming into inert form – silica, which quickly moves out from the organism (Voronkov *et al.*, 1978; Iler, 1979; D'yakov *et al.*, 1990).

5.8 General Conclusion

Silicon increases the level of resistance of living systems to **any** stressor(s), and does not toxically influence the organism. With active silicon forms is possible purposefully with low cost and without ***interfering*** the structure of genetic apparatus, to create conditions, which are necessary for optimal use of the Program of Realization of Genetic Information (Biel *et al.*, 1990; Biel and Fomina, 2005).

Our proposition can be one of the directing criteria for detection of the way(s) "to switch on" the silicon-protective function in cell with using natural materials and internal reserves of organism.

Acknowledgments

We thank Dr. Elena A. Bocharnikova (Institute of Physical and Chemical Soil-Science Problems, Russian Academy of Science, Pushchino, Moscow Region, Russia) for participation in some experiments.

References

Ahmad R., Zaheer S., Ismail S. Role of silicon in salt tolerance of wheat (*Triticum aestivum L.*), *Plant Sci.*, 1992, Vol. 85, p. 43–50.

Al-Wajeeh K. *Studies on the Microbiology of Silicon*, Ph.D. Thesis, University of Sheffield, England, 1999.

Aleshin N.E. Silicon content in rice RNA, *Soviet Agr. Sci.*, 1982, Vol. 6, p. 6–7. [in Russian]

Aleshin N.E., Avakyan E.P., Dyukunchak S.A., Aleshin E.P., Barushok V.P., Voronkov M.G. The role of silicon in protection of rice against diseases, *Proc. Acad. Sci. USSR*, 1987, Vol. 291, No. 2, p. 217–219.

Alvarez R., Sparks D.L. Polymerization of silicate anion in solution at low concentrations, *Nature*, 1985, Vol. 318, p. 649–651.

Anokhin S.M., Panichev A.M., Bgatov V.I. Basics of adaptation of animal organisms to alternative geochemical environment, *Natural Materials for Man's Benefit*, Novosibirsk, 1997, p. 141–142. [in Russian]

Banerjee A.K., Laya Mimo M.S., Vera Vegas W.J. Silica gel in organic synthesis, *Russian Chemical Reviews*, 2001, Vol. 70, No. 11, p. 971–990.

Bashkin V.N. *Biogeochemistry*, Nauchnii Mir, Moscow, 2005, 584 p. [in Russian]

Bazilevich N.I. 1993 *Biological Productivity of Ecosystems in Northern Eurasia*, Nauka, Moscow, 1993, 293 p. [in Russian]

Belanger R.R. The role silicon in plant-pathogen interaction: toward universal model, *Proc. 3rd Silicon Agricultural Conf.*, Korndorfer G.H. ed., October 22–26, 2005, Universidade Federal de Uberlândia, Uberlândia, 2005, p. 34–40.

Bgatov V.I., Motovilov K.Y., Speshilova M.A. Functions of natural minerals in metabolic processes of the farming poultries, *Agr. Biol.*, 1987, No. 7, p. 98–102. [in Russian]

Biel K.Y., Fomina I.R. Self-regulation in plant living systems: methanol as example of volatile mediators providing stress-tolerance, hypothesis, *The Complex Systems under Extreme Conditions*, Khlebopros R.G., Soukhovolsky V.G. eds., Krasnoyarsk, Russia, 2005, p. 116–133.

Biel K.Y., Kolmakov P.V., Lastivka A.N., Titlyanov E.A. Photosynthetic carbon metabolism in ontogenesis of green algae *Ulva fenestrate*, *Marine Biol.*, 1990, Vol. 5, p. 52–58. [in Russian]

Biel K.Y., Matichenkov V.V., Bocharnikova E.A., Fomina I.R. Location and redistribution of silicon within different organs of *Distichlis spicata* and *Hordeum vulgare* under high salinities, 2018. [in preparation]

Biel K.Y., Matichenkov V.V., Fomina I.R. Protective role of silicon in living systems, *Functional Foods for Chronic Diseases. Advances in the Development of Functional Foods*, Martirosyan D.M. ed., D&A, Inc. 580 W Arapaho Rd., Suite 130, Richardson, TX 75080, USA, 2008, Vol. 3, p. 208–231. http://www.functionalfoodscenter.net

Birchall J.D., Chappel J.S. The chemistry of aluminium and silicon in relation to Alzheimer's disease, *Clin. Chem.*, 1988, Vol. 23, p. 154–156.

Bocharnikova E.A., Matichenkov V.V. Influence of plant associations on the silicon cycle in the soil-plant system, *Appl. Ecol. Environ. Res.*, 2012, Vol. 10, No. 4, p. 547–560.

Bocharnikova E.A., Matichenkov V.V., Pinsky D.L. The influence of soluble silica acids on behavior of heavy metals in soil and natural waters, *Pollution in Large Cities*, Proc. World-Wide Symp. Venice-Padova, Italy, 1995, p. 43–50.

Bocharnikova E.A., Matichenkov V.V., Snyder G.H. The management of heavy metal behavior and mobility in the soil-plant system, *Proc. 31st Mid-Atlantic Industrial and Hazardous Waste Conf.*, Nikolaidis N., Erkey C., Smets B.F. eds., June 1999, Technomic Publ. Comp., Inc., Lancaster, Pennsylvania, 1999, p. 621–630.

Bocharnikova E.A., Matichenkov V.V., Matichenkov I.V. Silicon fertilizers and soil amendments: history, theory and practice, *Agrochem.*, 2011, Vol. 7, p. 84–96. [in Russian]

Brown C.A. The use of silicon gel for treating children's burn scars in Saudi Arabia: a case study, *Occup. Ther. Inst.*, 2002, Vol. 9, No. 2, p. 121–130.

Brummer F. Living inside a glass box - silica in *Diatoms, Silicon Biomineralization*, Muller W.E.G. ed., Springer, Berlin, 2003, p. 3–10.

Camilleri C., Markich S.J., Noller B.N., Turley C.J., Parker G., van Dam R.A. Silica reduces the toxicity of aluminium to a tropical freshwater fish (*Mogurnda mogurnda*), *Chemosphere*, 2003, Vol. 50, No. 3, p. 355–364.

Carlisle E.M. Silicon. Essential element for the chick, *Science*, 1972, Vol. 178, p. 619–621.

Carlisle E.M. Silicon, *Biochemistry of the Essential Ultratrace Elements*, Frieden E. ed., Plenum, New York, 1984, p. 257–291.

Carlisle E.M. Silicon, *Handbook of Nutritionally Essential Mineral Elements*, O'Dell B.L., Sundle R.A. eds., Marcel Dekker, Inc., New York, 1997, p. 603–618.

Carlisle E.M., McKeague J.A., Siever R., van Soest P.J. Silicon, *Trace Elements Related to Health and Disease*, Hopps H.C. ed., Geochem. and Environ., Natl. Acad. Sci., Washington, 1977, p. 54–72.

Cha J.N., Shimizu K., Zhou Y., Christianssen S.C., Chmelka B.F., Stucky G.D., Morse D.E. Silicate in filaments and subunits from marine sponge direct the polymerization of silica and silicones *in vitro*, *Proc. Natl. Acad. Sci. USA*, 1999, Vol. 96, p. 361–365.

Cherif M., Asselin A., Belanger R.R. Defense responses induced by soluble silicon in cucumber roots infected by *Pythium* spp., *Phytopathology*, 1994, Vol. 84, p. 236–242.

Cid M.S., Detling J.K., Brizuela M.A., Whicker A.D. Patterns in grass silicification: response to grazing history and defoliation, *Oecologia*, 1988, Vol. 80, p. 268–271.

Datnoff L.E., Raid R.N., Snyder G.H., Jones D.B. Effect of calcium silicate slag on blast and brown spot intensities and yields of rice, *Plant Dis.*, 1991, Vol. 75, p. 729–732.

Davies S. Silica in streams and ground water, *Am. J. Sci.*, 1964, Vol. 262, p. 870–876.

Dietzel M. Dissolution of silicates and the stability of polysilicic acid, *Geochim. Cosmochim. Acta*, 2000, Vol. 64, No. 19, p. 3275–3281.

Dietzel M. Interaction of polysilicic and monosilicic acid with mineral surfaces, *Water-Rock Interaction*, Stober I., Bucher K. eds., Kluwer, Dordrecht, 2002, p. 207–235.

Dobrovolsky G.V., Bobrov A.A., Gol'eva A.A., Shoba S.A. Opalescent phytholites in taiga biogeocenosis of middle taiga, *Biol. Sci.*, 1988, Vol. 2, p. 96–101. [in Russian]

D'yakov V.M., Matichenkov V.V., Chernyshev E.A., Ammosova Y.M. *Using of Silicon Compounds in Agriculture. Actual Problems of Chemical Science, Technology, and Preservation of the Environment* (Issue 7), Publ NIITEKhIM, Moscow, 1990, 32 p. [in Russian]

Elliot C.L., Snyder G.S. Autoclave-induced digestion for the colorimetric determination of silicon in rice straw, *J. Agric. Food Chem.*, 1991, Vol. 39, p. 1118–1119.

Emadian S.F., Newton R.J. Growth enhancement of loblolly pine (*Pinus taeda* L.) seedlings by silicon, *J. Plant Physiol.*, 1989, Vol. 134, No. 1, p. 98–103.

Epstein E. Silicon, *Annu. Rev. Plant Physiol. Plant Mol. Biol.*, 1999, Vol. 50, p. 641–664.

Exley C., Pinnegar J.K., Taylor H. Hydroxyaluminosilicates and acute aluminium toxicity in fish, *J. Theor. Biol.*, 1997, Vol. 189, No. 2, p. 133–139.

Fotiev V.A. About nature of aqueous humus, *Dokl. AN SSSR (Proc. Acad. Sci. USSR)*, 1971, Vol. 1199, No. 1, p. 198–201. [in Russian]

Frolov Y.G. *Course of Colloid Chemistry. Surface Phenomena and Dispersion Systems*, Alliance Publ., Moscow, 2004, 464 p. [in Russian]

Gol'eva A.A. *Phytholites and Their Informational Role in Investigation of Natural and Archeological Objects*, Moscow–Syktyvkar–Elista, 2001, 120 p. [in Russian]

Gol'eva A.A. Biogeochemistry of amorphous silica in plants and soils, *Soils, Biogeochemical Cycles, and Biosphere*, Glazovsky N.F. ed., Tov. Nauch. Izd. KMK, Moscow, 2004, p. 137–159. [in Russian]

Harrison C.C. Evidence for intramineral macromolecules containing protein from plant silicas, *Phytochemistry*, 1996, Vol. 41, p. 37–42.

Hodson M.J. Silicon deposition in the roots, culm and leaf of *Phalaris canariensis* L., *Ann. Bot.*, 1986, Vol. 58, No. 2, p. 167–177.

Hodson M.J., Sangster A.G. X-ray microanalysis of the seminal root of sorghum bicolor with particular reference to silicon, *Ann. Bot.*, 1989, Vol. 64, No. 6, p. 659–675.

Hodson M.J., White P.J., Mead A., Broadley M.R. Phylogenetic variation in the silicon (Si) composition of plants, *Ann. Bot.*, 2005, Vol. 96, p. 1027–1046.

Iler R.K. *The Chemistry of Silica*, Wiley, New York, 1979.

Jugdaohsingh R., Anderson S.H., Tucker K.L., Elliot H., Kiel D.P., Thompson R.P., Powell J.J. Dietary silicon intake and absorption, *Am. J. Clin. Nutr.*, 2002, Vol. 75, No. 5, p. 887–893.

Jurkic L.M., Cepanec I., Pavelic S.K., Pavelic K. Biological and therapeutic effect of ortho-silicic acid and ortho-silicic acid-releasing compounds: New perspectives for therapy, *Nutrition & Metabolism*, 2013, Vol. 10, p. 2–12.

Kaufmann K. *Silica: the Amazing Gel*, Alive Books, 1995*a*, 208 p.

Kaufmann K. *Silica: the Forgotten Nutrient*, Alive Books, 1995*b*, 106 p.

Kovda V.A. Mineral composition of plants and forming of soil, Soil Sci., 1956, Vol. 1, p. 6–38. [in Russian]

Kovda V.A. *Soil Biogeochemistry*, Nauka, Moscow, 1985, 263 p. [in Russian]

Kuzin N.M., Davani S.A., Kashevarov S.B., Leont'eva M.S., Koroleva I.M., Gorbunov A.S., Guznov I.G., Zainiddinov F.A. Using of adjustable silicon bandage in horizontal gastroplasty for patients with morbid adiposity, *Khirurgiya (Surgery)*, 2000, Vol. 10, p. 16–19. [in Russian]

Lanning F.C., Eleuterius L.N. Silica and ash in several marsh plants, *Gulf Research Reports*, 1981, Vol. 7, p. 47–52.

Leibigh J. *Chemistry in Its Application to Agriculture and Plant Physiology* (From the manuscript of the author by Lyon Playfair), Taylor and Walton, London, 1840, 345 p.

Ma J.F. *Studies on Beneficial Effects of Silicon on Rice Plants* (PhD Thesis), Kyoto University, Kyoto, 1990.

Ma J.F. 2003 Function of silicon in higher plants, *Progress in Molecular and Subcellular Biology*, Muller W.E.G. ed., Springer-Verlag, Berlin, Heidelberg, Vol. 33, p. 127–147.

Ma J.F., Goto S., Tamai K., Ichii M. Role of root hairs and lateral roots in silicon uptake by rice, *Plant Physiol.*, 2001, Vol. 127, p. 1773–1780.

Ma J.F., Takahashi E. *Soil, Fertilizer, and Plant Silicon Research in Japan*, Elsevier, The Netherlands, 2002, 281 p.

Ma J.F., Tamai K., Yamaji N., Mitani M., Konishi S., Katsuhara M., Ishiguro M., Murata Y., Yano M. Silicon transporter in rice, *Nature*, 2006, Vol. 440, p. 688–691.

Martin-Jezequel V.., Lopez P.J. Silicon – a general metabolite for diatom growth and morphogenesis, *Silicon Biomineralization*, Muller W.E.G. ed., Springer, Berlin, Germany, 2003, p. 99–126.

Matichenkov V.V. *The Role of Soluble Silicon Compounds in Plant and Soil-Plant System*, Dr. Sci. Dissertation, Institute of Basic Biological Problems of the Russian Academy of Sciences, Pushchino, 2008, 313 p. [in Russian]

Matichenkov V.V., Ammosova Y.M. Effects of amorphous silica on some properties of Sody-Podzolic soils, *Soil Sci.*, 1994, Vol. 7, p. 52–61. [in Russian]

Matichenkov V.V., Ammosova Y.M., Bocharnikova E.A. A method for measuring of the available for plants silicon forms in soil, *Agrochem.*, 1997, Vol. 1, p. 76–84. [in Russian]

Matichenkov V.V., Ammosova Y.M., Bocharnikova E.A. Influence of silicon fertilizers on plants and soil, *Agrochem.*, 2001*a*, Vol. 12, p. 30–38. [in Russian]

Matichenkov V.V., Bocharnikova E.A. Total and partial biogeochemistry cycle of Si in various ecosystems, *Sustainable Development: The View from the Less Industrialized Countries*, Monge-Najera J., San Jose eds., UNED, Costa Rica, 1994, p. 467–481.

Matichenkov V.V., Bocharnikova E.A. The relationship between silicon and soil physical and chemical properties, *Silicon in Agriculture*, Datnoff L.E., Snyder G.H., Korndorfer G.H. eds., Elsevier, Amsterdam, The Netherlands, 2001, p. 209–219.

Matichenkov V.V., Bocharnikova E.A., eds. *Silicon in Food, Agriculture and Environment*, Proc. Int. Conf., 2–5 August 2004, Pushchino, Russia, 2004*a*, 64 p.

Matichenkov V.V., Bocharnikova E.A. Si in horticultural industry, *Production Practices and Quality Assessment of Food Crops*, Vol 2, *Plant Mineral Nutrition and Pesticide Management*, Dris R., Jain S.M. eds., Kluwer Academic Publ., Amsterdam, 2004*b*, p. 217–228.

Matichenkov V.V., Bocharnikova E.A., Kosobryukhov A.A., Biel K.Y. Mobile forms of silicon in plants, *Dokl. Biol. Sci.*, 2008, Vol. 418, No. 2, p. 39–40.

Matichenkov V.V., Calvert D.V., Snyder G.H. Comparison study of soil silicon status in sandy soils of south Florida, *Soil Crop Sci. Soc. Fla. Proc.*, 1999*a*, Vol. 59, p. 132–137.

Matichenkov V.V., Calvert D.V., Snyder G.H. Silicon fertilizers for citrus in Florida, *Proc. Fla. State Hort. Soc.*, 1999*b*, Vol. 112, p. 5–8.

Matichenkov V.V., Calvert D.V., Snyder G.H. Prospective silicon fertilization for citrus in Florida, *Soil Crop Sci. Soc. Fla. Proc.*, 2000, Vol. 59, p. 137–141.

Matichenkov V.V., Calvert D.V., Snyder G.H. *Minimizing Nutrient and Pollutants Leaching from Sandy Agricultural Soils and Optimization of Plant Nutrition* (Final Report), DEP Direct Contract Fort Pierce, IRREC, University of Florida, FL, USA, 2005, 494 p.

Matichenkov V.V., Calvert D., Snyder G.H., Bocharnikova E.A. Effect of Si fertilization on growth and P nutrition of bahiagrass, *Soil Crop Sci. Soc. Fla. Proc.*, 2001, Vol. 60, p. 30–36.

Matichenkov V.V., Korenevsky A.A., Beveridge T.J. Adsorption of soluble silicon compounds by *Pseudomonas aeruginosa*, *Proc. V.M. Goldschmidt Conf.*, May 20–24, 2001, Hot Springs, Virginia, USA, 2001*b*, p. 3831.

Matichenkov V.V., Kosobryukhov A.A. Si effect on the plant resistance to salt toxicity, *Proc. 13th Int. Soil Conservation Organization Conf. (ISCO)*, 4-9 July, 2004, Brisbane, Australia, 2004, p. 287–295.

Matichenkov V.V., Pinsky D.L., Bocharnikova E.A. Influence of soil mechanical solidity on the state and forms of available silicon, *Soil Sci.*, 1994, Vol. 11, p. 71–76. [in Russian]

Matichenkov V.V., Snyder G.H. Mobile silicon compounds in some soils of south Florida, *Soil Sci.*, 1996, Vol. 12, p. 1448–1453. [in Russian]

Matichenkov V.V., Wei X., Liu D., Bocharnikova E.A. Theory, practice and prospection of Si fertilizer, *Agricultural Science & Technology*, 2013, Vol. 14, No. 3, p. 498–502.

Mayras Y., Riberi P., Cartier F., Allan P. Increase in blood silicon concentration in patients with renal failure, *Biomedicine*, 1980, Vol. 33, No. 7, p. 228–230.

McNaughton S.A., Bolton-Smith C., Mishra G.D., Jugdaohsingh R., Powell J.J. Dietary silicon intake in post-menopausal women, *Br. J. Nutr.*, 2005, Vol. 94, No. 5, p. 813–817.

Medvedev S.S. *Plant Physiology*, Publ. St. Petersburg State University, St. Petersburg, 2004, 336 p. [in Russian]

Milnes A.R., Twidale C.R. An overview of silicification in Cainozoic landscapes of arid central and southern Australia, *Aust. J. Soil Res.*, 1983, Vol. 21, No. 4, p. 387–410.

Mohanty B.K., Gosh S., Mishara A.K. The role of silicon in *Bacillus licheniformis*, *J. Appl. Bacteriol.*, 1990, Vol. 68, p. 55–60.

Olier K. *Erosion*, Nedra, Moscow, 1990, 347 p. [in Russian]

Orlov D.S. *Soil Chemistry*, Publ. Moscow State University, Moscow, 1985, 376 p. [in Russian]

Pellenard P. Use of silica gel and microorganisms for growing *Niphargus virei (hypogeal Crustacea Amphipoda)* in artificial media, *Ann. Nutr. Aliment*, 1969, Vol. 23, No. 4, p. 283–288. [in French]

Perelman A.I. *Geochemistry of Landscape*, Visshaya Shkola, Moscow, 1975, 341 p. [in Russian]

Perez-Grandos A.M., Vaquero M.P. Silicon, aluminium, arsenic and lithium: essentiality and human health implications, *J. Nutr. Health Aging*, 2002, Vol. 6, No. 2, p. 154–162.

Perry C.C., Keeling-Tucker T. Biosilicification: the role of the organic matrix in structure control, *J. Biol. Inorg. Chem.*, 2000, Vol. 5, p. 537–550.

Polynov B.B. *Geochemical Landscapes*, Geographgiz, Moscow, 1952, 400 p. [in Russian]

Price C.T., Koval K.J., Langford J.R. Silicon: A review of its potential role in the prevention and treatment of postmenopausal osteoporosis, *Int. J. Endocrinol.*, 2013, Vol. 2013, Article ID 316783, 6 p. http://dx.doi.org/10.1155/2013/316783

Rascher U., Bobich E.G., Lin G.H., Walter A., Morris T., Naumann M., Nichol C.J., Pierce D., Bil' K., Kudeyarov V., Berry J.A. Functional diversity of photosynthesis during drought in a model tropical rainforest – the contributions of leaf area, photosynthetic electron transport and stomatal conductance to reduction in net ecosystem carbon exchange, *Plant Cell Environ.*, 2004, Vol. 27, p. 1239–1256.

Reimers N.F. *Nature Management*, Mysl, Moscow, 1990, 640 p. [in Russian]

Savant N.K., Snyder., G.H., Korndorfer G.H. Silicon management and sustainable rice production, *Advance Agronomy*, 1997, Vol. 58, p. 151–199.

Siegel B.Z., Siegel S.M. The chemical composition of algal cell walls, *CRC Crit. Rev. Microbiol.*, 1973, Vol. 3, No. 1, p. 1–26.

Snyder G.H., Matichenkov V.V. Datnoff L.E. Silicon, *Handbook Plant Nutrition*, Massachusetts University, Massachusetts, 2006, p. 551–568.

Sokolova T.A. *Clay Minerals in Soils of Humid Zones of the USSR*, Nauka, Novosibirsk, 1985, 252 p. [in Russian]

Soomro F.M. *Effect of Silicon Compounds on Microbial Transformation in Soil*, Ph.D. Thesis, University of Sheffield, England, 2000.

Sripanyakorn S., Jugdaohsingh R., Elliot H., Walker C., Mehta P., Shoukru S., Thompson R.P., Powell J.J. The silicon content of beer and its bioavailability in healthy volunteers, *Br. J. Nutr.*, 2004, Vol. 91, No. 3, p. 403–409.

Strelko V.V., Gushchin P.P., Vysotsky Z.Z. About interaction of some amino compounds with dehydrated silica gel, *Dokl. AN SSSR (Proc. Acad. Sci. USSR)*, 1963, Vol. 153, No. 3, p. 619–621. [in Russian]

Stumm W., Morgan J.J. *Aquatic Chemistry - Chemical Equilibria and Rates in Natural Waters* (3rd Ed., Wiley-Interscience, New York, 1996, 1022 p.

Trevors J.T. Bacterial evolution and silicon, *Antonie Van Leeuwenhoek*, 1997, Vol. 71, p. 271–276.

Tuner R.E., Rabalais N.N., Justic D. Predicting summer hypoxia in the northern Gulf of Mexico: riverine N, P, and Si loading, *Mar. Pollut. Bull.*, 2006, Vol. 52, No. 2, p. 139–148.

Van Dyck K., Robbercht H., Van Cauwenbergh R., Van Vlasaer V., Deelstra H. Indication of silicon essentiality in humans: serum concentrations in Belgian children and adults, including pregnant women, *Biol. Trace Elem. Res.*, 2000, Vol. **77**, No. 1, p. 25–32.

Varshal G.V., Dracheva I.N., Zamkina N.S. About forms of silicic acids and the methods of their determination in natural waters, *Chemical Analysis Marine Sediments*, Nauka, 1980, p. 156–186. [in Russian]

Velikova V., Cociasu A., Popa L., Boicenco L., Petrova D. Phytoplankton community and hydrochemical characteristics of the Western Black Sea, *Water Sci. Technol.*, 2005, Vol. 51, No. 11, p. 9–18.

Vladimirov V.L., Kirilov M.P., Fantin V.M. 1998 Metabolism and productive properties of bullheads at feeding by mixed fodder with zeolite, *Russian Agr. Sci.*, 1998, No. 4, p. 38–40. [in Russian]

Voronkov M.G., Zelchan G.I., Lukevits A.Y. *Silicon and Life*, Zinatne, Riga, 1978, 578 p. [in Russian]

Vysotsky Z.Z., Danilov V.I., Strelko V.V. Properties of the polysilicic acid gels and conditions of the beginnings and development of life, *Biol. Bull. Rev.*, 1967, Vol. 63, No. 3, p. 362–379. [in Russian]

Wainwright M. Silicon and microbiology aspects - overview, *Proc 3rd Silicon Agricultural Conf*, Korndorfer G.H. eds., October 22-26, 2005, Universidade Federal de Uberlândia, Uberlândia, 2005, p. 19–22.

Wainwright M., Al-Wajeeh K., Grayston S.I. Effect of silicic acid and other silicon compounds on fungal growth in oligotrophic and nutrient rich media, *Mycol. Res.*, 1997, Vol. 101, p. 933–938.

Waterkeyn L., Bientait A., Peeters A. Callose et silice epidermiques rapports avec la transpiration culticulaire, *La Cellule*, 1982, Vol. 73, p. 263–287.

Yatsynin N.L. *Genesis of the Residual Solonetzs*, Nauka, Alma-Ata, 1989, 125 p. [in Russian]

Yatsynin N.L. *Colloidal-High-Polymeric Systems of Solonetzs in the Northern Kazakhstan*, Abstr. Doctor Sci. Dissertation, Tashkent, 1994, 37 p. [in Russian]

Yoshida S. The physiology of silicon in rice, *Tech. Bull.*, Food Fert Tech Center, Taipei, Taiwan, 1975, No. 25.

Yoshon T. Growth accelerating effect of silicon on *Pseudomonas aeruginosa*, *Saitama Med. School*, 1990, Vol. 17, p. 189–198.

6
Methanol as Example of Volatile Mediators Providing Plants' Stress Tolerance

Karl Y. Biel[1,2,*], **Irina R. Fomina**[1,2,*]

[1]*Institute of Basic Biological Problems, Russian Academy of Sciences, Pushchino, Moscow Region 142290, Russia*
[2]*Biosphere Systems International Foundation, Tucson, Arizona 85755, USA*

...the idea is to try to give all of the information to help others to judge the value of your contribution; not just the information that leads to judgment in one particular direction or another, –
Richard P. Feynman (as told to Ralph Leighton)

Abstract
The present brief review considers the rules of self-regulation in living plant systems from the view point of mediatory function of special volatile matter formed by plants, evaporated and passed from one plant to another. Methanol seems to be an important mediator providing stress-tolerance in plant organisms. According to the literature, methanol forms in plant cells through the tetrahydrofolate pathway, pectin and protein methylesterase reactions, protein repair, cell-wall loosening during cell expansion, and some other biochemical processes. Formation of methanol is activated during some stages of ontogenesis and under stress conditions. We speculate that the methanol formed in plants partly evaporates, but that which remains can be utilized in the plant cells by many metabolic pathways, including the involvement in self-regulation through DNA-methylation altering gene expression, and activation

[*] *Corresponding authors*: karlbiel@hotmail.com; irafomi@rambler.ru

of the functions necessary for plant survival. The evaporated methanol can penetrate the cells of neighboring plants and thus serves to either signal other plants of coming stress conditions and / or as a substrate for other self-regulatory reactions. On the bases of literature and some of our observations the hypothesis about mediatory role of methanol in plant living systems is described.

Keywords: Methanol, Pectin methylesterases, Formate dehydrogenase, Self-regulation

6.1 Introduction

Plants, like any living system, can self-regulate within the limits of the genotype under the action of internal or external factors. Special chemical mediators (usually phytohormones) provide the initiation of this self-regulating process (Dörffling, 1982; Srivastava, 2002). These mediators, formed during either ontogenesis or under stress conditions, activate features of the genotype. Concurrently, various mediators are the sources for the synthesis of different biochemical compounds, which are necessary for realization of the genetic information encoding proteins involved into acceleration or stress-adaptation. One of the known hormones, ethylene, is a volatile organic compound. Mediatory role of ethylene in self-regulation of plant growth and development is well described (see Rakitin, 1967; Dörffling, 1982; Muromtsev *et al.*, 1987; Chang and Stadler, 2001; Wang *et al.*, 2002; De la Torre *et al.*, 2006; Dong *et al.*, 2010; Zhao and Guo, 2011; and others). Besides regulatory functions, the volatile nature allows ethylene to pass from the plants, where it was formed, to other ones. Such a feature should be important to synchronize regulatory processes in plant communities.

The mediatory role of other volatile organic compounds, such as low-molecular-weight alcohols (methanol, ethanol, and others), terpenoids, phenylpropanoids / benzenoids, fatty acid derivatives, amino acid derivatives, *etc.*, produced by plant organisms (see Roshchina and Roshchina, 1993, 2012; Guenther *et al.*, 1995; Nemecek-Marshall *et al.*, 1995; Pichersky and Lewinsohn, 2011; Dorokhov *et al.*, 2012; Dudareva *et al.*, 2013; and others), is less studied. Methanol and its derivatives, however, seem to be an important mediator(s), providing self-regulatory response to stress conditions in plants (Benson and Nonomura, 1992; Nonomura and Benson, 1992*a*, 1992*b*, 1992*c*, 1997, 2012, 2013; Nishio *et al.*, 1993, 1996, 1999; Nonomura *et al.*, 1998, 2011; 2012; Biel *et al.*, 2003, 2010, 2018; Biel and Fomina, 2005; Kosobryukhov *et al.*, 2004; and others).

In plants, methanol is primarily formed during the demethylation of pectin constituents in the cell walls, in particular, in the meristomatic cells during their differentiation, in damaged leaf tissues to repair them, and in other cases (McCann *et al.*, 1990; Oberdorf *et al.*, 1990; Levy and Staehelin, 1992; Gaffe *et al.*, 1992, 1994; Nemecek-Marshall et al., 1995; Galbally and Kirstine, 2002; Yang and Yi, 2006; in details, see Emission of Methanol from Plants bellow). Alternatively, other authors (Ander *et al.*, 1985) expressed the view that methanol can be formed from methyl ether groups of lignin by the action of methylotrophic bacteria and fungi in the plant cell wall (Corpe and Rheem, 1989; Holland and Polacco, 1994).

The emission of methanol from plants is activated during some of the stages of ontogenesis and by environmental conditions (Nemecek-Marshall *et al.*, 1995). During formation the methanol partly evaporates, but that which remains is utilized in the plant cells in various pathways (Cossins, 1964; 1987; in greater details, see Utilization of Methanol by Plants bellow), including the reactions involved into self-regulation (see DNA Methylation in Plants below as well). It is reasonably to conclude that the evaporated methanol can penetrate into the cells of neighboring plants and thereby serve as either a signal to other plants on the stress conditions and / or a substrate for other self-regulation reactions. This hypothesis is actively developed in the field of plant virus-host interaction (Dorokhov *et al.*, 2012, 2014; Komarova *et al.*, 2014*a*). The authors convince that the emitted methanol is quickly perceived by intact leaves of affected plants and neighboring plants, promoting the emergence of bacterial immunity, but increases sensitivity to viral penetration.

The biochemistry of the processes of methanol utilization in plant cells actively investigated in the point of view their common biological importance.

In the present chapter we summarized the knowledge about the reactions of methanol formation and utilization in plant cells to develop an understanding of the plant adaptation mechanisms and survival during stress conditions.

6.2 Methanol Application for the Regulation of Productivity

Professors Arthur M. Nonomura and Andrew A. Benson (1992*a*, 1992*b*, 1992*c*) discovered that foliar methanol sprays could increase crop productivity and water use efficiency in plants. Tests on a variety of sun grown C_3 plant species (*e.g.*, *Lycopersicum esculentum* Mill., *Brassica oleracea*, *Gossypium hirsutum*, *Triticum aestivum* L., *Citrullus lanatus*, and others) showed that methanol treatment increases the turgidity, water use efficiency, and crop yield which increased by up to 100% (Nonomura and Benson, 1992*a*). An increase in tomato fruit sugar was also observed; and the increase in sugar production was attributed to a decrease in photorespiration (Benson and Nonomura, 1992).

A number of studies (Nonomura and Benson, 1997, 2012, 2013; Nishio *et al.*, 1993, 1996, 1999; Devlin *et al.*, 1994, 1995; Rowe *et al.*, 1994; Ferguson, 1995; Li *et al.*, 1995; Dorcus and Vivekanandan, 1996; Nonomura *et al.*, 1998, 2011, 2012; Makhdum *et al.*, 2002; Biel *et al.*, 2003, 2010, 2018; Zbieć *et al.*, 2003; Kosobryukhov *et al.*, 2004; Yang and Yi, 2006; Nadali *et al.*, 2010; Yoosefi *et al.*, 2011; and others) have corroborated the original findings of Nonomura and Benson (1992*a*, 1992*b*, 1992*c*).

Mentioned studies have shown that treatment with methanol, especially under stress conditions, causes the activation of plant photosynthesis and increases the rate of transpiration, stomatal conductance and antioxidant activity, accelerates flowering and induces changes in other physiological functions. This assures that the methanol in plants may serve not only as a source of carbon, but also a mediator of regulatory processes.

Consistency stimulation with methanol, nevertheless, is a problem (Hartz *et al.*, 1994; Mitchell *et al.*, 1994; McGiffen *et al.*, 1995; McGiffen and Manthey, 1996; Rajala *et al.*, 1998; Ramírez *et al.*, 2006; and others) and the conditions under which it works are not completely understood.

It is possible that the effect of methanol on plants, as suggested (Yang and Yi, 2006), is a complex of interactions between improvement of CO_2 availability via photosynthesis and non-photochemical energy-dissipative processes. But this may not be the only explanation for the action of mediator, as is known for its effects on gene expression (see Dorokhov *et al.*, 2012; Komarova *et al.*, 2014*a*; and others).

Moreover, these authors warn that when considering the use of methanol or other volatile organic compounds for plant protection, we must keep in mind the potential sensitization of plants to viral infections.

6.3 Emission of Methanol from Plants

Plants produce a large variety of volatile organic compounds (Isidorov *et al.*, 1985; Winer *et al.*, 1992; Guenther *et al.*, 1993, 1994, 1995) that are known to play various roles in plant physiology, signaling, plant-herbivore relationships, and defense against pathogens (Roshchina and Roshchina, 1993, 2012).

Emission of the volatile matter could be significant and may represent a considerable loss of carbon from the plants. Global emissions of isoprene and monoterpenes from plants were measured to be about 5×10^{14} and 1.3×10^{14} g C / year, respectively (Guenther *et al.*, 1995; Nemecek-Marshall *et al.*, 1995). Methanol emission was first found in citrus (Lund *et al.*, 1981). Subsequently methanol emission rates were determined from 11 different plant species, and the value range was from 0.6 to 17 μg C / h / g dry weight (MacDonald and Fall, 1993; Fall and Benson, 1996). Young leaves were found to emit greater amount of methanol than fully expanded ones, sometimes 40 and above μg C / h / g dry weight (Guenther *et al.*, 1995).

Dr. M. Nemecek-Marshall with colleagues (1995) showed, on many higher plants, the correlation between methanol fluxes from leaf surfaces and stomatal distribution and conductance. An increase in the methanol emission from the leaves was also observed with an increase in the intensity of light (Nemecek-Marshall *et al.*, 1995), which is thought to be due to the effects of light intensity on stomatal conductance rather than the direct influence of the light itself or the effect of light through photosynthesis. This conclusion is consistent with the presence of methanol emission in the dark when the stomata are open spontaneously (MacDonald and Fall, 1993).

In addition, it was shown (Nemecek-Marshall *et al.*, 1995) that most of the methanol emitted from hypostomatous leaves occurred from the abaxial leaf surfaces and methanol was emitted at the same rate both from adaxial and abaxial leaf surfaces of amphistomatous leaves. These results are consistent with the linkage of methanol fluxes to stomatal conductance. Direct evidences of decreased methanol emission rates with decreased stomatal conductance were obtained when stomata were induced to close by excision of leaf petioles or by influence of exogenous abscisic acid (Nemecek-Marshall *et al.*, 1995).

According to modern concepts (Galbally and Kirstine, 2002; Hüve *et al.*, 2007) free methanol contained in the leaf airspace, comes out with transpiration of water vapor through the stomata, and methanol emission though the cuticular surface is much lower (MacDonald and Fall, 1993).

6.3.1 Factors Affecting the Methanol Emission

Stomatal conductance in plant leaves is, as noted above, a main internal factor controlling the rate of methanol emission over the short-term. It is likely that the external and / or internal factors (temperature, leaf age and others) affect the rate of methanol synthesis by altering the balance between the internal concentration of methanol and the degree of opening of stomata. This in turn would have long-term effects on methanol emission. From this point of view, it is interesting that, for example, in a south-eastern pine forest in the U.S.A., methanol was detected as a major volatile organic compound in the air, both in the daytime and in the night, during the summer (Fehsenfeld *et al.*, 1992).

Stage of the leaf development has been found (Nemecek-Marshall *et al.*, 1995) to affect the intensity of methanol emission, and a variety of plants showed different patterns of methanol emission at developing leaves. Other factors that may influence methanol emission rates include: a) methanol pool size, which may vary with changing metabolism; and b) populations of leaf methylotrophic bacteria.

Methanol pool size. Because the plants can metabolize methanol, its volume emitted from the leaves may represent only a fraction of what is produced in the leaves. Moreover, methanol metabolism (its formation and utilization) might be changed during leaf development, resulting in changes in methanol pool size. For example, the determination of the content of methanol in leaf extracts (in cold perchloric acid by gas-chromatography) demonstrated that free methanol in bean leaves may vary from 10 to 27 μg / g fresh weight (Nemecek-Marshall *et al.*, 1995). By other words, the pool of free methanol may vary, resulting in changed methanol flux.

Methylotrophs. The leaves of most, if not all, plants are a harbor of methylotrophic bacteria that utilize methanol as their sole carbon source (Corpe and Rheem, 1989). Sometimes, as in the cases of bush beans and soybeans, such bacteria are the dominant species of viable bacteria isolated from the leaves (Hirano and Upper, 1991; Holland and Polacco, 1994). It is likely, but not proven, that leaf methylotrophs consume methanol, formed in the leaf. As a result, methanol fluxes from the leaves might be mediated by the population density and metabolic activity of these bacteria (Nemecek-Marshall *et al.*, 1995), *i.e.* the rate of methanol utilization in the leaves could be a sum of the rates of its utilization by leaf cells and by the leaf methylotrophs.

On the other hand, methylotrophic bacteria and fungi may produce methanol during lignin degradation of the plant cell wall (Ander *et al.*, 1985) and excrete other compounds (Ivanova *et al.*, 2000, 2001; Lidstrom and Chistoserdova, 2002), some of which presumably regulate plant metabolism and stomatal resistance by affecting the cell-genome.

Plausible model to explain the range of methanol fluxes from fully expanded leaves of different ages, at a given leaf temperature, is that flux is primary a function of the methanol pool size and stomatal conductance (Nemecek-Marshall *et al.*, 1995). It was suggested that a large fraction of leaf methanol constantly emitted when the stomata are open. On the other hand, since methanol has a large water / air partition coefficient, *i.e.* about 3700 at 34 °C (Jones *et al.*, 1990), it seems likely that only a small portion of the aqueous methanol pool will move into the leaf airspace and be emitted.

6.3.2 *Methanol Sources in Plants*

Measurements of the accumulation of free methanol in maturing soybean seeds (Oberdorf *et al.*, 1990) showed methanol production during seed growth and maturation. It was suggested that pectin methylesterases in cell walls produce methanol. The methanol was found to be a product of pectin demethylation by pectin methylesterases in cell walls of roots, stems, leaves, and fruits of various plants (see next section Pectin Methylesterases).

Other possible sources for methanol formation could be fungal degradation of lignin in plant secondary cell walls (Ander *et al.*, 1985), carbon-one intermediates of the tetrahydrofolate pathway (Cossins, 1964; 1987), and demethylation of DNA, proteins and other compounds, for example protein repair reactions involving methyltransferases (Mudgett and Clarke, 1993), that occur in all plant tissues.

Methanol formation in the leaves as has been shown (Knox *et al.*, 1990; McCann *et al.*, 1990, 1994; McCann and Roberts, 1991; Levy and Staehelin, 1992), associated with cell-wall loosening during cell expansion and formation of intercellular air spaces in both cases via pectin demethylation. It is possible that some part of the methanol pool in the leaf has a high concentration in the area of cell walls at the site of pectin methylesterase action (Knox *et al.*, 1990; Levy and Staehelin, 1992), where there is lower water content and a relatively high surface contact with the intercellular air space (Nemecek-Marshall *et al.*, 1995). Such a conclusion is qualitatively supported by our experiments, which showed that only 6–9% of ^{14}C-methanol spayed on the surface of the leaf, remained in the leaf after exposure to light (Biel and Fomina, 2005; Biel *et al.*, 2010, 2018).

6.3.3 *Pectin Methylesterases*

Pectin methylesterases (EC 3.1.1.11) are known as cell wall enzymes catalyzing de-esterification of galactosyluronic acid units in the pectin compounds (Figure 6.1). The activities of these enzymes were found in various organs and tissues of various higher plant species (Pressey and Avants, 1972; Evans and McHale, 1978; Goldberg, 1984; Gaffe *et al.*, 1992; and others). A large number of studies showed involvement of pectin methylesterases in a wide range of physiological processes, including microsporogenesis, pollen tube growth, seed germination, root development, polarity of leaf growth, stem elongation, fruit ripening, abscission and senescence (Bryan and Newcomb, 1954; Pressey and Avants, 1982; Sexton and Roberts, 1982; Yamaoka and Chiba, 1983; Goldberg, 1984;

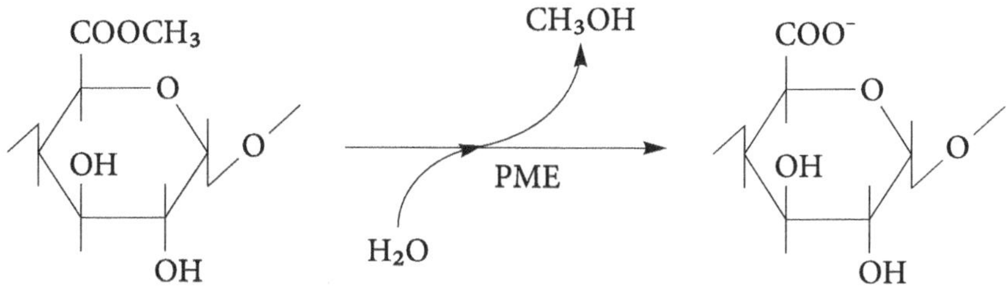

Fig. 6.1. Methanol formation in the de-esterification of galactosyluronic acid units of the pectin (adapted from Galbally and Kirstine, 2002; Yang and Yi, 2006)

Ricard and Noat, 1986; Northcote, 1986; Shea *et al.*, 1989; Fisher and Bennett, 1991; Moustacas *et al.*, 1991; Nari *et al.*, 1991; Liners and Van Cutsem, 1992; Wen *et al.*, 1999; Ren and Kermode, 2000; Pilling *et al.*, 2000, 2004; Micheli, 2001; Komarova *et al.*, 2014*b*; and others).

In the last quarter of the 20th century, several different isoforms of pectin methylesterase were found in higher plants, for example in tomato (*Lycopersicum esculentum*) (Pressey and Avants, 1972; Delincee, 1976; Tucker *et al.*, 1982; Gaffe *et al.*, 1994) and citrus (*Citrus*) fruits (Evans and McHale, 1978; Versteeg *et al.*, 1978), mung bean (the seed of *Vigna radiate*) hypocotyls (Goldberg, 1984; Goldberg *et al.*, 1992), and flax (*Linum usitatissimum*) callus (Gaffe *et al.*, 1992). Some of pectin methylesterases are encoded by different genes (Harriman *et al.*, 1991; Recourt *et al.*, 1992) while some others probably represent posttranslational modifications of the same pectin methylesterase protein.

Today the crucial role of pectin methylesterases in plant development is emphasized by the 79 isoforms found in *Arabidopsis* (Arabidopsis Genome Initiative, 2000).

As noted above, pectin methylesterases also act in defense mechanisms of plants' pathogens (Bateman and Millar, 1966) by releasing pectin fragments that act as messengers (Collmer and Keen, 1986) and by enhancing RNA silencing (Dorokhov *et al.*, 2006).

Recently the knowledge about the role of pectin methylesterases in plant immunity was summarized (Dorokhov *et al.*, 2012, 2014; Komarova *et al.*, 2014*a*) in the model of the effects of methanol emitted by the damaged plant. Mechanical damage to the leaves that often precedes the penetration of the leaf by a pathogen, leads to an increase in the expression level of pectin methylesterase, and induction of the release of gaseous methanol. Emitted methanol causes priming of adjacent leaves and neighboring plants, including activation of the methanol-induced genes, the emergence of bacterial immunity, the opening of plasmadesmata, and increased sensitivity to the virus penetration and intercellular distribution of viral pathogens.

It should be also noted that bacterial pectin methylesterase is in turn an enzyme that catalyzes the essential first step in bacterial invasion of plant tissues (Fries *et al.*, 2007). The authors provided a mechanism for the activity of pectin methylesterase which drives the enzyme along the pectin molecule and accounts for the sequential pattern of demethylation produced by both bacterial and plant pectin methylesterases (Fries *et al.*, 2007).

Quite a long time it was believed (see Gaffe *et al.*, 1994) that the main roles of pectin methylesterases in plants include: a) localized reduction in pH due to demethoxylation of pectins, resulting in cell extension and growth (Lamport, 1981; Ricard and Noat, 1986); b) de-esterification of pectins, making them more susceptible to degradation by pectinases and enhancing cell separation during fruit ripening, senescence, and abscission (Fisher and Bennett, 1991; Liners and Van Cutsem, 1992); and c) generation of free COO^-, which facilitates formation of new cross-linkages in cell walls (Fry, 1986). A change in cell-wall structure was shown to be also important for plant adaptation to abiotic stressors such as salt stress (Bigot and Binet, 1978).

Today, it is more and more clear that the methanol formed by demethylation of pectin is not a waste product of the reaction, but an important regulator of cell-to-cell communication.

6.3.4 *Utilization of Methanol by Plants*

Since the middle of the last century, a large body of information was obtained about the path of oxidized carbon-one compounds, such as carbon dioxide, in plants (Calvin and Benson, 1949; Benson, 1951; Bassham, Benson *et al.*, 1954; Stiller, 1962; Walker, 1962, and others). Much less, however, is known about the metabolism in plants of more reduced carbon-one compounds such as methanol (see ref. in: Gout *et al.*, 2000; Biel and Fomina, 2005; Yang and Yi, 2006; and others).

Work carried out on the animal tissues (Alexander and Greenberg, 1955) showed significant role of methanol, formaldehyde and formic acid in the synthesis of serine and methionine. Purine synthesis, as found (Buchanan, Flaks *et al.*, 1957), usually includes an intermediary formation of a complex with tetrahydrofolate. Methanol metabolism in *Pseudomonas* AM 1 (Large *et al.*, 1961, 1962) and *Pseudomonas gazotropha* Z-1156 (Romanova *et al.*, 1978) was examined in more detail. Both strains demonstrated the ability to grow using either CO_2 or methanol as a sole carbon source.

Labeled substrates quickly assimilated into the pathway leading to the formation of serine, glycine, malate, and aspartate. When the products of methanol utilization were degraded (Large *et al.*, 1962), the distribution of radioactivity in serine was consistent with a hydromethylation of glycine to yield serine. Malate could then arise by carboxylation serine. These results, as well as the results obtained on animal tissues (Alexander and Greenberg, 1955), indicate that methanol plays an important role in the biosynthesis of serine.

6.3.5 Ethanol-Water-Soluble Fraction in Different Parts of Plants

Eukaryotic algae and higher plants are capable to use methanol as the source of carbon in photosynthesis (Benson, 1951; Bassham, Benson *et al.*, 1954; Quayle, Fuller *et al.*, 1954; Doman and Romanova, 1962; Cossins, 1964). The rates of $^{14}CO_2$ fixation in cells of *Chlorella* and *Scenedesmus* were comparable with the rates of fixation of ^{14}C-methanol (Calvin and Benson, 1949).

Nonetheless, methanol was detected not only in leaves (Franzen, 1921) but also in fruits (White, 1950) and there is evidence that this compound may be subjected to interconversion in various plant tissues. For example, in a pulp of the ripening fruits was shown the ability to utilize methanol (Rakitin *et al.*, 1957).

Professors Nikolai G. Doman and Alla K. Romanova (1962) supplied a vapor of ^{14}C-methanol to the leaves of bean (*Vicia faba*) and barley (*Hordeum vulgare*) in the light and in the dark. The authors interpreted their data as indicating that carbon-one compounds such as methanol, formaldehyde, and formate are converted mainly to carbon dioxide, which is then involved in photosynthetic assimilation. Certain other transformations, which lead to the formation of serine and other compounds, have also been suggested (Doman and Romanova, 1962).

Special experiments (Cossins, 1964) showed that the carrot tissues (*Daucus carota* L. subsp. *sativus* (Hoffm.)), endosperm of castor bean (*Ricinus communis*), germinating pea (*Pisum sativum* L.), soybean seedlings (*Glycine max* (L.) Merr.), and mature beet tissues (*Beta vulgaris*) possess ability to metabolize micromolar amount supplied ^{14}C-methanol. In carrot tissues, more than 90% of supplied radioactivity was recovered in the products of methanol metabolism. In other tissues, a smaller, though significant, amount of supplied ^{14}C-methanol was utilized during the experimental period. As mentioned above, in our experiments (Biel *et al.*, 2018) with sugar beet (*Beta vulgaris*), not more than 9% of ^{14}C-methanol as a foliar spray, were utilized in the leaves.

Production of labeled formate and carbon dioxide during the exposure to ^{14}C-methanol indicated oxidation of the alcohol in all tissues examined. Methanol oxidation pathway in higher plants, believed to include the sequential formation of formaldehyde, formic acid and carbon dioxide (Colas des Francs-Small *et al.*, 1993).

Nevertheless, enzymes, oxidizing methanol to formaldehyde, remain unclear because higher plants are likely to lack the alcohol oxidase (EC 1.1.3.13) found in microorganisms (see Gout *et al.*, 2000; Ozimek *et al.*, 2005; Finlay, 2007). It was suggested (Holland and Polacco, 1994) that the pink-pigmented facultative methylotrophs are responsible for the activity of methanol dehydrogenase (or alcohol oxidase) and the process of oxidization of methanol in higher plants.

Next enzyme NAD^+-dependent formate dehydrogenase (EC 1.2.1.2), which performs a redox transformation between formaldehyde and formate, is widely found in plants (Davison, 1949, 1951; Mazelis, 1960; Sinha, 1964; Alekseeva *et al.*, 2011; and others). Plant formate dehydrogenase is localized primarily in the mitochondria, particularly of non-photosynthetic tissues (Colas des Fransis-Small *et al.*, 1993). Localization of formate dehydrogenase, both in mitochondria and chloroplasts, was evidently shown in transgenic plants (Herman *et al.*, 2002).

Plant's NAD^+-dependent formate dehydrogenase belongs to stress proteins (Hanson and Nelsen, 1978; Hourton-Cabassa *et al.*, 1998; Dubos and Plomion, 2003; Mimami *et al.*, 2003; Andreadeli *et al.*, 2009; David *et al.*, 2010; and others) which synthesis is induced by methanol or formaldehyde to a greater degree than by the formic acid (Fukusaki *et al.*, 2000).

Some authors (Olson *et al.*, 2000) believe that methanol stimulates the synthesis of formate dehydrogenase transcripts, while its oxidized modifications (formaldehyde, formate) can act as signaling molecules. The origin of formate in cells of the plants exposed to stress is still discussed (see Alekseeva *et al.*, 2011): photorespiration, methanol metabolism, and glyoxylate shift of the Krebs cycle were hypothesized as a possible way to form formate.

Further oxidation of formate produces CO_2. As noted above, the results obtained by N.G. Doman and A.K. Romanova (1962) showed that methanol is metabolized in the leaves of beans and barley, preferably after oxidation to CO_2. It is clear therefore, that methanol metabolism in these tissues might involve the reactions of photosynthetic or / and dark CO_2 fixation.

In Dr. E.A. Cossins's (1964) experiments, the tissues were incubated with ^{14}C-methanol in darkness, wherein any fixation of $^{14}CO_2$ might result in labeled organic acids. Castor bean endosperm and pea cotyledons fixed $^{14}CO_2$ in the dark, forming a ^{14}C-malate (Stiller *et al.*, 1958; Cossins, 1964; 1987). However, fixation of ^{14}C-methanol in the endosperm of castor bean and pea and soybean cotyledons does not lead to significant accumulation of labeled malate (Cossins, 1964).

From these studies, it appears unlikely that CO_2 fixation represents an important stage of methanol metabolism in the tissues examined. Furthermore, a considerable amount of labeled carbon was included in serine, methionine, methionine sulfone, and methionine sulfoxide (Cossins, 1964). Since the latter two compounds are the products of methionine oxidation, it is possible that their carbon skeletons are derived directly from methionine (Cossins, 1964).

Extensive synthesis of serine and methionine during the methanol fixation in higher plants agrees with the data obtained on *Pseudomonas* (Large *et al.*, 1961, 1962; Romanova *et al.*, 1978) and animal tissues (Alexander and Greenberg, 1955).

Thus, apart from the oxidation to CO_2, methanol may be used in plant tissues for the biosynthesis of amino acids, serine and methionine. Glyceric acid, which is an important product of serine metabolism in higher plants, has also been labeled after ^{14}C-methanol feeding (Cossins, 1964).

Biosynthetic pathways of serine and methionine in plant tissues involve transmethylation reactions, in which methanol presumably acts as a source of carbon-one units (Cossins, 1964; see also Figure 6.3). Studies with animal tissues have provided considerable evidence that an "active formyl" radical (formyl tetrahydrofolate) may readily arise from formaldehyde (Sakami, 1955) and then transfer activated carbon-one unit to glycine and homocysteine in the reactions leading to the formation of serine and methionine, respectively (Cossins, 1987).

According to Dr. E.A. Cossins (1964) the methanol metabolic pathways in plants may be as follows:

$$^{*}CH_3OH \rightarrow H^{*}CHO \rightarrow H^{*}COOH \rightarrow {}^{*}CO_2$$

H*CHO → "*active formyl" + homocysteine → *methionine

H*CHO → "*active formyl" + glycine → *serine

*denotes carbon atoms derived from methanol.

More detail schema (Gout *et al.*, 2000) of the proposed metabolic pathways involved in methanol assimilation by plants is presented in Figure 6.2.

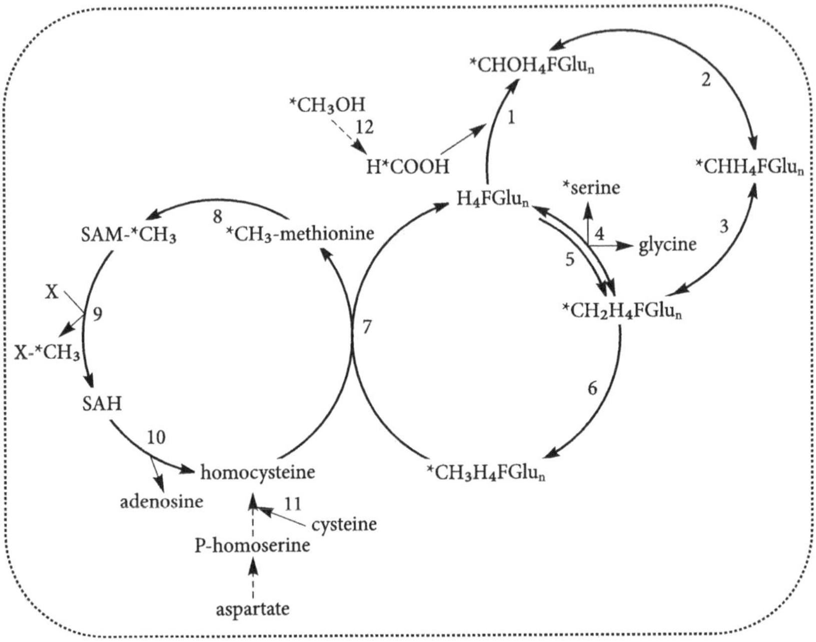

Fig. 6.2. Proposed metabolic pathway involved in methanol assimilation by plant cells (the origin of methyl-β-D-glucopyranoside is not indicated) (adapted from Gout *et al.*, 2000)

Note: *, (^{13}C) carbon. Enzymes: 1, formyltetrahydrofolate synthetase; 2, methenyl H_4Pte-Glu_n cyclohydrolase; 3, methylenetetrahydrofolate dehydrogenase; 4, Ser hydroxymethyltransferase; 5, Gly decarboxylase; 6, methylenetetrahydrofolate reductase; 7, cobalamin-independent Met synthase; 8, *S*-adenosyl-Met synthetase; 9, methyltransferase; 10, *S*-adenosylhomo-Cys hydrolase; 11, cystathionine γ-synthase and cystathionine β-lyase; 12, methanol oxidase and formaldehyde dehydrogenase. Compounds: H_4FGlu_n, 5,6,7,8 tetrahydropteroylpoly-Glu (tetrahydrofolate); $CHOH_4FGlu_n$, 10-formyltetrahydrofolate; CHH_4FGlu_n, 5,10-methenyltetrahydrofolate; $CH_2H_4FGlu_n$, 5,10-methylenetetrahydrofolate; $CH_3H_4FGlu_n$n, 5-methyltetrahydrofolate; SAM, *S*-adenosyl-Met; SAH, *S*-adenosylhomo-Cys

6.3.6 *Ethanol-Water-Insoluble Fractions in Plants*

Besides ethanol-water-soluble compounds, radioactive carbon of ^{14}C-methanol included in various insoluble substances, such as DNA, pectin, cellulose, protein, *etc.* (Cossins, 1964; Biel *et al.*, 2010, 2018). In pulse-chase experiments with *Beta vulgaris*, we found (Biel *et al.*, 2010, 2018), that after foliar spraying of ^{14}C-methanol or, for example, β-D-glucose ^{14}C-(U)-methyl-β-D-glucopyranoside (^{14}C-MeG) on a single leaf of a plant, ^{14}C was detected in all the neighboring leaves, and even in the roots. Redistribution of ^{14}C-methanol or ^{14}C-MeG within the plant organs takes place both in the light and in the dark. Label was detected in the ethanol-water soluble fraction, as well as in lipids, pectin, cellulose, starch and protein (Biel *et al.*, 2010, 2018). It is important to note, that incorporation of ^{14}C from ^{14}C-methanol into insoluble fractions displayed a strong dependence on tissue development. In one-day-old pea cotyledons, radioactivity of the insoluble fraction was 2.3% of the total; in storage tissue of mature beets it was 90.2% (Cossins, 1964).

It is likely that expressions of the genes encoding enzymes for methylation of high-molecular matter are determined by the course of plant tissue development and specialization. On the other hand, methylation of high-molecular-weight matter, especially DNA (Finnegan *et al.*, 1998), may influence, in its turn, the genes controlling plant development.

6.3.7 DNA Methylation in Plants

The control mechanisms for flowering, and possibly maturation, are the result of products or regulatory activities of many different genes. Some of these genes may affect the level of plant-growth regulators; others may affect the ability of the plant to respond to them, while still others may affect flowering by mechanisms that do not involve plant growth regulators at all.

Although Dr. P.F. Wareing (1980) suggested that root-producing plant-growth regulators affect the maturation, their role in this process is confusing. For example, β-auxin is implicated in both the promotion and reversal of maturation in conifers, and gibberellins promote flowering in conifers, but reverse the mature phase of *English ivy* (Zimmerman *et al.*, 1985; Greenwood, 1987). While gibberellins promote flowering in plants (conifers), there are some evidences that their application cannot offset the genetic disposition of an individual tree to flower (Chalupka, 1979; Webber *et al.*, 1985).

So, the hypothesis that the flowering or maturation can be controlled by only the plant-growth regulators does not include the role of other factors. In particular, methanol applied into fruit pulp stimulates ripening, and conversion of methanol in fruits has been observed at the same time (Rakitin *et al.*, 1957). It was suggested that stimulation of ripening by methanol may be associated with DNA methylation, especially nuclear DNA.

Higher plants, in addition to the nuclear DNA contain mitochondrial DNA and chloroplast DNA. The total amount of chloroplast DNA in plants is about 5×10^{-15} g or $95–100 \times 10^6$ Daltons per chloroplast. From these data it follows that there should be about 20–30 copies of circular DNA in each chloroplast (Sedzher, 1975). For chloroplast DNA is characteristic to be in contact with the membrane system of the chloroplasts. The same is true for the mitochondrial and bacterial DNA (Odintsova, 1976) and is considered important for the vector-valued matrix syntheses.

The presence in the molecules of DNA minor bases such as 5-methylcytosine (Hotchkiss, 1948; Wyatt, 1950) and N^6-methyladenine (Dunn and Smith, 1958) has been known since the middle of the last century. However, functional significance of DNA methylation was unknown for a long period of time. Then, in a series of molecular biological research, enzymetic DNA methylation has been shown to play an essential role in controlling global gene expression (see Gruenbaum *et al.*, 1981; Razin and Szyf, 1984; Razin and Cedar, 1991; Finnegan *et al.*, 1998; Bird, 2002; Vanyushin, 2005, 2007; and others). Some of these studies should be noted.

One of the first DNA nucleotide sequences methylated in bacterial DNA was deciphered in *Bacillus brevis* (Vanyushin and Dobritsa, 1975). The character of DNA methylation in bacteria was found to be species and strain specific (Vanyushin *et al.*, 1968), and associated with cell differentiation in microorganisms (see Vanyushin, 2005, 2007). Bacterial DNA of *Escherichia coli* showed one type of methylation pattern: all methylatable sites are methylated (Razin and Szyf, 1984). Chloroplast DNA in *Chlamydomonas* maternal gametes was also found to be hypermethylated (Nishiyama *et al.*, 2002). In contrast, the nuclear DNAs of eukaryotic organisms have shown diverse methylation patterns, in which some methylatable sites may remain unmethylated (Razin and Szyf, 1984).

Among the studies with higher plants should be mentioned the work of Dr. P.V. Boldtad and Dr. W.J. Libby (1982), in which changes in DNA methylation has been proposed as a basis for the maturation of woody plants. The authors suggested that maturation in trees might be associated with the varying of DNA methylation, because increased levels of methylation are usually associated with decreased gene expression at the transcriptional level (Razin and Riggs, 1980). Methylcytosine in the nuclear DNA of juvenile and mature foliage since DNA methylation can affect gene expression. Analysis of cDNA libraries made from poly (A+) RNA from expanding shoots of juvenile and mature plants showed that there are some differences in gene expression (Hutchison *et al.*, 1987).

Eukaryotic DNA contains 5-methylcytosine, as a sole modified base, as up to 30% of total cytosine residues (cited by Ito *et al.*, 2003). These methylated cytosine residues appear exclusively at CpG sequences in animal DNA (Razin and Szyf, 1984). In plants, methylation was shown to occur in CpG and CpXpG sites (X stands for A, T, or C) (Gruenbaum *et al.*, 1981), but one study showed that cytosines in CpXpX are also methylated (Oakeley and Jost, 1996).

The available data in the literature suggest a loss of methyl groups during the differentiation process (Razin and Szyf, 1984).

Two different activities were suggested: *de novo* methylation and maintenance methylation (Holliday and Pugh, 1975; Riggs, 1975). The *de novo* activity is responsible for the methylation of sites that are not methylated on both DNA strands, whereas maintenance methylation is involved in methylating of hemimethylated sites formed during DNA synthesis (Creusot *et al.*, 1982).

Three types of DNA methyltransferases (METs) have been identified in plants: MET1 for maintenance of CpG methylation (Finnegan and Kovac, 2000), chromomethylase for *de novo* methylation of CpXpG (Bartee *et al.*, 2001), and domain-rearranged MET, also for *de novo* methylation (Cao *et al.*, 2000). Multiplicity of DNA-methyltransferases allows us to distinguish between replicative and post-replicative DNA methylations in plants; these types of DNA methylation are different in site specificity and regulation by phytohormones and inhibitors (see Vanyushin, 2007).

It is believed that methylation of DNA inhibits transcription through changes in DNA structure caused by 5-methylcytosine. Two mechanisms were proposed: direct repression by methylation of the promoter regions, acting as the "locking mechanism" (Razin and Szyf, 1984; Bird, 2002), and indirect repression due to changes in the structure of nucleosomes, affecting chromatin conformation (Bird and Wolffe, 1999).

Among proteins that complex DNA to form chromatin, the proteins possessing a methyl-CpG-binding domain were found to play a critical role in transcription repression in animal cells (Bird and Wolffe, 1999; Bird, 2002). Methyl-CpG-binding domain protein with specifically bind to 5-methylcytosine in CpG and CpXpG was isolated from *Arabidopsis* (Ito, Koike *et al.*, 2003).

The genes expressed transiently, such as hormone-inducible genes (Wilks *et al.*, 1982; Burch and Weintraub, 1983; Gerber-Huber *et al.*, 1983) or ones encoding growth factors (including oncogenes), are found to be fully methylated. Studies where tissue-specific genes introduced into cells may be expressed only in the appropriate cells (Razin and Szyf, 1984) indicate that there is a requirement in the cellular factors that mediate tissue specificity. Because this tissue specificity factor may be a gene product, the expression of such a gene may still be dependent on its methylation pattern (Razin and Szyf, 1984).

Thus, due to modern view (see Vanyushin, 2005, 2007), DNA methylation controls all genetic functions and is the mechanism of cell differentiation and gene silencing, which plays a crucial role in the manifestation of the key elements of life, such as ontogenesis, origin of species, evolution, and aging. Regulatory function of DNA methylation is important to provide plant survival at stress conditions (Feng *et al.*, 2012) and even to save stress memory in plants (Pecinka *et al.*, 2010).

Besides DNA methylation, methylation and / or demethylation of other high-molecular matter such as pectin, as described above, and protein may also play a significant role in plant development and stress tolerance. Histone methylation at lysine and arginine residues is known as one of the most complex and important covalent modifications in eukaryotic organisms (Bannister and Kouzarides, 2005) including plants (see Niu *et al.*, 2007). These modifications are involved in the regulation of ontogenesis (Niu *et al.*, 2007) and stress response (Dahan *et al.*, 2011).

Methylation of L-isoaspartyl residues occurs during protein repair (Mudgett and Clarke, 1993).

Methylation of several alkaloids found among the pathways of formic acid utilization (Byerrum *et al.*, 1954; Battersby and Harper, 1958) seems to be important for alkaloid regulatory capacities.

Other regulatory compounds may also increase their biological activity through methylation. This was shown (Abrams, 1997, cited by Quatrano *et al.*, 1997) on methylated analog of abscisic acid such as (+)-8'-methylene-ABA in comparison with natural ABA.

6.4 Hypothesis of Methanol Influence on Different Levels of Cell Metabolism in C_3 Plants

Within their genotypic limits plants self-regulate their metabolism to adapt to the prevailing internal and external environmental factors. Special chemical mediators initiate the self-regulation process. The mediators are formed during either ontogenesis and / or under stress conditions and follow their specific genotype. Concurrently, various mediators are also used as sources for the synthesis of different biochemical compounds.

Methanol seems to be an important mediator providing self-regulation in plant organisms. In plant cells, methanol is primarily formed during the demethylation of pectin constituents in the cell walls. Tetrahydrofolate pathway, protein methylesterase reactions, and other processes may also contribute to the formation of methanol.

The formation of methanol is activated during some of the stages of ontogenesis and under stress conditions. During formation the methanol partly evaporates, but that which remains is utilized in the plant cells in various pathways including the reactions involved in the regulation of gene expression by DNA methylation.

The evaporated methanol can penetrate into the cells of neighboring plants and thereby serve as either a signal to other plants on the stress conditions and / or a substrate for other self-regulation reactions.

The biochemistry of the processes of methanol utilization in plant cells is only partly investigated. Some of these reactions described above have common biological importance. Complete knowledge about the reactions of methanol utilization in plant cells will help to develop an understanding of the plant adaptation mechanisms and survival under stress conditions.

The details of our hypothesis about methanol pathways in plant cells are presented in Figure 6.3.

One part of methanol, after the reaction with free water interstitial cavities, may be oxidized to formaldehyde by methanol oxidase of methylotrophic bacteria (Holland and Polacco, 1994). Formaldehyde may be involved in the metabolism of folic acid, synthesis of a serine, purines and other reactions or converted to formic acid and then to carbon dioxide.

Another part of the methanol enters a subcuticular space of cell walls and is released to the outside through the stomata (Yang and Yi, 2006). It is suggested that methanol formed in plant cell (or penetrated into cells from atmosphere) may be oxidized to CO_2, and then reduced in photosynthesis or by heterotrophic fixation. Another pathway of methanol oxidation is the methionine and serine biosynthesis.

Our data (Biel *et al.*, 2010, 2018) indicates that labeled carbon from exogenous ^{14}C-methanol fluxes into pectin, cellulose, protein, lipids, and starch. These pathways of methanol utilization have not yet been studied, but seem to be less important for the initiation of self-regulation responses to stress.

Regarding the mediatory function of methanol, we suggest that the principal pathway of methanol utilization should be the methylation of nuclear DNA, where up to 30% could be methylated in processes associated with regulation of gene expression, as described above (also see Razin and Szyf, 1984; Finnegan *et al.*, 1998; Bird, 2002). The fact that hormone-inducible genes or genes encoding growth factors are fully methylated (Wilks *et al.*, 1982; Burch and Weintraub, 1983; Gerber-Huber *et al.*, 1983) indicates involvement of DNA methylation into self-regulation response to stress conditions.

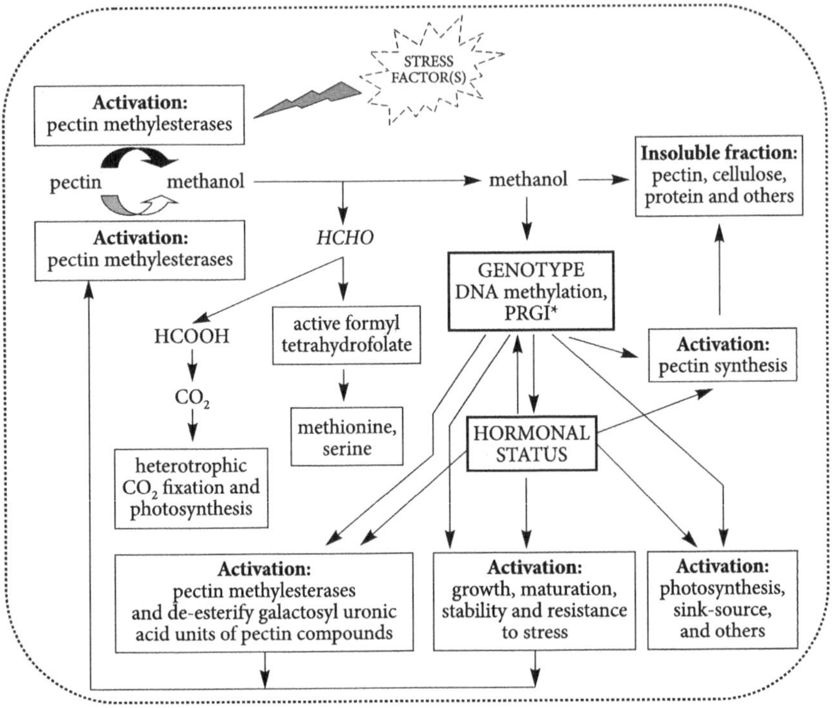

Fig. 6.3. Hypothetical scheme of stress resistance and plant productivity after methanol "vaccination" (modified from Biel and Fomina, 2005; we also used information from Doman and Romanova 1962; Cossins 1964; Razin and Szyf 1984; Nonomura and Benson 1992*a*; Gaffe *et al.*, 1994; Finnegan et al 1998; Bird 2002; Biel *et al.*, 1990; 2003; 2010, 2018; Gout *et al.*, 2000; Kosobryukhov *et al.*, 2004; and others)
Note: *PRGI – program realization of genetic information, *i.e.* the chain of changes such as repression of genes-repressors, de-repression of stress-inducible genes, primary the genes encoding hormone synthesis, then the hormone-inducible genes, *etc.*

This response is induced by methylation of the regulatory genes and alterations in the whole chromatin conformation. Presumably, some of these methylatable regulatory genes are the repressors for the genes encoding enzymes catalyzing synthesis of hormones, especially ABA, and / or synthesis of other inductors. De-repression of such genes leads to increase in hormonal level. Hormones, in their turn, induce hormone-inducible genes. In particular, ABA is produced in plants under water stress and low temperature, and ABA-dependent and ABA-independent signal pathways were shown to induce stress-inducible genes (Shinozaki and Yamagushi-Shinozaki, 2000; Zho, 2002).

There is a large body of other works in literature, demonstrated expression of the groups of genes by different stressors.

We suppose that methanol due to DNA methylation plays mediatory role between stress-receptors and genome, causing to repression of some genes-repressors. We also guess that de-repression of stress-inducible genes occurs due to their partial demethylation, or they are expressed when totally methylated similarly to *Xenopus* estrogen-inducible vitellogenin (Gerber-Huber *et al.*, 1983).

Alterations in DNA methylation patterns lead to alterations in cell functional activities, which are necessary to survive plant organism during stress. Various regulatory processes are initiated by DNA methylation, including the optimization of the Benson-Bassham-Calvin cycle and the synthesis of amino acids in photorespiration reactions and the efficiency of water use by increasing the number of roots and / or activity of the water pump.

At the same time, methanol reduces respiration rate, and maintains the sustaining respiration, as well as the coupling between photosynthesis and respiration, and increases resistance of photosystem II and respiration assistance of photosynthesis.

So, the methanol should also be included in the activation of alternative carbon flow pathways:

a) *Fd*-depended CO_2 carboxylation ($CO_2 \xrightarrow{Fd_{red}} formate \rightarrow tetrahydrofolate\ pool$), CO_2 fixation

b) in the inverted Krebs cycle ($CO_2 \xrightarrow{Acetyl\ CoA} converting\ 3\text{-}carbonic\ cycle$), and some other reactions.

Thus, the initial response to stress in plants should be the activation of methanol formation. This is likely due to activation of pectin methylesterases. Pectin demethylation produces methanol as a byproduct, and has been described to accompany the acceleration of plant ontogenesis (see above). The stimulation of stress tolerance by exogenous methanol also suggests a key role of methanol formation in the initiation of stress response.

Some new data that support such a conclusion were obtained in our experiments (Biel *et al.*, 2003; Kosobryukhov *et al.*, 2004) where methanol foliar spray did not stimulate considerably CO_2 assimilation in developing sugar beet (*Beta vulgaris*) leaves, but did increase the tolerance to drought stress.

In contrast, aging leaves sprayed by methanol, demonstrated more than the two-fold increase in CO_2 assimilation which helped them to survive drought stress.

These results indicate that developing leaves have a high methanol pool size, likely because of more active pectin demethylation in their cell walls, and do not need the external methanol to induce stress tolerance.

The second stage of the response to stress is methylation of genes-repressors and de-repression of stress-inducible genes, as noted above.

The third stage involves the changes in plant functional activities. Our experiments (Biel *et al.*, 2003; 2018; Kosobryukhov *et al.*, 2004) with sugar beet (*Beta vulgaris*) plants foliar sprayed by methanol, showed a long-term (14 days) increase in drought resistance regarding their photosynthetic apparatus.

This resistance was accompanied by: an increase in carbon fixation per leaf area, activation of antioxidant system, a decrease in lipid peroxidation, and other functional changes. These changes in metabolic pathways allowed survival of the plants.

In addition, a possible "feed-back" mechanism may exist wherein the induced pectin methylesterases release additional free mediator methanol. This speculation could be tested experimentally. It is appropriate that plants have a self-regulation system that allows them to endure stress with minimal negative profit.

It is known that healthy organisms endure stress better. We suppose that healthy plants produce free methanol as a volatile mediator, not only for themselves but also for the "survival" of neighboring plants. Conceivably, other volatile matter may be emitted by plants, as well as some soluble matter excreted by plant roots and absorbed by the roots of neighboring plants to play similar roles for a more effective plant self-regulation system.

Regulatory roles of the volatile hormone ethylene, and a few other volatile compounds (terpenes, isoprene, *etc.*), for plant growth and development, plant signaling, and in other plant relationships along with their biotic and abiotic environment, were described in a series of studies (see above). Still, the biochemical mechanisms by which mediators influence the plant genome, hormonal status, and functional activities are not completely understood.

Ethylene was shown to regulate biosynthesis of stress-hormone ABA (Ronen and Mayak, 1981).

Here the knowledge and suppositions about formation, utilization, and mechanisms of methanol regulatory action in plants are summarized based on the hypothesis that methanol is an important volatile mediator to induce plant development, productivity and survive stress tolerance.

6.5 Conclusion

During ontogenesis and / or under stress conditions plants may produce specific metabolic products (and most likely as by-products) in different organs (leaves, fruits, stems, roots). Some of these products are low-molecular weight compounds that function as self-regulation mediators. To mediate means to influence the genome, correcting program of the realization of genetic information. The mediator starts a chain of metabolic changes, which lead to an increase in stress tolerance or to an accelerated maturation. Such behavior may in turn lead to species survival. The mediators, which can be passed from one plant to others, increase weak plants potential for survival, as a possible result of group selection. As such it may also lead to synchronized plant community development and response to stress.

Methanol is shown to be important volatile mediator to induced plant productivity, maturation, and stress tolerance (especially to drought stress). The mediatory action of methanol is based primarily on the methylation of special clusters in nuclear DNA, and the ease of methanol formation in plant cells due to pectin methylesterase reactions in cell walls in response to internal or external factors.

Young tissues produce methanol more intensively than aging tissues therefore their metabolic reactions, which are necessary for stress tolerance, are better prepared to respond to stressful events.

Among the compounds emitted or excreted by plants, it may be possible to find out unknown mediators with regulatory roles like the known mediators (methanol, ethanol, ethylene, isoprene and others). Such mediators should be produced in small amounts as by-products of cell metabolism, and display a high biological activity. Under stress conditions the pool sizes of mediators whose roles are similar to methanol, should be increased.

Acknowledgments

Many thanks to Professors Andrew A. Benson (Scripps Institute of Oceanography, La Jolla, California, USA; deceased), John N. Nishio (Biocompatible Plant Research Institute, Chico, California 95973, USA), and Arthur M. Nonomura (Carbon Reactions of Photosynthesis Sector, BRANDT iHammer, Powell, Ohio 43065, USA) for encouraging our interest in methanol and its applications in plant physiology.

Abbreviations

ABA - abscisic acid
Fd_{ox}, Fd_{red} - oxidized and reduced ferredoxin
MET - methyltransferase

References

Alekseeva A.A., Savin S.S., V.I. Tishkov V.I. NAD^+-dependent formate dehydrogenase from plants, *Acta Naturae*, 2011, Vol. 3, No. 4(11), p. 38–54.

Alexander N., Greenberg D.M. Studies of the biosynthesis of serine, *J. Biol. Chem.*, 1955, Vol. 214, No. 2, p. 821–837.

Ander P., Eriksson M.E.R., Eriksson K.-E. 1985 Methanol production from lignin-related substances by *Phanerochaete chrysosporium*, *Physiol. Plant.*, 1985, Vol. 65, p. 317–321.

Andreadeli A., Flemetakis E., Axarli I., Dimou M., Udvardi M.K., Katinakis P., Labrou N.E. Cloning and characterization of *Lotus japonicus* formate dehydrogenase: A possible correlation with hypoxia, *Biochim. Biophys. Acta*, 2009, Vol. 1794, p. 976–984.

Arabidopsis Genome Initiative, Analysis of the genome sequence of the flowering plant *Arabidopsis thaliana*, *Nature*, 2000, Vol. 408, No. 6814, p. 796–815.

Bannister A.J., Kouzarides T. Reversing histone methylation, *Nature*, 2005, Vol. 436, p. 1103–1106.

Bartee L., Malagnac F., Bender J. Arabidopsis *cmt3* chromomethylase mutations block non-CG methylation and silencing of an endogenous gene, *Genes Dev.*, 2001, Vol. 15, p. 1753–1758.

Bassham J.A., Benson A.A., Kay L.D., Harris A.Z., Wilson A.T., Calvin M. The path of carbon dioxide in photosynthesis. XXI. The cyclic regeneration of carbon dioxide acceptor, *J. Am. Chem. Soc.*, 1954, Vol. 76, p. 1760–1770.

Bateman D., Millar R.L. Pectic enzymes in tissue degradation, *Annu. Rev. Phytopathol.*, 1966, Vol. 4, p. 119–146.

Battersby A.P., Harper B.J.T. Origin of methyl groups in morphine, codeine, and thebaine, *Chem. Ind.*, 1958, Vol. 12, p. 365–377.

Benson A.A. Identification of ribulose in $^{14}CO_2$ photosynthesis products, *J. Am. Chem. Soc.*, 1951, Vol. 73, p. 2971–2972.

Benson A.A., Nonomura A.M. The path of carbon in photosynthesis: methanol inhibition of glycolic acid accumulation, *Photosynth. Res.*, 1992, Vol. 34, p. 196, Abstr. P-522.

Biel K.Y, Fomina I.R. Self-regulation in plant living systems: methanol as example of volatile mediators providing stress-tolerance, hypothesis, *Complex Systems under Extreme Conditions*, Khlebopros R.G., Soukhovolsky V.G. eds., Krasnoyarsk, Russia, 2005, p. 116–133.

Biel K.Y., Fomina I.R., Nishio J.N. The protective role of alcohol with low-molecular-weight in stress resistance, productivity and rate of sugar beet development under increasing water deficiency in soil, 2018. [in preparation]

Biel K.Y., Kolmakov P.V., Lastivka A.N., Titlyanov E.A. Photosynthetic carbon metabolism of the green algae *Ulva fenestrata* in ontogenesis, *Russian Marine Biol.*, 1990, 5, p. 52–58. [in Russian]

Biel K.Y., Kosobryukhov A.A., Fomina I.R., Balakhnina T.I., Ivanov A.A., Shabnova N.I., Kreslavski V.G., Nazarova G.N., Nishio J.N. Protective role of methanol in sugar beet plants at water deficiency in soil, *New and Nontraditional Plants and Prospects of Their Utilization,* Proc. 5th Int. Symp., Russian Univ. Intern. Friendship, Moscow-Pushchino, Russia, 2003, Vol. 3, p. 18–21. [in Russian]

Biel K.Y., Nonomura A.M., Benson A.A., Nishio J.N. The path of carbon in photosynthesis. XXVI. Uptake and transport of methylglucopyranoside throughout plants, *J. Plant Nutr.*, 2010, Vol. 33, No. 6, p. 902–913.

Bigot J., Binet P. Modification de l'activite pectinesterase sous l'action du sel. Relations avec la succulence, *Soc. Bot. Fr.*, 1978, No. 3–4, p. 229–238.

Bird A.P. DNA methylation patterns and epigenetic memory, *Genes Dev.*, 2002, Vol. 16, p. 6–21.

Bird A.P., Wolffe A.P. Methylation-induced repression: belts, braces, and chromatin, *Cell*, 1999, Vol. 99, p. 451–454.

Boldstad P.V., Libby W.J. Comparison of radiate pine cuttings of hedge and tree-form origin after 7 growing seasons, *Silvae Genet.*, 1982, Vol. 31, p. 9–13.

Bryan W., Newcomb E.H. Stimulation of pectin methylesterase activity cultured tobacco pith by indoleacetic acid, *Physiol. Plant.*, 1954, Vol. 7, p. 290–297.

Buchanan J.M., Flaks J.G., Hartman S.C., Levenberg B., Lukens L.N., Warren L. 1957 The enzymatic synthesis of inosinic acid *de novo*, *Ciba Foundation Symposium – Chemistry and Biology of Purines,*

Wolstenholme G.E.W., O'Connor C.M. eds., Ciba Foundation, cited by John Wiley and Sons, Ltd., Chichester, UK, 2008, Chapter 19, p. 233–255.

Burch J.B.E., Weintraub H. Temporal order of chromatin structural changes associated with activation of the major chicken vitellogenin gene, *Cell*, 1983, Vol. 33, p. 65–76.

Byerrum R.U., Hamill U.R.L., Ball C.D. 1954 The incorporation of glycine into nicotine in tobacco plant metabolism, *J. Biol. Chem.*, Vol. 210, p. 645–650.

Calvin M., Benson A.A. The path of carbon in photosynthesis, *Science*, 1949, Vol. 109, p. 140–142.

Cao X., Springer M.N., Muszunski M.G., Philips R.L., Kaeppler S., Jacobsen S.E. Conserved plant genes with similarity to mammalian *de novo* DNA methyltransferases, *Proc. Natl. Acad. Sci. USA*, 2000, Vol. 97, p. 4979–4984.

Chalupka W. Effect of growth regulators on flowering of Norway spruce (*Picea abies* L Karst) grafts, *Silvae Genet.*, 1979, Vol. 28, p. 125–127.

Chang C., Stadler R. Ethylene hormone receptor action in *Arabidopsis*, *BioEssays*, 2001, Vol. 23, No. 7, p. 619–627.

Colas des Francs-Small C., Ambard-Bretteville F., Small I.D., Remy R. Identification of a major soluble protein in mitochondria from non-photosynthetic tissues as NAD-dependent formate dehydrogenase, *Plant Physiol.*, 1993, Vol. 102, p. 1171–1177.

Collmer A., Keen N.T. The role of pectin enzymes in plant pathogenesis, *Annu. Rev. Phytopathol.*, 1986, Vol. 24, p. 383–409.

Corpe W.A., Rheem S. Ecology of methylotrophic bacteria on living leaf surfaces, *FEMS Microbiol. Ecol.*, 1989, Vol. 62, p. 243–250.

Cossins E.A. The utilization of carbon-1 compounds by plants. 1. The metabolism of methanol-C^{14} and its role in amino acid biosynthesis, *Canadian J. Biochem.*, 1964, Vol. 42, p. 1793–1802.

Cossins E.A. Folate biochemistry and the metabolism of one-carbon units, *Biochemistry of Plants*, Davis D.D. ed., Academic Press, San Diego, California, USA, 1987, Vol. 11, p. 317–353.

Creusot F., Acs G., Christmang J.K. Inhibition of DNA methyltransferase and induction of Friend erythroleukemia cell differentiation by 5-azacytidine and 5-aza-2'-deoxycytidine, *J. Biol. Chem.*, 1982, Vol. 257, p. 2041–2048.

Dahan J., Koen E., Dutartre A., Lamotte O., Bourque S. Post-translational modifications of nuclear proteins in the response of plant cells to abiotic stresses, *Abiotic Stress Response in Plants - Physiological, Biochemical and Genetic Perspectives*, Shanker A.K., Venkateswarlu B. eds., InTech, Rijeka, Croatia, 2011, p. 77–113.

David P., Colas des Francs-Small C., Sevignac M., Thareau V., Macadre C., Langin T., Geffroy V. Three highly similar formate dehydrogenase genes located in a cluster of NBS-LRR are differentially expressed under biotic and abiotic stresses in *Phaseolus vulgaris*, *Theor. Appl. Genet.*, 2010, Vol. 121, p. 87–103.

Davison D.C. The distribution of formic acid and alcohol dehydrogenases in the higher plants, with particular reference to their variation in the pea plant during its life cycle, *Proc. Linn. Soc. NS Wales*, 1949, Vol. 74, p. 26–36.

Davison D.C. Studies on formate dehydrogenases, *Biochem. J.*, 1951, Vol. 49, p. 520–526.

De la Torre F., Del Carmen R.-G.M., Matilla A.J. How ethylene works in the reproductive organs of higher plants: a signaling update from the third millennium, *Plant Signal Behav.*, 2006, Vol. 1, No. 5, p. 231–242.

Delincee H. Thin layer isoelectric focusing of multiple forms of pectinsterases, *Phytochemistry*, 1976, Vol. 175, p. 903–906.

Devlin R.M., Bhowmik P.S., Karczmarczyk S.J. Influence of methanol on plant growth, *PGRSA Quarterly*, 1994, Vol. 22, No. 4, p. 102–108.

Devlin R.M., Karczmarczyk S.J., Lopes P.R. The effect of methanol on the growth and pigment content of bachelor's button (*Centaurea cyanus*) and geranium (*Pelargonium hortorus*), *PGRSA Quarterly*, 1995, Vol. 23, No. 2, p. 127–136.

Doman N.G., Romanova A.K. Transformations of labeled formic acid, formaldehyde, methanol, and CO_2 absorbed by bean and barley leaves from air, *Plant Physiol.*, 1962, Vol. 37, No. 6, p. 833–840.

Dong C.H., Jang M., Scharein B., Malach A., Rivarola M., Liesch J., Groth G., Hwang I., Chang C. Molecular association of the *Arabidopsis* ETR1 ethylene receptor and a regulator of ethylene signaling, RTE1, *J. Biol. Chem.*, 2010, Vol. 285, No. 52, p. 40706–40713.

Dorcus D., Vivekanandan M. Methanol-induced growth, biomass, and economic productivity in *Hibiscus esculentus, Vigna radiata,* and *V. catjung* in tropics, *Appl. Biochem. Biotech.*, 1996, Vol. 56, p. 111–114.

Dörffling K *Das Hormonsystem der Pflanzen*, Georg Thieme Verlag, Stuttgart, New York, 1982, Referred by Russian translation: Dörffling K. *Plant Hormones: System Approach*, Kefely V.I. ed., Mir, Moscow, 1985, 304 p.

Dorokhov Y.L., Frolova O.Y., Skurat E.V., Ivanov P.A., Gasanova T.V., Sheveleva A.A., Ravin N.V., Makinen K.M., Klimyuk V.I., Skryabin K.G., Gleba Y.Y., Atabekov J.G. A novel function for a ubiquitous plant enzyme pectin methylesterase: the enhancer of RNA silencing, *FEBS Lett.*, 2006, Vol. 580, p. 3872–3878.

Dorokhov Y.L., Komarova T.V., Petrunia I.V., Kosorukov V.S., Zinovkin R.A., Shindyapina A.V., Frolova O.Y., Gleba Y.Y. Methanol may function as a cross-kingdom signal, *PLoSONE*, 2012, Vol. 7, No. 4, e36122, doi:10.1371/journal.pone.0036122.

Dorokhov Y.L., Komarova T.V., Sheshukova E.V. Volatile organic compounds and plant virus-host interaction, *Plant Virus Host Interaction*, Gaur R.K., Hohn T., Sharma P. eds., Elsevier, Amsterdam, 2014, Chapter 13, p. 241–262.

Dubos C., Plomion C. Identification of water-deficit responsive genes in maritime pine (*Pinus pinaster* Ait.) roots, *Plant Mol. Biol.*, 2003, Vol. 51, p. 249–262.

Dudareva N. Klempien A., Muhlemann J.K., Kaplan I. Biosynthesis, function and metabolic engineering of plant volatile organic compounds, *New Phytol.*, 2013, Vol. 198, p. 16–32.

Dunn D.B., Smith J.D. The occurrence of 6-methylamino-purine in deoxyribonucleic acids, *Biochem J.*, 1958, Vol. 68, p. 627–636.

Evans R., McHale D. Multiple forms of pectinesterase in limes and oranges, *Phytochemistry*, 1978, Vol. 17, p. 1073–1075.

Fall R., Benson A.A. Leaf methanol – the simplest natural product from plants, *Trends Plant Sci.*, 1996, Vol. 1, 296–301.

Fehsenfeld F., Calvert C., Fall R., Goldan P., Guenther A.B., Hewitt C.N., Lamb B., Liu S., Trainer M., Westberg H., Zimmerman P. Emissions of volatile organic compounds from vegetation and the implications for atmospheric chemistry, *Global Biogeochem. Cycles*, 1992, Vol. 6, p. 389–430.

Feng Q., Yang C., Lin X., Wang J., Ou X., Zhang C., Liu Y.C.B. Salt and alkaline stress induced transgenerational alteration in DNA methylation of rice (*Oryza sativa*), *AJCS*, 2012, Vol. 6, No. 5, p. 877–883.

Ferguson T.U. Methanol and the productivity of tropical plants, *Methanol*, Proc. 1995 World Conf., Phoenix, Arizona, USA, 1995, p. 107–115.

Finlay C.J. *Pectic Methyl and Non-Methyl Esters and the Environmental Implications of Methanol Emissions from Plants* (Ph.D. Thesis), University of Glasgow, Glasgow, 2007, 239 p. http://theses.gla.ac.uk/176/1/2007finlay1phd.pdf

Finnegan E.J., Gender R.K., Peacock W.J., Dennis E.S. DNA methylation in plants, *Annu. Rev. Plant Physiol. Plant Mol. Biol.*, 1998, Vol. 49, p. 223–247.

Finnegan E.J., Kovac K.A. Plant DNA methyltransferases, *Plant Mol. Biol.*, 2000, Vol. 43, p. 189–201.

Fisher R.L., Bennett A.B. Role of cell wall hydrolases in fruit ripening, *Annu. Rev. Plant Physiol.*, 1991, Vol. 42, p. 675–703.

Franzen H Über die chemischen Bestand-teile grüner pflanzen. XII Mitt, *Z. Physiol. Chem.*, 1921, Vol. 112, p. 301–316.

Fries M., Ihrig J., Brocklehurst K., Shevchik V.E., Pickersgill R.W. Molecular basis of the activity of the phytopathogen pectin methylesterase, *EMBO J.*, 2007, Vol. 26, p. 3879–3887.

Fry S.C. Cross-linking of matrix polymers in the growing cell walls of angiosperms, *Annu. Rev. Plant Physiol.*, 1986, Vol. 37, p. 165–186.

Fukusaki E., Ikeda T., Shiraishi T., Nishikawa T., Kobayashi A. Formate dehydrogenase gene of *Arabidopsis thaliana* is induced by formaldehyde and not by formic acid, *J. Biosci. Bioeng.*, 2000, Vol. 90, p. 691–693.

Gaffe J., Morvan C., Jauneau A., Demarty M. 1992 Partial purification of flax cell wall pectin methylesterase, *Phytochemistry*, 1992, Vol. 31, p. 761–765.

Gaffe J., Tieman D.M., Handa A.K. Pectin methylesterase isoforms in tomato (*Lycopersicon esculentum*) tissues. Effects of expression of a pectin methylesterase antisense gene, *Plant Physiol.*, 1994, Vol. 105, p. 199–203.

Galbally I.E., Kirstine W. The production of methanol by flowering plants and the global cycle of methanol, *J. Atmos. Chem.*, 2002, Vol. 43, p. 195–229.

Gerber-Huber S., May F.E.B., Westley B.R., Felber B.K., Hosbach H.A., Andres A.C., Ryffel G. In contrast to other *Xenopus* genes the estrogen-inducible vitellogenin genes are expressed when totally methylated, *Cell*, 1983, Vol. 33, p. 43–51.

Goldberg R. Changes in the properties of cell wall pectin methylesterase along the *Vigna radiate* hypocotyls, *Physiol. Plant.*, 1984, Vol. 61, p. 58–63.

Goldberg R., Pierron M., Durand L., Mutaftschiev S. *In vitro* and *in situ* properties of cell wall pectin-methylesterases from mung bean hypocotyls, *J. Exp. Bot.*, 1992, Vol. 43, p. 41–46.

Gout E., Aubert S., Bligny R., Rébeillé F., Nonomura A.R., Benson A.A., Douce R. Metabolism of methanol in plant cells. Carbon-13 nuclear magnetic resonance studies, *Plant Physiol.*, 2000, Vol. 123, p. 287–296.

Greenwood M.S. Rejuvenation of forest trees, *Plant Growth Regul.*, 1987, Vol. 6, p. 1–12.

Gruenbaum Y., Naveh-Many T., Cedar H., Razin A. Sequence specificity of methylation in higher plant DNA, *Nature*, 1981, Vol. 292, p. 860–862.

Guenther A., Hewitt C.N., Erickson D., Fall R., Geron C., Graedel T., Harley P., Klinger L., Lerdau M., McKay W.A., Pierce P., Scholes B., Steinbrecher R., Tallamraju R., Taylor J., Zimmerman P. A global model of natural volatile organic compound emissions, *J. Geophys. Res.*, 1995, Vol. 100, p. 8873–8892.

Guenther A., Zimmerman P.R., Harley P.C., Monson R.K., Fall R. Isoprene and monoterpene emission rate variability: model evolutions and sensitivity analyses, *J. Geophys. Res.*, 1993, Vol. 98, p. 12609–12617.

Guenther A., Zimmerman P.R., Wildermuch M. Natural volatile organic compound emission rates for U.S. woodland landscapes, *Atmos. Environ.*, 1994, Vol. 28, p. 1197–1210.

Hanson A.D., Nelsen C.E. Betaine accumulation and [^{14}C]-formate metabolism in water-stressed barley leaves, *Plant Physiol.*, 1978, Vol. 62, p. 305–312.

Harriman R.W., Tieman D.M.G., Handa A.K. Molecular cloning of tomato pectin methylesterase and its expression in Rutgers, ripening inhibitor, nonripening, and never ripe tomato fruits, *Plant Physiol.*, 1991, Vol. 97, p. 80–87.

Hartz T.K., Mayberry K.S., McGiffen M.E., LeStrange M., Miyao G., Baameur A. Foliar methanol application ineffective in tomato and melon production, *Hort. Sci.*, 1994, Vol. 29 No. 9, p. 1087–1092.

Herman P.L., Ramberg H., Baack R.D., Markwell J., Osterman J.C. Formate dehydrogenase in *Arabidopsis thaliana*: overexpression and subcellular localization in leaves, *Plant Sci.*, 2002, Vol. 163, p. 1137–1145.

Hirano S.S., Upper C.D. Bacterial community dynamics, *Microbial Ecology of Leaves*, Andrews J.H., Hirano S.S. eds., Springer-Verlag, New York, 1991, p. 271–294.

Holland M.A., Polacco J.C. PPFMs and other covert contaminants: is there more to plant physiology than just plant? *Annu. Rev. Plant Physiol. Plant Mol. Biol.*, 1994, Vol. 45, 197–209.

Holliday R., Pugh J.E. DNA modification mechanisms and gene activity during development, *Science*, 1975, Vol. 187, p. 226–232.

Hotchkiss R.D. The quantitative separation of purines, pyrimidines and nucleosides by paper chromatography, *J. Biol. Chem.*, 1948, Vol. 175, p. 315–332.

Hourton-Cabassa C., Ambard-Bretteville F., Moreau F., Davy de Virville J., Remy R., Colas des Francs-Small C. Stress induction of mitochondrial formate dehydrogenase in potato leaves, *Plant Physiol.*, 1998, Vol. 116, p. 627–635.

Hutchison K.W., Singer P.B., Greenwood M.S. Molecular genetic analysis of development and maturation in larch, *Proc. 2nd IU-FRO Working Group on Molecular Genetics*, S2.04.06, June 16–18, Petawawa National Forestry Research Institute, Chalk River, Ontario, Canada, 1987, 26–33.

Hüve K., Christ M.M., Kleist E., Uerlings R., Niinemets Ü., Walter A., Wildt J. Simultaneous growth and emission measurements demonstrate an interactive control of methanol release by leaf expansion and stomata, *J. Exp. Bot.*, 2007, Vol. 58, p. 1783–1793.

Isidorov V.A., Zenkevich I.G., Ioffe B.V. Volatile organic compounds in the atmosphere of forests, *Atmos. Environ.*, 1985, Vol. 19, p. 1–8.

Ito M., Koike A., Koizume N., Sano H. Methylated DNA-binding proteins from *Arabidopsis*, *Plant Physiol.*, 2003, Vol. 133, p. 1747–1754.

Ivanova E.G., Doronina N.V., Shepeliakovskaya A.O., Laman A.G., Brovko F.A., Trotsenko Y.A. Facultative and obligate aerobic methylobacteria synthesize cytokinins, *Mikrobiologiya*, 2000, Vol. 69, 764–769. [in Russian]

Ivanova E.G., Doronina N.V., Trotsenko Y.A. Aerobic methylobacteria are capable of synthesizing auxins, *Microbiology*, 2001, Vol. 70, p. 392–397.

Jones A.W., Skagerberg S., Yonekura J., Sato A. Metabolic interaction between endogenous methanol and exogenous ethanol studied in human volunteers by analysis of breath, *Pharmacol. Toxicol.*, Vol. 66, p. 62–65.

Komarova T.V., Sheshukova E.V., Dorokhov Y.L. Cell wall methanol as a signal in plant immunity, *Frontiers Plant Sci.*, 2014*a*, Vol. 5, article 101.

Komarova T.V., Pozdyshev D.V., Petrunia I.V., Sheshukova E.V., Dorokhov Y.L. Pectin methylesterase-generated methanol may be involved in tobacco leaf growth, *Biochemistry (Moscow)*, 2014*b*, Vol. 79, p. 102–110.

Kosobryukhov A.A., Bil' K.Y., Nishio J.N. Sugar beet photosynthesis under conditions of increasing water deficiency in soil and protective effects of a low-molecular-weight alcohol, *Appl. Biochem. Microbiol.*, 2004, Vol. 40, No. 6, p. 581–587.

Knox J.P., Linstead P.J., King J., Cooper C., Roberts K. Pectin esterification is spatially regulated both within cell walls and between developing tissues of root apices, *Planta*, 1990, Vol. 181, p. 512–521.

Lamport D.T.A., Catt J.W. Glycoproteins and enzymes of the cell wall, *Encyclopedia of Plant Physiology, New Series, Vol 13B: Plant Carbohydrates II*, Tanner W., Loewus F.A. eds., Springer-Verlag, Berlin-Heidelberg-New York, 1981, p. 132–165.

Large P.J., Peel D., Quayle J.R. Microbial growth on C_1 compounds. 2. Synthesis of cell constituents by methanol- and formate-grown *Pseudomonas* AM 1, and methanol-grown *Hyphomicrobium vulgare*, *Biochem. J.*, 1961, Vol. 81, p. 470–480.

Large P.J., Peel D., Quayle J.R. Microbial growth on C_1 compounds. 3. Distribution of radioactivity in metabolites of methanol-grown *Pseudomonas* AM 1 after incubation with ^{14}C-methanol and ^{14}C-bicarbonate, *Biochem. J.*, 1962, Vol. 82, p. 483–488.

Levy S., Staehelin L.A. Synthesis, assembly and function of plant cell wall macromolecules, *Curr. Opin. Cell Biol.*, 1992, Vol. 4, p. 856–862.

Li Y., Gupta G., Joshi J.M., Siyumbano A.K. Effect of methanol on soybean photosynthesis and chlorophyll, *J. Plant Nutr.*, 1995, Vol. 18, No. 9, p. 1875–1880.

Lidstrom M.E., Chistoserdova L. Plants in the pink: Cytokinin production by *Methylobacterium*, *J. Bacteriol.*, 2002, Vol. 184, No. 7, p. 1818–1818.

Liners F., Van Cutsem P. Distribution of pectic polysaccharides throughout walls of suspension-cultured carrot cells, *Protoplasma*, 1992, Vol. 170, p. 10–21.

Lund E.D., Kirkland C.L., Shaw P.E. Methanol, ethanol and acetaldehyde contents of citrus products, *J. Agr. Food Chem.*, 1981, Vol. 29, p. 361–366.

MacDonald R.C., Fall R. Detection of substantial emissions of methanol from plants to the atmosphere, *Atmos. Environ.*, 1993, Vol. 27A, p. 1709–1713.

Makhdum M.I., Malik M.N.A., Din S.U., Ahmad F., Chaudhry F.I. Physiological response of cotton to methanol foliar application, *J. Res. Sci.*, 2002, Vol. 13, p. 37–43.

Mazelis M. Formate oxidation by particulate preparations from higher plants, *Plant Physiol.*, 1960, Vol. 35, p. 386–391.

McCann M.C., Roberts K. Architecture of the primary cell wall, *The Cytoskeletal Basis of Plant Growth and Form*, Lloyd C.W. ed., Academic Press, San Diego, California, USA, 1991, p. 109–129.

McCann M.C., Shi J., Roberts K., Carpita N.C. Changes in pectin structure and localization during the growth of unadapted and NaCI-adapted tobacco cells, *Plant J.*, 1994, Vol. 5, p. 773–780.

McCann M.C., Wells B., Roberts K. Direct visualization of cross-links in the primary plant cell wall, *J. Cell Sci.*, 1990, Vol. 96, p. 323–334.

McGiffen M.E., Green R.L., Manthey J.A., Faber B.A., Downer J., Sakovich N.J., Aguiar J. Field tests of methanol as a crop yield enhancer, *Hort. Sci.*, 1995, Vol. 30, p. 1225–1228.

McGiffen M.E., Manthey J.A. The role of methanol in promoting plant growth: a current evaluation, *Hort. Sci.*, 1996, Vol. 31, p. 1092–1096.

Micheli F. Pectin methylesterases: cell wall enzymes with important roles in plant physiology, *Trends Plant Sci.*, 2001, Vol. 6, p. 414–419.

Mimami A., Nagao M., Arakawa K., Fujikawa S., Takezawa D. Abscisic acid-induced freezing tolerance in the moss *Physcomitrella patens* is accompanied by increased expression of stress-related genes, *J. Plant Physiol.*, 2003, Vol. 160, p. 475–483.

Mitchell A.R., Crow F.J., Butler M.D. Plant performance and water use of peppermint treated with methanol and glycine, *J. Plant Nutr.*, 1994, Vol. 17, p. 1955–1962.

Moustacas A.M., Nari J., Borel M., Noat G., Ricard J. Pectin methylesterase, metal ions and plant cell-wall extension. The role of metal ions in plant cell-wall extension, *Biochem. J.*, 1991, Vol. 279, p. 351–354.

Mudgett M.B., Clarke S. Characterization of plant L-isoaspartyl methyltransferases that may be involved in seed survival: purification, cloning, and sequence analysis of the wheat germ enzyme, *Biochemistry*, 1993, Vol. 32, p. 11100–11111.

Muromtsev G.S., Chkanikov D.I., Kulaeva O.N., Gamburg K.Z. *The Basis of Chemical Regulation of Plant Growth and Productivity*, Agropromizdat, Moscow, 1987, 384 p. [in Russian]

Nadali I., Paknejad F., Moradi F., Vazan S., Tookalo M., Jami Al-Ahmadi M., Pazoki A. Effects of methanol on sugar beet (*Beta vulgaris*), *AJCS*, 2010, Vol. 4, No. 6, p. 398–401.

Nari J., Noat G., Ricard J. Pectin methylesterase, metal ions and plant cell-wall extension: hydrolysis of pectin by plant cell-wall pectin methylesterase, *Biochem. J.*, 1991, Vol. 279, p. 343–350.

Nemecek-Marshall M., MacDonald R.C., Franzen J.J., Wojciechowski C.L., Fall R. Methanol emission from leaves. Enzymatic detection of gas-phase methanol and relation of methanol fluxes to stomatal conductance and leaf development, *Plant Physiol.*, 1995, Vol. 108, p. 1359–1368.

Nishio J.N., Bil' K.Y., Kosobryukhov A.A., Fomina I.R. Photosynthesis and stress tolerance are improved by methanol and other compounds, *Abstr. 8th Western Regional Photosynthesis Conf.*, Asilomar Conf Center, Pacific Grove, California, 1999, p. 13.

Nishio J.N., Huang S., Winder T.L., Brownson M.P., Ngo L. Physiological aspects of methanol feeding to higher plants, *Plant Growth Regulator Society of America*, Proc. 20th Annual Meeting, Ferguson L. ed., St Louis, USA, 1993, p. 8–13.

Nishio J.N., Nonomura A.M., Benson A.A., Bil' K.Y. Methanol and other compounds stimulate higher plant photosynthesis, growth, and water use efficiency, *Bioenergetic of Photosynthesis*, Abstr. Int. Conf., 17–21 June 1996, Pushchino, Russia, 1996, p. 23.

Nishiyama R., Ito M., Yamagushi Y., Koizumi N., Sano H. A chloroplast-resident DNA methyltransferase is responsible for hyper-methylation of chloroplast genes in *Chlamydomonas* maternal gametes, *Proc. Natl. Acad. Sci. USA*, 2002, Vol. 99, p. 5925–5930.

Nonomura A.M., Benson A.A. The path of carbon in photosynthesis: improved crop yields with methanol, *Proc. Natl. Acad. Sci. USA*, 1992*a*, Vol. 89, p. 9794–9798.

Nonomura A.M., Benson A.A. The path of carbon in photosynthesis: stimulation of crop yields with methanol, *Photosynth. Res.*, 1992*b*, Vol. 34, Abstr. P-589.

Nonomura A.M., Benson A.A. The path of carbon in photosynthesis: methanol and light, *Research in Photosynthesis*, Proc. IX Int. Congr., Murata N. ed., Kluwer Acad. Publ., Dordrecht, The Netherlands, 1992*c*, Vol. 3, p. 991-914.

Nonomura A.M., Benson A.A. Methods and compositions for enhancing carbon fixation in plants, *United States Patent*, number 55974, 1997.

Nonomura A.M., Benson A.A. The path of carbon in photosynthesis. XXIX. Glass microbeads, *J. Plant Nutr.*, 2012, Vol. 35, p. 1896–1909.

Nonomura A.M., Benson A.A. The path of carbon in photosynthesis. XXX. α-mannosides, *Photosynthesis*, Dubinsky Z. ed., InTech, Janeza Trdine 9, 51000 Rijeka, Croatia, 2013, Chapter 1, p. 3–22.

Nonomura A.M., Benson A.A., Biel K.Y. The path of carbon in photosynthesis. XXVII. Sugar-conjugated plant growth regulators enhance productivity, *J. Plant Nutr.*, 2011, Vol. 34, No. 5, p. 653–664.

Nonomura A.M., Cullen B.A., Benson A.A. The path of carbon in photosynthesis. XXVIII. Response of plants to polyalkylglucopyranose and polyacylglucopyranose, *Advances in Photosynthesis - Fundamental Aspects*, Najafpour M.M. ed., InTech, Janeza Trdine 9, 51000 Rijeka, Croatia, 2012, Chapter 13, p. 259–272.

Nonomura A.M., Nishio J.N., Benson A.A. Methods and compositions for enhancing plant growth with p-amino- or p-nitro-benzoic acids, *United States Patent*, number 5846908 A, 1998.

Northcote D.H. Control of pectin synthesis and deposition during plant cell growth, *Chemistry and Function of Pectin*, Fishman M.L., Jen J.J. eds., American Chemical Society, Washington DC, 1986, p. 134–140.

Niu L., Lu F., Pei Y., Liu C., Cao X. Regulation of flowering time by the protein arginine methyltransferase AtPRMT10, *EMBO Reports*, 2007, Vol. 8, p. 1190–1195.

Oakeley E.J., Jost J.-P. Nonsymmetrical cytosine methylation in tobacco pollen DNA, *Plant Mol. Biol.*, 1996, Vol. 31, p. 927–930.

Oberdorf R.L., Koch J.L., Gorecki R.J., Amable R.A., Aveni M.T. Methanol accumulation in maturing seeds, *J. Exp. Bot.*, 1990, Vol. 41, p. 489–495.

Odintsova M.C. DNA of chloroplasts and mitochondria, *Overall Results of Science and Technique*, Biological Chemistry, VINITI, Moscow, 1976, 177 p. [in Russian]

Olson B.J., Skavdahl M., Ramberg H., Osterman J.C., Markwell J. Formate dehydrogenase in *Arabidopsis thaliana*: characterization and possible targeting to the chloroplast, *Plant Sci.*, 2000, Vol. 159, p. 205–212.

Ozimek P., Veenhuis M., van der Klei I.J. Alcohol oxidase: A complex peroxisomal, oligomeric flavoprotein, *FEMS Yeast Res.*, 2005, Vol. 5, p. 975–983.

Pecinka A., Dinh H.Q., Baubec T., Rosa M., Lettner N., Mittelsten Scheid O. Epigenetic regulation of repetitive elements is attenuated by prolonged heat stress in *Arabidopsis*, *Plant Cell*, 2010, Vol. 22, p. 3118–3129.

Pichersky E., Lewinsohn E. Convergent evolution in plant specialized metabolism, *Annu. Rev. Plant Biol.*, 2011, Vol. 62, p. 549–566.

Pilling J., Willmitzer L., Bucking H., Fisahn J. Inhibition of a ubiquitously expressed pectin methyl esterase in *Solanum tuberosum* L. affects plant growth, leaf growth polarity, and ion partitioning, *Planta*, 2004, Vol. 219, p. 32–40.

Pilling J., Willmitzer L., Fisahn J. Expression of a *Petunia inflata* pectin methyl esterase in *Solanum tuberosum* L. enhances stem elongation and modifies cation distribution, *Planta*, 2000, Vol. 210, p. 391-399

Pressey R., Avants J.K. Multiple forms of pectinesterase in tomatoes, *Phytochemistry*, 1972, Vol. 11, p. 3139–3142.

Pressey R., Avants J.K. Solubilization of cell walls by tomato polygalacturonase: effects of pectin methylesterase, *J. Food Biochem.*, 1982, Vol. 6, p. 57–74.

Quatrano R.S., Bartels D., Ho T.-H.D., Pages M. New insights into ABA-mediated processes, *Plant Cell*, 1997, Vol. 9, No. 4, p. 470–475.

Quayle J.R., Fuller R.C., Benson A.A., Calvin M. 1954 Enzymatic carboxylation of ribulose diphosphate, *J. Am. Chem. Soc.*, 1954, Vol. 76, p. 3610–3611.

Rajala A., Kärkkäinen J., Peltonen J., Peltonen-Sainio P. Foliar applications of alcohols failed to enhance growth and yield of C_3 crops, *Indust. Crops Prod.*, 1998, Vol. 7, No. 2-3, p. 129–137.

Rakitin Y.V. Ethylene as high-effective defoliant, *Soviet Plant Physiol.*, 1967, Vol. 14, p. 936–951. [in Russian]

Rakitin Y.V., Krylov A.V., Tarakanova G.A. Conversions of methanol in fruit when applied for stimulation of ripening, *Proc. USSR Acad. Sci.*, 1957, Vol. 11, p. 874–877.

Ramírez I., Dorta F., Espinoza V., Jiménez E., Mercado A., Peña-Cortés H. Effects of foliar and root applications of methanol on the growth of *Arabidopsis*, tobacco, and tomato plants, *J. Plant Growth Regul.*, 2006, Vol. 25, No. 1, p. 30–44.

Razin A., Cedar H. DNA methylation and gene expression, *Microbiol. Rev.*, 1991, Vol. 55, No. 3, p. 451–458.

Razin A., Riggs A.D. DNA methylation and gene function, *Science*, 1980, Vol. 210, p. 604–610.

Razin A., Szyf M. DNA methylation patterns: formation and function, *Biochim. Biophys. Acta*, 1984, Vol. 782, p. 331–342.

Recourt K., Ebelelaar M.E.M., Barbisan P., Laats M.M., Wichers H.W. Accession X68028, X68029, X68030, *EMBL Databank*, 1992.

Ren C., Kermode A.R. 2000 An increase in pectin methylesterase activity accompanies dormancy breakage and germination of yellow cedar seeds, *Plant Physiol.*, 2000, Vol. 124, p. 231–242.

Ricard J., Noat G. Electrostatic effects and the dynamics of enzyme reactions at the surface of plant cells. 1. A theory of the ionic control of a complex multi-enzyme system, *Eur. J. Biochem.*, 1986, Vol. 155, p. 199–202.

Riggs A.D. X-inactivation, differentiation and DNA-methylation, *Cytogenet. Cell Genet.*, 1975, Vol. 14, p. 9–25.

Romanova A.K., Vedenina I.Y., Zykalova K.A., Ermolenko Z.M., Semenova L.R. Pathways of assimilation of carbon in methylotrophy of *Pseudomonas gazotropha*, *Microbiology*, 1978, Vol. 47, No. 4, p. 492–498.

Ronen M., Mayak S. Interrelationships between abscisic acid and ethylene in the control of senescence processes in carnation flowers, *J. Exp. Bot.*, 1981, Vol. 32, p. 759–765.

Roshchina V.V., Roshchina V.D. *The Excretory Function of Higher Plants*, Springer-Verlag, Berlin, Heidelberg, 1993, 314 p.

Roshchina V.V., Roshchina V.D. *The Excretory Function of Higher Plants*, 3rd Ed., LAP Lambert Acad. Publ. GmbH, Saarbrucken, 2012, 476 p. [in Russian]

Rowe R.N., Farr D.J., Richards R.A.J. Effect of foliar and root applications of methanol or ethanol on the growth of tomato plants (*Lycopersicon esculentum* Mill), *New Zealand J. Crop Hort. Sci.*, 1994, Vol. 22, p. 335–337.

Sakami W. *Handbook of Isotope Tracer Methods. Department of Biochemistry*, Western Reserve University, Cleveland, Ohio, 1955.

Sedzher P. *Cytoplasmic Genes and Organelles*, Mir, Moscow, 1975, 423 p. [in Russian]

Sexton R., Roberts J.A. Cell biology of abscission, *Annu. Rev. Plant Physiol.*, 1982, Vol. 33, p. 133–162.

Shea E.M., Gibeaut D.M., Carpita N.C. Structural analysis of the cell walls regenerated by carrot protoplasts, *Planta*, 1989, Vol. 179, p. 293–308.

Shinozaki K., Yamagushi-Shinozaki K. Molecular responses to dehydration and low temperature: differences and cross-talk between two stress signaling pathways, *Curr. Opin. Plant Biol.*, 2000, Vol. 3, p. 217–223.

Sinha S.K. *Role of Carbon-One and Carbon-Two Compounds in the Biosynthesis of Amino Acids in Higher Plant Tissues* (PhD Thesis), University of Alberta, Edmonton, Alberta, 1964, 233 p.

Srivastava L.M. *Plant Growth and Development: Hormones and Environment*, Academic Press, Amsterdam, 2002, 772 p.

Stiller M. The path of carbon in photosynthesis, *Annu. Rev. Plant Physiol.*, 1962, Vol. 13, p. 151–170.

Stiller M.L., Neal G.E., Beevers H. CO_2 fixation during the conversion of fat to carbohydrate in the castor bean, *Plant Physiol.*, 1958, Vol. 33, suppl., p.34.

Tucker G.A., Robertson N.G., Grierson D. Purification and changes in activities of tomato pectinesterase isoenzymes, *J. Sci. Food Agric.*, 1982, Vol. 33, p. 396–400.

Vanyushin B.F. Enzymatic DNA methylation is an epigenetic control for genetic functions of the cell, *Biochemistry (Moscow)*, 2005, Vol. 70, p. 488–499.

Vanyushin BF A View of an elemental naturalist at the DNA world (base composition, sequences, methylation), *Biochemistry (Moscow)*, 2007, Vol. 72, No. 12, p. 1289–1298.

Vanyushin B.F., Belozersky A.N., Kokurina N.A., Kadirova D.X. 5-methylcytosine and 6-methylaminopurine in bacterial DNA, *Nature*, 1968, Vol. 218, p. 1066–1067.

Vanyushin B.F., Dobritsa A.P. On the nature of the cytosine-methylated sequence in DNA of *Bacillus brevis* var. G.-B., *Biochim. Biophys. Acta*, 1975, Vol. 407, p. 61–72.

Versteeg C., Rombouts F.M., Pilnik W. Purification and some characteristics of two pectinesterase isoenzymes from orange, *Lebensm. Wiss. Technol.*, 1978, Vol. 11, p. 267–274.

Walker D.A. Pyruvate carboxylation and plant metabolism, *Biol. Rev. Cambridge Phil. Soc.*, 1962, Vol. 37, p. 215–256.

Wang K.L., Li H., Ecker J.R. Ethylene biosynthesis and signaling networks, *Plant Cell*, 2002, Vol. 14, Suppl., S131–S151.

Wareing P.F. Root hormones and shoot growth, *Control of Shoot Growth in Trees*, Proc. IUFRO Workshop, July 2-24, 1980, Little C.H.A. ed., Fredericton, NB, Canada, 1980, p. 237–256.

Webber J.E., Ross S.D., Pharis R.P., Owens J.N. Interaction between gibberellin $A_{4/7}$ and root-pruning on the reproductive and vegetative process in *Douglas fir*. II. Effects on shoot elongation and its relationship to flowering, *Can. J. For. Res.*, 1985, Vol. 15, p. 248–353.

Wen F., Zhu Y., Hawes M.C. Effect of pectin methylesterase gene expression on pea root development, *Plant Cell*, 1999, Vol. 11, p. 1129–1140.

White J.W. Composition of a volatile fraction in apples, *Food Res.*, 1950, Vol. 15, p. 68–78.

Wilks A.F., Cozens P.J., Mattaj I.W., Jost J.P. Estrogen induces a demethylation at the 5' end region of the chicken vitellogenin gene, *Proc. Natl. Acad. Sci. USA*, 1982, Vol. 79, p. 4252–4255.

Winer A.M., Arey J., Atkinson R., Aschmann S.M., Long W.D., Morrison C.L., Olszyk D. Emission rates of organics from vegetation in California's central valley, *Atmos. Environ.*, 1992, Vol. 26A, p. 2647–2659.

Wyatt G.R. Occurrence of 5-methylcytosine in nucleic acid, *Nature*, 1950, Vol. 166, p. 237–238.

Yamaoka T., Chiba N. Changes in coagulating ability of pectin during growth of soybean hypocotyls, *Plant Cell Physiol.*, 1983, Vol. 24, p. 1281–1290.

Yang Y., Yi X. Emission and utilization of methanol in higher plants, *Ecol. Environ.*, 2006, Vol. 15, No. 6, p. 1258–1263.

Yoosefi S.M.M., Vishekaei M.N.S., Noormohammadi G., Niyaki S.A.N. Effect of foliar spraying by methanol on growth of common bean and snap bean in Rasht, North of Iran, *Res. J. Biol. Sci.*, 2011, Vol. 6, p. 47–50.

Zbieć I., Karczmarczyk S Podsiadło C. Response of some cultivated plants to methanol as compared to supplemental irrigation, *Electronic J. Polish Agricult.* University, Agronomy, 2003, Vol. 6, No. 1.

Zhao Q., Guo H.W. Paradigms and paradox in the ethylene signaling pathway and interaction network, *Mol. Plant*, 2011, Vol. 4, p. 626–634.

Zho J.K. Salt and drought stress signal transduction in plants, *Annu. Rev. Plant Biol.*, 2002, Vol. 53, p. 247–273.

Zimmerman R.H., Hackett W.P., Pharis R.P. 1985 Hormonal aspects of phase change and precocious flowering, *Encyclopedia of Plant Physiology* (New Series), Pharis R.P., Reid D.M. eds., Springer-Verlag, Heidelberg, 1985, Vol. 11, p. 79–115.

EXPERIMENTS

7
Patterns of Carbon Metabolism within Leaves

Karl Y. Biel[1,2,*], Irina R. Fomina[1,2,*], Galina N. Nazarova[1], Vladislav G. Soukhovolsky[3,4], Rem G. Khlebopros[3], John N. Nishio[5,*]

[1]*Institute of Basic Biological Problems, Russian Academy of Sciences, Pushchino, Moscow Region 142290, Russia*
[2]*Biosphere Systems International Foundation, Tucson, Arizona 85755, USA*
[3]*International Scientific Centre for Organism Extreme States Research, Krasnoyarsk Scientific Centre, Siberian Branch of the Russian Academy of Sciences, Krasnoyarsk 660036, Russia*
[4]*Institute of Forest, Siberian Branch of the Russian Academy of Sciences, Krasnoyarsk 660036, Russia*
[5]*Biocompatible Plant Research Institute, Chico, California 95973, USA*

Leaves are heterogeneous. Grinding a whole leaf to measure biochemical activity to understand leaf metabolism is like grinding a whole frog to do the equivalent for frog metabolism, –
Professor John N. Nishio

Abstract
The localization of the key photoreductive and oxidative processes and some stress-protective reactions within the leaf of mesophytic C_3 plants were investigated. The role of light in determining the profile of Rubisco, glutamate oxaloacetate transaminase, catalase, fumarase and cytochrome-*c*-oxidase across spinach leaves was examined in leaves exposed to full illumination by adaxial or abaxial surface. Oxygen evolution in fresh paradermal leaf sections and CO_2 gas exchange in whole leaves under

* *Corresponding authors:* karlbiel@hotmail.com; irafomi@rambler.ru; nishio@biocompatible.org

adaxial or abaxial illumination was also examined. The results showed that the palisade mesophyll is responsible for the midday depression of photosynthesis in spinach leaves. The photosynthetic apparatus was more sensitive to the light environment than the respiratory apparatus. Additionally, examination of the paradermal leaf sections by optical microscopy allowed describing two new types of parenchyma in spinach (*Spinacia oleracea*) - pirum mesophyll and pillow spongy mesophyll. A hypothesis about the role of oxaloacetic acid in the biochemical protection of the leaf-top mesostructure from destructive influence of active oxygen forms is presented. The application of the mathematical modeling shows that pattern of enzymatic ensemble distribution goes by principle of maximal ecological utility. Light regulation of carbon metabolism across leaves is discussed.

Keywords: *Spinacia olerace*a, Carbon metabolism, Enzyme activity, Ecological utility

7.1 Introduction

The focus of the present chapter is carbon metabolism within mesophytic leaves. The chapter is divided into two parts. The first part focuses on the complex interactions among light, leaf anatomy, the distribution of metabolic activity, and environmental stress tolerance across leaves. After a brief literature review of leaf anatomy, light absorption within leaves, and metabolic specialization across leaves, we will turn our attention to how we used the direction of light on a mesophytic leaf to provide important information about the complexity of the metabolic specialization within leaves. The second part presents an updated model of optimal photosynthesis within a mesophytic leaf is presented.

7.2 Interactions among Light, Leaf Anatomy, the Metabolic Activity, and Environmental Stress Tolerance across Leaves

Light is the ultimate source of energy for photosynthetic organisms. However, excessive light may result in production of active oxygen species during photosynthesis and cause photoinhibition (Osmond *et al.*, 1997). The compromise between efficient absorption and utilization of light under non-saturating light conditions and avoiding photodamage under over-saturating light conditions is likely linked to leaf anatomical adaptations, as well as physical, biochemical, and biophysical adaptations. Many different leaf anatomies (Clements, 1905; Fahn, 1990; Biel, 1988, 1993); modes of photosynthesis, including C_3 or reductive pentose phosphate Benson–Bassham-Calvin cycle (Benson and Calvin, 1950*a*, 1950*b*; Benson, 1952; Bassham, Benson *et al.*, 1954; Bassham and Calvin, 1957; Bassham and Buchanan, 1982; Benson, 2002; Biel and Fomina, 2015; see also Chapter 11 in this Book, and others), C_4 or cooperative photosynthesis (Hartt *et al.*, 1954; Nezgovorova, 1956, 1957; Karpilov, 1960; Tarchevsky and Karpilov, 1963; Kortschak *et al.*, 1965; Hatch and Slack, 1970; see also Nickell, 1993; Bil', 2000, and others), CAM (Osmond and Allaway,1974; Kluge and Ting, 1978; Osmond, 1978; Monson, 1989, and others), C_3 / C_4 photosynthesis (Monson *et al.*, 1984), and C_4 - CAM photosynthesis (Bil' *et al.*, 1981; Biel *et al.*, 1983; Nishio and Ting, 1987; 1993; Biel, 1988, 1993); and avenues for dealing with high light (see ref. in Sun *et al.*, 1996*a*; Nishio, 2000) are found in nature. Such adaptations attest to the many different successful strategies utilized by higher plants for survival under highly varied environmental conditions.

A steep gradient of light from the top to the bottom of leaves (Terashima and Saeki, 1983; Cui *et al.*, 1991; Vogelmann, 1993) and epidermal focusing of light (Vogelmann, 1993) means that cells and chloroplasts within a bifacial leaf can experience extremely different light conditions. Since light is an important regulator of plant development (Tobin and Silverthorne, 1985; Tobin and Kehoe, 1994), cells and organelles may develop with different metabolic activity depending on position within the leaf. For example, individual leaves contain a gradient of "sun" and "shade"-type chloroplasts (Outlaw, 1987; Terashima, 1989; Nishio *et al.*, 1993). In *Camellia* leaves, the spongy mesophyll is highly light scattering (Terashima and Saeki, 1983), so the light gradient is accentuated when leaves are illuminated abaxially. It was shown unequivocally that in intact *Spinacia oleracea* leaves labeled with $^{14}CO_2$ under illumination directed towards the abaxial surface, the majority of carbon fixation occurs in the spongy mesophyll, and that relatively little fixation occurs in the palisade mesophyll (Sun and Nishio, 2001). Such a pattern of fixation under abaxial illumination must be due to light scattering and light trapping in the spongy mesophyll (Sun and Nishio, 2001). On the other hand, the palisade mesophyll allows light to pass to the spongy mesophyll (Vogelmann and Martin, 1993).

Thus, leaves developed originally under adaxial illumination will experience profound changes in the light microenvironment within the leaves, when the leaves are inverted. Illumination of the abaxial surface of leaves during leaf expansion inverted the intra-leaf gradient of the thylakoid number in chloroplasts (Terashima *et al.*, 1986).

It was also found that CO_2 fixation rates, Rubisco content, and Rubisco activity across *Spinacia oleracea* leaves illuminated on the adaxial leaf surface increase from the adaxial leaf surface to a maximum midway through the palisade mesophyll, about 20–40% deep within the leaf, and gradually decreases towards the abaxial leaf surface (Nishio *et al.*, 1993; Fomina, Biel *et al.*, 2001*a*, 2001*b*; Sun and Nishio, 2001; Bil' *et al.*, 2003; Bil', Fomina *et al.*, 2003; Biel, Fomina *et al.*, 2008, 2010). The role of light in shaping the distribution profiles of Rubisco and other enzymes across leaves is not known.

The distribution of chlorophyll across leaves is similar to that of Rubisco, in that the maximum chlorophyll content is in the middle of the leaves. However, a direct correlation between chlorophyll and Rubisco does not always exist (Nishio *et al.*, 1993). Consequently, the maximum Rubisco in the middle of leaves is not necessarily linked to the absorptive photosynthetic pigments, and the distribution of Rubisco across leaves is likely optimal for photosynthesis in bifacial leaves, such as *Spinacia oleracea* (Sun *et al.*, 1996*a*, 1996*b*) and could be the result of optimizing light utilization across the whole leaf.

One of our working hypotheses is that the lower CO_2 fixation rates and Rubisco content at the top of the leaves, compared to midway through the leaves, is due to higher light fluxes that can be toxic; whereas the lower CO_2 fixation rates and Rubisco content towards the bottom of the leaf are due to light limitations. The basic hypothesis is supported by the distribution of photosynthetic activity, such as that of the nitrogen-metabolizing enzyme, glutamate oxaloacetate transaminase, which follows the profile of Rubisco due to the high level of metabolism associated with Rubisco, as discussed below. We also discuss in this chapter how enzymes involved in the detoxification of active oxygen species, such as catalase exhibit high activity at the illuminated leaf surface, because of the possibility of excess irradiation (Cui *et al.*, 1991; Nishio *et al.*, 1994; Sun *et al.*, 1996*a*; Nishio, 2000).

By inverting leaves, the light gradient is reversed; however, leaf inversion of mature leaves does not alter leaf anatomy. Unlike leaf anatomy, the activity of enzymes was distinctly changed by leaf inversion. It will be discussed here how a prolonged leaf inversion affects the distribution profiles of Rubisco, glutamate oxaloacetate transaminase (GOT), catalase, NAD- and NADP-malate dehydrogenase (NAD- and NADP-MDH), fumarase, and cytochrome-*c*-oxidase in *Spinacia oleracea* leaves. We also present data on the effect of leaf inversion on oxygen evolution in fresh paradermal leaf sections and whole leaf gas exchange.

Finally, an overall discussion of the complex interactions among light, leaf anatomy, the distribution of metabolic activity, and environmental stress tolerance across leaves is presented.

7.2.1 Anatomy and Pattern of Enzymes within the Leaf of Spinacia oleracea

7.2.1.1 Leaf Anatomy

Leaves are composed of many different tissues. A mesophytic, bifacial C_3 leaf, such as that of spinach, has an upper and lower epidermis, with or without stomata. Underlying the upper epidermis is palisade mesophyll, which may be one to several cell layers thick and is composed of rod shaped cells, with the long axis perpendicular to the leaf surface. Below the palisade mesophyll is the spongy mesophyll, which is composed of loosely packed, irregularly shaped cells. The palisade mesophyll and spongy mesophyll are the main photosynthetic tissues. The spongy mesophyll is typified by large air cavities, and the palisade mesophyll has a large surface area exposed to air. The vascular tissue, intercalated between the palisade mesophyll and spongy mesophyll, is cytoplasmically dense and metabolically very active (Sun *et al.*, 1996*b*). For experiments we typically utilize the 5th–8th true leaves of *Spinacia oleracea*.

Under the used growth conditions, the palisade mesophyll from such leaves consists of two layers of cylindrical cells with the long axis perpendicular to the leaf surface (illustrated in Figure 7.1). As mentioned earlier, the internal light gradient in leaves will be greatly altered by the direction of light entering leaves, so inverting leaves could lead to significant changes in the leaves. Compared to control leaves, we could detect no anatomical differences, either in leaf cross sections or in paradermal leaf sections from leaves inverted for 15 days (not shown). In the enzyme section, many of the changes that have been observed in enzyme activity after leaf inversion are discussed.

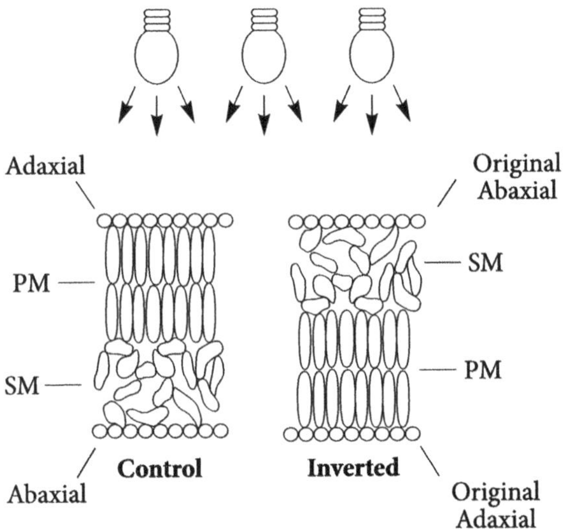

Fig. 7.1. Experimental conditions for leaf inversion experiments. A cross section of a control leaf is illustrated on the left, and the inverted leaf on the right. The leaf surfaces are labeled as adaxial and abaxial for top and bottom in the controls. In the inverted leaves, the abaxial leaf surface is exposed to the incident light. The illuminated leaf surface of the inverted leaves is referred to as the "original abaxial" leaf surface, and the bottom surface is referred to as the "original adaxial" leaf surface. The reversal of spongy mesophyll from the bottom to the top in the inverted leaves is clearly illustrated

In freshly cut paradermal leaf slices, prepared using a lab-built microvibratome (Skupchenko, 1979), the palisade mesophyll viewed from the bottom at two different magnifications appear as round cells (Figure 7.2). The arrangement of chloroplasts along the wall, with the central part from the top to the bottom free of chloroplasts make the cylindrical cells appear as "light tunnels" (Figure 7.2A). Chloroplasts were not observed at the top or bottom of palisade mesophyll cells in light-adapted leaves or in leaves maintained in the dark for 48 hours. Additionally, chloroplasts did not move to the open space at the top and bottom of the palisade mesophyll after leaves were inverted for 12–15 days. At lower magnification, the light-transmitting quality of the palisade mesophyll cells is further illustrated. The more extensive green areas are due to the surface along the long axis of palisade mesophyll cells that are not vertical and / or attached broken cells (Figure 7.2B).

Panels 7.2A and 7.2B illustrate the minimal contact between the long axes of the palisade mesophyll cells and consequently the large surface exposed area of the palisade mesophyll. The minimal contact between the long axes of palisade mesophyll cells was confirmed in a separate experiment. In thin paradermal slices of palisade mesophyll without the upper epidermis, the palisade mesophyll cells separated from each other as groups of cells during the preparation of the slices (not shown).

The tissue between the columnar palisade mesophyll and spongy mesophyll, that contained the vascular tissue, exhibited pear-shaped cells with strong connections to each other and to the idioblasts (Figure 7.2C and 7.2D). The tissue was one cell thick, and like the spongy mesophyll, contained large air spaces around the cells. In *Glycine max*, similar cells are referred to as paraveinal mesophyll (Fisher, 1967). Based on cross-sections of *Cucumis sativus* leaves, the veinal cells were referred to as the "intermediate layer of parenchyma" (Chugunova *et al*., 1976), and Dr. G. Haberlandt (1914) referred to them as "funnel-cells" or "collecting-cells".

We call the special layer of palisade cells, the pirum mesophyll. The chloroplasts were equally distributed along the whole surface of pirum mesophyll cells; however, in some pirum mesophyll cells, the areas clearly lacking chloroplasts were observed (Figure 7.2C). These areas are likely the direct connection points with the overlying palisade mesophyll cells (see Haberlandt, 1914). *Spinacia oleracea* leaf idioblasts are characterized by their spherical form and calcium oxalic acid crystal inside the cell. Each idioblast was in close contact with surrounding pirum mesophyll cells that appeared like spokes of a wheel (Figure 7.2C and 7.2D). Figure 7.2C illustrates three distinctly different sizes of oxalic acid crystals, which likely are related to the length of time that oxalic acid has been accumulating. The

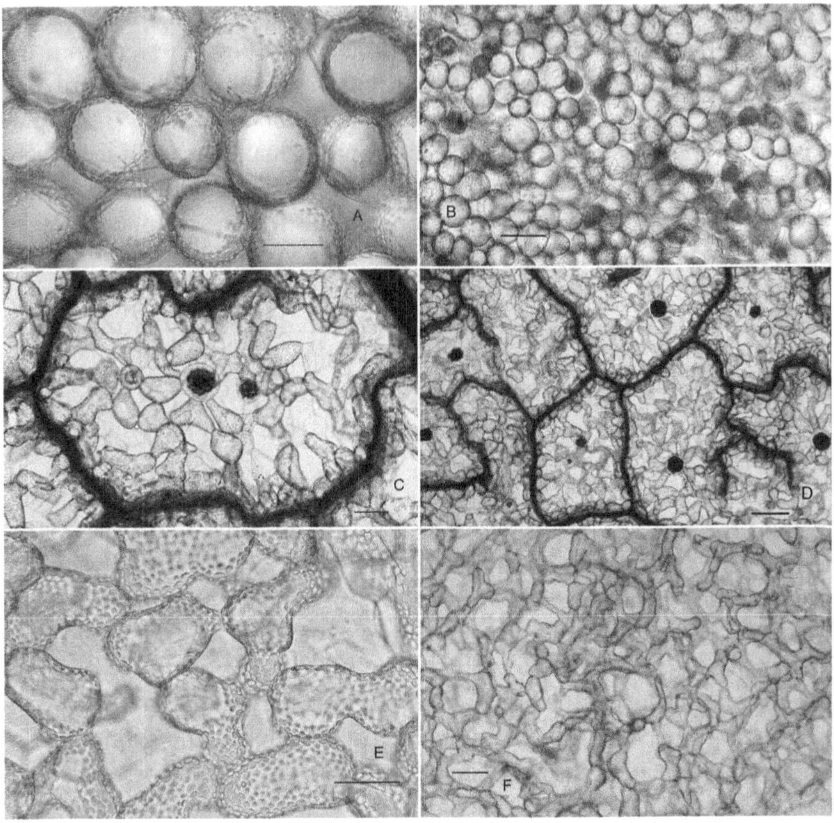

Fig. 7.2. Light microscopic photomicrographs of *Spinacia oleracea* leaf fresh paradermal sections.

A. Palisade mesophyll (line = 50 μm). The image illustrates the chloroplasts localized along the long surfaces of the cells, but not at the top of the cells.

B. Some of the palisade mesophyll cells are not perpendicular to the leaf surface, as seen as the darker green (line = 100 μm). The darker green is also due to attached broken cells.

C. Minor vascular tissue containing specialized palisade mesophyll cells shaped like pears, referred to as pirum mesophyll (line = 100 μm). Idioblasts containing crystals of calcium oxalic acid are occurred in the pirum mesophyll. Three different developmental stages of idioblasts are shown.

D. The pirum mesophyll cells appear as spokes of a wheel around the idioblasts (line = 200 μm). The idioblasts are relatively evenly spaced within the tissue layer.

E. Pillow spongy mesophyll cells underlay the vascular tissue (line = 50 μm).

F. Typical spongy mesophyll cells from the lower portion of the spongy mesophyll were curved and tightly connected to one another (line = 100 μm). Chloroplasts were distributed along the whole surface of the cells minor vascular tissue was surrounded by a layer of tightly connected green cells with a cubical form. Little airspace exists between the cells, and they can be identified as bundle sheath cells (see also Figure 7.2D).

The spongy mesophyll exhibited large air spaces and consisted of 2-3 layers of multifaceted cells with large surface area contacts with each other (Figure 7.2E and 7.2F). Unlike the palisade mesophyll cells, the spongy mesophyll cells did not separate from each other during preparation of thin paradermal slices of spongy mesophyll without lower epidermis. The chloroplasts were equally distributed along the whole surface of the cell (Figure 7.2E).

Two distinct forms of spongy mesophyll exist in *Spinacia oleracea*. The spongy mesophyll cells that underlay the vascular tissue were more condensed and "pillow" like (Figure 7.2E). This layer of spongy mesophyll we call the pillow spongy mesophyll. Spongy mesophyll shown in Figure 2F was typical of the spongy mesophyll cells from the abaxial surface connected to the epidermis. The cells are elongated and curved–like groups of interconnected, curved fingers (Figure 7.2F). Figures 7.2E and 7.2F illustrate the spongy characteristic of the spongy mesophyll and the overlapping quality of cells that would lead to high light scattering characteristics (Shibata, 1958).

In summary, there were seven main morphologically different tissue types in the mesostructure of the examined *Spinacia oleracea* leaves.

From the top to the bottom there were: 1) two layers of rod shaped palisade mesophyll cells; 2) one layer of pirum mesophyll, which are specialized palisade mesophyll associated with the vascular tissue; 3) spherical idioblasts containing crystals of calcium oxalic acid; 4) various vascular tissue cells; 5) cubical bundle sheath cells; and two forms of spongy mesophyll – 6) pillow spongy mesophyll, the tissue below the vascular tissue and 7) typical spongy mesophyll cells that are the most abaxial cells associated with the lower epidermis. Again, the above tissues are protected by epidermal cells.

7.2.1.2 What Are the Roles of the Different Cells?

The pirum mesophyll, a modified palisade cell, that forms at the bottom of the typical palisade mesophyll, appear to form a network of cells that connect directly to the bundle sheath cells and the pillow spongy mesophyll. They were also associated with idioblasts containing Ca-oxalate crystals (Zindler-Frank, 1976). Presumably, glycolate (or glyoxylate, or oxalate), formed during photorespiration in the upper palisade mesophyll, can be transported from the palisade mesophyll to the pirum mesophyll, where the oxalate can be "detoxified" as Ca-oxalate crystals.

Whether transport is symplastic or apoplastic remains to be examined. The structural network suggests that transport is symplastic. However, since oxalate is toxic, apoplastic transport would minimize accumulation in the non-idioblastic cells. It was shown recently (Keates *et al.*, 2000; Kostman *et al.*, 2001) that the complete set of enzymes for oxalate formation from isocitrate exists in the idioblasts of *Pistia stratiotes*. The contribution of that pathway vs. direct synthesis from other routes should be examined. As mentioned below, the rate of $^{14}CO_2$ incorporation into oxalate, suggests that the direct route may play a significant role. [^{35}S]-methionine incorporation into a specific subset of polypeptides occurred in the vascular region of *Spinacia oleracea* leaves (Sun *et al.*, 1996*b*). Using microautoradiography, we concluded that the majority of the synthesis was occurring in the phloem (Sun *et al.*, 1996*b*). It is possible that the pirum mesophyll also plays a direct role in the high rates of protein incorporation observed in the vascular tissue region in *Spinacia oleracea* (Sun *et al.*, 1996*b*), because of the anatomical structures presented here and the physiological and biochemical data in *Glycine max* and other plants, that suggest the paraveinal tissue may be storage cells, as well as transport cells (Tranbarger *et al.*, 1991; Brubaker and Lersten, 1995; Lansing and Franceschi, 2000).

The exact role of Ca-oxalate crystals in plants is not known (Zindler-Frank, 1976; Franceschi and Horner, 1980; Webb, 1999). Oxalate oxidation is involved in active oxygen formation in pathogen responses in plants (Bolwell and Wojtaszek, 1997), and it may play a role in moderating the Ca^{2+} concentration in plants (Zindler-Frank, 1976; Franceschi and Horner, 1980; Webb, 1999; Keates *et al.*, 2000). Ca-oxalate crystals and oxalic acid alone may also perform a protective role against predation or other roles (Franceschi and Horner, 1980; Finley, 1999). Its role in structure has also been investigated (Franceschi and Horner, 1980; Türk *et al.*, 1995).

There is a negative correlation between starch accumulation and Ca-oxalate crystals accumulation in *Portulaca* (Alexandrov and Shanidze, 1925). In *Oxalis corniculata*, oxalate is labeled with $^{14}CO_2$ in the light within 5 seconds (Seal and Sen, 1970). In *Amaranthus*, Ca-oxalate is labeled with $^{14}CO_2$ in the light within 1-2 minutes after labeling (Karpilov, Pristupa et *al.*, 1970). Clearly there is some relationship between photosynthesis and oxalate, and it is likely formed via the glycolate → glyoxylate → oxalate pathway (Osmond, 1963; Osmond and Avadhani, 1968; Seal and Sen, 1970; see also Osmond, 2014). Alternate pathways contribute significantly to oxalate formation and are localized to the idioblasts in *Pistia stratiotes* (Keates *et al.*, 2000). The possibility that Ca-oxalate crystals can dissolve during stress conditions to provide CO_2 (Karpilov,

Pristupa *et al.*, 1970; Karpilov *et al.*, 1978), when the stomata are closed has never been shown, but remains a logical and tantalizing hypothesis.

Thus, in *Spinacia oleracea*, the pirum mesophyll appears to be the final transport cells from the palisade mesophyll to the vascular tissue, and the storage area for Ca-oxalate crystals. Formation of oxalic acid and Ca-oxalate crystals remains an active area of research (Millerd *et al.*, 1963*a*, 1963*b*; Mostafa and Ulrich, 1976; Franceschi and Horner, 1979; Franceschi and Loewus, 1995; Webb, 1999; Keates *et al.*, 2000).

Interestingly, the Ca-oxalate crystals in *Spinacia oleracea* is stored in the region of the leaf that also contains the highest amount of Rubisco. In *Pepperomia spp.*, it has been shown that the Ca-oxalate crystals are in the palisade mesophyll (Franceschi and Horner, 1980). It has previously been shown that the majority of Rubisco in *Pepperomia camptotricha*, a plant with leaf architecture similar to the *Pepperomia spp.* described above, is in the palisade mesophyll (Nishio and Ting, 1987). At least in the case of *Spinacia oleracea* and apparently for some species of *Pepperomia*, the Ca-oxalate crystals are in the part of the leaf near the site of maximal Rubisco carboxylation. Such a location would make sense if Ca-oxalate crystals played a role in stress metabolism and provided CO_2 for carboxylation when stomata are closed.

Figure 7.3 shows how Ca-oxalic acid could supply CO_2. The hypothesis remains to be tested. Oxalic acid crystals are also found in wood, however (Horner and Wagner, 1995), and in the epidermal tissue of certain plants. When they are in the upper portion of the leaf, they may serve light blockers, since they are normally associated with succulents or xerophytic leaves. In succulents, however, internal CO_2 could be limiting, when the stomata are closed. Often succulents have no stomata on the upper surface.

Thus, if the CO_2 were released under stressful conditions, the CO_2 would have to traverse the leaf thickness before exiting through the stomata, if the stomata are "leaky". Additionally, the presence of Ca-oxalate crystals near the middle of the leaf, where Rubisco is maximal, may be a coincidence of being by the vascular tissue where transported Ca^{2+} is released.

Calling the paraveinal cells in *Spinacia oleracea* pirum mesophyll has structural significance. First of all, the cells are shaped like pears. Secondly, pears have a high concentration of sclerenchyma cells, which can be Ca-oxalate containing cells.

Besides storage and transport, another possible role can be added to the list of physiological possibilities for the specialized cells – that of stress tolerance (see Figure 7.3). Paraveinal mesophylls in other plants are known to contain idioblasts (without oxalic acid crystals). Pirum mesophyll seems to be a good name for paraveinal cells in other plants, as well.

The pillow spongy mesophyll was the uppermost spongy mesophyll found directly below the pirum mesophyll. The pillow spongy mesophyll was chloroplast rich and denser than the typical spongy mesophyll. The pillow spongy mesophyll exhibited close contact with one another. Unlike the palisade mesophyll, the chloroplasts were distributed completely around the cell, though at cell junctions there may be decreased chloroplast populations (as in the palisade mesophyll), which needs to be examined more carefully. The distribution of the chloroplasts and cellular structure of the pillow spongy mesophyll suggests that it is the highest light scattering layer of photosynthetic tissue, and was the region that exhibited the highest rates of Rubisco activity in the inverted leaves. The pillow spongy mesophyll would contribute significantly to carbon fixation and is adjacent to the vascular tissue, which exhibited high NAD-MDH activity (Figure 7.4) and protein synthesis (Sun *et al.*, 1996*b*).

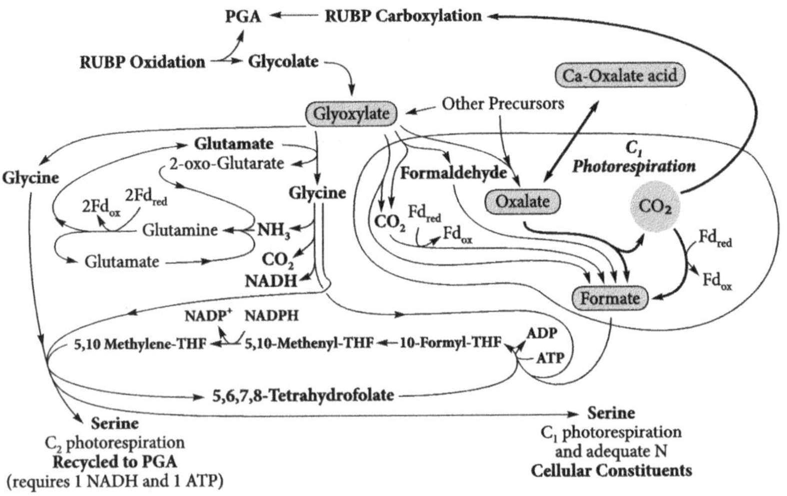

Fig. 7.3. Calcium oxalic acid crystals could supply CO_2 when plants are under stress. Stomata may close under extreme stress, and oxalic acid crystals could dissolve to supply CO_2. The relationship among the photorespiratory pathways is highlighted (adapted from Biel, Fomina *et al.*, 2010)

Although the importance of metabolism in the leaf function has been shown regardless of leaf anatomy, there is a question: "What function does the anatomy of the palisade mesophyll and spongy mesophyll actually serve?" The structure of the palisade mesophyll cells is most likely related to a function of allowing light to reach the inner part of the leaf (Vogelmann and Martin, 1993); as leaves that have a vertical orientation, tend to have palisade mesophyll on both leaf surfaces, and leaves that are grown in high light tend to have thicker palisade mesophyll (Clements, 1905; Chugunova *et al.*, 1976). It has also been suggested that the structure of the palisade mesophyll is related to transport of fixed carbon toward the vascular tissue of the leaf (Thoday, 1931; Kursanov, 1976; Karpilova *et al.*, 1982).

In contrast to the palisade mesophyll cells, the spongy mesophyll cells act like a light trap and minimize light loss from the bottom of the leaf (Shibata, 1958; Terashima and Saeki, 1983; Terashima, 1989; Sun and Nishio, 2001). A more lucid view of the structure and function of the palisade mesophyll and spongy mesophyll will continue to develop, as we learn more about the specific metabolic activity of the different cell layers across leaves.

7.2.2 *Enzyme Activity*

For enzyme analysis we routinely use tissue sectioned on a freezing microtome. Therefore, the effect of freezing on enzyme activity must be considered. To test the effect of freezing, we froze fresh tissue preparations in liquid nitrogen, which had little effect on the enzymes measured (Table 7.1). Total activities of Rubisco, GOT, and NADP-MDH were not significantly changed by freezing. Catalase activity expressed per kg protein was decreased about 30 ± 16% by freezing; total activity of NAD-MDH on a leaf areal basis was increased 29 ± 10% by freezing. The increase in NAD- and NADP-MDH activities was in the soluble fraction, and membrane activities decreased. Since the changes in enzyme activities after freezing were small (and we presume consistent for the different tissue layers), frozen tissue taken from the freezing microtome were used for analysis of enzyme activity across leaves.

7.2.2.1 *How Does Inverting the Leaves Alter the Distribution of Enzyme Activity within Spinacia oleracea Leaf?*

Rubisco activity in control leaves on an area basis doubles from the top of the leaf to the maximum activity approximately 200 μm deep (Figure 7.4 Rubisco A). $^{14}CO_2$ fixation exhibits the same profile (Nishio *et al.*, 1993). Twelve to 21 days after leaf inversion, the maximum Rubisco activity shifts to approximately 200 μm from the bottom of the leaf (Figure 7.4 Rubisco A). The soluble protein strongly correlates with Rubisco activity in both leaf types (Figure 7.4 Rubisco C). In contrast, the specific activity of Rubisco is constant across control and inverted leaves (Figure 7.4 Rubisco B). Thus, 12 days after inverting leaves, the pattern of Rubisco across the leaf completely reverses.

Table 7.1. Effect of freezing in liquid nitrogen on *Spinacia oleracea* leaf enzyme activity

Enzyme	Fraction	Areal activity, µmol substrate m^{-2} s^{-1} (for catalase, mmol m^{-2} s^{-1})		Specific activity, mmol substrate kg^{-1} protein s^{-1} (for catalase, mol kg^{-1} protein s^{-1})	
		Control	Frozen	Control	Frozen
Rubisco	Total[1]	34.1±4.8	32.5±2.1	13.6±1.6	14.2±1.3
Catalase	Total	15.8±3.2	12.3±1.9	5.9±0.7	4.1±0.6
GOT	Total	14.8±1.8	11.9±1.4	5.6±1.3	5.9±0.6
NADP-MDH	Soluble[2]	17.1±3.2	23.4±1.5	8.7±1.5	9.4±0.5
NADP-MDH	Membrane[3]	3.9±0.7	1.8±0.7	8.6±0.9	5.8±0.2
NADP-MDH	Total[4]	21.0±3.3	25.2±1.7	8.7±1.4	9.0±0.7
NAD-MDH	Soluble	280±10	412±29	144±8	167±13
NAD-MDH	Membrane	53.8±1.7	19.9±4.2	118±10	85±15
NAD-MDH	Total	334±10	432±30	138±4	154±11

Note: Leaf disks were frozen in liquid nitrogen for 30–40 min. [1]Supernatant after centrifugation at 745 × g for 10 min; [2]Supernatant from 8300 × g spin; [3]Pellet (8300 × g) contained chloroplasts and mitochondria + 0.3% Triton X-100; [4]Sum of soluble + membrane. Rubisco activity was measured as $^{14}CO_2$ incorporation (Perchorowitz *et al.*, 1982), Catalase, GOT, and MDH activities were measured spectrophotometrically (Hatch and Mau, 1973; Hatch *et al.*, 1975; Worthington, 1988) with minor modifications (for GOT and MDH) in the reaction mediums (Bil', Fomina *et al.*, 1990). Soluble protein, in the samples prepared for enzyme measurements, was determined by Bradford's method (Bradford, 1976). Data points represent the mean ± SD; n = 3

The glutamate oxaloacetate transaminase (GOT) activity profile across control leaves on an area or volume basis (Figure 7.4 GOT A) essentially mimics that of Rubisco activity (Figure 7.4 Rubisco A). Ten to fifteen days after leaf inversion, the maximum GOT activity was about 280 µm from the leaf bottom compared to 200 µm for Rubisco activity. The soluble protein profiles reversed after inversion (Figure 7.4 GOT C) and were similar to the Rubisco experiment (Figure 7.4 Rubisco C). The specific activity was relatively constant across control and inverted leaves, though near the vascular tissue, activity was higher in inverted leaves (Figure 7.4 GOT B). Therefore, maximal GOT activity shifted toward the sponge mesophyll in inverted leaves.

Catalase activity in control leaves is highest at the top of the leaf, and decreases exponentially toward the bottom of the leaf (Figure 7.4 Catalase A). Three to eight days after leaf inversion, the maximum catalase activity shifts to the bottom of the leaf (Figure 7.4 Catalase A). The maximum rate in the inverted leaf is lower than the maximum rate at the top of the control leaf. Catalase activity from the 600 μm depth is statistically different (*t*-test) from the activity measured in the 40–360 μm samples ($p < 0.05$), except the 280 μm sample where $p = 0.062$. The profile of catalase specific activity across the control leaf is similar to that on an area basis (Figure 7.4 Catalase B), but it is "U"-shaped in inverted leaves.

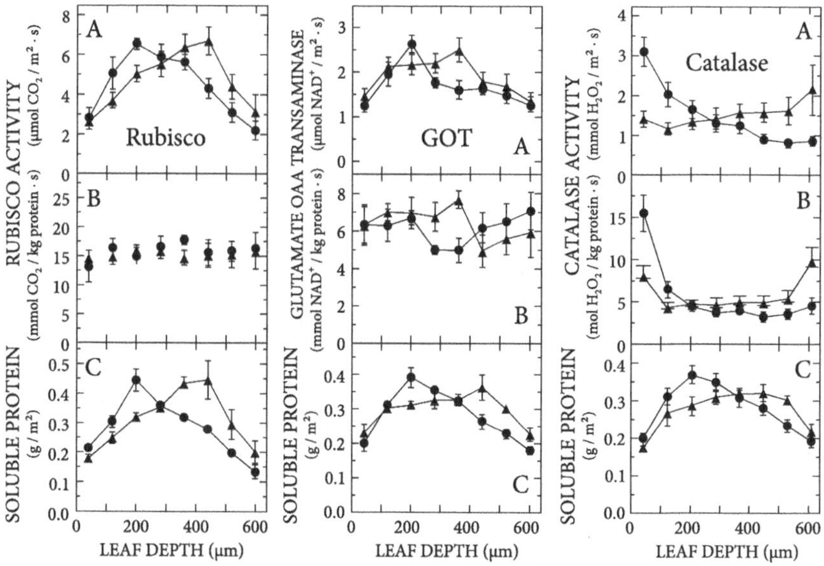

Fig. 7.4. Long-term effect of leaf inversion on the profiles of Rubisco, glutamate oxaloacetate transaminase (GOT), and catalase activities across *Spinacia oleracea* leaf

Note: A, Activity per area; B, Specific activity; C, Soluble protein per area. Measurements were made on leaf slices from the control (•) and inverted (▲) leaves. Leaf inversion was 12–21 days for Rubisco measurements, 10–15 days for GOT, and 3–8 days for catalase measurements. Age of plants from sow date was 60–74 days. Activities of the enzymes were measured as described in Note of Table 7.1. Each measurement was made on a pooled sample of 5–7 leaf discs. Data points represent the mean ± SD. Control, $n = 3 \div 6$; inverted, $n = 3 \div 5$

The soluble protein profile of the inverted and control leaves is reversed (Figure 7.4 Catalase C), as in the Rubisco and GOT experiments; however, the extent of the reversal is smaller, because leaves had been inverted only 3–8 days for the catalase measurements compared to 10–21 days for the Rubisco and GOT experiments.

NADP-malate dehydrogenase (NADP-MDH) and NAD-malate dehydrogenase (NAD-MDH) activities. NADP-MDH activity (mainly chloroplastic) is relatively flat across control leaves (Table 7.2), with an apparent maximum in the region of the vascular bundles, between 300 and 400 μm deep within the leaf. NAD-MDH activity (cytosolic and mitochondrial enzyme) clearly exhibits a peak in the region of the vascular tissue (Table 7.2).

The specific activity of both enzymes is highest at the top of the leaves (Table 7.2). The specific activity of NADP-MDH is also higher in the bottom of the leaf, with the profile appearing as a "U". The profile of the soluble protein follows that of the other experiments (data not shown). MDH activity was not measured in inverted leaves.

Table 7.2. Profiles of NADP- and NAD-MDH activities across *Spinacia oleracea* leaves

	NADP-MDH		NAD-MDH	
Leaf depth (from top to bottom), μm	Areal activity (μmol substrate m^{-2} s^{-1})	Specific activity (mmol substrate kg^{-1} protein s^{-1})	Areal activity (μmol substrate m^{-2} s^{-1})	Specific activity (mmol substrate kg^{-1} protein s^{-1})
40	2.09±0.32	13.8±2.67	22.6±3.4	184±27
120	2.04±0.26	9.62±1.98	27.9±0.7	140±17
200	2.32±0.31	6.74±0.50	31.6±5.8	92±21
280	2.34±0.27	7.16±1.06	36.5±7.3	123±13
360	2.60±0.06	9.15±0.28	39.9±5.7	141±15
440	2.31±0.24	8.38±0.75	31.2±7.2	113±28
520	2.12±0.35	9.64±1.86	23.1±1.0	105±22
600	2.09±0.24	11.5±2.56	22.1±2.5	121±19

Note: Age of plants from sow date was 60–65 days. Each measurement was made on a pooled sample from 5–7 leaf discs. Data points represent the mean ± SD. NADP-MDH, $n = 4$; NAD-MDH, $n = 3$

Table 7.3. Distribution of fumarase and cytochrome-*c*-oxidase activities in control and inverted *Spinacia oleracea* leaves

	Fumarase (μmol substrate m^{-2} s^{-1})		Cytochrome-*c*-oxidase (μmol substrate m^{-2} s^{-1})	
Leaf tissue	Control leaf	Inverted leaf	Control leaf	Inverted leaf
Palisade mesophyll*	0.23±0.05	0.33±0.01	2.16±0.26	2.57±0.14
Spongy mesophyll**	1.15±0.26	1.21±0.35	3.65±0.89	4.34±0.30

Note: *Palisade mesophyll with upper epidermis; **Spongy mesophyll with lower epidermis. Leaf inversion was 14 days; plants sampled 60–65 days from sow date. Each measurement was made on fresh leaf slices (80–120 μm) from a pooled sample of 6–7 leaf discs. Data are mean ± SD; n = 3÷4

Fumarase and cytochrome-c-oxidase activities were higher in spongy mesophyll cells as compared to palisade mesophyll cells (Table 7.3). Inverting of the leaf did not alter the profile of these enzymes.

7.2.2.2 Summary of Enzyme Activity across Leaves

Light is a key factor controlling the profile of photosynthetic function across the bifacial leaves of *Spinacia oleracea*, a C_3 plant. Maximum carbon fixation occurs midway through the palisade mesophyll of leaves illuminated adaxially (Nishio *et al.*, 1993), and Rubisco activity across leaves correlates with the distribution of carbon fixation (Figure 7.4; see also Sun and Nishio, 2001).

This chapter reviews the data showing that leaf inversion, so that the leaves are illuminated on the original abaxial leaf surface, does not alter leaf anatomy, but reverses the distribution profile across leaves of the following enzymes: Rubisco, the main carboxylating enzyme in photosynthetic organisms; GOT, an enzyme of nitrogen metabolism; and catalase, a protective enzyme from oxidative damage (Fomina, Biel *et al.*, 2001*a*; 2001*b*; Bil', Fomina *et al.*, 2003; Biel, Fomina *et al.*, 2010).

The reversal in the enzyme activity profiles across leaves is likely due to both up-regulation and down-regulation of protein synthesis in different parts of the leaf, as the profile of total soluble protein is also reversed. Carbon fixation per Rubisco is relatively flat across spinach leaves, when they are illuminated on the adaxial leaf surface (Sun and Nishio, 2001). When leaves are illuminated on the abaxial leaf surface, carbon fixation per Rubisco is decreased in the palisade mesophyll (as well as total carbon fixation), because of light limitations, due to light trapping and scattering in the spongy mesophyll cells (Sun and Nishio, 2001), as the specific activity across leaves is not changed by leaf inversion (Figure 7.4) (Bil', Fomina *et al.*, 2003, Biel, Fomina *et al.*, 2008, 2010).

Figure 7.2 illustrates that the palisade mesophyll may pass light through the interior of the cells and act as light tunnels that allow light into the spongy mesophyll cells. The results further illustrate the importance of leaf structure and the distribution of photosynthetic pigments, Rubisco, and other enzymes across leaves for optimum light utilization, carbon fixation, and metabolism in plants (Nishio *et al.*, 1993; Sun *et al.*, 1996*a*; 1998; Fomina, Biel *et al.*, 2001*a*, 2001*b*; Sun and Nishio, 2001; Bil', Fomina *et al.*, 2003; Biel, Fomina *et al.*, 2008, 2010).

7.2.2.3 Functional Significance to Profiles of Enzyme Activity across Spinacia oleracea Leaves

The results support our hypothesis that protective enzyme activity is maximal at the top of the leaf. Catalase activity, a protective enzyme against oxidative stress, decreased exponentially from the top of the leaf to the bottom on an area basis (Figure 7.4) and chlorophyll basis (not shown). Inverting the leaves completely reverses the profile, so maximum catalase shifts to the original abaxial leaf surface (Figure 7.4) (Fomina, Biel *et al.*, 2001*a*; 2001*b*; Bil', Fomina *et al.*, 2003; Biel, Fomina *et al.*, 2008, 2010).

By other words, the distribution of catalase activity across leaves matched the light gradient across *Spinacia oleracea* leaves (Cui *et al.*, 1991), and was highest where photon fluxes were also the highest.

As just mentioned, the top of the leaf may be exposed to excess light, and therefore, a high concentration of superoxides and over-reduction is possible at the top of the leaves and could lead to photoinhibition. At the bottom of the leaf, light, in particular red and blue light may become limiting. Maximum carbon fixation and Rubisco activity occurred about 20–40% deep into the leaf from the illuminated leaf surface. Closer to the top of the leaves, where over-reduction is likely and more than 40% deep within the leaf, where light may become limiting, the amount of Rubisco (Nishio *et al.*, 1993) and Rubisco activity (Figure 7.4, see also Fomina, Biel *et al.*, 2001*a*; Sun and Nishio, 2001; Bil', Fomina *et al.*, 2003; Biel, Fomina *et al.*, 2008, 2010) are decreased.

Therefore, in normally oriented *Spinacia oleracea* leaves, maximum photosynthesis midway through the palisade mesophyll likely represents a region of the leaf that is most optimal for photosynthesis and accounts for the contradictory problem of over-excitation and non-saturating light conditions that normally exist in nature. The same argument would apply for inverted leaves, where the maximum amount of photosynthesis shifted to the region of the spongy mesophyll near the vascular tissue. Besides avoiding the toxicity due to superoxides and over-reduction, another intriguing possibility is that the distribution of Rubisco is also related to the high amount of electron transport at the surface directly irradiated.

Professor I. Terashima showed that maximal electron transport on a chlorophyll basis occurs at the top of *Spinacia oleracea* leaves (Terashima and Inoue, 1985; Terashima, 1989). Maximal carbon fixation per chlorophyll also occurs at the top of leaves (Nishio *et al.*, 1993; Sun *et al.*, 1996*a*). The light gradient across *Spinacia oleracea* leaves is very steep with maximum light fluxes and absorption occurring at the top of leaves (Terashima and Saeki, 1983; Terashima, 1989; Cui *et al.*, 1991; Nishio, 2000). High electron transport activity at the top of the leaves would contribute to elevating the oxygen concentration due to water-splitting. Since oxygen is toxic to Rubisco carboxylase activity (Somerville and Ogren, 1982; Chen and Spreitzer, 1992), placing the maximum activity of Rubisco more deeply into the leaf decreases the O_2 / CO_2 at the region that exhibits maximal carboxylation by Rubisco. Possibly more nitrate reduction and / or some other reductive process(es) occur(s) in the upper part of leaves.

The activity of GOT mainly mimics that of Rubisco across control leaves, but appears to have less relative activity in the spongy mesophyll (Figure 7.4) (Fomina, Biel *et al.*, 2001*a*, 2001*b*). After inversion, the activity shifts to the spongy mesophyll and region of the vascular tissue, where carbon fixation and likely synthetic activities were maximal. The GOT activity in the spongy mesophyll and palisade mesophyll of *Spinacia oleracea* (Figure 7.4) were remarkably different from that in *Vicia faba* (Outlaw *et al.*, 1976). In *Vicia faba*, the GOT activity was reported to be 9-times higher on a chlorophyll basis, and 4-times higher on a protein basis, in the spongy mesophyll than in the palisade mesophyll (Outlaw *et al.*, 1976).

There was no difference in the specific activity of GOT in the palisade mesophyll and spongy mesophyll of *Spinacia oleracea* leaves (Figure 7.4) (Bil', Fomina *et al.*, 2003; Biel, Fomina *et al.*, 2008, 2010). The reason for such large species differences is a matter of speculation that may be related to the source of nitrogen, as *Vicia faba* is a legume that has symbiotic nitrogen fixers. It would be interesting to examine characteristics of carbon and nitrogen metabolism across leaves of nitrogen-fixing and non-nitrogen-fixing plants grown under similar conditions.

One of the main hypotheses presented is that the distribution of metabolic activity in leaves is partly due to dealing with high light. Clearly, the distribution of catalase (Figure 7.4) supports that hypothesis. Rubisco does not follow the light gradient. The model that follows discusses that issue directly and also supports our hypothesis.

Figure 7.5 illustrates some of the detoxifying pathways that may play a role and may be distributed across the leaf to reflect their possibly protective nature. The figure also incorporates Ca-oxalate crystals. The detoxifying enzymes and metabolites can follow the distribution of light, catalase, or photosynthetic electron transport more than that of Rubisco.

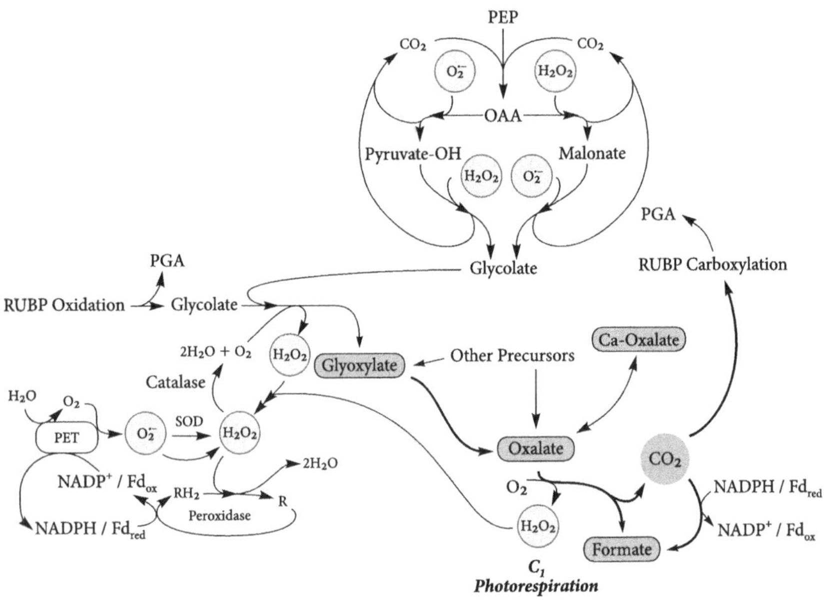

Fig. 7.5. Stress-related metabolism should concentrate at sites most needed. Reactive oxygen species are associated with photosynthetic electron transport, high light, and stress that can cause over-reduction. Such activity will likely be higher at the surface of leaves directly exposed to light. The diagram also illustrates how oxalate metabolism can contribute directly to the reactions by production of hydrogen peroxide (adapted from Biel, Fomina *et al.*, 2010)

NADP-MDH activity is relatively flat across *Spinacia oleracea* leaves, but NAD-MDH exhibits maximal activity near the vascular tissue in control leaves (Table 7.2) (Bil', Fomina *et al.*, 2003; Biel, Fomina *et al.*, 2008, 2010). The shifting of maximal GOT activity to the vascular tissue reasonably coincides with maximal Rubisco activity in inverted leaves. That NAD-MDH activity was already high towards the bottom of the leaf suggests that there is a great deal of respiratory metabolic activity in the vascular and spongy mesophyll of control leaves. The respiration rates (not shown) as well the activity of the key respiratory enzymes (Table 7.3) (Fomina, Biel *et al.*, 2001*a*, 2001*b*; Bil' *et al.*, 2003; Bil', Fomina *et al.*, 2003; Biel, Fomina *et al.*, 2008, 2010) were significantly higher in the spongy mesophyll compared to the palisade mesophyll in control and inverted leaves.

Summarizing, it can be concluded that leaf inversion does not appear to alter the respiratory activity as much as it does the photosynthetic enzymes and activity. As was shown earlier, high amounts of [^{35}S]-methionine-incorporation occurs near the vascular tissue (Sun *et al.*, 1996*b*). We did not measure NADP-MDH and NAD-MDH in inverted leaves, nor did we measure cytosolic vs. mitochondrial activity across leaves, which should be very enlightening.

7.2.3 CO_2 / O_2 Gas Exchange

7.2.3.1 CO_2 gas exchange

Figure 7.6 IA shows that CO_2 gas exchange in the control *Spinacia oleracea* leaves is higher when the adaxial surface is illuminated, as compared to illumination of the abaxial surface. During 2-week experiments, leaf inversion causes the CO_2 gas exchange rate during illumination of the original adaxial leaf surface to decrease (Figure 7.6 IB). In contrast, the CO_2 gas exchange rate increased from the control, when the original abaxial leaf surface of inverted leaves was illuminated (Figure 7.6 IB).

The increase was statistically significant ($p < 0.05$) seven days after leaf inversion. Interestingly, during the course of the experiment, conductance is relatively unchanged, whereas transpiration more than doubled on the seventh day of the experiment. The difference in CO_2 assimilation, when leaves are illuminated either adaxially or abaxially, is due to light (Sun and Nishio, 2001), not conductance, so a strong correlation between conductance and CO_2 assimilation when the leaves are irradiated abaxially is not necessarily expected. The lack of correspondence between transpiration and conductance (Figure 7.6 IC-F) is puzzling and may be related to the age of the leaves.

Short-term experiments provide further insight into the role of leaf anatomy in carbon fixation. Carbon dioxide (CO_2) gas exchange dramatically decreases in the first two hours, when inverted leaves are illuminated on the original adaxial leaf surface (Figure 7.6 IIB). Monitoring gas

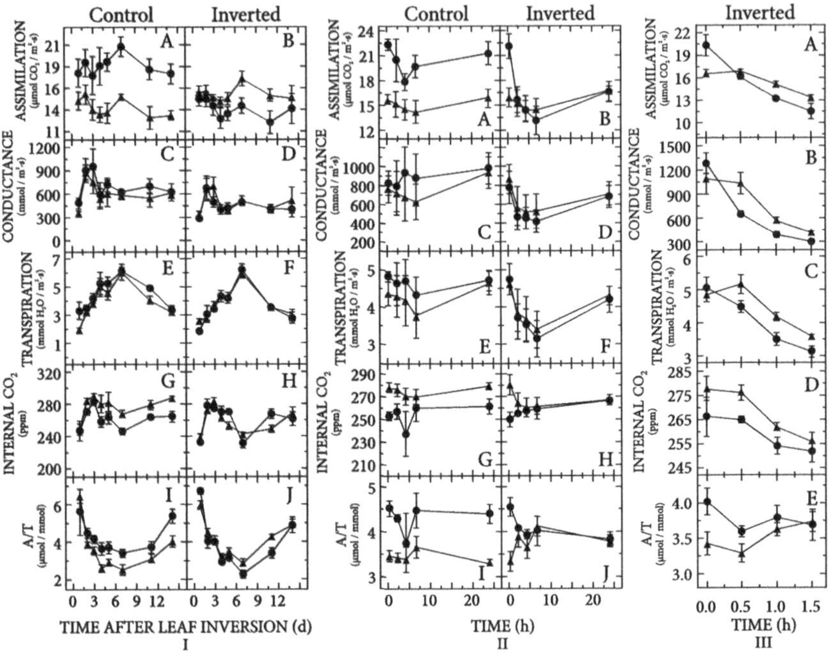

Fig. 7.6. Effect of leaf inversion on gas exchange parameters of *Spinacia oleracea* leaves (adapted from Biel, Fomina *et al.*, 2010)

I. Long-term effect during the two weeks following leaf inversion. Measurements were begun 1 day after leaf inversion and were made (between 10:00–12:00 hours) with the light directed either to the adaxial (•) or abaxial (▲) leaf surface for control leaves, either to the original adaxial (•) or original abaxial (▲) leaf surface for inverted leaves. A, B: CO_2 gas exchange; C, D: conductance; E, F: transpiration; G, H: Ci (internal CO_2 concentration); I, J: A/T (assimilation / transpiration)

II. Middle-term effect during the 24 hours following leaf inversion. The arrow indicates when the leaves were inverted. 0 = Time experiment was started, 9:00 hours. Other symbols as in I

III. Short-term effect during the 1.5 hours following leaf inversion. The data obtained on the inverted leaves are presented. 0 = Time experiment was started, 11:00 hours. A: CO_2 gas exchange; B: conductance; C: transpiration; D: Ci; E: A/T. Other symbols as in I and II

Note: Age of plants from sow date was 35–46 days after sowing. Data points represent the mean ± SD; $n = 3 \div 6$

exchange in a more short-term experiment shows that in less than 30 minutes, the CO_2 gas exchange, in inverted leaves irradiated on the original adaxial leaf surface, decreases to the level of control leaves irradiated on the abaxial leaf surface (Figure 7.6 IIIA). At the same time CO_2

gas exchange, in inverted leaves illuminated on the original abaxial surface, starts to increase only after a few days and never reaches (during the 2-week experiment) the level of gas exchange in leaves irradiated on the adaxial leaf surface of control leaves (Figure 7.6 IA and IB).

When the leaves are inverted and illuminated on the original adaxial leaf surface, conductance, transpiration, and Ci measured and estimated simultaneously by CO_2 gas exchange, decrease within 0.5–1.5 hours (Figure 7.6 IIIB-D). However, in the two-week experiments the same parameters in inverted leaves do not change significantly as compared to control leaves (Figure 7.6 IC-H).

Thus, CO_2 gas exchange is dramatically changed by leaf inversion, unlike the conductance, transpiration, and Ci.

Although the profile of Rubisco is reversed in the leaves, the maximum rate of gas exchange in the inverted leaves illuminated on the original abaxial leaf surface never reaches the rate of controls illuminated adaxially (Figure 7.6). It is no surprise that inverting leaves and illuminating them on the abaxial leaf surface decreased whole leaf photosynthesis in *Spinacia oleracea* leaves (Figure 7.6), since the palisade mesophyll acts like light tunnels and the spongy mesophyll is light scattering (see Sun and Nishio, 2001). Such decreases in photosynthesis due to abaxial illumination have been previously observed in other species (Moss, 1964; Oya and Laisk, 1976; Leverenz and Jarvis, 1979; Syvertsen and Cunningham, 1979; Turner and Singh, 1984; Terashima, 1986). Inverting leaves decreased whole leaf CO_2 gas exchange, when the original abaxial leaf surface was illuminated, compared to when the adaxial leaf surface was illuminated in control leaves.

The decrease in photosynthesis in inverted leaves occurred within 30 minutes (Figure 7.6), and the changes appear to be directly linked to stomatal changes. Over the longer time course of the experiments, however, conductance, transpiration, and Ci in the inverted leaves were not significantly changed from the controls, but CO_2 gas exchange remained different from the controls. This is reasonable, since the stomata can readjust rapidly, compared to alterations in enzyme activity across leaves. Furthermore, it suggests that the concentration of CO_2 within the leaves may not be a major factor in determining the distribution of Rubisco across leaves, and that the photon flux within the palisade mesophyll of inverted leaves illuminated on the original abaxial leaf surface never reaches that within the spongy mesophyll of control leaves illuminated adaxially.

During the course of the experiments, we observed a significant phenomenon in *Spinacia oleracea*. The palisade mesophyll cells are responsible for the midday depression in photosynthesis. Figure 7.6 IIA shows that CO_2 gas exchange in control leaves illuminated on the adaxial leaf surface exhibits a midday depression. However, the same leaves do not exhibit a midday depression when CO_2 gas exchange is measured when leaves are illuminated on the abaxial leaf surface (Figure 7.6 IIA). Conductance, though it exhibited a large standard deviation, actually increased after 5 hours, so the decrease in assimilation was not due to stomatal limitations.

The measurements were made on whole leaves, so each measurement has a component due to the underlying tissue. In the case of the adaxially illuminated leaves, the spongy mesophyll contributes significantly to carbon fixation (Nishio *et al.*, 1993), and therefore to CO_2 gas exchange, as well. On the other hand, when leaves are illuminated on the abaxial leaf surface, the light is mainly trapped in the spongy mesophyll, and the palisade mesophyll contributes a significantly smaller proportion to the overall carbon fixation rate (Sun and Nishio, 2001). Therefore, the gas exchange measured under adaxial light reflects both palisade mesophyll and spongy mesophyll contributions, whereas under abaxial light the rate is due mainly to the spongy mesophyll. Since the palisade mesophyll contributes less to the overall carbon fixation rate under abaxial illumination, the slight depression in gas exchange observed under abaxial light may be due mainly to the contribution of the palisade mesophyll.

In the light of the above, it can be concluded that the palisade mesophyll was responsible for the midday depression in *Spinacia oleracea*, a plant with typical bifacial leaves.

7.2.3.2 HCO_3^--Dependent Oxygen Evolution

The rate of oxygen evolution in the spongy mesophyll of inverted leaves increased during the two-week experiment, but the rate never reaches that of the palisade mesophyll in the control leaves (Figure 7.7). Maximum oxygen (O_2) evolution in leaf slices from the palisade mesophyll of control *Spinacia oleracea* leaves is more than double that in leaf slices from the spongy mesophyll (Figure 7.7A and C). After 15-19 days of leaf inversion, leaf slice photosynthesis of both tissue types exhibits the same P_{max} (Figure 7.7B and D). The maximum rate of oxygen evolution in leaf slices from the palisade mesophyll decreases, whereas oxygen evolution of spongy mesophyll leaf slices increases (Figure 7.7). The same trend is observed on an area or chlorophyll basis, as inverting the leaves caused little change in pigments (data not shown). Respiration rates are 35% higher in the spongy mesophyll compared to the palisade mesophyll of inverted leaves and about 41% higher in the control leaves (data not shown).

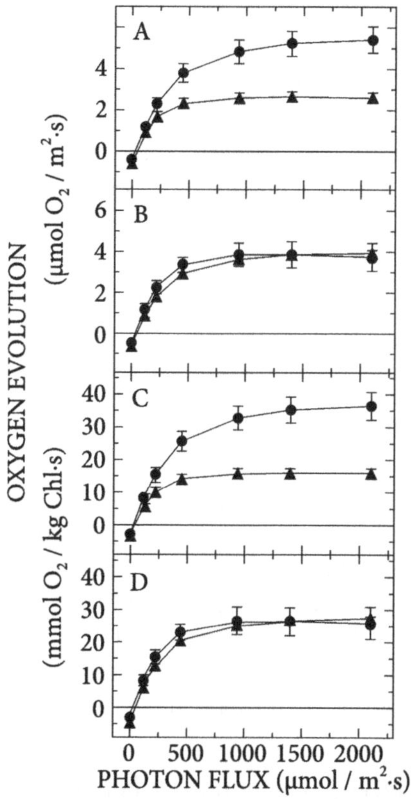

Fig. 7.7. Long-term effect of leaf inversion on the photosynthetic capacity of different tissues of *Spinacia oleracea* leaves (adapted from Biel, Fomina *et al.*, 2010). Light saturation curves of oxygen evolution were measured in fresh paradermal leaf slices from the palisade mesophyll (•) and spongy mesophyll (▲) of control leaves (A, C) and leaves inverted for 15–19 days (B, D)
Note: Chlorophylls *a* and *b* were determined using the extinction coefficients of Lichtenthaler (1987). Age of plants from sow date was 38–42 days. Data points represent the mean ± SD. Control, *n* = 5; inverted, n = 4

To our knowledge, the measurements of HCO_3^--dependent oxygen evolution in fresh paradermal leaf sections of palisade mesophyll and spongy mesophyll are the first of their kind (Bil' *et al.*, 2003; Bil', Fomina *et al.*, 2003; Biel, Fomina *et al.*, 2008, 2010). The capacity for electron transport increases in the spongy mesophyll after leaf inversion, whereas it decreases in the palisade mesophyll (Figure 7.7B). The capacity for electron transport in the spongy mesophyll never reaches that of the control palisade mesophyll. However, it increases to the same rate to which the palisade mesophyll decreases (Figure 7.7B). That the capacity for electron transport never reaches that of the palisade mesophyll may be due to high light. Interestingly, the sum of electron transport capacities of the two tissue types in the inverted leaves is equivalent to that of the sum in the controls on both an area and chlorophyll basis (see Figure 7.7). Such a result suggests that there may be some sort of whole leaf control on the total photosynthetic capacity due to the coordination of the anabolic and catabolic reactions of the two tissue types.

Thus, the interactions between the palisade mesophyll and spongy mesophyll and the control of total leaf photosynthesis are an intriguing question. Experiments altering the source / sink relations or the incident light on the leaves after leaf inversion should provide pertinent information about such a fundamental question.

The decrease in Rubisco activity in the palisade mesophyll, after leaf inversion, is likely due to a decrease in light and change in light quality. In contrast, whole leaf photosynthesis in inverted leaves increases, when the original abaxial leaf surface is illuminated compared to when the abaxial leaf surface is illuminated in control leaves. The increase in Rubisco activity (Figure 7.4) and photosynthetic electron transport capacity (Figure 7.7) in the spongy mesophyll after leaf inversion is likely linked to increases in the light microenvironment within the spongy mesophyll and pillow spongy mesophyll.

7.2.4 Enzyme Activity, Carbon Metabolism, and Stress Tolerance across Spinacia oleracea Leaves

While the distribution of photosynthetic enzymes across leaves, such that there is a maximum midway through the palisade mesophyll, may be related to optimal light environments, we further hypothesize that different modes of carbon metabolism occur at different depths within the leaves, and that such metabolism is related to the known stress tolerance at the top of leaves (Sun *et al.*, 1996*a*). For example, a higher rate of photorespiration at the surface of the leaf may be unavoidable, or it may serve a protective function (Wu *et al.*, 1991; Osmond and Grace, 1995). It can be also predicted that protective enzymes and compounds, such as antioxidants, SOD, catalase, peroxidase (Halliwell and Gutteridge, 1989) are more active in the top of bifacial leaves, as shown for the activity of catalase (Figure 7.4). Additionally, the activity of the enzyme was higher in the spongy mesophyll of leaves that had been inverted (Figure 7.4).

Figure 7.8 schematically summarizes literature and our data (modified from Biel, Fomina *et al.*, 2010; see also Bassham, Benson *et al.*, 1954; Bassham and Calvin, 1957; Whiteley *et al.*, 1959; Doman and Romanova,

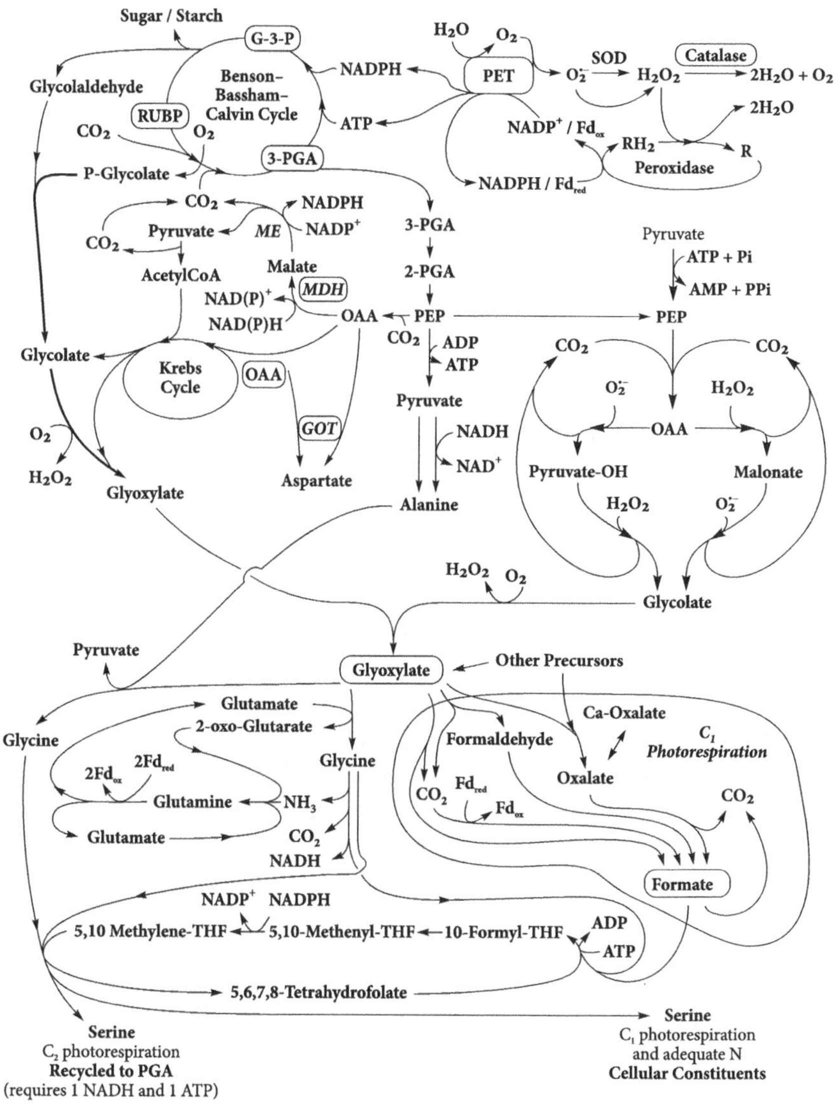

Fig. 7.8. Interrelationships among various metabolic activities in a leaf of C_3 plant (adapted from Biel, Fomina *et al.*, 2010). The metabolic activities related to stress tolerance are likely more active in the upper part of the leaf that is directly exposed to light. C_1 / C_2 photorespiration represents an efficient pathway of photorespiration that may be induced by stress in C_3 plants. Various mechanisms and pathways for dealing with active oxygen forms are shown. The various enzymes and pathways are discussed in the text

1962; Tarchevsky 1964, 1982; Hiatt, 1965; Kent, 1972; Zelitch, 1973; Ramaswamy *et al.*, 1976; Grodzinski, 1978; Lyubimov *et al.*, 1978; Lyubimov and Karpilov, 1978; Beevers, 1979; Oliver, 1979; Cossins, 1980; 2000; Tolbert, 1955, 1981; Bassham and Buchanan, 1982; Magomedov, 1982, 1998; Nasyrov, 1982; Nonomura *et al.*, 1998; Benson, 2002; Ramberg *et al.*, 2002; Bil' *et al.*, 2003; Bil', Fomina *et al.*, 2003; and others) that illustrate some of the relationships between photosynthetic carbon metabolism, photosynthetic electron transport, the Krebs cycle and a number of metabolic pathways that could contribute to environmental stress tolerance. As an example, production of superoxides at the top of the leaf can be detoxified by superoxide dismutase and catalase, as well as non-enzymatic decarboxylation of oxaloacetate to pyruvate-OH and malonate and finally to glycolate (Zelitch, 1973; Lyubimov *et al.*, 1978; Lyubimov and Karpilov, 1978).

Glycolate from oxygenation of RuBP and other sources could be metabolized via C_2 photorespiratory pathways (Tolbert, 1981) or alternative C_2 / C_1 photorespiratory pathways (Nonomura *et al.*, 1998). Many of the reactions involved in C_1 photorespiration have been previously reviewed (Cossins, 1980; Cossins, 2000). Oxidation of glycolate forms a key compound, glyoxylate, the branch point for C_2 or C_1 metabolism. Glyoxylate can also be formed directly from isocitrate via the glyoxylate pathway (Beevers, 1979; Tolbert, 1981). The C_1 photorespiratory pathway allows direct formylation of tetrahydrofolate (Whiteley *et al.*, 1959; Hiatt, 1965) and incorporation of formate directly into serine (Tolbert, 1955; Doman and Romanova, 1962). The formate can originate from any number of sources. Some examples include glyoxylate (Tolbert *et al.*, 1949; Zelitch, 1972; Grodzinski, 1978; Oliver, 1979) and oxalate decarboxylation (Millerd *et al.*, 1963*a*; see also Franceschi and Horner, 1980), and direct reduction of CO_2 to formate (Kent, 1972; Ramaswamy *et al.*, 1976). In the presence of adequate nitrogen, the C_1 photorespiratory pathway would eliminate one amination of glyoxylate, and therefore decrease the amount of nitrogen-cycling during photorespiration.

As discussed above in the description of the leaf anatomy, namely, idioblasts containing calcium oxalate crystals, oxalate may be a component of C_1 photorespiration.

Carbon dioxide from glyoxylate and oxalate decarboxylation and formate oxidation (Mazelis, 1960; Halliwell, 1974; Oliver, 1981) can be refixed via RuBP carboxylation, PEP carboxylation, or directly into (back to) formate for incorporation into the C_1 pathway via tetrahydrofolate synthetase. A direct fixation of CO_2 into formate (Kent, 1972; Ramaswamy *et al.*, 1976) followed by incorporation into the tetrahydrofolate-C_1

metabolite pool, while being an extremely efficient mode of CO_2 fixation appears to represent only a fraction of the carbon flux in plants. It is possible, however, that under stressful conditions, C_1 photorespiration accounts for a greater portion of the carbon flux in plants, and C_1 photorespiration may be important in stress tolerance.

Under drought conditions, the percentage of total carbon fixed into alanine doubles from 10% to 20% (Tarchevsky, 1964, 1982) apparently due to a decrease in the ATP / NADPH ratio that concomitantly decreases the export of trioses from the Benson-Bassham-Calvin cycle. Consequently, a higher percentage of 3-PGA is converted to alanine, rather than to sucrose under drought conditions. The phenomenon has been referred to as the "alanine effect" (Tarchevsky, 1964, 1982).

Figure 7.9 illustrates that alanine formed by transamination could be used to aminate glyoxylate under stressful conditions and provide glycine for photorespiratory recycling of glycolate.

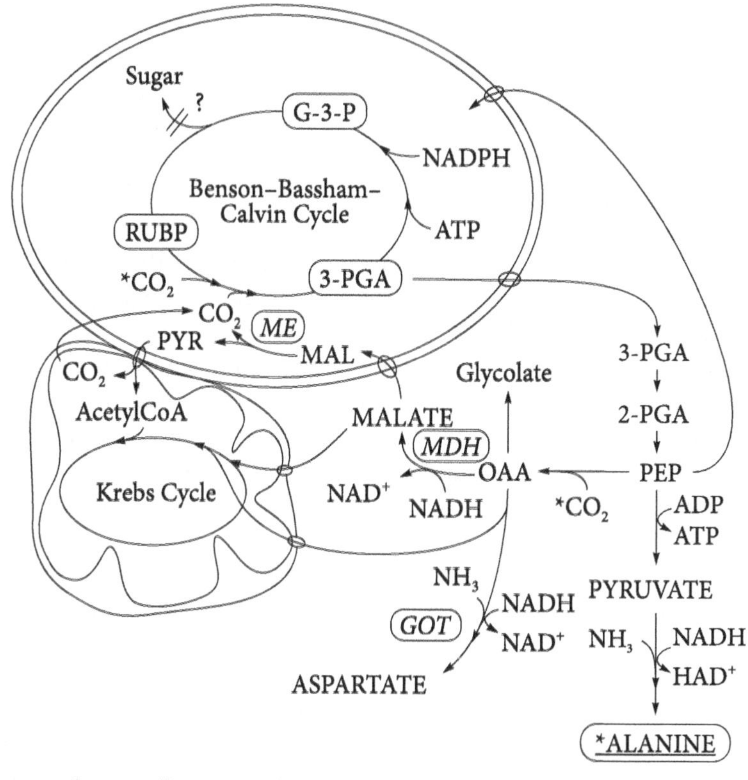

Fig. 7.9. Alanine effect. Drought treatment results in increased $^{14}CO_2$ labeling of alanine (Tarchevsky, 1964, 1982). Other related reactions are illustrated (adapted from Biel, Fomina *et al.*, 2010)

The oxaloacetate-malate shuttle shown in Figure 7.8 illustrates a possible chloroplast CO_2-enriching mechanism in a C_3 plant, as has been previously suggested (Nasyrov, 1982; Magomedov, 1982). The source of oxaloacetate is from PEP, that may form from 3-PGA diverted from the Benson-Bassham-Calvin cycle or glycolysis (not shown in Figure 7.9), which may operate in the light under stressful conditions. The oxaloacetate can be reduced to malate in the chloroplast or in the cytoplasm and be transported to the chloroplast, where it is decarboxylated by NADP-ME. Under stressful conditions, the CO_2 released can be refixed. Deamination and decarboxylation of aspartate in chloroplasts (Magomedov, 1998) could play a similar role as the oxaloacetate-malate shuttle shown.

The illustrated CO_2-enriching pathway in a C_3 plant represents a futile cycle, because the PEP comes from 3-PGA or pyruvate. The product of PEP carboxylation, oxaloacetate is reduced to malate and then decarboxylated to pyruvate. The pyruvate is either converted directly back to PEP or converted to 3-PGA, which can reenter the RuBP pool. In either case there is no net gain in carbon, unless the CO_2 / O_2 ratio is significantly increased. Since the contribution of a CO_2-enriching pathway would be insignificant (enzyme limitations, for example) under non-stressful conditions, such a pathway would only be useful under stressful conditions, when stomata are likely to be closed. The CO_2 released in the chloroplast would increase the CO_2 / O_2 ratio. If the other product of decarboxylation (pyruvate) is used to directly generate PEP, then ATP is utilized. If the pyruvate is used to regenerate RuBP, then 4 ATP's and 2 NADPH's are consumed. Consuming redox potential could prevent over-reduction under stressful conditions. The consumption of ATP utilizes the electrochemical potential across the thylakoids to produce more ATP, and decreases inhibition of photosynthetic electron transport due to the buildup of a high electrochemical potential. Inhibition of photosynthetic electron transport could lead to photodamage in the light.

It will be interesting to examine the distribution profiles across leaves of energy dissipating mechanisms (Demmig-Adams and Adams, 1992), membrane composition (Havaux and Gruszecki, 1993; Mishra *et al.*, 1994; Richter *et al.*, 1994; Schindler *et al.*, 1994; Moon *et al.*, 1995; Somerville, 1995), and various metabolic pathways, such as an oxaloacetate-malate shuttle and C_2 / C_1 photorespiration (Nonomura *et al.*, 1998), as they may all contribute to environmental stress tolerance at the top of leaves.

7.2.5 *Light Regulation of Photosynthetic Enzyme Activity across Leaves*

Since the capacity for gene expression and protein synthesis must initially be the same in all cells across the leaf, the observed distribution of Rubisco across leaves may be the result of down regulation of transcription or translation of the genes encoding Rubisco subunits or increased Rubisco degradation caused by some factor or factors in the upper portion of the leaves. At the bottom of the leaf, the same or other factors may result in lower amounts of Rubisco. The reversal of the Rubisco profile across mature leaves after inversion supports such a notion. It was shown previously that the capacity for carbon fixation in the palisade mesophyll was significantly reduced in leaves illuminated unilaterally on the abaxial leaf surface, because of light trapping in the spongy mesophyll (Sun and Nishio, 2001).

It is likely that the light gradient across bifacial leaves illuminated abaxially is steeper than when illuminated adaxially. Nonetheless, the Rubisco profile completely reversed after leaf inversion. The results suggest that photon flux alone cannot be responsible for the Rubisco profile across leaves, and that the controlling factors in Rubisco synthesis are much more sensitive than carbon fixation alone. The controlling factors, while not light flux alone, may be related to light flux and light quality, which impact cellular redox potential, ATP / ADP ratio, concentration of active oxygen species and sugars, for example. Concentration of carbon dioxide, temperature, and / or humidity gradients and the like may also play a role.

In summary it can be assumed that the light microenvironment within the leaf is a key factor in "shaping" the optimal leaf, and factors both similar and different to those controlling Rubisco synthesis play a role in controlling the distribution of other enzymes and pigments across leaves.

Because of their sessile nature, plants must have many adaptations to deal with stress. Leaves are exposed to highly variable light environments, temperatures, and physical perturbations. We have been investigating the relations among leaf anatomy, biochemistry, and influence of light on the bifacial leaves of *Spinacia oleracea* for over two and half decades. The bifacial leaf is exquisitely designed for its function. Besides the primary function of light harvesting and carbon reduction, different cell types are metabolically specialized to aid in plant survival. For example, the palisade mesophyll is also specialized to pass light and photosynthate to the tissue below, and it exhibits higher stress tolerance than the underlying tissue. The pirum mesophyll appears to specialize in the storage of Ca-oxalate crystals and the apparent transfer of photosynthates from the upper palisade mesophyll to the vascular tissue. They may be active in protein synthesis, as well. The spongy mesophyll acts like a light trap and efficiently captures light. Additionally, the spongy mesophyll is a metabolically active region that likely specializes in respiration and synthetic reactions. Inverting leaves showed that it is possible to reverse certain photosynthetic enzymes across the leaves, but not those involved in respiration.

Despite the shift in photosynthetic enzyme activity, total productivity could not be reached after more than two weeks of leaf inversion, which illustrates the importance of the leaf anatomy in bifacial leaves.

The selective forces acting over the past 500 million years shaped the many forms of leaves and modes of photosynthesis in higher plants, and the bifacial leaf is only one strategy for survival in a particular environment. The stimuli that result in the formation of specific structural and functional leaf features may emerge by generalization of the experimental data using the unified theory of ecological utility.

7.3 Model of Optimal Photosynthesis within a Mesophytic Leaf

In this section, the hypothesis that the distribution of carbon fixation across leaves is optimal and-an updated model of optimal photosynthesis within a mesophytic leaf that was developed on the basis of the principle of maximal ecological utility are described (Soukhovolsky *et al.*, 2001, 2002; Soukhovolsky, 2004). We assumed that the pattern of metabolic activity within the leaf is relatively optimal not only for utilization of light energy, but also for protection from photooxidation. For describing the function of biological systems by the model of ecological utility (Soukhovolsky, 1996, 2004), the function of partial ecological utility is entered for each defined parameter of the system. It is possible to use any limited monomial function to write the functions of partial utility (Hicks, 1939).

At optimal functioning of a system, the general ecological utility as a product of partial ecological utilities (in the case of their independence from each other) should be maximized. We have used this approach (Soukhovolsky *et al.*, 2001; 2002) on a hypothetical model of a mesophytic leaf, where the leaf is considered to be a flat plate, and its surface is illuminated by light ($\boldsymbol{I_o}$; $\boldsymbol{v}$ in the PAR range). The light is absorbed by chlorophyll ($a/b = 3/1$; see Nishio *et al.*, 1993) in a layer of phototrophic cells at different depths within the plate; the plate itself has the same absorption spectrum. In such a leaf the function, $\mathbf{r(h)}$, partial ecological utility of protection from photooxidation, and the function, $\mathbf{s(h)}$, partial ecological utility of light absorption for photosynthesis, depend on the relative depth, $\mathbf{h}$, of the photosynthetic layer within the leaf. The relative leaf depth, variable $\mathbf{h}$, can be incorporated by standardized leaf depth, variable z, where $z = \mathbf{h} \cdot (1 - \mathbf{h})^{-1}$. When $\mathbf{h} \to 0, z \to 0$; when $\mathbf{h} \to 1, z \to \infty$.

After such incorporation the functions $\mathbf{r(z)}$ and $\mathbf{s(z)}$ can be written as limited monomial functions:

$$r(z) = r_{min} + (r_{max} - r_{min}) \cdot (1 - \exp(-\alpha(\nu)z)), \quad (1)$$

$$s(z) = s_{min} + (s_{max} - s_{min}) \, exp(-\beta(\nu)z), \quad (2)$$

where r_{max}, s_{max} and r_{min}, s_{min} are the values of partial utility at $z = 0$ and $z \to \infty$; $\alpha(\nu)$, $\beta(\nu)$ are the parameters proportional, respectively, either to the uptake of O_2 with production of active forms of oxygen (Mehler reaction), or to the photosynthesis at leaf depth, z, at fixed I_o, ν.

The function of whole ecological utility of the photosynthetic layer at the standardized leaf depth is a product of (1) and (2), as follows:

$$F(z,\nu) = [r_{min} + (r_{max} - r_{min}) \cdot (1 - \exp(-\alpha(\nu)z)] \times [s_{min} + (s_{max} - s_{min}) \exp(-\beta(\nu)z)]. \quad (3)$$

Equation 3 shows that the function $\mathbf{F(z,\nu)}$, whole ecological utility of the photosynthetic layer at the standardized leaf depth, is a skewed Gaussian curve (Figure 7.10). The maximum of this curve occurs at the leaf depth h_{opt} (after returning from z to $\mathbf{h}$). In the model, h_{opt} depends on the parameters $\alpha(\nu)$ and $\beta(\nu)$. It is assumed that under the considered conditions (I_o is near saturating; natural [CO_2] and [O_2]), the quantum yield of the Mehler reaction does not change with leaf depth, and $\alpha(\nu)$ is proportional to the absorption coefficient $\mathbf{k(\nu)}$ of the photosynthetic pigments. Under the same conditions, the quantum yield of photosynthesis $\mathbf{q(\nu)}$ changes with leaf depth (Evans, 1999), but in the first approximation, $\beta(\nu)$ may be considered as proportional to the product of $\mathbf{k(\nu)}$ and the mean of $\mathbf{q(\nu)}$ (over the leaf depth).

To test the assumptions, model curves of the whole ecological utility of the photosynthetic layers in a model leaf exposed to the light of different wavelengths were constructed (Figure 7.10; Table 7.4) on the basis of experimental profiles of CO_2 fixation in *Spinacia oleracea* leaves exposed to light of 450, 560, and 650 nm (Sun *et al.*, 1998; Nishio, 2000).

Table 7.4. Parameters of the whole ecological utility model (WEU-model) in a model leaf for different spectral ranges (see Figure 7.10)

Light	WEU-model parameters					
	s_{max}	r_{max}	$\alpha(v)$	$\beta(v)$	s_{min}	r_{min}
450 nm	1	1	5	1.9	0.05	0.10
560 nm	1	1	1.6	0.8	0.15	0.10
650 nm	1	1	4	1.9	0.10	0.005
White	1	1	3.2	1.4	0.3	0.05

Note: $\alpha(v)$, $\beta(v)$, r_{max}, s_{max}, r_{min}, and s_{min} are defined in the text.

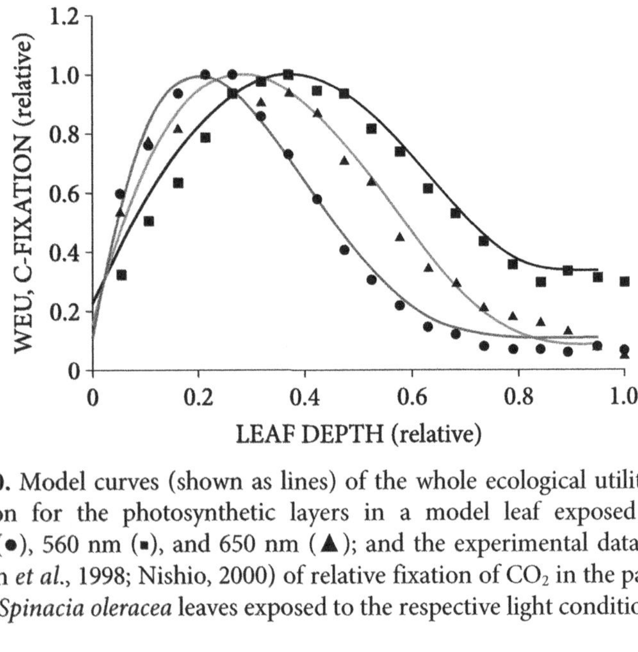

Fig. 7.10. Model curves (shown as lines) of the whole ecological utility (WEU) prediction for the photosynthetic layers in a model leaf exposed to light: 450 nm (●), 560 nm (▪), and 650 nm (▲); and the experimental data (adapted from Sun *et al.*, 1998; Nishio, 2000) of relative fixation of CO_2 in the paradermal slices of *Spinacia oleracea* leaves exposed to the respective light conditions

Calculations using coefficients of extinction of chlorophyll *a* and *b* (Mackinney, 1941) (the ratio *a*/*b* = 3/1 was used for calculations) and the data on the quantum yield of photosynthesis in *Spinacia oleracea* leaves (Sun *et al.*, 1998), showed that $\boldsymbol{\alpha(450)}$ and $\boldsymbol{\alpha(650)}$ of the model curves are proportional to the corresponding absorption coefficients $\boldsymbol{k(v)}$ of chlorophyll (*a* + *b*). Furthermore, $\boldsymbol{\beta(450)}$ and $\boldsymbol{\beta(650)}$ are proportional to the corresponding products of $\boldsymbol{k(v)}$ and quantum yield of photosynthesis $\boldsymbol{q(v)}$ (Table 7.5). The relative values of $\boldsymbol{\alpha(560)}$ and $\boldsymbol{\beta(560)}$ proved much higher than calculated (Table 7.5), probably because of absorption of green light by carotenoids, which was not considered in the calculations.

Table 7.5. Parameters $\alpha(\nu)$ and $\beta(\nu)$ of the whole ecological utility model (adapted from Table 7.4) and the calculated relative values: $k(\nu)$, absorption coefficient of chlorophyll ($a + b$) at chlorophyll $a/b = 3/1$; and $k(\nu) \times q(\nu)$, product of absorption coefficient and quantum yield of photosynthesis

Light	$\alpha(\nu)$, relative	$k(\nu)^a$, relative	$\beta(\nu)$, relative	$k(\nu) \times q(\nu)^b$, relative
450 nm	1	1	1	1
560 nm	0.32	0.12	0.42	0.16
650 nm	0.80	0.78	1	1

Note: [a]Coefficients of extinction for chlorophyll *a* and *b* in 80% acetone (Mackinney, 1941) and the data on [b]quantum yield of photosynthesis in *Spinacia oleracea* leaves (Sun *et al.*, 1998) were used for calculations; "1" is relative value of the parameters for blue light

Scattering of green light is also relatively high within leaves compared to red and blue light. Characteristically, the values of $\alpha(\nu)$ and $\beta(\nu)$ for white light in the model curve (Figure 7.11A), constructed on the basis of profiles of Rubisco activity, were about the means of the corresponding values for 450, 560 and 650 nm (Table 7.4).

Since the value of h is based on the depth within the leaf in relation to light flux, the curve of the whole ecological utility of the photosynthetic layers in a model leaf inverted 180° should be reversed compared to the initial curve (Figure 7.11A), and $h_{opt,\,invert} = 1 - h_{opt,\,norm}$. This aspect of the model was experimentally verified by measuring Rubisco activity in *Spinacia oleracea* leaves (Figure 7.4 Rubisco A; Figure 7.11A). In the inverted leaves, the profile of Rubisco activity reversed after 12-21 days, and the model ratio $h_{opt,\,invert} = 1 - h_{opt,\,norm}$ was achieved with high accuracy, though no correction for different absorbencies of palisade mesophyll and spongy mesophyll cells due to light scattering were applied.

At the bottom of the leaf, light, in particular red and blue, may become limiting. The elevated Rubisco activity predicted by the model curve compared to the experimental data in the bottom of the model leaf (Figure 7.11A) may be due to the regulatory role of red and blue light for Rubisco synthesis (Voskresenskaya, 1965; Tobin and Silverthorne, 1985; Tobin and Kehoe, 1994), which was not taken into consideration. Accordance between the model curves and experimental data for leaf depths close to the bottom of the leaf is improved (Figure 7.11B) with modifications to equation (3) that take the ecological utility of the absorption of regulatory light into account:

$$W(z,\nu) = F(z,\nu) \cdot [w_{min} + (w_{max} - w_{min}) \cdot \exp(-\gamma(\nu)z)], \qquad (4)$$

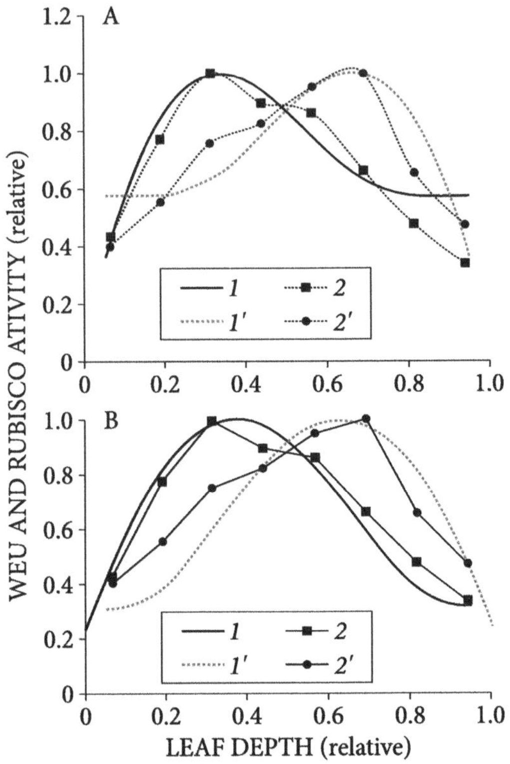

Fig. 7.11. Whole ecological utility prediction (A) and modified the whole ecological utility prediction (B) for the photosynthetic layers in a model leaf exposed to white light by adaxial (1) or abaxial (1′) surface and the experimental profiles of the Rubisco activity (adapted from Figure 7.4) in *Spinacia oleracea* leaves: control (2); inverted for 12–21 days (2′)

where w_{max} and w_{min} are the values of the whole ecological utility at $z = 0$ and $z \to \infty$; $\gamma(v)$ is the parameter proportional to absorption of regulatory light in the leaf depth, z, at fixed I_o, v. The improvement in correspondence at the bottom of the leaf resulted in an under prediction in activity at the top of leaves illuminated abaxially. Again, this may be due to the light scattering characteristics of the spongy mesophyll.

Our results described in section 1 of the chapter and data from earlier work lead us to assume that, after hundreds of millions of years of selection, the typical mesophytic leaf structure and the pattern of metabolic activity within it are relatively optimal, not only for utilization of light energy, but also for protection from photooxidation.

It has been suggested that photosynthesis simply follows the light absorption gradient (Evans, 1999). However, light absorption decreases from the top, illuminated surface to the bottom of the leaf (Cui *et al.*, 1991; Nishio, 2000), while the maximal CO_2 fixation occurs in the middle of palisade mesophyll (Nishio *et al.*, 1993). The results imply that a significant amount of absorbed light energy may not be utilized directly for CO_2 fixation at the top of the leaf. Clearly, the low carbon fixation is related to the low amount of Rubisco. The control of Rubisco accumulation is not known. It is possible that low Rubisco content in the upper layer of palisade is due to excess illumination, which induces the formation of active oxygen species; oxygen uptake under saturated light might reach 50% from its maximal evolution (Voskresenskaya, 1965).

The modeling (Equations 1–3) shows that optimal CO_2 fixation within a bifacial mesophytic leaf of a C_3 plant conforms to the rule of maximal ecological utility when only the simple factors of maximization of light energy utilization and minimization of photodestruction are applied (Soukhovolsky *et al.*, 2001; 2002; Soukhovolsky, 2004). Curves generated by the model show a good correlation with experimental profiles of carbon fixation in *Spinacia oleracea* leaf under different light conditions (Figure 7.10), as well as with the profiles of Rubisco activity in control and inverted *Spinacia oleracea* leaves (Figure 7.11).

Thus, the leaf depth optimum for photosynthesis is apparently determined, in part, by the internal light conditions within the leaf. Of course, such a distribution of Rubisco must be related to the corresponding chlorophyll distribution, which mainly controls the light gradient in *Spinacia oleracea* (Nishio, 2000). The same argument would apply for inverted leaves, where the maximum amount of photosynthesis shifted to the region of spongy mesophyll near the vascular tissue.

7.4 General Conclusion

Most physiological stress caused by environmental factors must result from interactions of the abiotic factors with light (also an abiotic factor), leaf anatomy, and a broad range of plant metabolic activities within the leaf. The interaction among light, leaf anatomy, and cellular metabolism within leaves is an important area of research that will provide fundamental knowledge about the operation of the integrated leaf.

While we have shown many aspects of metabolic distribution within leaves and how such distribution may be related to stress tolerance, there is still little knowledge about the integrated functioning of whole leaves. For example, there are a number of ^{35}S-methionine incorporating polypeptides that accumulate in the vascular tissue (mainly in the phloem)

(Sun *et al.*, 1996*b*). One of the highly incorporating polypeptides has been identified as a ribosomal protein, based on a peptide sequence (unpublished). Its identity, and that of the other highly incorporating polypeptides, can be confirmed by mass spectrometry.

The present chapter describes how the maximal amount of Rubisco across a leaf occurs midway through the palisade mesophyll, and that the profile of Rubisco across leaves can be reversed by leaf inversion, whereas the measured respiratory enzymes do not reverse. As mentioned, the physiological reason for such a distribution is not known, but we have speculated that the lower content of Rubisco at the top of leaves is due to high light, and that photoprotective mechanisms must be elevated at the leaf surface exposed to light.

Support for this hypothesis comes from the rules of maximal ecological utility, which were successfully applied to model the gradients. The changes that occur to carbon metabolism within leaves during various abiotic stresses require refinement and reexamination. Basic knowledge of gene expression and protein expression within leaves is lacking. Applying genomic microarray approaches in combination with proteomic approaches (Gómez, Nishio *et al.*, 2002; Gómez, Bil' *et al.*, 2003; Whitelegge, Laganowsky *et al.*, 2006; Laganowsky *et al.*, 2009) to paradermal leaf slices will provide necessary insight for further understanding whole leaf metabolism.

Acknowledgments

Many thanks for the valuable comments on the manuscript from Academician Professor Andrew A. Benson (Scripps Institute of Oceanography, La Jolla, California, USA; deceased), Professor Edwin A. Cossins (University of Alberta, Edmonton, Canada), and Professor William H. Outlaw Jr. (Florida State University, Tallahassee, Florida, USA. We also thank Dr. Vincent R. Franceschi (Electron Microscopy Center, Washington State University, Pullman, Washington, USA; deceased) for useful discussions about calcium oxalic acid.

References

Alexandrov V.G., Shanidze M.I. Effect of Ca-oxalate crystals on the function of chloroplasts, *J. Russian Bot. Soc.*, 1925, Vol. 10, p. 165–167. [in Russian]

Bassham J.A., Benson A.A., Kay L.D., Harris A.Z., Wilson A.P., Calvin M. The path of carbon dioxide in photosynthesis. XXI. The cyclic regeneration of carbon dioxide acceptor, *J. Am. Chem. Soc.*, 1954, Vol. 76, p. 1760–1770.

Bassham J.A., Buchanan B.B. Carbon dioxide fixation pathways in plants and bacteria, *Photosynthesis: Development, Carbon Metabolism, and Plant Productivity*, 1st edn., Govindjee ed., Academic Press, New York, 1982, Vol. **2**, p. 141–189.

Bassham J.A., Calvin M. *The Path of Carbon in Photosynthesis*, Prentice-Hall, Inc., Englewood Cliffs, New Jersey, 1957, 104 p.

Benson A.A. Mechanism of biochemical photosynthesis, *Zeitschr. Electrochem.*, 1952, Vol. 56, p. 848–854.

Benson A.A., Calvin M. Carbon dioxide fixation by green plants, *Annu. Rev. Plant Physiol.*, 1950*a*, Vol. 1, p. 25–42.

Benson A.A., Calvin M. The path of carbon in photosynthesis. VII. Respiration and photosynthesis, *J. Exp. Bot.*, 1950*b*, Vol. 1, p. 63–68.

Benson A.A. Following the path of carbon in photosynthesis: a personal story, *Photosynth. Res.*, 2002, Vol. 73, p. 29–49.

Beevers H. Microbodies in higher plants. Glyoxysomes, leaf peroxisomes, microbodies from other tissues, *Annu. Rev. Plant Physiol.*, 1979, Vol. 30, p. 159–193.

Biel K.Y. *Phototrophic Carbon Metabolism in Plants with Differential Organization of Photosynthetic Apparatus*, Dr. Sci. Dissertation, Pushchino, Russia, 1988, 543 p. [in Russian]

Biel K.Y. *Ecology of Photosynthesis*, Nauka, Moscow, 1993, 221 p. [in Russian]

Biel K.Y., Fomina I.R. Benson-Bassham-Calvin cycle contribution to the organic life on our planet, *Photosynthetica*, 2015, Vol. 53, No. 2, p. 161–167.

Biel K.Y., Lyubimov V.Y., Demidova R.N., Gedemov T. The phenomenon of cooperative function in a leaf with CAM and C_4 photosynthesis, *Photosynthetic Carbon Metabolism*, Mokronosov A.T. ed., Publ. Ural. Univer., Sverdlovsk, 1983, 47–57. [in Russian]

Biel K.Y., Fomina I.R., Nazarova G.N., Soukhovolsky V.G., Khlebopros R.G., Nishio J.N. Untangling metabolic and spatial interactions of stress tolerance in plants. 1. Patterns of carbon metabolism within leaves, *Protoplasma*, 2010, Vol. 245, p. 49–73.

Biel K.Y., Fomina I.R., Nishio J.N., Soukhovolsky V.G. Optimization of photosynthesis within leaves of C_3 and C_4 plants. Example of complex system self-regulation when energy source is also stress inductor, *System Identification and Control Problems*, Proc VII Int. Conf. SICPRO '08, Moscow, 2008, p. 566–593. ISBN 978-5-91450-002-0

Bil' K.Y. Yuri S. Karpilov: the man, the scientist, and the discoverer of cooperative photosynthesis (1931-1977), *Russ. J. Plant Physiol.*, 2000, Vol. 47, No. 4, p. 565–568.

Bil' K.Y., Fomina I.R., Nazarova G.N., Nishio J.N. Light dependent organization of carbon metabolism within spinach leaves, *Physiol. Biochem. Cult. Plants*, 2003, Vol. 35, No. 5, p. 392–402.

Bil' K.Y, Fomina I.R., Nishio J.N. Stress tolerance and light dependent organization of carbon metabolism within spinach leaves, *New and Nontraditional Plants and Prospects of Their Utilization*, Proc. 5th Int. Symp., Publ. Russian Univer. Int. Friendship, Moscow–Pushchino, Russia, 2003, Vol. 3, p. 24–27.

Bil' K.Y., Fomina I.R., Shabnova N.I., Terekhova I.V., Tsenova E.N. The influence of ammonium nitrogen nutrition and conditions of illumination on activity of enzymes of photosynthetic carbon metabolism in corn of different genotypes, *Biochemistry-USSR*, 1990, Vol.54, p. 920–925.

Bil' K.Y., Lyubimov V.Y., Demidova R.N., Gedemov T. Assimilation of CO_2 by plants of the family *Chenopodiaceae* with three types of autotrophic tissues in the leaves, *Soviet Plant Physiol.*, 1981, Vol. 28, p. 808–815.

Bolwell G.P., Wojtaszek P. Mechanism for the generation of reactive oxygen species in plant defense – A broad perspective, *Physiol. Mol. Plant Pathol.*, 1997, Vol. 51, p. 347–366.

Bradford M.M. A rapid and sensitive method for the quantitation of microgram quantities of protein utilizing the principle of protein-dye binding, *Anal. Biochem.*, 1976, Vol. 72, p. 248–254.

Brubaker C.L., Lersten N.R. Paraveinal mesophyll: review and survey of the subtribe *Erythrininae* (*Phaseoleae, Papilionoideae, Leguminosae*), *Plant Syst. Evol.*, 1995, Vol. 196, p. 31–62.

Chen Z., Spreitzer R.J. How various factors influence the CO_2 / O_2 specificity of ribulose-1,5-bisphosphate carboxylase / oxygenase, *Photosynth. Res.*, 1992, Vol. 31, p. 157–164.

Chugunova N.G., Bil' K.Y., Chermnykh L.N. The structure and photosynthesis of cucumber leaves at different temperatures in the root zone, *Soviet Plant Physiol.*, 1976, Vol. 22, p. 584–591.

Clements E.S. The relation of leaf structure to physical factors, *Trans. Am. Micr. Soc.*, 1905, Vol. 26, p. 19–102.

Cossins E.A. One-carbon metabolism, *The Biochemistry of Plants*, P.K. Stumph and E.E. Conn eds., Academic Press, London, 1980, Vol. 2, p. 365–418.

Cossins E.A. The fascinating world of folate and one-carbon metabolism, *Can. J. Bot.*, 2000, Vol. 78, p. 691–708.

Cui M., Vogelmann T.C., Smith W.K. Chlorophyll and light gradients in sun and shade leaves of *Spinacia oleracea, Plant Cell Environ.*, 1991, Vol. 14, p. 493–500.

Demmig-Adams B., Adams W.W. III. Photoprotection and other responses of plants to high light stress, *Annu. Rev. Plant Physiol. Plant Mol. Biol.* 1992, Vol. 43, p. 599–626.

Doman N.G., Romanova A.K. Transformations of labeled formic acid, formaldehyde, methanol, and CO_2 absorbed by bean and barley leaves from air, *Plant Physiol.*, 1962, Vol. 37, p. 833–840.

Evans J.R. Leaf anatomy enables more equal access to light and CO_2 between chloroplasts, *New Phytol.*, 1999, Vol. 143, p. 93–104.

Fahn A. *Plant Anatomy*, 4th edn., Pergamon Press, Oxford–New York, 1990, 588 p.

Finley D.S. Patterns of calcium oxalate crystals in young tropical leaves: a possible role as an anti-herbivory defense, *Rev. Biol. Trop.*, 1999, Vol. 47, No. 1, p. 27–31.

Fisher D.B. An unusual layer of cells in the mesophyll of the soybean leaf, *Bot. Gaz.*, 1967, Vol. 128, p. 215–218.

Fomina I.R., Biel K.Y., Nazarova G.N., Nishio J. Light dependent organization of carbon metabolism within leaf of C_3 plants, *Vestnik Bashkirskogo Universiteta* (Bull. Bashkirian Univer.), 2001*a*, Vol. 2, No. 1, p. 82–84. [in Russian]

Fomina I.R., Biel K.Y., Nazarova G.N., Nishio J. Light dependent organization of carbon metabolism within spinach leaf, *New and Nontraditional Plants and Prospects of Their Utilization*, Proc. 4th Int. Symp., NCBI Press, Pushchino, Russia, 2001*b*, Vol. 2, p. 549–551. [in Russian]

Franceschi V.R., Horner H.T. Use of *Psychotria punctata* callus in study of calcium oxalate crystal idioblast formation, *Z. Pflanzenphysiol.*, 1979, Vol. 92, p. 61–75.

Franceschi V.R., Horner H.T. Calcium oxalate crystals in plants, *Bot. Rev.*, 1980, Vol. 46, p. 361–427.

Franceschi V.R., Loewus F.A. Oxalate biosynthesis and function in plant and fungi, *Calcium Oxalate in Biological Systems*, S.R. Khan ed., CRC Press, Boca Raton, FL, 1995, p. 113–130.

Gómez S.M., Bil' K.Y., Aguilera R., Nishio J.N., Faull K.F., Whitelegge J.P. Transit peptide cleavage Sites of integral thylakoid membrane proteins, *Mol. Cell Proteomics,* 2003, Vol. 2, p. 1068–1085.

Gómez S.M., Nishio J.N., Faull K.F., Whitelegge J.P. The chloroplast grana proteome defined by intact mass measurements from LC-MS, *Mol. Cell Proteomics,* 2002, Vol. 1, p. 46–59.

Grodzinski B. Glyoxylate decarboxylation during photorespiration, *Planta,* 1978, Vol. 144, p. 31–37.

Haberlandt G. *Physiological Plant Anatomy,* Translated by Montagu Drummond, MacMillan and Co Ltd, London, 1914, 777 p.

Halliwell B. Oxidation of formate by peroxisomes and mitochondria from spinach leaves, *Biochem. J.,* 1974, Vol. 138, p. 77–85.

Halliwell B., Gutteridge J.M.C. *Free Radicals in Biology and Medicine,* 2nd edn., Clarendon Press–Oxford University Press, Oxford–New York, 1989, 543 p.

Hartt C.E., Kortschak H.P., Burr G.O. Photosynthesis by sugar cane fed radioactive carbon dioxide, *Proc. Haw. Acad. Sci.,* 1954, Vol. 1953/1954, p. 13–14.

Hatch M.D., Mau S. Activity, location and role of aspartate aminotransferase and alanine aminotransferase isoenzymes in leaves with C_4 pathway of photosynthesis, *Arch. Biochem. Biophys.,* 1973, Vol. 156, p. 195–206.

Hatch M.D., Slack C.R. Photosynthetic CO_2-fixation pathways, *Annu. Rev. Plant Physiol.,* 1970, Vol. 21, p. 141–184.

Hatch M.D., Kagava T., Craig S. Subdivision of C_4 pathway species based on different C_4-acid decarboxylating systems and ultrastructural features, *Aust. J. Plant Physiol.,* 1975, Vol. 2, p. 110–128.

Havaux M., Gruszecki W.I. Heat and light-induced chlorophyll *a* fluorescence changes in potato leaves containing high or low levels of the carotenoid zeaxanthin: indications of a regulatory effect of zeaxanthin on thylakoid membrane fluidity, *Photochem. Photobiol.,* 1993, Vol. 58, p. 607–614.

Hiatt A.J. Formic acid activation in plants. I. Purification, properties and distribution of formyl-tetrahydrofolate synthetase, *Plant Physiol.,* 1965, Vol. 40, p. 184–188.

Hicks J.R. *Value and Capital,* Clarendon Press, Oxford, 1939, 450 p.

Horner H.T., Wagner B.L. Calcium oxalate formation in higher plants, *Calcium Oxalate in Biological Systems,* S.R. Khan ed., CRC Press, Boca Raton, 1995, p. 53–72.

Karpilov Y.S. Distribution of ^{14}C among the photosynthetic products of maize, *Tr. Kazan S-kh Inst.* (Works Kazan Agricultural Institute), 1960, Vol. 41, p. 15–24. [in Russian]

Karpilov Y.S., Pristupa N.A., Bil' K.Y., Brick P.L. The structure of the photosynthetic apparatus in the leaves of plants with cooperative photosynthesis, *Cooperative Photosynthesis of Xerophytes*, Proc. Moldavian Inst. Res. Irrigation Farming and Vegetable Growing, Publ. Agricult Ministry Moldova, Kishinev, 1970, Vol. 11, No. 3, p. 18–24. [in Russian]

Karpilov Y.S., Kuzmin A.N., Bil' K.Y. Distribution of glycolytic enzymes in assimilating tissues of C_4 plant leaves and their link with characteristics of photosynthetic and photorespiratory reactions, *Soviet Plant Physiol.*, 1978, Vol. 25, p. 891–897.

Karpilova I.F., Chugunova N.G., Bil' K.Y., Chermnykh L.N. Ontogenetic changes of chloroplast ultrastructure, photosynthates, and photosynthate outflow from the leaves in cucumber plants under conditions of reduced night temperature, *Soviet Plant Physiol.*, 1982, Vol. 29, p. 96–104.

Keates S.E., Tarlyn N.M., Loewus F.A., Franceschi V.R. L-ascorbic acid and L-galactose are sources for oxalic acid and calcium oxalate in *Pistia stratiotes*, *Phytochem.*, 2000, Vol. 53, p. 433–440.

Kent S.S. Photosynthesis in the higher plant *Vicia faba*. II. The non-Calvin cycle origin of acetate and its metabolic relationship to the photosynthetic origin of formate, *J. Biol. Chem.*, 1972, Vol. 247, p. 7293–7302.

Kluge M., Ting I.P. *Crassulacean Acid Metabolism: Analysis of an Ecological Adaptation*, Springer–Verlag, New York, 1978, 209 p.

Kortschak H.P., Hartt C.E., Burr G.O. Carbon dioxide fixation in sugarcane leaves, *Plant Physiol.*, 1965, Vol. 40, p. 209–213.

Kostman T.A., Tarlyn N.M., Loewus F.A., Franceschi V.R. Biosynthesis of L-ascorbic acid and conversion of carbons 1 and 2 of L-ascorbic acid to oxalic acid occurs within individual calcium oxalate crystal idioblasts, *Plant Physiol.*, 2001, Vol. 125, p. 634–640.

Kursanov A.L. *Transport of Assimilates in Plants*, Nauka, Moscow, 1976, 649 p. [in Russian]

Laganowsky A., Whitelegge J.P., Nishio J.N. Hydroponics on a Chip: analysis of the Fe deficient *Arabidopsis* thylakoid membrane proteome, *J. Proteomics*, 2009, Vol. 72, p. 397–415.

Lansing A.J., Franceschi V.R. The paraveinal mesophyll: a specialized path for intermediary transfer of assimilates in legume leaves, *Aust. J. Plant Physiol.*, 2000, Vol. 27, p. 757–767.

Leverenz J.W., Jarvis P.G. Photosynthesis in Sitka spruce. VIII. The effects of light flux density and direction on the rate of net photosynthesis and the stromal conductance of needles, *J. Appl. Ecol.*, 1979, Vol. 16, p. 919–932.

Lichtenthaler H.K. Chlorophylls and carotenoids: pigments of photosynthetic biomembranes, *Meth. Enzymol.*, 1987, Vol. 148, p. 350–382.

Lyubimov V.Y., Biel K.Y., Karpilov Y.S. The peculiarities of the glycolate formation in assimilation tissues of C_4 plants, *Plant Physiology*, Materials 6th National Conf. Bulgarian Acad. Sci., Sofia, Bulgaria, 1978, Vol. 4, No. 1, p. 41–56.

Lyubimov V.Y., Karpilov Y.S. Influence of org anic acids on the photoinduced absorption of oxygen by corn chloroplasts, *Proc. Acad. Sci. USSR*, 1978, Vol. 237, p. 142–144.

Mackinney G. Absorption of light by chlorophyll solutions, *J. Biol. Chem.*, 1941, Vol. 140, p. 315–322.

Magomedov I.M. *Physiology and Biochemistry of Photosynthesis in Plants Exhibiting Primary Carboxylation of Phosphoenolpyruvate,* Doctor Biol. Sci. Thesis, Publ. Leningrad State University, Leningrad, 1982, 418 p. [in Russian]

Magomedov I.M. Modern concepts concerning the role of organic acids in photosynthesis, *Vestnik St. Petersburg. Univer.* (Herald St. Petersburg Univer.), Biol. Ser., 1998, Vol. 3, p. 52–63. [in Russian]

Mazelis M. Formate oxidation by particulate preparations from higher plants, *Plant Physiol.*, 1960, Vol. 35, p. 386–391.

Millerd A., Morton R.K., Wells J.R.E. Enzymic synthesis of oxalic acid in *Oxalis pes-caprae, Biochem. J.*, 1963*a*, Vol. 88, p. 281–288.

Millerd A., Morton R.K., Wells J.R.E. Oxalic acid synthesis in shoots of *Oxalis pes-caprae.* The precursors of glycolic acid and glyoxylic acid, *Biochem. J.*, 1963*b*, Vol. 88, p. 276–281.

Mishra R.K., Mishra N.P., Gauri S.S. Protection of photosystem II complex from strong visible light and high temperature stress in aqueous dispersions of phosphatidylcholines, *J. Plant Physiol.*, 1994, Vol. 144, p. 248–250.

Monson R.K. On the evolutionary pathways resulting in C_4 photosynthesis and Crassulacean acid metabolism (CAM), *Adv. Ecol. Res.*, 1989, Vol. 19, p. 57–110.

Monson R.K., Edwards G.E., Ku M.S.B. C_3-C_4 intermediate photosynthesis in plants, *BioScience*, 1984, Vol. 34, p. 563–574.

Moon B.Y., Higashi S.-I., Gombos Z., Murata N. Unsaturation of the membrane lipids of chloroplasts stabilizes the photosynthetic machinery against low-temperature photoinhibition in transgenic tobacco plants, *Proc. Natl. Acad. Sci. USA*, 1995, Vol. 92, p. 6219–6223.

Moss D.N. Optimum lighting of leaves, *Crop. Sci.*, 1964, Vol. 4, p. 131–136.

Mostafa M.A.E., Ulrich A. Absorption, distribution, and form of Ca in relation to Ca deficiency (tip burn) of sugarbeet, *Crop. Sci.*, 1976, Vol. 16, p. 27–30.

Nasyrov Y.S. Genetical regulation of formation and activity of photosynthetic apparatus, *Physiology of Photosynthesis*, A.A. Nichiporovich ed., Nauka, Moscow, 1982, p. 146–164. [in Russian]

Nickell L.G. A tribune to Hugo P. Kortschak: the man, the scientist and the discoverer of C_4 photosynthesis, *Photosynth. Res.*, 1993, Vol. 35, p. 201–204.

Nezgovorova L.A. To a question about the products of photosynthesis, *Soviet Plant Physiol.*, 1956, Vol. 3, p. 497–507. [in Russian]

Nezgovorova L.A. Influence of water regime of plants on the receipt and distribution of carbon in photosynthesis, *Soviet Plant Physiol.*, 1957, Vol. 4, p. 440–449.

Nishio J.N. Why are higher plants green?: Evolution of the higher plant photosynthetic pigment complement, *Plant Cell Environ.*, 2000, Vol. 23, p. 539–548.

Nishio J.N., Sun J., Vogelmann T.C. Carbon fixation gradients across spinach leaves do not follow internal light gradients, *Plant Cell*, 1993, Vol. 5, p. 953–961.

Nishio J.N., Sun J., Vogelmann T.C. Photoinhibition and the light environment within leaves, *Photoinhibition of Photosynthesis: from Molecular Mechanisms to the* Field, 1st edn., Baker N.R., Bowyer J.R. eds., Bios. Scientific Publ., Oxford, 1994, p. 221–237.

Nishio J.N., Ting I.P. Carbon flow and metabolic specialization in the tissue layers of the Crassulacean Acid Metabolism plant *Peperomia camptotricha*, *Plant Physiol.*, 1987, Vol. 84, p. 600–604.

Nishio J.N., Ting I.P. Photosynthetic characteristics of the palisade mesophyll and spongy mesophyll in the CAM / C_4 intermediate plant *Peperomia camptotricha*, *Bot. Acta*, 1993, Vol. 106, p. 120–125.

Nonomura A.M., Nishio J.N., Benson A.A. *Methods and Compositions for Enhancing Plant Growth with ρ-Amino- or ρ-Nitro-Benzoic Acids*, 1998, United States of America Patent 5846908.

Oliver D.J. Mechanism of decarboxylation of glycine and glycolate by isolated soybean cells, *Plant Physiol.*, 1979, Vol. 64, p. 1048–1052.

Oliver D.J. Formate oxidation and oxygen reduction by leaf mitochondria, *Plant Physiol.*, 1981, Vol. 68, p. 703–705.

Osmond C.B. Oxalates and ionic equilibria in Australian saltbushes (*Atriplex*), *Nature*, 1963, Vol. 98, p. 503–505.

Osmond C.B. Crassulacean acid metabolism: a curiosity in context, *Annu. Rev. Plant Physiol.*, 1978, Vol. 29, p. 379–414.

Osmond C.B. Our eclectic adventures in the slower eras of photosynthesis: from New England down under to Biosphere 2 and beyond, *Annu. Rev. Plant Biol.*, 2014, Vol. 65, p. 1–32.

Osmond C.B., Allaway W.G. Pathway of CO_2 fixation in the CAM plant *Kalanchoe daigremontiana.* I. Patterns of $^{14}CO_2$ fixation in the light, *Aust. J. Plant Physiol.*, 1974, Vol. 1, No. 4, p. 503–511.

Osmond C.B., Avadhani P.N. Acid metabolism in *Atriplex.* II. Oxalate synthesis during acid metabolism in the dark, *Aust. J. Biol. Sci.*, 1968, Vol. 21, p. 917–927.

Osmond B., Badger M., Maxwell K., Bjorkman O., Leegood R. Too many photons: photorespiration, photoinhibition and photooxidation, *Trends Plant Sci.*, 1997, Vol. 2, p. 119–121.

Osmond C.B., Grace S.C. Perspectives on photoinhibition and photorespiration in the field: quintessential inefficiencies of the light and dark reactions of photosynthesis, *J. Exp. Bot.*, 1995, Vol. 46, p. 1351–1362.

Outlaw W.H.J. A minireview: comparative biochemistry of photosynthesis in palisade cells, spongy cells, and guard cells of C_3 leaves, *Progress in Photosynthesis Research*, Biggins J. ed., Martinus Nijhoff, Dordrecht, 1987, Vol. 4, p. 265–272.

Outlaw W.H.J., Schmuck C.L., Tolbert N.E. Photosynthetic carbon metabolism in the palisade parenchyma and spongy parenchyma of *Vicia faba* L., *Plant Physiol.*, 1976, Vol. 58, p. 186–189.

Oya V.M., Laisk A.K. Adaptation of the photosynthesis apparatus to the light profile in the leaf, *Soviet Plant Physiol.*, 1976, Vol. 23, p. 381–386.

Perchorowitz J.T., Raynes D.A., Jensen R.G. Measurement and preservation of the *in vivo* activation of ribulose-1,5-bisphosphate carboxylase / oxygenase proteins in leaf extracts, *Plant Physiol.*, 1982, Vol. 69, p. 1165–1168.

Ramaswamy N.K., Behere A.G., Nair P.M. A novel pathway for synthesis of solanidine in the isolated chloroplast from greening potatoes, *Euro. J. Biochem.*, 1976, Vol. 67, p. 275–280.

Ramberg H.A., Olson B.J.S.C., Nishio J.N., Markwell J., Osterman J.C. The role of methanol in promoting plant growth: an update, *Rev. Plant Biochem. Biotechnol.*, Goyal A., Mehta S.L., Lodha M.L. eds., 2002, Vol. 1, p. 113–126.

Richter M., Goss R., Bothin B., Wild A. Zeaxanthin dependent and zeaxanthin independent changes in non-photochemical energy dissipation, *J. Plant Physiol.*, 1994, Vol. 143, p. 495–499.

Schindler C., Reith P., Lichtenthaler H.K. Differential levels of carotenoids and decrease of zeaxanthin cycle performance during leaf development in a green and an aurea variety of tobacco, *J. Plant Physiol.*, 1994, Vol. 143, p. 500–507.

Seal S.N., Sen S.J.P. The photosynthetic production of oxalic acid in *Oxalis corniculata, Plant Cell Physiol.*, 1970, Vol. 11, p. 119–128.

Shibata K. Spectrophotometry of intact biological materials; absolute and relative measurements of their transmission, reflection and absorption spectra, *J. Biochem. (Tokyo)*, 1958, Vol. 45, p. 599–623.

Skupchenko V.B. The vibrating microtome of the soft tissues. The new scientific methods, *The New Scientific Methods*, Kalinin E.P. ed., Publ. Acad. Sci. USSR, Komi Branch, Syktyvkar, 1979, Vol. 2, p. 1–56. [in Russian]

Somerville C. Direct tests of the role of membrane lipid composition in low-temperature-induced photoinhibition and chilling sensitivity in plants and cyanobacteria, *Proc. Natl. Acad. Sci. USA*, 1995, Vol. 92, p. 6215–6218.

Somerville C.R., Ogren W.L. Genetic modification of photorespiration, *Trends Biochem. Sci.*, 1982, Vol. 7, p. 171–174.

Soukhovolsky V.G. Free competition for resource, *Mathematical Problems of Ecology*, Nauka, Novosibirsk, 1996, p. 151–154. [in Russian]

Soukhovolsky V.G. *Economics of Living Things: Optimizing Approach in the Description of Processes in Ecological Communities and Systems*, Khlebopros R.G. ed., Nauka, Novosibirsk, 2004, 139 p. [in Russian]

Soukhovolsky V.G., Fomina I.R., Biel K.Y., Nishio J., Khlebopros R.G. Optimization model of photosynthetic leaf. Optimization of photosynthetic CO_2 fixation within mesophytic leaf of C_3plants, *Biological Resources and Sustainable Development*, Materials Int. Conf., NIA-Priroda, Pushchino, Russia, 2001, p. 209–210. [in Russian]

Soukhovolsky V.G., Fomina I.R., Bil' K.Y., Nishio J., Khlebopros R.G. An optimization model of the photosynthetic leaf: the model of optimal photosynthetic CO_2 fixation within leaves of mesophytic C_3 plants, *Dokl. Biol. Sci.*, 2002, Vol. 382, p. 28–30.

Sun J., Nishio J.N. Why abaxial illumination limits photosynthetic carbon fixation in spinach leaves, *Plant Cell Physiol.*, 2001, Vol. 42, p. 1–8.

Sun J., Nishio J.N., Vogelmann T.C. High-light effects on CO_2 fixation gradients across leaves, *Plant Cell Environ.*, 1996*a*, Vol. 19, p. 1261–1271.

Sun J., Nishio J.N., Vogelmann T.C. ^{35}S-methionine incorporates differentially into polypeptides across leaves of spinach (*Spinacia oleracea*), *Plant Cell Physiol.*, 1996*b*, Vol. 37, p. 996–1006.

Sun J., Nishio J.N., Vogelmann T.C. Green light drives CO_2 fixation deep within leaves, *Plant Cell Physiol.*, 1998, Vol. 39, p. 1020–1026.

Syvertsen J.P., Cunningham G.L. The effects of irradiating adaxial or abaxial leaf surface on the rate of net photosynthesis of *Perezia nana* and *Helianthus annuus, Photosynthetica*, 1979, Vol. 13, p. 287–293.

Tarchevsky I.A. *Photosynthesis and Drought*, Publ. Kazan. Univer., Kazan, Russia, 1964. [in Russian]

Tarchevsky I.A. Mechanisms of drought effects on photosynthetic absorption of CO_2, *Physiology of Photosynthesis*, Nichiporovich A.A. ed., Nauka, Moscow, 1982, p. 118–129. [in Russian]

Tarchevsky I.A., Karpilov Y.S. On the nature of products of short-term photosynthesis, *Soviet Plant Physiol.*, 1963, Vol. 10, p. 229–231. [in Russian]

Terashima I. Dorsiventrality in photosynthetic light response curves of a leaf, *J. Exp. Bot.*, 1986, Vol. 37, p. 399–405.

Terashima I. Productive structure of a leaf, *Photosynthesis*, Proc. CS French. Symp., Briggs W.R. ed., Alan R. Liss, New York, 1989, p. 207–226.

Terashima I., Inoue Y. Vertical gradients in photosynthetic properties of spinach chloroplasts dependent on intra-leaf light environment, *Plant Cell Physiol.*, 1985, Vol. 26, p. 781–785.

Terashima I., Saeki T. Light environment within a leaf. I. Optical properties of paradermal sections of *Camellia* leaves with special reference to differences in the optical properties of palisade and spongy tissues, *Plant Cell Physiol.*, 1983, Vol. 24, p. 1493–1501.

Terashima I., Sakaguchi S., Hara N. Intra-leaf and intracellular gradients in chloroplast ultrastructure of dorsiventral leaves illumination from the adaxial or abaxial side during their development, *Plant Cell Physiol.*, 1986, Vol. 27, p. 1023–1031.

Thoday D. The significance of the reduction in the size of leaves, *J. Ecol.*, 1931, Vol. 19, p. 297–303.

Tobin E.M., Kehoe D.M. Phytochrome regulated gene expression, *Semin. Cell Biol.*, 1994, Vol. 5, p. 335–346.

Tobin E.M., Silverthorne J. Light regulation of gene expression in higher plants, *Annu. Rev. Plant Physiol.*, 1985, Vol. 36, p. 569–593.

Tolbert N.E. Formic acid metabolism in barley leaves, *J. Biol. Chem.*, 1955, Vol. 217, p. 27–34.

Tolbert N.E. Metabolic pathways in peroxisomes and glyoxysomes, *Annu. Rev. Biochem.*, 1981, Vol. 50, p. 133–157.

Tolbert N.E., Clagett C.O., Burris R.H. Products of the oxidation of glycolic acid and L-lactic acid by enzymes from tobacco leaves, *J. Biol. Chem.*, 1949, Vol. 181, p. 905–914.

Tranbarger T.J., Franceschi V.R., Hildebrand D.F., Grimes H.D. The soybean 94-kilodalton vegetative storage protein is a lipoxygenase that is localized in paraveinal mesophyll cell vacuoles, *Plant Cell*, 1991, Vol. 3, p. 973–987.

Türk S., Kuhn A.J., Schimansky C., Schröder W.H., Fink S. Uptake and fate of Ca in Norway spruce seedlings (*Picea abies* L. Karst.), *Int. Union Forest Research Organizations XX World Congr.*, Tampere, Finland, 1995, Poster 106, Abstract http://www.metla.fi/iufro/iufro95abss

Turner N.C., Singh D.P. Responses of adaxial and abaxial stomata to light and water deficits in sunflower and sorghum, *New Phytol.*, 1984, Vol. 96, p. 187–195.

Vogelmann T.C. Plant tissue optics, *Annu. Rev. Plant Physiol. Plant Mol. Biol.*, 1993, Vol. 44, p. 233–251.

Vogelmann T.C., Martin G. The functional significance of palisade tissue: penetration of directional versus diffuse light, *Plant Cell Environ.*, 1993, Vol. 16, p. 65–72.

Voskresenskaya N.P. Oxygen uptake by plants under the light, *Biochemistry and Biophysics of Photosynthesis*, Krasnovsky A.A. ed., Nauka, Moscow,1965, p. 219–235. [in Russian]

Webb M.A. Cell-mediated crystallization of calcium oxalate in plants, *Plant Cell*, 1999, Vol. 11, p. 751–761.

Whitelegge J.P., Laganowsky A., Nishio J.N., Souda P., Zhang H., Cramer W.A. Sequencing covalent modifications of membrane proteins, *J. Exp. Bot.*, 2006, Vol. 57, p. 1515–1522.

Whiteley H.R., Osborn M.J., Huennekens F.M. Purification and properties of the formate-activating enzyme from *Micrococcus aerogenes, J. Biol. Chem.*, 1959, Vol. 234, p.1538–1543.

Worthington C.C. Catalase, *Worthington Enzyme Manual: Enzymes and Related Biochemicals*, C.C. Worthington ed., Worthington Biochemical Corp, Freehold, New Jersey, 1988, p. 72–75.

Wu J., Neimanis S., Heber U. Photorespiration is more effective than the Mehler reaction in protecting the photosynthetic apparatus against photoinhibition, *Bot. Acta.*, 1991, Vol. 104, p. 283–291.

Zelitch I. The photooxidation of glyoxylate by envelope-free spinach chloroplasts and its relation to photorespiration, *Arch. Biochem. Biophys.*, 1972, Vol. 150, p. 698–707.

Zelitch I. Alternate pathways of glycolate synthesis in tobacco and maize leaves in relation to rates of photorespiration, *Plant Physiol.*, 1973, Vol. 51, p. 299–305.

Zindler-Frank E. Oxalate biosynthesis in relation to photosynthetic pathway and plant productivity–a survey, *Z. Pflanzenphysiol*, 1976, Vol. 80, p. 1–13.

8
4-Hydroxyphenethyl Alcohol and Dihydroquercetin Increase Adaptive Potential of Barley Plants under Soil Flooding Conditions

Tamara I. Balakhnina*

Institute of Basic Biological Problems, Russian Academy of Sciences, Pushchino, Moscow Region 142290, Russia

Abstract

Increase of plant adaptation to unfavorable factors under changing environmental conditions is one of the most important aims of the plant physiologists and ecologists' investigations. Soil flooding, like soil drought, salinity and other negative stress factors often takes place in nature and results in a decrease of crop production. The role of pre-sowing seed treatment with physiologically active substances of different nature in enhancing of plant resistance to soil flooding is presented here. As active factors were used cytokinin-like substance - 4-hydroxyphenethyl alcohol (4-HPEA) and bioflavonoid - dihydroquercetin (DHQ). The data on possible involvement of 4-HPEA and DHQ in increasing energy seeds germination, further growth of seedlings, biomass accumulation, and plant adaptive potential under optimal soil watering and flooding conditions are analyzed. The adaptive potential of the plants was estimated through the intensity oxidative processes, tested by the content of thiobarbituric acid reactive substances (TBARs) and the activities of antioxidant enzymes guaiacol-dependent peroxidase (GPX) and ascorbate peroxidase (AsP).

Keywords: Ascorbate peroxidase, Dihydroquercetin, 4-Hydroxyphenethyl alcohol, Flooding, Growth, Guaiacol peroxidase, Lipid peroxidation, Soil hypoxia

* *Corresponding authors:* tbalakhnina@rambler.ru

8.1 Introduction

8.1.1 Effect of Soil Flooding on Plants

One of the essential negative consequences of soil flooding is oxygen deficiency in the tissues of submerged plant organs. Oxygen deficiency affects the intensity and direction of a number of physiological and biochemical reactions and induces oxidative stress in plant cells (Balakhnina, 2015; Dat *et al.*, 2004; Grichko and Glick, 2001; Yan *et al.*, 1996). Under oxygen deficiency conditions experienced by plant roots during the period of soil flooding, generalized adaptive response takes place in leaves, which were in normal aerobic conditions (Balakhnina *et al.*, 2004, 2010a, 2010b, 2012; Yan *et al.*, 1996; Zakrzhevsky *et al.*, 1996). The root hypoxia results in an increase of the value of stomatal diffusive resistance, controlling the tissue turgor and the rate of CO_2 fixation (Balakhnina, 2015; Bennicelli *et al.*, 1998). Protection of plants from oxidative destruction is associated with active functioning of low molecular antioxidants (carotenoids, ascorbic acid, glutathione *etc.*), as well as antioxidant enzymes such as superoxide dismutase (SOD), ascorbate peroxidase (APX), guaiacol dependent peroxidase (GPX) and others, detoxifying the reactive oxygen species (ROS) (Balakhnina, 2015; Elstner, 1982; Halliwell, 1984; Noctor *et al.*, 2002). Besides APX, GPX is also involved in scavenging of soluble hydroperoxides in plants (Prasad *et al.*, 1995). Soil flooding induced SOD and APX in plant leaves (Balakhnina *et al.*, 2009; 2010a; Kalashnikov *et al.*, 1994; 1985) but GPX activity did not increase significantly (Kalashnikov *et al.*, 1994). A dependence of the membrane lipid peroxidation intensity on the activity of antioxidant enzymes has been shown repeatedly in roots and leaves of cultural plants (barley, maize, pea, and wheat) under flooding conditions (Balakhnina *et al.*, 2009; 2010; Bennicelli *et al.*, 1998, Kalashnikov *et al.*, 1985; 1994; 1999; Zakrzhevsky *et al.*, 1996). Physiological biochemical reactions of damaging and adaptation of the investigated plants were found to depend on both the dose of acting stress factors and the evolution-formed tolerance of these cultures to soil flooding and oxygen deficiency in the root zone (Balakhnina, 2015).

Nowadays, in connection with the growing ecological effort, search for and development of new technologies for increasing the adaptive potential of cultural plants is actual. One of such approaches may be the treatment of seeds with physiologically active substances, for example, with natural or synthetic cytokinins (Bessonova *et al.*, 1985; Chernyad'ev, 2005). The beginning of investigations on substances of cytokinin nature is connected with the name of D. Letham (Letham and Palni, 1983) who isolated zeatin from unripe maize corns and decoded its structure. At present, there are enough data confirming that these phytohormones play a role in the regulation of seed development, dormancy, and germination (Brault and Maldiney, 1999; Dungey *et al.*, 1980; Khan, 1975; 1982; Webb *et al.*, 1973). Taking into consideration that seeds in deep dormancy usually contain indeterminable small amounts of cytokinins (Dungey *et al.*, 1980; Webb *et al.*, 1973), it can be expected that the use of exogenous substances with cytokinin activity can contribute to increasing germinating energy. Taking into account that seeds in deep dormancy usually contain undetectable small amounts of cytokinins, it can be expected that the use of exogenous substances with cytokinin activity can contribute to increasing germinating energy.

The growth acceleration in the initial stage of development should take a special significance for regions with critical climatic and ecological conditions. 4-Hydroxyphenethyl alcohol (4-HPEA) is a phenolic compound, which was isolated from the phototrophic bacterium *Rhodospirillum rubrum* and characterized as a nonpurine cytokinin-like substance (Serdyuk *et al.*, 2000). It has been shown in a few studies that 4-HPEA stimulates plant growth in the juvenile phase (Ivanova *et al.*, 2003).

On the other hand, frequent natural flooding of important agricultural lands can lead to mold growth on seeds, and consequently, a decrease in their germination capacity, which ultimately leads to a decrease in crop yields. To eliminate this negative effect of soil flooding may be used treatment of the seeds with physiologically active natural substances such as polyphenolic compounds. Their antibiotic capacities to protect plants against pathogenic microorganisms and fungi are well known (Baraboy, 1976). It was demonstrated that plant tolerance to pathogens is correlated with the content of phenolic compounds in plant tissues (Baraboy, 1976; Morita, 1988). Antibiotic capacities of the polyphenols are connected with their antioxidant activity, which is shown to be high in quercetin, rutin, DHQ and some other bioflavonoids (Rice-Evans *et al.*, 1997; Yamasaki *et al.*, 1997). Antioxidant activity of DHQ is determined by its ability to donate hydrogen or electron. In this way DHQ is either oxidized, reducing other compounds, or prevents their oxidation and inhibits lipid peroxidation (LPO) (Baraboy, 1976; Rice-Evans *et al.*, 1997). It was shown (Teselkin *et al.*, 1996) that DHQ, similarly to the liposoluble antioxidants such as α-tocopherol and β-carotene, inhibits oxidation of liposomes induced by iron sulphate or the Fe^{2+}-ascorbate system. In another work (Kostyuk and Potapovich, 1998), DHQ, as well as rutin and quercetin, was found to be effective protector of phagocytic cells against injury caused by asbestos; at that DHQ inhibited TBARs formation due to its capacity to scavenge superoxide anion. Antioxidant and antibiotic activities of DHQ have been widely used in medicine, pharmacology and dietetics (Kandaswami *et al.*, 1992; Haraguchi *et al.*, 1996; Uminsky *et al.*, 2007), whereas application of DHQ in agriculture has been recently started in the field experiments and showed protective effect of DHQ against mold on the seeds (Biokhimzashchita, 2002).

8.2 Effect of 4-Hydroxyphenethyl Alcohol on Growth and Adaptive Potential of Barley Plants at Optimal Soil Watering and Flooding

Barley *Hordeum vulgare* L. cv. "Rudzik" seeds were soaked in distilled water, or in aqueous 10^{-6} M 4-HPEA, (+4-HPEA)-treatment and then germinated on a moist filter paper at 25 ± 2 °C. Energy of germination was estimated by the percentage of the full sprouted seeds 72 hours after the beginning of germination. The seedlings were grown in 1 dm^3 plastic pots filled with a brown loess soil. Soil pH was 7.3 in distilled water and contained 1.69% C_{org}, 29% of 1–0.1 mm fraction, 67% of 0.1–0.002 mm fraction, and 4% of clay. Each pot had 0.75 kg of soil (dry weight) packed to the bulk density of 1.35 Mg m^{-3}. All the seedlings were maintained under optimal conditions of aeration and water availability during the first 5 days of growth. Then, on the 5th day of growth, the pots were divided into four groups. In this way, half of the pots with (-4-HPEA) and (+4-HPEA) plants were maintained at optimal soil watering; the other pots were flooded with 1-cm of water layer stagnating on the soil surface for 7 days. The adaptive potential of the plants was estimated by the concentration of thiobarbituric acid reactive substances (TBARs) and the activity of guaiacol-dependent peroxidase (GPX). It was shown that the 4-HPEA had a stimulatory action on seeds germination, shoot and root growth and biomass production, expressed in a greater extent at the early stage of plant development. Pre-sowing soaking of barley seeds in aqueous 4-HPEA solution increased plant tolerance to the effect of soil flooding, too.

8.2.1 Plant Reactions

8.2.1.1 Seed Germination

The percentage of full sprouted seeds pre-soaked with water was 64 ± 5% 72 hours after the beginning of germination, but in the case of pre-soaking in aqueous 10^{-6} M 4-HPEA the percentage of full sprouted seeds was 85 ± 7%. This result showed that 4-HPEA stimulated seed germination by about 20%.

8.2.1.2 Plant Growth

The investigation of growth processes on the basis of shoots and roots elongation showed that on the 3rd day of growth the average length of the (-4-HPEA)-barley seedling shoots was about 4.6 cm, and then reached 18.2 cm on the 12th day of growth under optimal soil moisture conditions (Figure 8.1a).

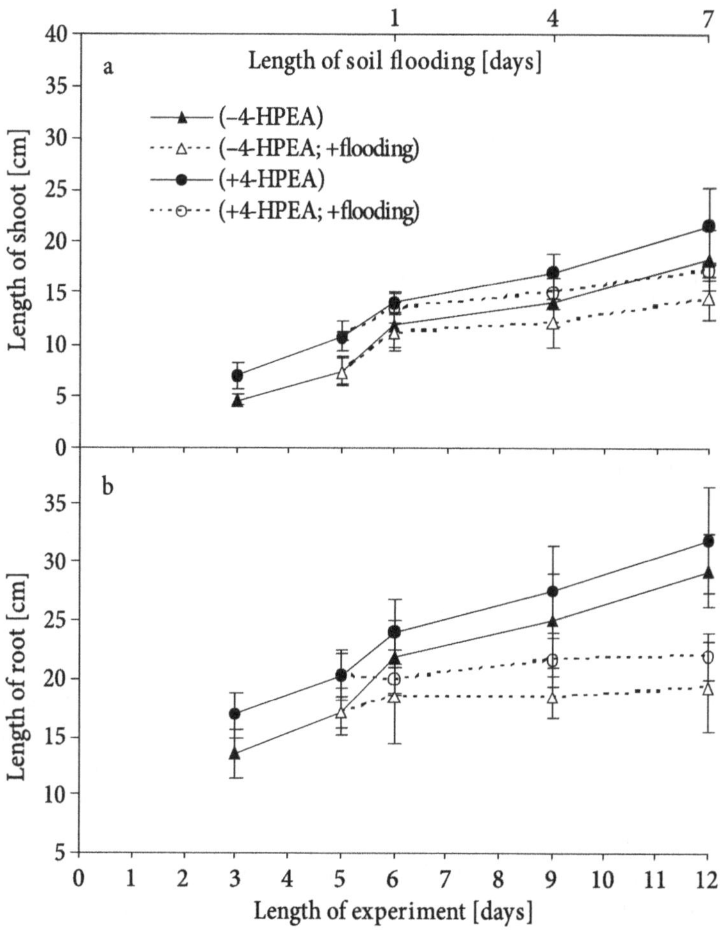

Fig. 8.1. The effects of pre-sowing treatment of *Hordeum vulgare* seeds with 10^{-6} M 4-hydroxyphenethyl alcohol (4-HPEA) on the shoot (a) and root (b) length of the seedlings grown under optimal soil watering or flooding conditions (adapted from Balakhnina *et al.*, 2010*b*)

Table 8.1. Statistical significance of differences in shoot and root length among seedlings germinated from the seeds treated (+4-HPEA) and non-treated (-4-HPEA) by 4-HPEA and grown under optimal soil conditions (-flooding) and under soil flooding (+flooding) started at the 5th day of growth (calculations were made on the basis of 95% LSD method)

Treatments which are compared	Time (days)				
	3	5	6	9	12
	Time of soil flooding (days)				
	–	–	1	4	7
	Difference				
	Length of shoot				
(+4-HPEA) / (-4-HPEA)	2.3*	3.36*	2.17*	2.67*	3.29*
(+4-HPEA) / (+4-HPEA;+flooding)	–	–	0.57	1.85*	4.33*
(-4-HPEA) / (-4-HPEA;+flooding)	–	–	0.81	2.06*	3.71*
(+4-HPEA;+flooding) / (-4-HPEA;+flooding)	–	–	2.41*	2.88*	2.67*
(+4-HPEA) / (-4-HPEA;+flooding)	–	–	2.98*	4.73*	7.0*
(+4-HPEA;+ flooding) / (-4-HPEA)	–	–	1.6*	0.82	−1.04
	Length of root				
(+4-HPEA) / (-4-HPEA)	3.4*	3.17*	2.19*	2.5*	2.65*
(+4-HPEA) / (+4-HPEA;+flooding)	–	–	3.97*	5.86*	9.93*
(-4-HPEA) / (-4-HPEA;+flooding)	–	–	3.34*	6.59*	10.0*
(+4-HPEA;+flooding) / (-4-HPEA;+flooding)	–	–	1.56	3.23*	2.72*
(+4-HPEA) / (-4-HPEA;+flooding)	–	–	5.53*	9.09*	12.65*
(+4-HPEA;+ flooding) / (-4-HPEA)	–	–	−1.79	−3.36*	−7.28*

*denotes a statistically significant difference.

The average length of the (+4-HPEA)-seedling shoots was about 6.9 and 21.5 cm on the 3rd and 12th day of growth, respectively. The data set in Figure 8.1a demonstrate more intensive growth of the (+4-HPEA)-seedling shoots, especially during the first 3–5 days of growth when the difference between the treatments reached 50%. Significant increase in shoots length of the (+4-HPEA)-plants, in comparison with (-4-HPEA)-ones, was observed during the whole period of the experiment (Table 8.1).

The shoots length of (+4-HPEA; +flooding)-plants, one day after the beginning of the overwatering, was significantly greater than the shoots length of (-4-HPEA;+flooding)-plants. This positive influence, though not significant, lasted during 4 days after the beginning of the soil flooding

(Table 8.1). It means that soaking of barley seeds with aqueous 4-HPEA solution led to stronger adaptive response in the leaves under oxygen deficiency conditions, even compared with well aerated soil. Soaking of barley seeds in the 4-HPEA solution induced root growth, too, but to a lesser degree than growth of the shoots. The length of roots in 3- and 5-day (+4-HPEA)-plants exceeded significantly the length of roots in the (-4-HPEA)-seedlings, by 5 and 19%, respectively (Table 8.1, Figure 8.1b).

Soil hypoxia resulted in a tendency to inhibit shoots elongation and in a reliable inhibition (by 15-17%) of the root elongation both in the (-4-HPEA; +flooding) and (+4-HPEA;+flooding)-plants, after one day of soil flooding action (Figure 8.1a,b). During ulterior development of the hypoxic stress, the elongation of shoots decreased reliably by 14–20% in all of the plants, but differences between the (-4-HPEA; +flooding) and (+4-HPEA;+flooding)-plants were maintained (Figure 8.1a). The effect of soil hypoxia on the root elongation was more pronounced. At the end of the flooding experiment the roots in the (-4-HPEA; +flooding) and in (+4-HPEA; +flooding)-seedlings were significantly shorter (by 34 and 31%) than in (-HPEA) and in (+4-HPEA) seedlings, respectively (Figure 8.1b). Soaking of barley seeds with 4-HPEA induced stimulation of shoot and root elongation under flooding compared with non-treated by 4-HPEA flooded plants (Table 8.1).

The examination of biomass production showed that shoot biomass of the (-4-HPEA) and (+4-HPEA)-seedlings averaged 0.054 and 0.079 g, respectively, on the 3rd day of growing, and reached 0.174 g (-4-HPEA) and 0.189 (+4-HPEA) on the 12th day of growing (Figure 8.2a). The data presented in Figure 8.2a show that the effect of 4-HPEA on biomass accumulation was more significant during the first 3–5 days of growing, when the differences between the variants reached 46 and 41%, respectively. The promoting effect of 4-HPEA on the root biomass was similar to the phytohormonal effect on the shoot biomass accumulation (Figure 8.2a,b). Under optimal growth conditions, the root biomass of the (+4-HPEA) seedlings exceeded the (-4-HPEA) value of this parameter on the 3rd and 5th days of growing by 35 and 26%, respectively. Then the differences between the variants declined and by the end of the experiment were about 6–8% (Figure 8.2b).

Under soil flooding conditions the growth of the shoot and root biomass decreased both in the (-4-HPEA;+flooding) and (+4-HPEA; +flooding) seedlings (Figure 8.2a,b). After 7 days of stress factor effects, the shoot biomass of the (-4-HPEA; +flooding) and (+4-HPEA; +flooding) seedlings declined by 20%; the root biomass of the same seedlings declined by 33 and 26%, respectively, compared to the (-4-HPEA) and (+4-HPEA) ones grown with optimal soil watering (Figure 8.2a,b). It is significant that under flooding conditions the seedlings grown from seeds treated with 4-HPEA kept the advantage in biomass accumulation, in comparison with untreated plants.

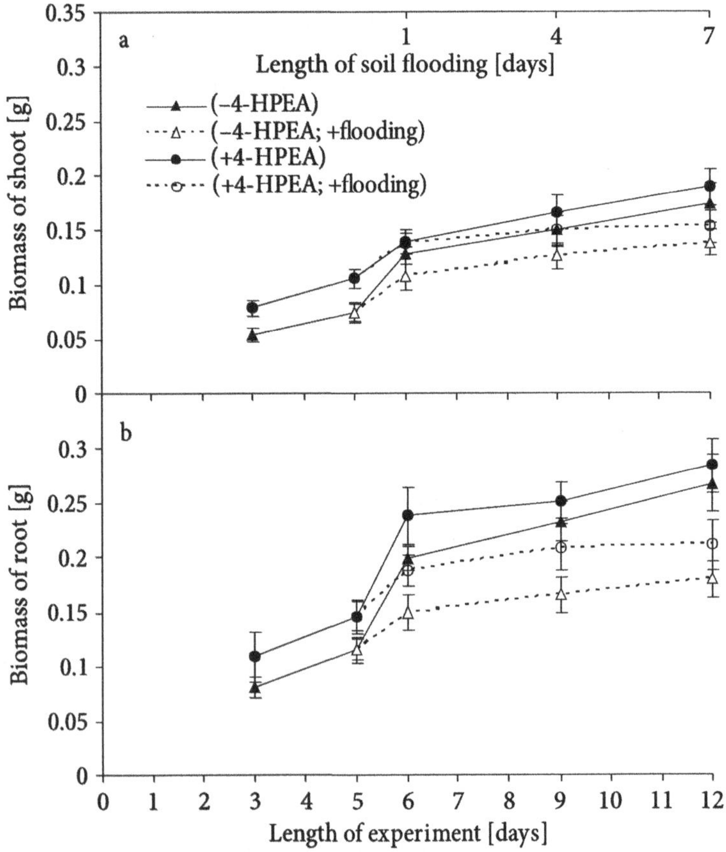

Fig. 8.2. The effects of pre-sowing treatment of *Hordeum vulgare* seeds with 10^{-6} M 4-hydroxyphenethyl alcohol (4-HPEA) on the shoot (a) and root (b) biomass production of the seedlings grown under optimal soil watering or flooding conditions (adapted from Balakhnina *et al.*, 2010*b*)

8.2.1.3 Lipid Peroxidation Intensity

The TBARs concentration (Figure 8.3) in the first leaves of 6-day seedlings of the (-4-HPEA) and (+4-HPEA) plants averaged 111 and 132 nmol g^{-1} FW, respectively, and was significantly varied (Table 8.2). During the next 6 days no significant decrease of the TBARs concentration was observed in the leaves of the (-4-HPEA) and (+4-HPEA) plants. As a result, the TBARs concentration in the leaves of seedlings of 9–12 days age, in both of the variants, was practically the same (Figure 8.3). The opposite effect on the TBARs concentration was observed in the (-4-HPEA; +flooding) and (+4-HPEA; +flooding) seedlings under soil hypoxia. An increase followed by a decrease of the TBARs concentration took place in the (-4-HPEA; +flooding) plants, and a significant decrease of the TBARs concentration with subsequent increase to the level of the control variant was noted in the plants (+4-HPEA;+flooding) (Table 8. 2, Figure 8.3).

8.2.1.4 Guaiacol Peroxidase Activity

GPX activity in the leaves of the (-4-HPEA) seedlings under optimal soil watering was monotone decrease during plant growth, from 7.55 μmol tetraquaiacol g^{-1} FW min^{-1} in 6-day seedlings to 3.48 μmol tetraquaiacol g^{-1} FW min^{-1} in 12-day seedlings (Figure 8.4). GPX activity in the leaves of 6- and 9-day (+4-HPEA) seedlings was significantly higher than in the (-4-HPEA) plants of the same age, by 81 and 130%, but after that it decreased and till the 12th day of growth was of about the same value as the (-4-HPEA) variant (Table 8.2, Figure 8.4).

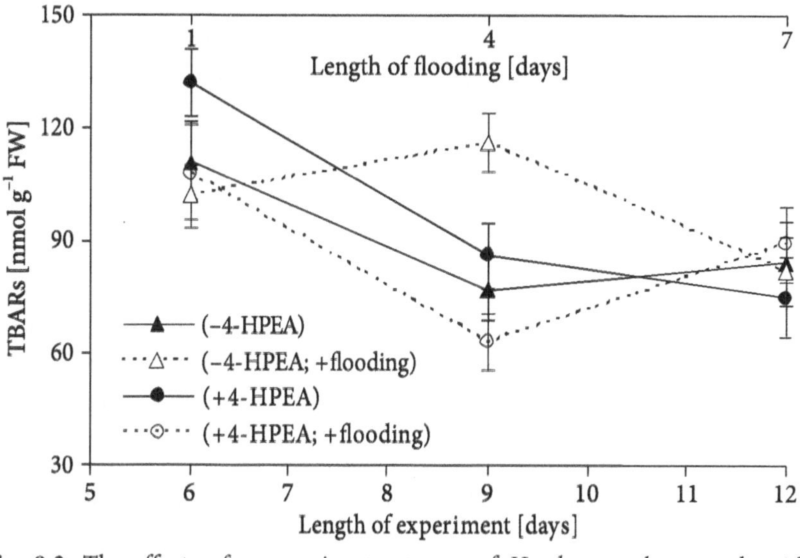

Fig. 8.3. The effects of pre-sowing treatment of *Hordeum vulgare* seeds with 10^{-6} M 4-hydroxyphenethyl alcohol (4-HPEA) on the contents of thiobarbituric acid reactive substances (TBARs) in the leaves of seedlings grown under optimal soil watering or flooding conditions (adapted from Balakhnina *et al.*, 2010b)

Table 8.2. Statistical significance of differences in TBARs concentration and GPX activity among seedlings germinated from the seeds treated (+4-HPEA) and non-treated (-4-HPEA) by 4-HPEA and grown under optimal soil conditions (-flooding) and under soil flooding (+flooding) started at the 5th day of growth (calculations were made on the basis of 95% LSD method)

Treatments which are compared	Time (days)				
	3	5	6	9	12
	Time of soil flooding (days)				
	–	–	1	4	7
	Difference				
	TBARs concentration				
(+4-HPEA) / (-4-HPEA)	–	–	21.0*	9.0	-9.0
(+4-HPEA) / (+4-HPEA; +flooding)	–	–	24.0*	23.0*	-14.0
(+4-HPEA) / (-4-HPEA; +flooding)	–	–	30.0*	-30.0*	-7.0
(-4-HPEA) / (+4-HPEA; + flooding)	–	–	3.0	14.0	-5.0
(-4-HPEA) / (-4-HPEA; +flooding)	–	–	9.0	-39.0*	2.0
(+4-HPEA; +flooding) / (-4-HPEA; +flooding)	–	–	6.0	-53.0*	7.0
	GPX activity				
(+4-HPEA) / (-4-HPEA)	–	–	6.13*	6.45*	0.84
(+4-HPEA) / (+4-HPEA; +flooding)	–	–	2.24*	3.27*	-4.23*
(+4-HPEA) / (-4-HPEA; +flooding)	–	–	5.76*	5.64*	-3.57*
(-4-HPEA) / (+4-HPEA; + flooding)	–	–	-3.89*	-3.18*	-5.07*
(-4-HPEA) / (-4-HPEA; +flooding)	–	–	-0.37	-0.81	-4.41*
(+4-HPEA; +flooding) / (-4-HPEA;+ flooding)	–	–	3.52*	2.37*	0.66

*denotes a statistically significant difference.

During the first 4 days of soil hypoxia action, the GPX activity in the leaves of the (-4-HPEA; +flooding) plants did not practically differ from the (-4-HPEA) variant (Figure 10.4). In contrast, the GPX activity in the leaves of (+4-HPEA; +flooding) plants, during the first 4 days of hypoxic stress, was lower than in (+4-HPEA) plants but significantly higher than in the (-4-HPEA) and (-4-HPEA; +flooding) plants (Table 8.2, Figure 8.4). On the 7th day the levels of GPX activities in the (-4-HPEA; +flooding) and (+4-HPEA; +flooding) plants. Moreover, the activity of this enzyme in all flooded plants was twice as high as that in the plants non-stressed by flooding (Figure 8.4).

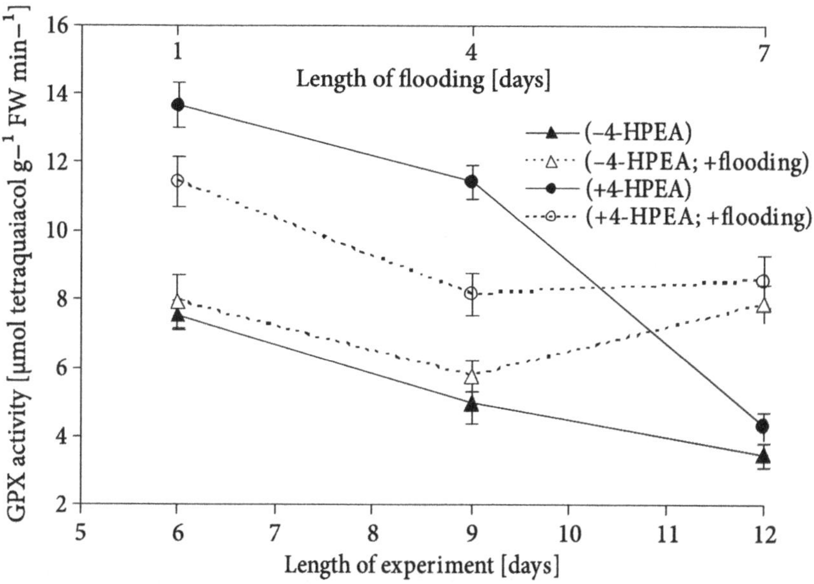

Fig. 8.4. The effects of pre-sowing treatment of *Hordeum vulgare* seeds with 10^{-6} M 4-hydroxyphenethyl alcohol (4-HPEA) on the quaiacol peroxidase (GPX) activity in the leaves of seedlings grown under optimal soil watering or flooding conditions (adapted from Balakhnina *et al.*, 2010*b*)

8.2.1.5 Discussion

The effect of natural and synthetic cytokinins on seed germination has been shown in numerous works (Khan, 1975; 1982; Webb *et al.*, 1973; Shishov *et al.*, 1990 and others). It is known that soaking of seeds in water with growth regulator solutions with antigibberellin and cytokinin activity before sowing increases the energy and rate of cauliflower seeds germination and the phenophasis replacement (Shishov *et al.*, 1990). It has also been found that cytokinins have a unique capability to eliminate different inhibitory effects on the seeds. The germination of seeds could be blocked by abscisic acid or by various phenolic compounds. Treatment of seeds with cytokinins removes or reduces the inhibitory effect of such compounds (Khan, 1982). It has been proposed that the regulation of dormancy and germination of seeds is realized through balance of their stimulators and inhibitors (Khan, 1975). The mechanism of withdrawal of seeds from the dormancy state is connected with activation by cytokinins of the RNA synthesis (Khan, 1975).

Our results showed that pre-treatment of seeds with 4-HPEA stimulated their germination by about 20%. Besides germination, shoots and roots growth induction and their biomass production were shown in the (+4-HPEA)-barley seedlings. These effects of 4-HPEA on the seedlings were significant at the earliest phase of plant development, with subsequent decrease at the end of the growth experiment. Moreover, the growth of (+4-HPEA)-seedlings was not suppressed by soil flooding to as significant a degree as in the (-4-HPEA)-plants. This result indicates that 4-HPEA plays a role in plant tolerance process.

There is little data in the literature about the cytokinin regulatory role in the formation of plant tolerance to unfavourable factors (Bessonova *et al.*, 1985; Chernyad'ev, 2005; Muromcev *et al.*, 1987) in comparison with a lot of data about cytokinin effect on the seed germination and seedling growth. In particular, the cytokinin influence on plant tolerance to drought has been shown on maize, horse bean and wheat plants (Chernyad'ev, 2005). It has also been established that kinetin diminished the effect of a higher dose of zinc on pea plant growth (Bessonova *et al.*, 1985). The cytokinin action on stress tolerance is associated with induction of "stress" protein biosynthesis (Nover *et al.*, 1984), activation of leaf growth, stimulation of chloroplast differentiation, and mesophyll cells growth and division (Muromcev *et al.*, 1987). The effect of cytokinin-like substance on the intensity of lipid peroxidation process and activity of antioxidant enzyme in the aspect of estimation of the adaptive potential of the plants was investigated by us for the first time.

Lipid peroxidation is known to be a parameter of stress response indicator in living organisms (Balakhnina, 2015). Usually, the intensity of lipid peroxidation is estimated by measuring TBARs concentration. The concentration of TBARs increased in (-4-HPEA)-barley seedlings after 4 days flooding. This increase indicates intensification of the oxidative destruction processes induced by flooding conditions. The (+4-HPEA)-seedlings did not show any increase in the concentration of TBARs but, in contrast, they exhibited a decreasing trend in TBARs concentration during flooding. We suppose that seed presoaking with 4-HPEA provides activation of antioxidant system in plants. Measurements of GPX activity supported this assumption. The GPX activity was increased significantly in the first leaves of (+4-HPEA)-plants. This increase was the biggest in the first stage of seedling growth, both under optimal and flooding conditions.

Thus, pre-soaking with 4-HPEA stimulates germination of barley seeds, growth and development of seedlings, and their stress tolerance. The adaptive potential in the 4-HPEA treated plants is higher than in the non-treated plants. Treatment of plant seeds with 4-HPEA could be used to increase plant tolerance to unfavourable factors.

8.3 Dihydroquercetin Protects Barley Seeds against Mold and Increases Seedling Adaptive Potential under Soil Flooding

Barley (*Hordeum vulgare* L.) seeds were soaked in aqueous 10^{-4} M dihydroquercetin (DHQ) to examine its influence on seed germination and further growth of seedlings under optimal soil watering and flooding conditions. The adaptive potential of the plants was estimated by the content of thiobarbituric acid reactive substances (TBARs) and the activity of ascorbate peroxidase (AsP). High-grade seeds were germinated evenly under (-DHQ)- and (+DHQ)-treatments. Low-grade seeds soaked in DHQ, showed no mold and twofold germination rate in comparison with the same seeds soaked in water. The seedlings grown from the similarly germinated seeds did not differ from each other in the shoot growth, independent of the DHQ-pretreatment. The root growth was higher in DHQ-pretreated plants. Soil flooding suppressed the shoot and root growth rates in non-pretreated and DHQ-pretreated plants, however TBARs content was lower in the roots and leaves of (+DHQ)-seedlings as compared to the (-DHQ)-ones. The activity of AsP increased more significantly in the (+DHQ)-plants. The ratio between TBARs content and the AsP activity was lower in the leaves of (+DHQ)-plants both under optimal soil conditions and flooding. Thus, the treatment of low-grade barley seeds with DHQ protects the seeds against mold and increases adaptive potential of the seedlings.

8.3.1 Plant Reactions

8.3.1.1 Seed Germination

The appearance of mold was detected on the surface of low-grade barley seeds soaked with distilled water and their germination was only 42% after 72 hours. At the same time the low-grade seeds, soaked with water solution of 10^{-4} M DHQ, remained clean and germinated up to 87% (Figure 8.5a). The high-grade seeds showed equal germination in the (-DHQ)- and (+DHQ)-variants of soaking (Figure 8.5b).

8.3.1.2 Growth Parameters

The seedlings, grown up from the low-grade seeds, showed intensification of the root growth in case of pre-sowing treatment with DHQ. After 3–5 days of growing under optimal watering, the main roots in (+DHQ)-plants were by 10–12% longer than that ones in (-DHQ)-plants (Figure 8.6a).

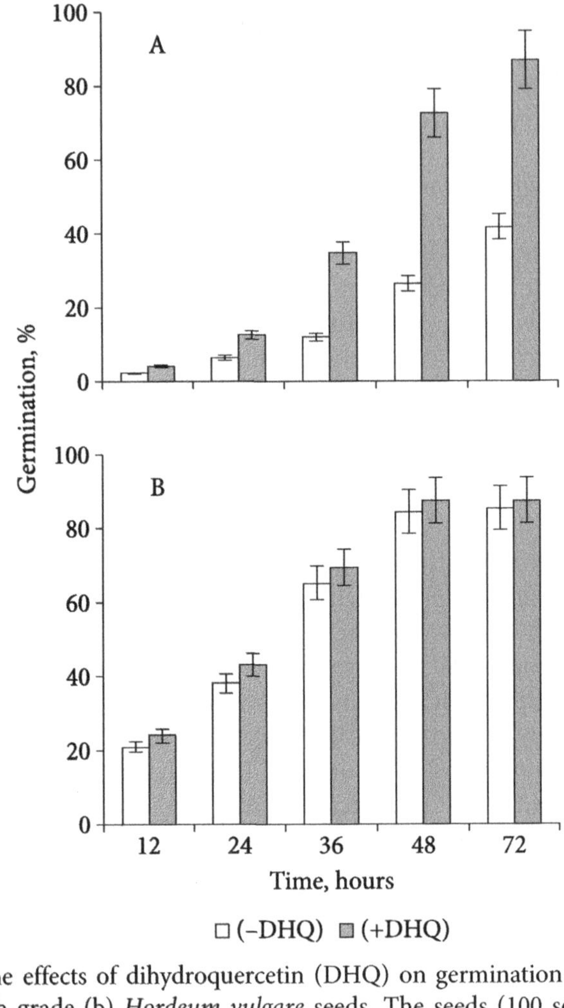

Fig. 8.5. The effects of dihydroquercetin (DHQ) on germination of low-grade (a) and high-grade (b) *Hordeum vulgare* seeds. The seeds (100 seeds per each replication) were soaked for 22 hours with distilled water or with water solution of 10^{-4} M DHQ, then germinated for 72 hours (adapted from Balakhnina *et al.*, 2009)
Note: Bars indicate standard deviations, $n = 6$

Visually, the roots of (+DHQ)-seedlings were darker and coarser than the roots of (-DHQ)-seedlings. Soil flooding suppressed the root growth uniformly in non-pretreated and pretreated with DHQ plants (Figure 8.6a). No differences in the shoot length between the (-DHQ)- and (+DHQ)-seedlings were detected neither under optimal soil watering nor under flooding conditions (Figure 8.6b).

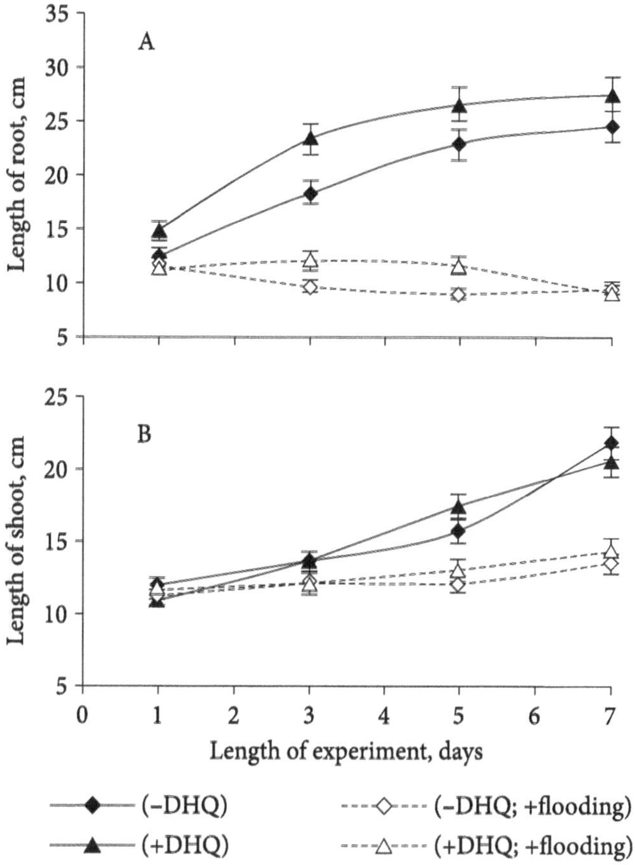

Fig. 8.6. The effects of pre-sowing treatment of *Hordeum vulgare* seeds with 10^{-4} M dihydroquercetin (DHQ) on the root (a) and shoot (b) length of the seedlings grown under optimal soil watering or flooding conditions (adapted from Balakhnina *et al.*, 2009)

Note: Bars indicate standard deviations, $n = 17$

The root biomass of (-DHQ)-plants increased during the 7 days of the experiment from 0.089 to 0.153 g per plant; during the same period, the (+DHQ)-seedlings showed the faster growth of root biomass, from 0.093 to 0.202 g per plant (Figure 8.7a). Hence, the difference in root biomass between the (+DHQ)- and (-DHQ)-variants was about 32% at the end of the experiment. Under soil flooding the growth of root biomass was suppressed equally in (+DHQ)- and (-DHQ)-plants (Figure 8.7a). Unlike the roots, the growth of shoot biomass did not depend on pre-sowing treatments, it was the same in (+DHQ)- and (-DHQ)-plants under optimal watering, decreasing evenly under flooding conditions (Figure 8.7b).

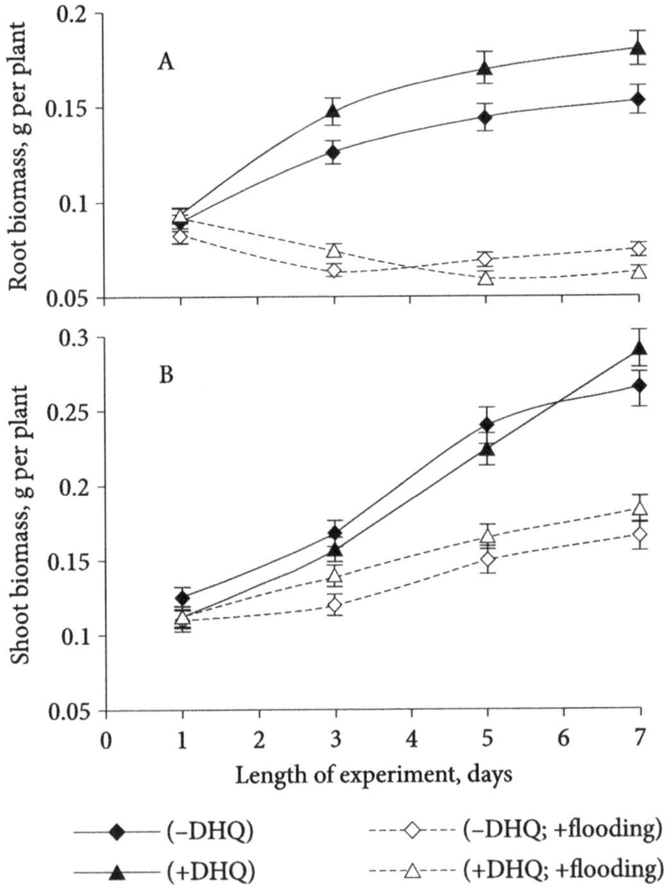

Fig. 8.7. The effects of pre-sowing treatment of *Hordeum vulgare* seeds with 10^{-4} M dihydroquercetin (DHQ) on the root (a) and shoot (b) biomass production of the seedlings grown under optimal soil watering or flooding conditions (adapted from Balakhnina *et al.*, 2009)
Note: Bars indicate standard deviations, $n = 17$

8.3.1.3 Intensity of Lipid Peroxidation

The changes of LPO intensity in the roots displayed an oscillatory character (Figure 8.8a). Among four treatments the highest concentration of TBARs was observed in the roots of the (-DHQ; +flooding)-seedlings. Under optimal soil watering the TBARs content in the roots of the (-DHQ)-seedlings was much lower than that under flooding (Figure 8.8a). Pre-sowing treatment with DHQ resulted in a decrease of LPO

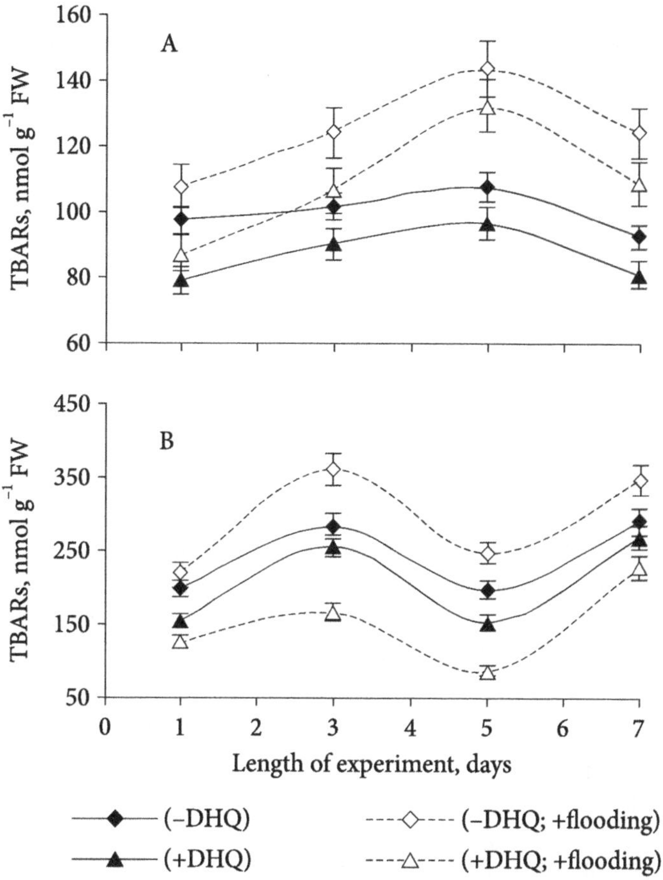

Fig. 8.8. The effects of presowing treatment of *Hordeum vulgare* seeds with 10^{-4} M dihydroquercetin (DHQ) on the contents of thiobarbituric acid reactive substances (TBARs) in the roots (a) and leaves (b) of the seedlings grown under optimal soil watering or flooding conditions (adapted from Balakhnina *et al.*, 2009)

Note: Bars indicate standard deviation, $n = 5$

intensity in the plant roots even under optimal soil watering. Under soil flooding the (+DHQ; +flooding)-plants kept lower level of LPO intensity during stress development as compared with the (-DHQ; +flooding)-plants (Figure 8.8a). The changes of LPO intensity in the leaves displayed oscillatory character as well (Figure 10.8b). In particular, the TBARs concentration in the leaves of (-DHQ)-seedlings changed during the plant growth from 199 to 290 nmol g^{-1} fresh mass (Figure 8.8b). The leaves of (+DHQ)-plants showed smaller values of TBARs concentration, which varied from 155 to 269 nmol g^{-1} fresh mass during the plant growth (Figure 8.8b). The soil flooding induced LPO intensity in the leaves of plants non-pretreated with DHQ (Figure 8.8b), namely, the percentage ratio of TBARs concentrations between (-DHQ; +flooding)- and (-DHQ)-plants increased from 111 to 128% during the experiment with flooding. In contrast, the leaves of plants pretreated with DHQ showed even a decrease in TBARs content under the flooding (Figure 8.8b). A comparison of the values of TBARs concentration in (+DHQ; +flooding)- and (-DHQ; +flooding)-plants also demonstrated that the pre-sowing treatment of seeds with DHQ resulted in a decrease of the leaf LPO intensity (Figure 8.8b).

8.3.1.4 Activity of Ascorbate Peroxidase

The activity of AsP in the leaves of (-DHQ)-seedlings decreased during the experiment of the seedlings' growth from 8.0 to 4.2 μmol g^{-1} min^{-1} (Figure 8.9). The leaves of (+DHQ)-seedlings also showed a decrease in AsP activity during the same time of the plants' growth from 11.0 to 6.0 μmol g^{-1} min^{-1}; however, the values of enzymatic activity were always higher in (+DHQ)-seedlings relative to the (-DHQ)-ones; the percentage ratio between these variants was 138, 157, 148 and 143%, respectively, after 1, 3, 5 and 7 days from the beginning of the experiment.

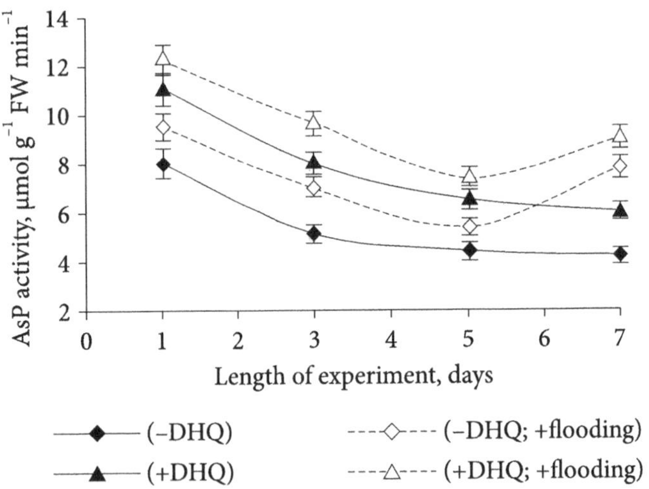

Fig. 8.9. The effects of presowing treatment of *Hordeum vulgare* seeds with 10^{-4} M dihydroquercetin (DHQ) on the ascorbate peroxidase (AsP) activity in the leaves of the seedlings grown under optimal soil watering or flooding conditions (adapted from Balakhnina *et al.*, 2009)
Note: Bars indicate standard deviation, $n = 5$

The soil flooding led to an increase in AsP activity both in the leaves of plants non-pretreated and pretreated with DHQ (Figure 8.9). In particular, the percentage ratio of AsP activities between the (-DHQ; +flooding)- and (-DHQ)-seedlings was 119% at the beginning of hypoxic stress development; at the end of flooding experiment this ratio reached 186%. Similarly, in the (+DHQ; +flooding)-seedlings an enzymatic activity was 111% as compared with the non-flooded (+DHQ)-plants at the beginning of flooding and reached 150% at the end of the experiment.

These data indicate that at the end of flooding experiment an increase of AsP activity was significant both in non-pretreated and pretreated with DHQ plants. Moreover, a comparison of AsP activities between the (+DHQ; +flooding)- and (-DHQ)-seedlings showed percentage ratio of 153% in the beginning of flooding and 214% at the end of the experiment. The ratios of AsP activities between (+DHQ; +flooding)- and (-DHQ; +flooding)-seedlings varied within the range of 115–137%. For detailed estimation of the relationship between the intensity of oxidative destruction and the AsP activity in the leaves, we calculated the ratios of TBARs content / AsP activity (Table 8.3).

Table 8.3. The ratios between the content of thiobarbituric acid reactive substances (TBARs) and the activity of ascorbate peroxidase (AsP) in the leaves of *Hordeum vulgare* due to seed treatments and seedling growth conditions

Length of experiment, days	Ratio of TBARs content / AsP activity (nmol / unit activity)[a]			
	(-DHQ)	(-DHQ; +flooding)	(+DHQ)	(+DHQ; +flooding)
1	25	23	14	11
3	55	52	32	18
5	45	46	24	12
7	69	44	45	25

[a]The data presented in Figures 8.8 and 8.9 were used for calculations

The values of this ratio increased in the leaf fresh mass of (-DHQ)-and (+DHQ)-seedlings during their growth, from 25 to 69 nmol TBARs / unit AsP activity and from 14 to 45 nmol TBARs / unit AsP activity, respecttively.

Under soil flooding conditions, the values of TBARs content / unit AsP activity in (-DHQ; +flooding)-plants were similar to these in the non-flooded ones, whereas plants pretreated with DHQ showed a decrease in this parameter (Table 8.3).

8.3.1.5 Discussion

The observations of germination of low- and high-grade barley seeds confirmed protective action of DHQ against mold. In field experiments the same effects were demonstrated using spring barley, winter wheat, potato, long-fibred flax, sugar beet and other plants (Biokhimzashchita, 2002). The dependence of plant tolerance to different pathogens on the concentration of phenolic compounds in plant tissues was shown in a number of works (see Baraboy, 1976; Morita, 1988). According to Morita's experiments (Morita, 1988) the oxygen radicals, generated during autooxidation of the polyphenols, are involved in bacteriophage inactivation. Antioxidant enzyme catalase and some low-molecular antioxidants inhibited biological injury of the phages by polyphenols (Morita, 1988). From this point of view, fungicide effect of DHQ may be explained by easy oxidation of bioflavonoid molecules with further formation of the radicals, which suppress, in their turn, mold fungus biological activity. The equal energies of germination observed in high-grade barley seeds treated and non-treated with DHQ supported this conclusion.

It is significant to note that in our experiments the pre-sowing treatment of low-grade seeds with DHQ suppressed mold growth on the seed surfaces but did not inhibit germination and seedling growth. Thus, the applied DHQ concentration of 10^{-4} M was really fatal for fungi without negative action on the growth processes in barley. Moreover, the root growth increased even by 10–12% in plants pretreated with DHQ.

It has long been known that phenolic compounds may inhibit growth of higher plants. In particular, the short-caulescent pea species were shown to contain considerably larger amount of phenolic compounds than the high-caulescent ones (Muzafarov and Ruzieva, 1985). Inhibitory effects of several phenolic compounds on seed germination and plant growth were directly demonstrated in a number of investigations (see Baraboy, 1976). In our experiments, we did not measure the concentration of DHQ transferred into barley seedlings from the seeds and could only assume that it was smaller than the required one for exhibition of inhibitory effect.

Besides protective action against mold, pretreatment of barley seeds with DHQ was useful for seedlings growing under soil flooding. The first symptoms of the development of hypoxic stress induced in plants by soil flooding are increasing stomatal diffusive resistance and wilting (Jackson, 1991). This seems to be a consequence of higher resistance to mass flow of water through the roots. Stomatal closure in leaves of flooded plants was considered to be caused by ABA, which was synthesized in the roots soon after flooding and then transported to leaves (Jackson, 1991). It is known that stomatal closure results in declining photosynthetic activity (Jackson and Drew, 1984). Direct measurements of stomatal diffusive resistance in the leaves of different cultural plants under soil flooding showed its increase with a subsequent decrease in photosynthesis and growth processes (Bennicelli *et al.*, 1998; Balakhnina *et al.*, 2004; Blanke and Cooke, 2004). Thus, another symptom of the plant hypoxic stress development is the inhibition of biomass production (Bennicelli *et al.*, 1998). The decreasing in plant growth under soil flooding is not only a consequence of reducing photosynthesis, but of lacking nitrogen or other major nutrients as well (Trought and Drew, 1980). Under the long-term development of the hypoxic stress, a decrease of the plants growth seems to be a result of accumulation of metabolic toxins (Jackson and Drew, 1984; Blanke and Cooke, 2004). In our experiments a decrease in the shoot and root growth during the development of hypoxic stress was steady in (+DHQ; +flooding)- and (-DHQ; +flooding)-plants. However, measurement of antioxidant potential in the seedlings showed its increase provided the seeds were presoaked with DHQ.

The nonspecific response of higher plants to soil hypoxia induced by soil over-moistening is an increasing concentration of reactive oxygen species (ROS) in plant tissues (Kalashnikov *et al.*, 1994; Yan *et al.*, 1996). An increasing of H_2O_2 was earlier demonstrated in barley leaves and roots under soil flooding (Kalashnikov *et al.*, 1994). The living cells of every organism contain antioxidant systems to neutralize ROS (Asada, 2006; Larson, 1988). Among low molecular antioxidants, which determine antioxidant potential of the plants, ascorbic acid is one of the most important. It is a strong reducing agent, which neutralizes H_2O_2 to H_2O and O_2 with participation of AsP (Asada, 2006). The oxidized molecules of ascorbic acid can be reduced in the plant cells by bioflavonoids (Melnikova and Yoffe, 2002). Reciprocal stabilization of ascorbic acid and polyphenol compounds was shown under model conditions (Melnikova and Yoffe, 2002). Bioflavonoid, such as polyphenol, can be in touch with copper-containing center of ascorbate oxidase and it protects ascorbic acid from oxidative destruction, inhibiting in this way the formation of dihydroascorbic acid. Then the flavonoid is oxidized up to orthodiquinone, which can be reduced either by NADP-dependent quinone oxidoreductase or by ascorbic acid with further formation of dihydroascorbic acid (Brown *et al.*, 1998; Melnikova and Yoffe, 2002).

Our results showed that soil flooding leads to an increase of AsP activity in barley leaves especially after 7 days of stress development. Presowing treatment of barley seeds with DHQ also led to increasing AsP activity in the leaves. As was mentioned above, antioxidant activity of DHQ allowed LPO to be inhibited (Baraboy, 1976; Rice-Evans *et al.*, 1997). Therefore, the fact of increasing AsP activity in the leaves of (+DHQ)- and (+DHQ; +flooding)-plants, as compared with (-DHQ)- and (-DHQ; +flooding)-plants, allows us to suppose that a part of DHQ can be transferred from the seeds into the leaves, where it can be involved both in the direct ROS neutralization and reduction of oxidized ascorbate. As an index reflecting adaptive potential in barley tissues (under our conditions) we suggest to use the ratio of the LPO intensity to antioxidant activity, expressed by nmol TBARs / unit AsP activity. The values of these ratios were significantly lower in the DHQ-pretreated seedlings relative to the non-pretreated ones both under optimal soil watering and flooding conditions (see Table 10.3). This observation demonstrates close relationship between the LPO intensity and the AsP activity in barley tissues and possible decrease of oxidative destruction processes by stimulation of antioxidant enzyme as a result of DHQ-pretreatment.

Thus, the technology of pre-sowing soak of low-grade seeds in aqueous 10^{-4} M DHQ causes suppression of the mold fungus biological activity improves seed germination and increases adaptive potential of the seedlings.

Acknowledgments

The author heartily thanks Dr. Irina R. Fomina (Institute of Basic Biological Problems, RAS, Pushchino, Moscow Region, Russia); Professor Teresa M. Włodarczyk, Dr. Magdalena Nosalewicz and Dr. Aneta Borkowska (Institute of Agrophysics, PAS, Lublin, Poland) for participating in some experiments and discussing obtained results.

Abbreviations

AsP – Ascorbate peroxidase
Chl – Chlorophyll
C_{org} – Organic carbon
DW – Dry weight
GR – Glutathione reductase
GPX – Guaiacol peroxidase
LPO – Lipid peroxidation
MDA – Malondialdehyde

ROS – Reactive oxygen species
SOD – Superoxide dismutase
TBARs – Thiobarbituric acid reactive substances

References

Asada K. Production and scavenging of reactive oxygen species in chloroplasts and their functions, *Plant Physiol.*, 2006, Vol. 141, p. 391–396.

Balakhnina T., Bennicelli R., Stêpniewska Z., Stêpniewski W. and Fomina I. Oxidative damage and antioxidant defense system in leaves of *Vicia faba major* L. cv. Bartom during soil flooding and subsequent drainage, *Plant Soil*, 2010a, Vol. 327, p. 293–301.

Balakhnina T., Bennicelli R., Stepniewska Z., Stepniewski W., Borkowska A., Fomina I. Stress responses of spring rape plants to soil flooding, *Int. Agrophys.*, 2012, Vol. 26, p. 347–353.

Balakhnina T., Borkowska A. Effects of silicon on plant resistance to environmental stresses: review, *Int. Agrophys.*, 2013, Vol. 27, p. 225–232.

Balakhnina T., Wlodarczyk T., Borkowska A., Nosalewicz M., Serdyuk O., Smolygina L., Ivanova E., and Fomina I. Effect of 4-hydroxyphenethyl alcohol on growth and adaptive potential of barley plants under optimal and soil flooding conditions, *Polish J. Environ. Studies*, 2010b, Vol. 19, No. 3, p. 565–572.

Balakhnina T.I. Plant responses to soil flooding, *Stress Responses in Plants, Mechanisms of Toxicity and Tolerance*, Tripathi B.N., Müller M. eds., Springer International Publishing AG Switzerland, 2015, p. 115–143.

Balakhnina T.I., Bennicelli R.P., Stêpniewska Z., Stêpniewski W. Oxygen stress In the root zone and plant response (some examples), *Physics, Chemistry and Biogeochemistry in Soil and Plant Studies*, Józefaciuk G. ed., Institute of Agrophysics PAS Press, Lublin, Poland, 2004.

Balakhnina T.I., Gavrilov A.B., Włodarczyk T.M., Borkowska A., Nosalewicz M., Fomina I.R. Dihydroquercetin protects barley seeds against mould and increases seedling adaptive potential under soil flooding, *Plant Growth Regul.*, 2009, Vol. 57, p. 127–135.

Baraboy V.A. *Biological Action of Plant Phenolic Compounds*, Naukova Dumka, Kiev, 1976, xxx p. [in Russian]

Bennicelli R.P., Stepniewski W., Zakrzhevsky D.A., Balakhnina T.I., Stêpniewska Z. The effect of soil aeration on superoxide dismutase activity, malondialdehyde level, pigment content and stomatal diffusive resistance in maize seedlings, *Environ. Exp. Bot.*, 1998, Vol. 39, p. 203–211, doi:10.1016/S0098-8472 (97)00044-0.

Bessonova V.P., Lizhenco I.I., Mikhailov O.F., Kulaeva O.N. Effect of kinetin on growth of pea germs and content of pigments under overage of zinc in the nutrient medium, *Fiziol. Rast.*, 1985, Vol. 32, p. 1149–1156. [in Russian].

Biokhimzashchita (Biochemical Protection). Larixin as Plant Growth Regulator with Fungicidal Activity to Complex of Mycotic, Bacterial and Viral Infections is Made from the Larch Wood, Institute of Cytology and Genetics SB RAS, Technical Certificate, 2002, 2449-001-55457436-02. [in Russian]

Blanke M.M., Cooke D.T. Effects of flooding and drought on stomatal activity, transpiration, photosynthesis, water potential and water channel activity in strawberry stolons and leaves, *Plant Growth Regul.*, 2004, Vol. 42, p.153–160, doi:10.1023/B:GROW.0000017489.21970.d4.

Brault M., Maldiney R. Mechanism of cytokinins action, *Plant Physiol. Biochem.* 1999, Vol. 37, p. 403–412.

Brown J.E., Khodr H., Hider R.C. Rice-Evans C.A. Structural dependence of flavonoid interactions with Cu^{2+} ions: implications for their antioxidant properties, *Biochem. J.*, 1998, Vol. 330, p. 1173–1178.

Chernyad'ev I.I. Effect of water stress on the photosynthetic apparatus of plants and the protective role of cytokinins, *Appl. Biochem. Microbiol.*, 2005, Vol. 41, p. 133–147.

Dat J.F., Capelli N., Folzer H., Bourgeade P., Badot P.M. Sensing and signaling during plant flooding, *Plant Physiol. Biochem.*, 2004, Vol. 42, p. 273–282.

Dungey N.O., Brown G., Pinfield N.J. The involvement of cytokinins in the removal of test-induced dormancy in *Acer* seeds, *Z. Pflanzenphysiol.*, 1980, Vol. 96, p. 233–237.

Elstner E.F. Oxygen activation and oxygen toxicity, *Annu. Rev. Plant Physiol.*, 1982, Vol. 33, p. 73–96.

Hallivell B. Oxidative damage, lipid peroxidation and antioxidant protection in chloroplasts, *Chem. Phys. Lipids*, 1984, Vol. 44, p. 327–340.

Haraguchi H., Mochida Y., Sakai S. *et. al.* Protection against oxidative damage by dihydroflavonols in *Engelhardtia chrysolepis*, *Biosci. Biotechnol. Biochem.*, 1996, Vol. 60, No. 6, p. 945–948.

Ivanova E.P., Serdyuk O.P., Smolygina L.D., Nazarova G.N., Kirillova L.L. Reduction of the "latent growth" phase in amaranth seedlings by the 4-hydroxyphenethyl alcohol, *New and Nontraditional Plants and Prospects of Their Utilization*, Materials of V Int. Symp., Moscow-Pushchino, 2003, Vol. 3, p.74–76. [In Russian]

Jackson M.B. Regulation of water relationships in flooded plants by ABA from leaves, roots and xylem sap, *Abscisic Acid: Physiology and Biochemistry*, Davies W.J., Jones H.G. eds., Bios Scientific, Oxford, 1991, p. 217–226.

Jackson M.B., Drew M.C. Effects of flooding on growth and metabolism of herbaceous plants, *Flooding and Plant Growth*, Kozlowski T.T. ed., Academic Press, Orlando, Florida, 1984, p. 47–128.

Kalashnikov Y.E., Balakhnina T.I., Bennicelli R.P., Stepniewski W., Stepniewska Z. Antioxidant activity and lipid peroxidation in wheat plants as related to cultivar tolerance to excessive soil moisture, *Russ. J. Plant Physiol.*, 1999, Vol. 46, p. 227–233.

Kalashnikov Y.E., Zakrzhevsky D.A., Balakhnina T.I. Effect of soil hypoxia on activation of oxygen and the system of protection from oxidative destruction in roots and leaves of *Hordeum vulgare* L., *Russ. J. Plant Physiol.*, 1994, Vol. 41, p. 583–588.

Kalashnikov Y.E., Zakrzhevsky D.A., Balakhnina T.I., Sheveleva E.V., Zastrizhnaya O.M. Effect of drought or water logging of soil on oxygen medium, *Fiziol. Rast.* 1985, Vol. 32, p. 1149–1157. [in Russian]

Kandaswami C., Perkins E., Drzewiecki G. *et. al.* Differential inhibition of proliferation of human squamous cell carcinoma, gliosarcoma and embryonic fibroblast-like lung cells in culture by plant flavonoids, *Anticancer Drugs,* 1992, Vol. 3, No. 5, p. 525–530, doi:10.1097/00001813-199210000-00013.

Khan A.A. Primary, preventive and permissive roles of hormones in plant systems, *Bot. Rev.*, 1975, Vol. 41, p. 391–420.

Khan A.A. The rest of seeds: conversion of the conception and theories, *Physiology and Biochemistry of the Resting and Growing of Seeds*, Kolos, Moscow, 1982, p. 47–67. [In Russian].

Kostyuk V.A., Potapovich A.I. Antiradical and chelating effects in flavonoid protection against silica-induced cell injury, *Arch. Biochem. Biophys.*, 1998, Vol. 355, No. 1, p. 43–48, doi:10.1006/abbi.1998.0708.

Larson R.A. The antioxidants of higher plants, *Phytochemistry*, 1988, Vol. 27, p. 969–978, doi:10.1016/0031-9422(88)80254-1.

Letham D.S., Palni L.M.S. The biosynthesis and metabolism of cytokinins, *Ann. Rev. Plant. Physiol.*, 1983, Vol. 34, p. 163–197.

Melnikova N.B., Yoffe I.D. Biocompatibility of dihydroquercetin with lipophilic fragments of biomembrane. Effects of metal ions and ascorbic acid, *Khim. Rast. Syr. (Chem. Plant Stuff)*, 2002, Vol. 2, p. 93–103. [in Russian]

Morita J. Inactivation of bacteriophages by polyphenols in the presence of cupric ion, *Agric. Biol. Chem.*, 1988, Vol. 52, No. 7, p. 1669–1673.

Muromcev G.S., Chkanikov D.I., Kulaeva O.N., Gamburg G.C. *The Basis of the Chemical Growth Regulation and Plant Productivity*, Agropromizdat, Moscow, 1987. [in Russian].

Muzafarov E.N., Ruzieva R.K. Phenolic compounds as a factor of photosynthesis and growth regulation, *Plant Growth and Its Regulation*, Kefely V.I., Toma S.I. eds., Shtiintsa, Kishinev, 1985, p. 210–218.

Noctor G., Gomez L., Vanacker H., Foyer C. Interactions between biosynthesis and transport in the control of glutathione homeostasis and signaling, *J. Exp. Bot.*, 2002, Vol. 53, p. 1283–1304.

Nover L., Yellmund D., Neumann D., Scharf K.D., Selfling E. The head shock response of eukaryotic cells, *Biol. Zentralblatt.*, 1984, Vol. 103, p. 357–435.

Prasad T.K., Anderson M.D., Steward C.R. Localization and characterization of peroxidases in the mitochondria of chilling-acclimated maize seedlings, *Plant Physiol.*, 1995, Vol. 108, p. 1597–1605.

Rice-Evans C.A., Miller N.J., Paganga G. Antioxidant properties of phenolic compounds, *Trends Plant Sci.*, 1997, Vol. 2, No. 4, p. 152–159, doi:10.1016/S1360-1385(97)01018-2.

Serdyuk O.P., Smolygina L.D., Ivanova E.P., Firsov A.P., Pogrebnoi P.V. 4-Hydroxyphenethyl alcohol – a new cytokininlike substance isolated from phototrophic bacterium *Rhodospirillum rubrum*. Exhibition of activity on plants and transformed mammalian cells, *Process Biochem.*, 2000, Vol. 36, No. 5, p. 475–479.

Shishov A.D., Matevosyan G.L., Sutulova V.I., Ivanova S.I. Influence of the phytoregulators upon the growth, development and productivity of the cauliflower, *Physiol Biochem. Cult. Plants*, 1990, Vol. 22, p. 360–366.

Teselkin Yu.O., Zhambalova B.A., Babenkova I. *et. al.* Dihydroquercetin antioxidant properties, *Biophys.*, 1996, Vol. 41, No. 3, p. 620–624.

Trought M.C.T., Drew M.C. The development of waterlogging in wheat seedlings (*Triticum aestivum* L.). I. Shoot and root growth in relation to changes in the concentrations of dissolved gases and solutes in the soil solution, *Plant Soil*, 1980, Vol. 54, p. 77–94, doi:10.1007/BF02182001.

Uminsky A.A., Havsteen B.H., Bakaneva B.F. *The Biochemistry and Medical Significance of Flavonoids*, Photon-vek, Pushchino, 2007. [in Russian]

Webb D.P., Van Staden J., Wareing P.F. Seed dormancy in *Acer*. Changes in endogenous citokinins, gibberellins and germinations inhibitors during the breaking of dormancy in *Acer saccharum*. Marsh., *J. Exp. Bot.*, 1973, Vol. 24, p. 105–116.

Yamasaki H., Sakihama Y., Ikehara N. Flavonoid-peroxidase reaction as a detoxification mechanism of plant cells against H_2O_2, *Plant Physiol.*, 1997, Vol. 115, p.1405–1412.

Yan N.B., Dai Q., Liu X. et. al. Flooding-induced membrane damage, lipid oxidation and activated oxygen generation in corn leaves, *Plant Soil* 1996, Vol. 179, p. 261–268, doi:10.1007/BF00009336.

Zakrzhevsky D.A., Balakhnina T.I., Stepniewski W., Stepniewska S., Bennicelli R.P., Lipiec J. Oxidation and growth processes in roots and leaves of higher plants at different oxygen availability in soil, *Russ. J. Plant Physiol.*, 1995, Vol. 42, p. 242–248.

9
Cooperation of Photosynthetic and Nitrogen Metabolism

Anatoly A. Ivanov*, **Anatoly A. Kosobryukhov***

Institute of Basic Biological Problems, Russian Academy of Sciences, Pushchino, Moscow Region 142290, Russia

Abstract
This chapter is devoted to the consideration of the interaction of carbon and nitrogen metabolism, which occupy a central place among the biochemical processes that support the growth and development of plants. In the first part, the main mechanisms for the interrelation and regulation of the reactions of the metabolism of carbon and nitrogen in the photoreduction of carbon dioxide and nitrite with the participation of ferredoxin are considered, and the processes do not compete for the reduction equivalents. The regulation of the assimilation rate of carbon dioxide and nitrogen in plant leaves depends on the inflow of substrates and the removal of the products of their assimilation. It has been shown that a complex network of metabolite-status-sensing systems coordinates supply and demand fluctuations with changes in plant metabolism and development. The second part of the review is devoted to the study of the interaction of carbon-nitrogen metabolism with changing environmental conditions. The mechanisms underlying the adaptation of plants with high and low carbon dioxide content in the atmosphere are shown, which leads to a change in the concentration and availability of the basic biogenic elements and the formation of a phenotype maximally adapted to certain conditions. Attention is paid to the strategy of adaptive plant changes in conditions of water and salt stress, which can contribute

* *Corresponding authors*: demfarm@mail.ru; kosobr@rambler.ru

to understanding the causal relationship between the processes of nitrogen and carbon metabolism and the mechanisms of their coordinated regulation.

Keywords: Carbon and nitrogen metabolism, Photosynthesis, Stress, Elevated and low carbon dioxide concentration, Drought, Salinity

9.1 Introduction

Optimal functioning and coordination of carbon and nitrogen metabolism plays a key role in plant growth and productivity, what is reflected in the works of many researchers (Zheng, 2009; Coruzzi and Zhou, 2001; Martin, 2002; Krapp and Truong, 2005). Carbon metabolism provides available energy (ATP) and carbon skeletons (α-ketoglutarate) to reduce nitrate for subsequent synthesis of amino acids (Lawlor, 2002*a*). At the same time, nitrogen metabolism competes with carbon metabolism for the assimilatory powers (NADH and NADPH), as well as certain photoreaction intermediates (ferredoxin), especially when CO_2 assimilation is limited (Lawlor, 2002*a*; Sánchez-Rodríguez *et al.*, 2011). In addition, high content of carbohydrates can increase the activity of enzymes related to nitrogen metabolism, promoting ammonium assimilation (Zhang *et al.*, 2011). Thus, there are many points of interdependent control between carbon and nitrogen metabolism, including at the level of gene expression and enzyme activity (Ghisi *et al.*, 2002; Commichau *et al.*, 2006).

Nitrogen is quantitatively one of the most important elements for the growth and development of plants (Chen *et al.*, 2003). Most of the plant nitrogen is accumulated in the leaves, and most of the leaf nitrogen is embedded in the photosynthetic apparatus (Makino and Osmond, 1991). A low nitrogen supply leads to suppression of protein biosynthesis, nucleic acids and other metabolites and, ultimately, to a decrease in biomass and leaf area (Sheehy Skeffington and Jeffrey, 1985; Wu *et al.*, 2008*a*; Brueck and Senbayram, 2009). The percentage of nitrogen in the leaves can be used to assess the photosynthetic capacity (Evans, 1983; Llorens *et al.*, 2003; Niu *et al.*, 2003; Takashima *et al.*, 2004; Xu and Zhou, 2005*b*; Chen *et al.*, 2005).

Nitrate reductase, which catalyzes the first stage of the nitrate assimilation process, up-regulated at both the transcriptional and post-translational levels by means of CO_2 products (Cheng *et al.*, 1992; Vincentz *et al.*, 1993; Larios *et al.*, 2001). Nitrate assimilation for the synthesis of amino acids depends on photosynthetic carbon assimilation, which supplies carbon skeletons. Assimilation of nitrate proceeds at a low rate in plants with a low carbohydrates level (Stitt and Schulze, 1994*b*). At the same time, increased CO_2 assimilation under elevated CO_2 concentration, and hence higher availability of carbohydrates, enhances the expression and activity of nitrate reductase and glutamine synthetase, which leads to a decrease in nitrate accumulation in leaves (Larios *et al.*, 2004).

Since the number of carbon and nitrogen resources varies with the environment, the plants have developed a very complex system for controlling the uptake, transport and assimilation of carbon and nitrogen (Gutiérrez *et al.*, 2007; Zhang, 2007; Vidal and Gutiérrez, 2008; Pant *et al.*, 2009; Cai *et al.*, 2010; Nunes-Nesi *et al.*, 2010; Vidal *et al.*, 2010). In stable environmental conditions, there is a balance between carbon and nitrogen metabolism, supported by a number of monitoring mechanisms, understanding the nature of which can contribute to the intensification of the growth of agricultural plants and increase their productivity. One way of understanding the regulating mechanisms can be the study of plant metabolism under stressful conditions, when a change in the carbon / nitrogen balance is observed. In this chapter, our aim is to show the interaction between the assimilation of carbon and nitrogen in plant leaves, and the imbalance of this interaction, both with a change in the availability of the basic substrates of the carbon and nitrogen metabolism, and in conditions of drought and salinity, with reference to the activity of key enzymes of carbon and nitrogen pathways.

9.2 Carbon Uptake and Rubisco

In higher plants, the assimilation of inorganic carbon occurs mainly through the CO_2 assimilation with the help of Rubisco (EC 4.1.1.39), associated with the photosynthetic apparatus. Rubisco is the most common enzyme in C_3 plants. It includes 27% of the total nitrogen in the leaves (Evans, 1989; Makino *et al.*, 1997*b*). Nevertheless, in C_3 plants at aCO_2, Rubisco activity is only 25% of the potential catalytic capacity (Sage and Pearcy, 1987; Makino *et al*, 1997b; Spreitzer and Salvucci, 2002; Jin *et al.*, 2006).

When saturating light intensity and aCO_2, the photosynthetic rate of C_3 plants depends on several limiting factors: the amount and activity of Rubisco (Makino *et al.*, 1997*b*), the ribulose-1,5-diphosphate (RuBP) regeneration rate, Rubisco carboxylation capacity, the synthesis of starch and sucrose as photosynthesis products and the inorganic phosphate regeneration rate for photophosphorylation (Farquhar *et al.*, 1980*a*; Sage, 1990; Manter and Kerrigan, 2004).

Rubisco activity is induced by chloroplastic CO_2, but the carboxylation capacity of Rubisco depends on the CO_2 concentration at the Rubisco catalytic site (C_c), which in C_3 plants is usually far from saturation due to low kinetic affinity for CO_2 and high resistance to CO_2 diffusion in the liquid phase of leaf tissues (Bernacchi *et al.*, 2002; Flexas *et al.*, 2006*a*), *i.e.* the photosynthetic rate can be limited by internal conductivity to supply CO_2 from substomatal cavities to the carboxylation sites, which often depends on the supply of nutrients. The rate of CO_2 diffusion from the atmosphere into the carboxylation site depends on the conductivities of the boundary layer, stomata and mesophyll (Bernacchi *et al.*, 2002). The consequence of this is that C_c is less than ambient CO_2 concentration due to a number of gas-phase and liquid-phase resistances (Aalto and Juurola, 2002). There is evidence that C_c can be much smaller than intercellular CO_2 concentration (C_i) (Evans *et al.*, 1986; Parkhurst and Mott, 1990; Epron *et al.*, 1995; Ethier and Livingston, 2004). In contrast, in C_4 plants, C_c for Rubisco is close to saturation due to the presence of the CO_2-concentrating mechanism, which results in a fully activated Rubisco (Sage *et al.*, 1987; Dai *et al.*, 1993).

Internal conductance to CO_2 (g_i) is a function of the photosynthetic rate (A) ($g_i = A / (C_i - C_c)$) and affects the interpretation of the photosynthesis response to intercellular CO_2 concentration in the biochemical model of photosynthesis by Farquhar *et al.* (1980*b*), according to which the rate of photosynthesis is limited by one of two processes, and this restriction passes from one to another as CO_2 concentration changes. At low CO_2 concentration photosynthesis is limited by maximum rate of carboxylation (V_{cmax}), while at high CO_2 concentration photosynthesis is limited by maximum rate of electron transport (J_{max}).

Carboxylation and RuBP-regeneration capacities determine light-saturated photosynthetic rate. With increasing the atmospheric CO_2 concentration, Rubisco carboxylation becomes more effective than regeneration of ribulose-1,5-diphosphate. To increase the photosynthetic nitrogen use efficiency, it is theoretically necessary to reduce the Rubisco share with respect to ribulose-1,5-diphosphate regeneration components at high CO_2 concentration (Sage, 1994; Medlyn, 1996; Makino *et al.*, 1997*a*). Some balance change occurs in the case of nutrient deficiency, but it is usually less than the optimally predicted, and often it is not (Sage, 1994; Medlyn, 1996). According to Stitt and Schulze (1994*a*), under constant conditions, the photosynthetic apparatus is not optimized to maximize photosynthesis, which allows the plant to direct resources to use changes that would be expected in a natural environment.

In the case of a balance between carboxylation and ribulose-1,5-diphosphate regeneration capacities, the optimum depends on temperature and photosynthetic photon flux density, which fluctuate too quickly to acclimate in the natural environment. A model was developed describing the balance of carboxylation and ribulose-1,5-diphosphate regeneration capacities as a function of the atmospheric CO_2 concentration, temperature, and photosynthetic photon flux density (Mitchell *et al.*, 2000). This model allows investigating the difference between the optimal balance of capacities under the measuring conditions of light-saturation and

20 °C and the optimal balance for a natural environment. A model was used to estimate the effects of fluctua in photosynthetic photon flux density and temperature in the growth environment on the optimal balance between these capacities. There is a tendency to over-invest in Rubisco at elevated atmospheric CO_2 concentration, so genetic manipulation to reduce Rubisco relative to ribulose-1,5-diphosphate regeneration capacity could benefit nitrogen use efficiency of photosynthesis under these conditions (Mitchell *et al.*, 2000).

It should also be borne in mind that the balance between capacities maximum rate of carboxylation and maximum rate of electron transport, which determines the CO_2-dependent photosynthetic rate, varies with the growth temperature, which affects the relative amounts of photosynthetic proteins and the CO_2 response of photosynthesis (Onoda *et al.*, 2005). CO_2 stimulation of photosynthesis should increase with increasing temperature, but this is not always observed in plants grown in seasonal climates (Teskey, 1997, Lewis *et al.*, 2001), which can be explained by temperature acclimatization (Onoda *et al.*, 2005).

9.2.1 Dependence of the Carbon Assimilation on Nitrogen Supply

A number of studies have shown the dependence of photosynthesis on nutrients supplay (see, *e.g.*, Walcroft *et al.*, 1997; Warren *et al.*, 2003), and positive correlation between internal conductance to CO_2 in leaves and the leaf nitrogen content per unit area (von Caemmerer and Evans, 1991; Poorter and Evans, 1998).

The amount of Rubisco increases with the increase in leaf nitrogen content (Ookawa *et al.*, 2004). However, this reduces the activation state Rubisco, which can lead to a decrease in photosynthetic nitrogen use efficiency and partially eliminates the benefits of a large amount of Rubisco for photosynthesis (Cheng and Fuchigami, 2000; Ray *et al.*, 2003). On the other hand, it should be borne in mind that with ambient CO_2 concentration, the amount of Rubisco is usually higher than required for photosynthesis (Quick *et al.*, 1991; Eichelmann and Laisk, 1999; Warren *et al.*, 2003) and Rubisco may also function as nitrogen-storage compound (Warren *et al.*, 2003).

However, with high leaf nitrogen content, despite the reduced activation state of Rubisco, the light-saturated photosynthetic rate (Reich *et al.*, 1995; Makino *et al.*, 2003) and the general activity of Rubisco (Evans, 1983; Cheng and Fuchigami, 2000), as a rule, are higher than in plants with low leaf nitrogen content. A higher net photosynthetic rate is usually associated with an increase in carboxylation efficiency. Nevertheless, as shown by Li Y. *et al.* (2009), carboxylation capacity of Rubisco in plants grown at different leaf nitrogen content was constant. This may be due to the fact that with a higher nitrogen supply, the chloroplast size increases, resulting in a decrease in CO_2 transport resistance in the liquid phase, which leads to higher mesophyll conductance and CO_2 concentration at the catalytic site of Rubsico (Li Y. *et al.*, 2009). Consequently, an increased rate of photosynthesis at high leaf nitrogen content is apparently associated with a higher CO_2 concentration in chloroplasts. Thus, CO_2 concentration at the catalytic site of Rubsico can be considered as one of the main factors limiting the photosynthetic efficiency at a higher nitrogen supply (Li Y. *et al.*, 2009).

The relative limitation of photosynthesis due to intercellular CO_2 concentration increases with increasing supply of nutrients and positively correlates with the nitrogen leaf content and Rubisco (Warren, 2004). This is due to the large positive response of photosynthesis to nutrient supply, while nutrients had a relatively small and contradictory effect on the internal conductance to CO_2 (von Caemmerer and Evans, 1991; Warren, 2004).

Trends in relationships of the leaf nitrogen content and the internal conductance to CO_2 or the Rubisco activity and the internal conductance to CO_2 are similar to relations of the net photosynthetic rate and the leaf nitrogen content (Field and Mooney, 1986). Coincidental relationships the internal conductance to CO_2 with leaf nitrogen content or Rubisco may reflect the main correlation with the anatomical and morphological features of the leaf (Evans *et al.*, 1994; Syvertsen *et al.*, 1995; Singsaas *et al.*, 2004), *i.e.* the internal conductance to CO_2 is largely constitutive and is determined by leaf structural features. On the other hand, leaf structural traits fail to explain the rapid responses of the internal conductance to CO_2 to environmental conditions (Bernacchi *et al.*, 2002; Centritto *et al.*, 2003), which can best be explained by the actions of carbonic anhydrase or aquaporins (Gillon and Yakir, 2000; Terashima and Ono, 2002). Apparently, leaf structural traits and enzymatic processes are both correlated with g_i.

In plants, there is a sugar-regulated expression of the enzymes associated with the carbon and nitrogen metabolism. Low carbohydrate levels up-regulate gene expression for photosynthesis and reserves mobilization, while high sugar levels enhance gene expression for carbon storage (Koch, 1996). Nitrate assimilation occurs at a low rate in plants with a low carbohydrate level (Stitt and Schulze, 1994*a*), while sucrose feeding leads to an increase in the nitrate assimilation rate and the accumulation of amino acids (Morcuende *et al.*, 1998), for the synthesis of which requires carbon skeletons obtained from photosynthetically synthesized carbohydrates (Huppe and Turpin, 1994). The addition of sugar in the dark to a carbohydrate depleted plant also enhances the expression of nitrate reductase genes (Cheng *et al.*, 1992).

9.3 Alternative Electron Acceptors in Photosynthesis

The processes of CO_2 and nitrite assimilation interact at the level of photosynthetic electron transport and the reference point for their reciprocal regulation is the reducing side of PS I (Eichelmann *et al.*, 2011). With linear transport of electrons from H_2O to $NADP^+$, 4 electrons (2 NADPH) co-transport 12 H^+, which, in turn, supports the synthesis of 3 ATP. This is sufficient to assimilate one CO_2 molecule in the Benson-Bassham-Calvin cycle (Laisk *et al.*, 2007). An additional amount of ATP for secondary metabolisms can be formed by transferring electron to acceptors other than CO_2 with the inclusion of metabolic pathways with less ATP consumption per electron (Eichelmann *et al.*, 2011). As an alternative electron acceptor can be nitrite, the photoreduction of which leads to the production of additional ATP by reducing the absorption of CO_2 (Noctor and Foyer, 1998). Mathematical modelling of nitrogen and carbon metabolism suggested that nitrogen limitation, common in natural communities, may proportionally limit CO_2 assimilation and help maintain a rather constant nitrogen / carbon ratio in plants (Laisk *et al.*, 2009).

For photoassimilation of one molecule of nitrite, 8 electrons (6 electrons for the reduction of nitrite to ammonium and 2 electrons for assimilation of ammonium into Glu) are required, whereas for assimilation of CO_2 only 4 electrons per molecule is required (see Figure 9.1). With available substrates, the enzymatic ability of nitrite photoreduction can be compared with the CO_2 reduction (Eichelmann *et al.*, 2011). However, despite the equal distribution of the electron flows between carbon and nitrogen reduction, under the conditions of elevated CO_2 concentration, the CO_2 recovery will dominate (Robinson, 1986). Moreover, in the absence of a substrates deficiency, the rate of CO_2 fixation is stimulated by the simultaneous nitrite photoreduction. Presumably, the nitrite reduction can help maintain a balance in the production / use of ATP and NADPH (Robinson, 1986; Eichelmann *et al.*, 2011) and promote the synthesis of ATP for alternative uses aimed at protein and starch synthesis in chloroplasts. However, these *in vitro* rates may differ from the actual rates *in vivo* due to differences in the availability of various substrates (Eichelmann *et al.*, 2011).

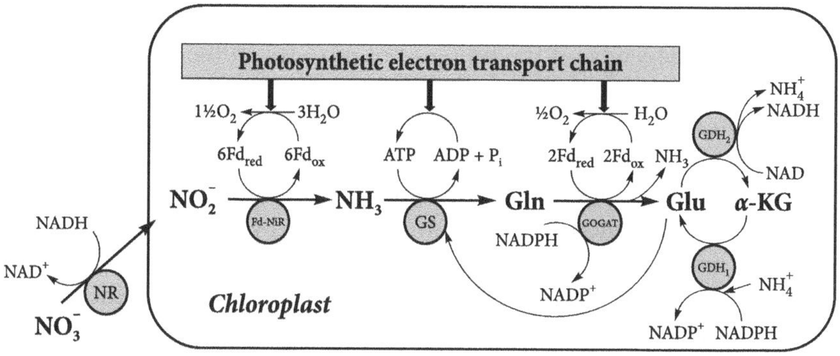

Fig. 9.1. A light-dependent pathway of primary assimilation of inorganic nitrogen into organic compounds
Note: NR – nitrate reductase, Fd-NiR– ferredoxin-dependent nitrite reductase, GS – glutamine synthetase, GOGAT – ferredoxin-dependent glutamate synthase, GDH – glutamate dehydrogenase, Fd – ferredoxin

The assimilation coefficient (AQ=CO_2/ O_2 flux ratio) is 1.0 when CO_2 is the only electron acceptor. Any decrease of this value below 1.0 means redirection of the electron flow to alternative acceptors (Cen *et al.*, 2001). The use of photosynthetic electrons with alternative acceptors is less than 10% in C_3 plants, but it is larger in C_4 plants (Badger *et al.*, 2000; Ruuska *et al.*, 2000; Siebke *et al.*, 2003). Nitrogen reduction increases with increasing light intensity, causing a decrease in AQ by 10 to 15% (Bloom *et al.*, 2002; Rachmilevitch *et al.*, 2004).

Eichelmann *et al.* (2011) have shown that the nitrite and CO_2 reduction competes for the same reductant pool with similar properties. However, the rate of nitrite reduction saturates at low light intensity, while CO_2 recovery continues to increase with increasing light intensity. Thus, ferredoxin-dependent-nitrite reductase successfully competes with ferredoxin-NADP$^+$ oxidoreductase (EC 1.18.1.2) for reduced ferredoxin at low light intensity, but the nitrite reduction has saturation and cannot compete with ferredoxin-NADP$^+$ oxidoreductase at high light intensities.

9.4 Nitrogen Metabolism

9.4.1 *Primary Assimilation of Inorganic Nitrogen*

The assimilation of inorganic nitrogen in higher plants is one of the determinants of plant growth and yield. Nitrates are the main source of nitrogen available in the soil (Lawlor, 2002*a*). The ability of plants to assimilate nitrates depends on the presence of nitrogen in the soil and the activity of the nitrogen cycle inside the plant, which includes both the uptake of nitrate from the soil by specific transporters, and the hydrolysis of the spare nitrogen nitrogen compounds, and the nitrogen reassimilation during photorespiration. Nitrate is partially transported to the leaves, where it is first reduced to nitrite in the cytosol by nitrate reductase and then to ammonium in chloroplasts by ferredoxin-dependent nitrite reductase, which is associated with reductants from photosynthetic electron transport (Lawlor, 2002*a*) (see Figure 9.1).

The resulting ammonium is first converted to glutamine and then to glutamate by the glutamine synthetase / ferredoxin-dependent glutamate synthase (GS / GOGAT) cycle or by an alternative glutamate dehydrogenase pathway (Ireland and Lea, 1999; Thomas and Hilker, 2000; Ziegler *et al.*, 2003; Robredo et. al., 2011; Sánchez-Rodríguez *et al.*, 2011). Combined action of glutamine synthetase and ferredoxin-dependent glutamate synthase is the main way for assimilation of ammonia. However, ammonium can directly be incorporated into glutamate in an amination reaction involving NADH-dependent glutamate dehydrogenase, especially at high concentrations of ammonium, but its function is the subject of debate (Ireland and Lea, 1999; Ziegler *et al.*, 2003). The NADH-dependent glutamate dehydrogenase amination function appears to be activated only when the glutamine synthetase / glutamate synthase cycle activity is inhibited under stress conditions (Oaks, 1995; Robredo et. al., 2011). On the other hand, since glutamate dehydrogenase reversibly deaminates glutamate (NAD-dependent glutamate dehydrogenase), the physiological role of this enzyme *in vivo* can consist in obtaining α-ketoglutarate and reducing equivalents for use in other metabolic pathways (Robredo et. al., 2011). Energy, reduction equivalents and carbon skeletons are used by the nitrate reductase and glutamine synthetase / glutamate synthase cycle to produce glutamate.

Glutamate acts as a precursor for the biosynthesis of other amino acids (Miflin and Lea, 1976), as well as practically all nitrogen containing compounds, including nucleic acids (Hodges, 2002; Ziegler *et al.*, 2003). Synthesis of amino acids in photosynthetic cells includes both carbon and nitrogen metabolism (Solomonson and Barber, 1990; Lam *et al.*, 1996). Amino acids can also act as feedback regulators when uptaking and assimilating nitrate (Walch-Liu *et al.*, 2005). Thus, photosynthesis and assimilation of nitrogen are regulated in many points of interdependent control. From this point of view becomes special interest study enzymes nitrogen metabolism (Lam *et al.*, 1996).

9.4.1.1 Nitrate Reductase

At biological assimilation of inorganic nitrogen in plants, nitrate is reduced to ammonium during the transfer of 8 electrons in two stages. The first stage catalyzed by nitrate reductase (NR; EC 1.7.1.1) performs two-electron reduction of nitrate to nitrite with NADH as an electron donor (Redinbaugh and Campbell, 1985; Campbell and Kinghorn, 1990). Nitrate reductase is the key enzyme of nitrogen metabolism, the activity of which is considered a stage limiting the rate of primary nitrogen assimilation.

In higher plants, nitrate reductase is localized in the cytoplasm of cells (Dalling *et al.*, 1972; Wallsgrove *et al.*, 1979; Hoff *et al.*, 1994; Crawford, 1995). There are three forms of nitrate reductase: NADH-dependent enzyme in higher plants and algae (EC 1.7.1.1), NAD(P)H-two-dependent enzyme in higher plants, algae and fungi (EC 1.7.1.2) and NADPH-dependent enzyme found only in mushrooms (EC 1.7.1.3) (Kleinhofs and Warner, 1990; Solomonson and Barber, 1990; Campbell and Kinghorn, 1990; Rouze and Caboche, 1992). In contrast to prokaryotes where ferredoxin-dependent nitrate reductase was detected (Mikami and Ida, 1984), in plants and algae there is exclusively NAD(P)H-dependent nitrate reductase (Mikami and Ida, 1984; Knaff and Hirasawa, 1991).

Nitrate reductase is a water-soluble, multi-centered redox enzyme and belongs to the class of flavin-dependent cytochrome b5 reductases (Campbell and Kinghorn, 1990; Karplus *et al.*, 1991; Correll *et al.*, 1992, 1993). Nitrate reductase is a homodimer with 100 kDa polypeptide subunits, each of which binds one equivalent of cofactors: Mo-pterin, heme-Fe, FAD and has one binding site for pyridine nucleotides (Ida and Mikami, 1983; Redinbaugh and Campbell, 1985; Campbell, 1989; Campbell and Kinghorn, 1990; Kleinhofs and Warner, 1990; Solomonson and Barber, 1990; Rouze and Caboche, 1992). Three functional fragments are arranged in the form of a linear array in the polypeptide chain (Lu *et al.*, 1994), therefore the catalytic reduction of nitrate in nitrite can be regarded as a mini electron transport system with electron flux NADH → FAD → cytochromes *b* → Mo-pterin → nitrate (Redinbaugh and Campbell, 1985; Campbell, Kinghorn, 1990; Kleinhofs and Warner, 1990; Solomonson and Barber, 1990; Rouze and Caboche, 1992). The Mo-pterin cofactor forms an active site for the nitrate reduction (Solomonson and Barber, 1986; Neame and Rarber, 1989; Campbell and Kinghorn, 1990). The Heme-Fe domain is in the middle of the chain and connects the FAD domain and the pyridine nucleotide binding site that reduced the flavin cofactor (Calza *et al.*, 1987; Campbell and Kinghorn, 1990; Lu G. *et al.*, 1992; Lu *et al.*, 1994).

Regulation of nitrate reductase genes expression is carried out by a number of factors, including nitrate, light, metabolites, phytohormones and stress. Regulation of the nitrate reductase genes expression by transcription factors is described in detail in a review by Yanagisawa (2014). The main modulator of nitrate reductase gene expression is nitrate, which functions as a signal transcription molecule.

At the post-translational level, nitrate reductase activity can be regulated by protein kinases, phosphatases, 14-3-3 proteins and proteases (Hoff *et al.*, 1994; Crawford, 1995; MacKintosh and Meek, 2001). The main factor in controlling nitrate reductase activity at the post-translational level is the phosphorylation / dephosphorylation process and the binding of the signaling 14-3-3 protein to the phosphorylated form of the enzyme (Bachmann *et al.*, 1996; Moorhead *et al.*, 1996). The phosphorylated form of nitrate reductase in the presence of 14-3-3 is inhibited by Mg^{2+}. Activity state nitrate reductase is defined as the nitrate reductase activity in the presence of Mg^{2+} as a percentage of nitrate reductase activity without Mg^{2+} (Mackintosh *et al.*, 1995). The splitting of nitrate reductase by specific proteinases *in vivo* can also be a way of controlling the activity of the enzyme, especially in plant leaves where nitrate reductase activity has diurnal rhythms (Campbell and Kinghorn, 1990).

One of the most important factors for increasing the nitrate reductase genes expression is light (Hoff *et al.*, 1994; Crawford, 1995; MacKintosh and Meek, 2001). Even though nitrate reductase in higher plants is not a light-dependent enzyme, the flux density of photosynthetic photons is important for regulating the nitrate reductase activity, since its inhibition leads to a drop in the activity state of nitrate reductase (Lillo *et al.*, 1996). Perhaps in the process of photosynthesis, a signal is generated that triggers the enzyme dephosphorylation mechanism (Provan and Lillo, 1999), followed by an increase in the activity state nitrate reductase. The activity of nitrate reductase may also depend on the redox state of chloroplasts (Knaff and Hirasawa 1991; Provan and Lillo, 1999).

Light signaling is directly involved in the nitrate reductase genes expression (Jonassen *et al.*, 2009). However, in higher plants, the reduction of nitrate (Beevers and Hageman, 1972, Hewitt, 1975, Guerrero *et al.*, 1981) can not be considered light-dependent (Aslam *et al.*, 1979; Aslam and Huffaker, 1982). The light-dependent induction of nitrate reductase gene expression may be indirectly associated with the activation of carbohydrate synthesis in photosynthesis (Hoff *et al.*, 1994; Crawford, 1995), since sucrose, glucose and fructose induce nitrate reductase gene expression and mimic the light induction of its genes (Crawford, 1995; Provan and Lillo, 1999).

Sucrose-induced expression of nitrate reductase genes may be part of the regulatory system responsible for activation of nitrate assimilation combined with activation of photosynthesis and maintenance of a balanced level of carbohydrates and amino acids. Conversely, nitrogen containing metabolites (probably mainly glutamine) acts as negative regulators of nitrate reductase genes through feedback regulation (Hoff *et al.*, 1994; Crawford, 1995). However, inhibition of nitrate reductase transcription with glutamine was not observed in low-ferredoxin-dependent glutamate synthase mutant plants accumulating glutamine *in vivo*. Perhaps the negative effect of glutamine on NR transcripts is removed at a high α-ketoglutarate level, and the relative amounts of α-ketoglutarate and glutamine are more important for transcription of nitrate reductase than the concentrations of each of them individually (Ferrario-Méry *et al.*, 2001).

For the photosynthetic activation of nitrate reductase, the activity of the Benson-Bassham-Calvin cycle is important, the intermediate products of which could act as signaling molecules for the regulation of nitrate reductase activity (Huber *et al.*, 1996). In mutants, devoid of the activity of the Benson-Bassham-Calvin cycle, there is no light activation of nitrate reductase. Moreover, the removal of CO_2 from the ambient air results in a decrease in the nitrate reductase activity state in the leaves (Kaiser and Brendle-Behnisch, 1991). On the other hand, according to Provan and Lillo (1999), the activity of the Benson-Bassham-Calvin cycle is not a prerequisite for the activation of nitrate reductase, since the high activity state of nitrate reductase in the light was not strictly associated with either the pool of reduced ferredoxin or the high activity of the Benson-Bassham-Calvin cycle. This indicates the complexity of the activation process and the possible involvement of various signals that affect nitrate reductase activity.

Light-dependent nitrate utilization can be associated with the stimulating role of carbohydrates in the synthesis of enzymes, as well as with the enhanced generation of reducing equivalents for enzymatic reactions. The reduction of nitrate to nitrite by nitrate reductase and nitrite to ammonium by nitrite reductase is a energy-dependent pathway that requires NAD(P)H and reduced ferredoxin as the energy source, respectively (Hoff *et al.*, 1994). Assimilation of ammonium in the GS / GOGAT cycle also requires ATP and NADH or reduced ferredoxin. Thus, light inducible expression of nitrate reductase genes is consistent with the requirement of light energy for the nitrogen assimilation.

Expression of nitrate reductase genes fluctuates during the day / night cycle due to transcriptional or post-transcriptional regulation (Hoff *et al.*,

1994; Crawford, 1995). Other environmental factors, such as low temperature (Jones *et al.*, 1998) and drought (Foyer *et al.*, 1998), also affect the expression of nitrate reductase genes, reducing mRNA levels. Phytohormones, cytokinins and abscisic acid antagonistically modulate the nitrate reductase mRNA level (Hoff *et al.*, 1994). However, despite the fact that nitrate promotes both the genes expression of cytokinin biosynthesis enzymes and their accumulation, nitrate inducible expression of nitrate reductase genes is largely independent of the accumulation of cytokinins, but is controlled directly by nitrate (Sakakibara, 2003).

9.4.1.2 Ferredoxin-Dependent Nitrite Reductase

Ferredoxin-dependent nitrite reductase (NiR; EC 1.7.7.1) is a water-soluble, multicenter redox enzyme (Campbell, Kinghorn, 1990) and catalyzes the second step of nitrogen assimilation (Knaff and Hirasawa, 1991) – 6-electron reduction of nitrite to ammonium. Ferredoxin-dependent nitrite reductase of higher plants is a monomeric protein with a molecular weight of 60 to 64 kD and contains siro heme and [4Fe–4S] clusters as prosthetic groups (Hilliard *et al.*, 1991; Knaff and Hirasawa, 1991) and uses reduced ferredoxin in as an electron donor (Knaff and Hirasawa, 1991). The interaction between nitrite reductase and ferredoxin is of an electrostatic nature, the negative charge ferredoxin interacts with the positive nitrite reductase charge (Hirasawa *et al.*, 1994). Thus, the reduction of nitrite through ferredoxin is functionally associated with the photosystem I (Neyra and Hageman, 1974; Plaut *et al.*, 1977). Ferredoxin-dependent nitrite reductase is located exclusively in the stroma of chloroplasts (Wallsgrove *et al.*, 1979), but is encoded by nuclear genes with the polypeptides synthesis in the cytoplasm (Lahners *et al.*, 1988). In fungi and bacteria, nitrite reductase is a pyridine nucleotide-dependent enzyme (Campbell and Kinghorn, 1990).

Reduction of nitrite in chloroplasts depends on the rate of ferredoxin photoreduction and the activity CO_2 fixing. The nitrite reduction very quickly reacts to light-dark transition and is stimulated by uncouplers, probably due to the amplification of electronic transport (Plaut *et al.*, 1977). Thus, the photosynthetic CO_2 reduction and the nitrite reduction consumes electrons from the photosynthetic electron transport chain through ferredoxin, therefore some regulatory or competitive interaction between these two processes should be expected, especially when limiting the regeneration of the reduced ferredoxin. Under physiological conditions, the reduction rate of CO_2 is much higher than the reduction rate of nitrite (Plaut *et al.*, 1977).

In vivo, photosynthetic carbon reduction cycle intermediates can to a certain extent regulate the reduction rate of nitrite to ammonium. The increase in photosynthetic activity can promote the use of ammonium, formed by the reduction of nitrite, together with the available carbon skeletons for the synthesis of amino acids (Plaut *et al.*, 1977).

9.4.1.3 Glutamine Synthetase / Glutamate Synthase (GS / GOGAT) cycle

In higher plants, ammonium is the only inorganic nitrogen compound that is directly used as a precursor for the biosynthesis of organic nitrogen-products (Lea, 1997). Glutamine synthetase is the key enzyme of nitrogen metabolism, which catalyzes the ATP-dependent synthesis of glutamine from glutamate and ammonium. Glutamine synthetase is often used as a marker of the nitrogen plant status (Suzuki and Knaff, 2005; Kichey *et al.*, 2006) and is rate-limiting enzyme during biomass production (Migge *et al.*, 2000).

Glutamine synthetase is responsible for the primary assimilation of ammonium generated by nitrite reductase (García-Calderón *et al.*, 2012), and participates in the re-assimilation of ammonium obtained from photorespiration and catabolic processes (Miflin and Lea, 1980; Lea *et al.*, 1990; Hirel and Lea, 2001; Tobin and Yamaya, 2001). The primary assimilation of ammonium into glutamine / glutamate is carried out by coordinated actions of glutamine synthetase and glutamate synthase (Temple *et al.*, 1993; Lam *et al.*, 1995; Ireland and Lea, 1999; Migge *et al.*, 2000). With the help of glutamine synthetase the ammonium first assimilated into glutamine. In the future, from glutamine and α-ketoglutarate is formed glutamate by means of ferredoxin-dependent glutamate synthase in chloroplasts or by NADH-dependent glutamate synthase in plastids of roots or etiolated shoots (Lam *et al.*, 1996; Temple *et al.*, 1996, 1998; Ireland and Lea, 1999; Tobin and Yamaya, 2001; Lancien *et al.*, 2002).

The GS / GOGAT cycle generates two glutamate molecules. One of them is used as a cycled back substrate for the assimilation of ammonium by the glutamine synthetase enzyme; the other is exported from the GS / GOGAT cycle to synthesize other types of amino acids (Ireland and Lea, 1999). Thus, glutamate and glutamine serve as the main nitrogen donor for the biosynthesis of other organic nitrogen compounds. Amino acids and presumably ammonium, produced in the mesophyll or imported through the xylem, can be loaded to sieve elements, and the phloem highly contributes to cycling of nitrogen molecules (Coschigano *et al.*, 1998; Bush, 1999; Feraud *et al.*, 2005).

9.4.1.3.1 Glutamine Synthetase

Glutamine synthetase (GS; EC 6.3.1.2) in higher plants is an octameric enzyme from 320 to 380 kD and has 2 isoenzymes, glutamine synthetase 1 and glutamine synthetase 2, which differ in their organ- and cell-specificity, where they perform various functions. In the leaves of C_3 plants, glutamine synthetase 1 is localized in the cytosol of the phloem companion cells (Hirel *et al.*, 1993) and appears to be involved in the formation of glutamine for transporting nitrogen (Kamachi *et al.*, 1992; Pereira *et al.*, 1992).

Glutamine synthetase 2 is expressed in photosynthetic cells and is found in the chloroplasts stroma (Mann *et al.*, 1979, Hirel and Gadal, 1980, Stewart *et al.*, 1980, McNally *et al.*, 1983), where it used to assimilate ammonium obtained by reduction of nitrate or from photorespiration (Wallsgrove *et al.*, 1987; Cren and Hirel, 1999). Glutamine synthetase 2 is the most abundant isoform (Cabello *et al.*, 1991).

Glutamine synthetase is a nuclear-encoded protein. In higher plants, there is a single nuclear gene glutamine synthetase 2 (Hayakawa *et al.*, 1990; Kamachi *et al.*, 1991; Ishiyama, 2004; Tabuchi, 2005; Bernard *et al.*, 2008; Deng *et al.*, 2010; Gadaleta *et al.*, 2011). Glutamine synthetase 1 encodes multiple homologous but distinct genes (Lightfoot *et al.*, 1988; Tingey *et al.*, 1988; Lam *et al.*, 1995; Keegstra and Cline, 1999; May and Soll, 1999). There is also a post-transcriptional control of synthesis and activity of glutamine synthetase 2 (Migge *et al.*, 2000). Chloroplast glutamine synthetase 2 is initially synthesized in the cytosol as a higher-molecular-mass precursor polypeptide and contains a transit peptide (Kozaki and Tabeka, 1996; Hoshida *et al.*, 2000). The transit peptide mediates routing to the inner side of the chloroplast and is then removed by treatment with stromal peptidases (Husted *et al.*, 2002; Chu *et al.*, 2003).

Expression of both forms of glutamine synthetase in leaves is induced by light and carbon metabolites, such as sucrose (Oliveira and Coruzzi, 1999, Larios *et al.*, 2004), while amino acids are antagonists of the induction of glutamine synthetase (Oliveira and Coruzzi, 1999). The enzyme activity is modulated by the rate of CO_2 assimilation involving photosynthesized sugars as regulatory metabolites (Larios *et al.*, 2004).

The stable level of transcription of glutamine synthetase mRNA depends on nitrogen supply (Edwards and Coruzzi, 1989; Cock *et al.*, 1991; Li *et al.*, 1993; Lam *et al.*, 1996; Finnemann and Schjoerring, 1999) and is important for normal plants development and maintaining the carbon / nitrogen metabolic balance (Bao *et al.*, 2015). Lower metabolic status caused by the decreased glutamine synthetase 2 mRNA level in co-suppressed plants demonstrated an important function of the glutamine synthetase 2 gene in pathways of metabolism of nitrogen and carbon, as well as synthesis of amino acids (Cai *et al.*, 2010). The increased activity of glutamine synthetase 2 correlates with a decrease in the ammonium pool in the leaves and with the increase in the level of some free amino acids, including glutamate and glutamine (Migge *et al.*, 2000). Overexpression of a glutamine synthetase 1 gene in transgenic plants can lead to an increase in biomass (Lam *et al.*, 1995). However, according to other studies, overexpression of glutamine synthetase 1 genes was not associated with increased protein synthesis or biomass production (Eckes *et al.*, 1989; Hirel *et al.*, 1992; Vincent *et al.*, 1997).

9.4.1.3.2 Ferredoxin-Dependent Glutamate Synthase

Ferredoxin-dependent glutamate synthase (Fd-GOGAT; EC 1.4.7.1) localized exclusively in chloroplasts (Becker *et al.*, 1993) and NADH-dependent glutamate synthase (NADH-GOGAT, EC 1.4.1.14), absent in green tissues and found preferably in the vascular bundles of unexpanted leaves (Temple *et al.*, 1998) or in the root nodules of legumes (Trepp *et al.*, 1999). NADPH-dependent glutamate synthase is found in non-photosynthetic bacteria and archaea. Ferredoxin-dependent glutamate synthase participates in the assimilation of ammonium, obtained during the restoration of nitrates in leaves or during photorespiration (Lea, 1997). The NADH-dependent glutamate synthase function may consist in co-acting in the glutamine synthetase 1 / NADH-dependent glutamate synthase cycle during nitrogen transport via the vascular bundles from the roots or senescent tissues (Sivasankar and Oaks, 1996).

Expression of glutamine synthetase 2 and ferredoxin-dependent glutamate synthase genes is co-regulated (McGrath and Coruzzi, 1991, Migge *et al.*, 2000), however, the expression of glutamine synthetase 2 and ferredoxin-dependent glutamate synthase genes is not interdependent. An increase in glutamine synthetase 2 activity in the leaves did not affect the content of the ferredoxin-dependent glutamate synthase protein. Apparently, more ferredoxin-dependent glutamate synthase is present in the leaves than is required to meet the requirements of primary ammonium assimilation and that there is no strong interdependence between glutamine synthetase 2 and the expression of the ferredoxin-dependent glutamate synthase protein (Migge *et al.*, 2000).

According to Feraud *et al.* (2005), GLU1, encoding ferredoxin-dependent glutamate synthase, was expressed in the darkgrown seedlings and dark-adapted leaves. White light, red light and sucrose served as the signals to induce GLU1 expression in mesophyll and vascular cells. The light-induction of both the transcripts and the protein for ferredoxin-dependent glutamate synthase led to the predominance of ferredoxin-dependent glutamate synthase over NADH-dependent glutamate synthase in the photosynthetic leaves. Glutamine synthetase / ferredoxin-dependent glutamate synthase cycle can assimilate ammonium both in light and dark. Ferredoxin-dependent glutamate synthase shares the subcellular localization in the phloem with NADH-dependent glutamate synthases, providing glutamate for long-distance transport from sources via vascular tissues to sink to maintain the nitrogen status necessary for plant growth and development (Bush, 1999).

9.4.1.4 Glutamate Dehydrogenase

Ammonium can also be incorporated into glutamate with NADH-dependent glutamate dehydrogenase (NADH-GDH; EC 1.4.1.2) located in the mitochondria (Lawlor, 2002a), especially at high concentrations of ammonium, but the role of glutamate dehydrogenase in assimilation of ammonium has not been proven *in vivo* (Ireland and Lea, 1999; Miflin and Habash, 2002; Ziegler *et al.*, 2003). It is generally accepted that GDH is not a significant alternative route for the formation of glutamate in plants (Suzuki and Knaff, 2005). Since glutamate dehydrogenase reversibly deaminates glutamate (NAD-glutamate dehydrogenase), the physiological role of this enzyme *in vivo* can consist in obtaining α-ketoglutarate and reducing equivalents for use in other metabolic pathways (Robredo *et. al.*, 2011). Perhaps NAD(P)H-glutamate dehydrogenase performs a compensatory role in stabilizing the glutamate pool in leaves when ferredoxin-dependent glutamate synthase is limited (Ferrario-Méry *et al.*, 2002).

It should also be mentioned that ammonium can be integrated into asparagine with cytosolic asparagine synthetase (AS, EC 6.3.5.4) using free ammonium or glutamine-amide nitrogen, but this path occurs mainly in the dark (Thum *et al.*, 2003).

9.4.2 Relationship of Photorespiration and Nitrtogen Metabolism

In addition to the primary assimilation of nitrogen soil, there are other processes in plants accompanied by the liberation of ammonium, among which the most important are photorespiration (Mattsson *et al.*, 1997), the transfer of nitrogen compounds between different compartments of plants and the decay of the protein during senescence (Joy, 1988; Lea *et al.*, 1992).

Leaf senescence, in which the key role belongs to glutamine synthetase 1 (Miflin and Habash, 2002) leads to a significant intracellular accumulation of ammonium and amino acids as a result of the breakdown of proteins (Simpson and Dalling, 1981). Senescence can also be caused in young plants in the dark, which leads to a decrease in glutamine synthetase activity and the accumulation of ammonium (Peeters and Van Laere, 1992; Mattsson and Schjoerring, 1997).

Photorespiratory metabolism in plants is a consequence of the oxygenase activity of Rubisco (Ogren, 1984; Husic *et al.*, 1987). Subsequent metabolic reactions include the work of glycine decarboxylase complex along with serine hydroxymethyltransferase in mitochondria catalyzing the conversion of two glycine molecules to one serine molecule, which is accompanied by the release of CO_2, ammonia and reduction NAD^+. Photorespiration pathway and subsequent loss of CO_2, reduces the growth rate and productivity of C_3 plants.

The relationship between the processes of photosynthesis and photorespiration can be manifested in the fact that there is a balance between stomatal conductance and the photorespiration rate in plants, which has been shown in plants deficient in photorespiratory enzymes. The photorespiration rate depends on intercellular CO_2 concentration, which is regulated by stomata, the closure of which reduces CO_2 concentration at the catalytic site of Rubsico, thereby increasing the photorespiration. The need to maintain CO_2 concentration in chloroplasts for photosynthesis requires an increased opening of stomata, *i.e.* photorespiration is involved in the regulation of intercellular / ambient CO_2 concentration ratio, both via effects on stomata and via discrimination of ^{13}C in the glycine decarboxylase reaction (Igamberdiev *et al.*, 2004). Perhaps, photorespiration is associated with the regulation of various metabolic processes, which include signal transduction leading to the closure of stomata.

Ammonium, generated by photorespiration, can be re-processed in the glutamine synthetase / glutamate synthase cycle (Somerville and Ogren, 1980; Kendall *et al.*, 1986; Blackwell *et al.*, 1987, 1988; Wallsgrove *et al.*, 1987). The importance of this reaction is emphasized by the fact that photorespiration produces at least 10 times higher ammonium flux than when nitrates are restored (Givan *et al.*, 1988). Reduction of glutamine synthetase capacity to reassimilate ammonium in mutant plants leads to an increase in ammonium concentration and a decrease in amino acid and protein content in leaves (Mattsson and Schjoerring, 1997). Such plants were not able to reassimilate photorespiratory ammonium and were viable only in an atmosphere with high CO_2 concentration or exogenous glutamine, *i.e.* in conditions of suppression of photorespiration (Blackwell *et al.*, 1987, 1988; Wallsgrove *et al.*, 1987; Melo *et al.*, 2003). In contrast, overexpression of glutamine synthetase 2 leads to an increase in photorespiration capacity without altering the parameters of plant growth (Husted *et al.*, 2002), as well as to increased high light intensity tolerance (Oliveira *et al.*, 2002) and an increase in salt tolerance (Hiei *et al.*, 1994; Hoshida *et al.*, 2000). However, with a strong fall in glutamine synthetase activity, can be called up a compensatory mechanism that limits the accumulation of ammonium in leaves (Hausler *et al.*, 1994*a*, *b*). Thus, the reaction catalyzed by glutamine synthetase 2 represents the rate-limiting step of photorespiration (Kozaki and Takeba, 1996). Similarly, ferredoxin-dependent glutamate synthase also participates in the reassimilation of photorespiratory ammonium, since the decrease in ferredoxin-dependent glutamate synthase activity in the leaves of transformed plants is proportional to the accumulation of ammonium, glutamine and α-ketoglutarate in them (Ferrario-Méry *et al.*, 2000).

The rates of photosynthesis and photorespiration are determined by the kinetic properties of Rubisco, the relative proportion of its alternative substrates, and competitive CO_2 and O_2 inhibitors (Woodrow and Berry, 1988). The possibility of regulating glutamine synthetase 2 in terms of the rate of photorespiration was analyzed in leaf of C_3 plants grown at elevated CO_2 concentration to suppress photorespiration. It has been suggested that photorespiration promotes the expression of the glutamine synthetase 2 gene, since the expression of the glutamine synthetase 2 gene in the leaves decreases with elevated CO_2 concentration in the atmosphere (Edwards and Coruzzi, 1989; Cock *et al.*, 1991). However, suppression of photorespiration under conditions of elevated CO_2 concentration did not influence the level of glutamine synthetase 2 transcripts in the leaves (Beckmann *et al.*, 1997; Migge *et al.*, 1997). Apparently, photorespiration does not exert a metabolic control on the expression of the glutamine synthetase 2 gene (Beckmann *et al.*, 1997), and a lower level of glutamine synthetase 2 mRNA was due to the metabolic acclimation of photosynthesis with prolonged plant exposure at elevated CO_2 concentration, leading to a decrease in the ability to absorb CO_2 (Drake *et al.*, 1997). During the short-time exposure of plants to elevated CO_2 concentration, when the photosynthesis does not acclimate, the expression of the glutamine synthetase 2 genes and the corresponding increase in enzyme activity is positively modulated by CO_2 assimilation through the formation of sugars, but not by high photorespiration. Glutamine synthetase is regulated by post-translationally by reversible phosphorylation (Finnemann and Schjoerring, 2000; Riedel *et al.*, 2001).

9.5 Relationship of Carbon and Nitrogen Metabolism in Stress Conditions

9.5.1 *High CO_2 Concentration in the Atmosphere*

9.5.1.1 *Plants' Growth*

The effect of CO_2 concentration in the atmosphere on photosynthesis and plant growth is shown in many studies (Ainsworth *et al.*, 2002; Long *et al.*, 2004; Ainsworth and Rogers, 2007; Leakey *et al.*, 2009*a*; Rogers *et al.*, 2009; Aranjuelo *et al.*, 2011*a*; Perez *et al.*, 2011).

Under natural conditions, the carboxylation capacity of Rubisco is far from saturation (Stitt and Schulze, 1994a; Long *et al.*, 2006), which allows the plant to use an increase in CO_2 concentration (eCO_2) in air to accumulate additional biomass (Pal *et al.*, 2004; Xu *et al.*, 2007; Carlisle *et al.*, 2012) and increasing the ratio of root biomass to shoots (Jach *et al.*, 2000; Prior *et al.*, 2003) to expand the nutrient absorption area from the soil. However, in many other studies, only a small change in biomass and its distribution between organs has been shown (Bernacchi *et al.*, 2000, Zhou and Shangguan, 2009; Newingham *et al.*, 2013), which indicates the ambiguous effect of the elevated CO_2 concentration in plants.

The distribution of biomass in plant organs at elevated CO_2 concentration also depends on the conditions of mineral nutrition (Xu *et al.*, 2007). In this case, the increase in the rate of growth processes can be limited by the plant's ability to transport and assimilate nutrients in an amount sufficient to correspond to an elevated level of carbon fixation for the growth of new organs and tissues (Ainsworth *et al.*, 2007). In many respects, plant growth is limited by the presence and availability of nitrogen (Stitt and Krapp, 1999; Oren *et al.*, 2001; Hungate *et al.*, 2003; Luo *et al.*, 2004). The increase in CO_2 concentration changes the ratio of carbon to nitrogen in fully developed and young, growing leaves, which results in a downstream effect on the growth and productivity of plants (Ainsworth *et al.*, 2007).

Metabolite balance of plant cells depends on the metabolic status of the tissue (Geiger *et al.*, 1998; Matt *et al.*, 2001; Nielsen and Stitt, 2001). The old leaves are dominated by assimilation and export pathways, and mainly soluble carbohydrates are synthesized, while in young leaves, the processes of biosynthesis and growth predominate, and more amino acids, proteins and cell wall material are synthesized (Masclaux *et al.*, 2000). Since carbohydrates are capable of down-regulating photosynthesis (Long *et al.*, 2004; Rogers and Ainsworth, 2006), to determine carbon /

nitrogen status of plants can be used the metabolic pools (Stitt and Krapp, 1999; Matt *et al.*, 2001; Foyer *et al.*, 2003; Kruse *et al.*, 2003*b*; Jeong *et al.*, 2004). In old leaves increases the content of hexoses, sucrose and starch (Ainsworth *et al.*, 2007). Young leaves have high concentrations of ureides and amino acids at relatively low concentrations of carbohydrates (Ainsworth *et al.*, 2007). Developing leaves at elevated CO_2 concentration have small pools of ureides as compared to developing leaves at ambient CO_2 concentration, which indicates that nitrogen assimilation in young leaves is improved by elevated CO_2 concentration.

The accumulation and distribution of biomass also varies depending on the plant species or functional group (C_3 and C_4) that may have different specialized strategies for regulating biomass distribution for maximum use of external resources (Auerswald *et al.*, 2012; Liu *et al.*, 2012; Poorter *et al.*, 2012; Vogan and Sage, 2012).

In plant cells, there may be a competitive interaction between CO_2 fixation and nitrate assimilation for reduced equivalents (Bloom *et al.*, 2002, 2010). According to Rachmilevitch *et al.* (2004), an increase in photosynthetic activity at elevated CO_2 concentration may reduce the amount of ferredoxin available for nitrite reduction. In addition, to restore nitrate to nitrite requires NADH, produced from malate and then transported to chloroplast. Deficiency of ATP and NADH, observed in some studies (Aranjuelo *et al.*, 2013), can limit the reduction of nitrate, which is the second largest sink for the utilization of photosynthetic energy (Bloom *et al.*, 2002).

9.5.1.2 Rubisco Content

Regulation of photosynthesis by changing the distribution of nitrogen throughout the plant is one of the important strategies for plant adaptation in the elevated CO_2 concentration condition (Makino *et al.*, 2000*a*). Specifically, with elevated CO_2 concentration, a decrease in the soluble protein content is observed, due mainly to a decrease in the amount of Rubisco (Bowes, 1993; Drake *et al.*, 1997).

The adaptation of the photosynthetic apparatus to elevated CO_2 concentration does not affect the synthesis of Rubisco (Andreeva *et al.*, 1979). The light curves of photosynthesis are the same in plants adapted to 0.04% (ambient CO_2 concentration) and 0.2% CO_2 (elevated CO_2 concentration). Moreover, with long-term growth of many cultures with eCO_2, a decrease in Rubisco activity occurs (Besford *et al.*, 1990; Wong, 1990; Rowland-Bamford *et al.*, 1991), which may be due to a change in the enzyme content or its specific activity. According to Drake *et al.*

(1997), a decrease in the Rubisco content by an average of 15% was observed in 12 plant species studied. However, according to Neales and Nichols (1978), in plants grown at aCO_2 and then placed in the atmosphere with an elevated CO_2 concentration, the Rubisco content in the leaves increased by 20%, but the enzyme activity decreased by 22%.

These examples illustrate two different mechanisms by which elevated CO_2 concentration affects photosynthesis: as a result of a change in the total Rubisco content and as a result of a change in the amount of enzyme per unit leaf area. These mechanisms form the basis of acclimation of plants with elevated CO_2 concentration. According to Ainsworth and Rogers (2007), redirecting excess nitrogen from Rubisco to other processes does not affect the rate of carbon fixation in plant leaves.

The photosynthetic rate under conditions of elevated CO_2 concentration is limited by the electron flow rate in the photosynthetic electron transport chain (Farquhar *et al.*, 1980*b*) and the presence of inorganic phosphate in chloroplasts for the synthesis of ATP (Sharkey, 1985), resulting in significantly reduced Rubisco (Sage, 1990). According to Eaton-Rye and Govinjee (1988), and also Klimov *et al.* (1995), the effect of CO_2 is manifested in the regulation of the electron transport rate at the level of photosystem II. The carbon dioxide binding sites are located on the 32 kDa protein as the secondary electron acceptor. The carbon dioxide binding sites are located on the 32 kD protein, near quinone (Q_B), as secondary electron acceptors. Trace amounts of CO_2 change the conformation of the protein in these centers and thus ensure a high activity of the photosynthetic electron transport chain in the area between photosystem II and photosystem I.

Physiological and biochemical features and molecular mechanisms of plant adaptation to the elevated CO_2 concentration at the Rubisco level are discussed in detail in the works of Romanova (1997, 2005).

9.5.1.3 Photosynthetic Acclimation

The photosynthetic rate depends both on the photosynthetic electron transport rate (Aranjuelo *et al.*, 2008), and Rubisco carboxylation capacity (Stitt and Krapp, 1999; Long *et al.*, 2004; Aranjuelo *et al.*, 2005). Initially, the incubation of plants in an atmosphere with an elevated CO_2 concentration leads to an increase in carboxylase activity of Rubisco due to the greater substrate availability and the suppression of competitive Rubisco oxygenation (Stitt, 1991; Aranjuelo *et al.*, 2005; Farquhar *et al.*, 1980*b*; Bowes, 1993; Martínez-Carrasco *et al.*, 2005; Pérez *et al.*, 2007; Alonso *et al.*, 2009; Gutiérrez *et al.*, 2009; Ellsworth *et al.*, 2004). However, after initial stimulation, the photosynthetic rate decreases during acclimation (Ainsworth *et al.*, 2004; Aranjuelo *et al.*, 2009), at which the photosynthetic activity is adjusted in accordance with the carbon-requirements for plants (Ainsworth *et al.*, 2002; Long *et al.*, 2004; Ainsworth and Rogers, 2007; Leakey *et al.*, 2009*a*; Rogers *et al.*, 2009).

Down-regulation of photosynthetic capacity may also be a consequence of closure of stomata and a decrease in transpiration rate observed with prolonged exposure to elevated CO_2 concentration (Stitt and Krapp, 1999; Pérez *et al.*, 2005; Taub and Wang, 2008; Gutiérrez *et al.*, 2013). In this case, the stomatal limitation of photosynthesis may result from the depletion of intercellular CO_2 concentration (Naumburg *et al.*, 2004) or the reduction of mesophyll conductance to diffusion of CO_2 to the carboxylation sites (Singsaas *et al.*, 2003; Flexas *et al.*, 2007; Evans *et al.*, 2009).

However, with long-term elevated CO_2 concentration, the decreased Rubisco carboxylation is likely to occur through two basic mechanisms: first, by reducing the content of Rubisco through the sugar sensing mechanism (Sheen, 1990, Moore *et al.*, 1999) or carbon source-sink relationships; secondly, as a result of a reduction in Rubisco content with a decrease in nitrogen content in the leaves or as a result of nitrogen allocation (Luo *et al.*, 1994; Nakano *et al.*, 1997).

At the increase in the photosynthetic rate at the beginning of the incubation of plants at elevated CO_2 concentration, many photoassimilates accumulate in the leaves, which cause suppression of the expression of genes encoding photosynthetic proteins, which results in down-regulation of photosynthetic capacity (Moore *et al.*, 1999; Jifon and Wolfe, 2002). An increased amount of hexose induces hexose kinase activity (Kaschuk *et al.*, 2010; Ainsworth and Bush, 2011), which initiates the transduction reaction by causing the inhibition of the Rubisco activity promoter of a small subunit (Jang and Sheen, 1994; Moore *et al.*, 1999; Long *et al.*, 2004; Ainsworth and Rogers, 2007). In this case, increasing the hexose concentration can serve as a source / sink signal of imbalance followed by down-regulation of Rubisco content (Moore *et al.*, 1999; Long *et al.*, 2004). In addition, for photosynthetic down-regulation, the balance between the synthesis of sucrose and starch is important (Moore *et al.*, 1999; Stitt and Krapp, 1999). Enhancement of starch synthesis at elevated CO_2 concentration is up-regulate the photosynthetic rate. Up-regulating properties are also possessed by glucose-6-phosphate isomerase, triose phosphate isomerase (Moore *et al.*, 1999) and germin-like proteins (Stitt *et al.*, 2010). All this indicates a close coordination of the processes of photosynthesis and sink utilization of carbohydrates (Ainsworth and Rogers, 2007).

In other words, when the amount of photosynthetic products exceeds the ability of sink organs to use them (Lewis *et al.*, 2002; Aranjuelo *et al*, 2009; Ainsworth *et al.*, 2004; Aranjuelo *et al.*, 2011*a*), these photosynthates promote to decrease photosynthetic activity to balance the carbon source activity and sink capacity (Thomas and Strain, 1991). Thus, the performance of plants depends on the sink strength, which is one of the key conditioning factors responsible for the responsiveness of plants to elevated CO_2 concentration (Ziska *et al.*, 2004; Alonso *et al.*, 2008; Uddling *et al.*, 2008; Hogy *et al.*, 2009, 2010; Aranjuelo *et al.*, 2011*a*; Perez *et al.*, 2011). Limiting sink activity can cause accumulation of carbohydrates in the leaves and altered carbon / nitrogen balance with subsequent inhibition of photosynthetic activity, which is especially pronounced for plants with low sink strength (Uddling *et al.*, 2008). As the source activity increases, the value of sink strength as a limiting factor becomes especially relevant. In this case, maintaining a high photosynthesis rate with elevated CO_2 concentration will depend on the ability of the plant to form new sinks or to support the growth of existing sinks, and to increase storage capacity. For example, in wheat at elevated CO_2 concentration, the main carbon and nitrogen sinks during the grain filling stage are not able to overcome the accumulation of carbohydrates in the leaves, resulting in photosynthetic down-regulation (Aranjuelo *et al.*, 2011*a*).

The degree of down-regulation varies depending on the stage of ontogenesis (Van Oosten and Besford, 1995, Miller *et al.*, 1997), growth habit (determinate or indeterminate) (Ceulemans *et al.*, 1997; Ainsworth *et al.*, 2004) and functional groups (C_3 or C_4 plants) (Ainsworth and Rogers, 2007).

9.5.1.4 Photosynthesis and Nitrogen Content

Reducing Rubisco in leaves can be associated with a reduction in the total nitrogen content of the leaf (Ellsworth *et al.*, 2004; Aranjuelo *et al.*, 2005*b*; Aranjuelo *et al.*, 2011*a*), and as a result of redistribution of nitrogen within the plant (Nakano *et al.*, 1997). The physiological consequences of reducing the nitrogen content in the leaves can be crucial for plant growth at elevated CO_2 concentration.

At first glance, an increase in carbon assimilation with short-term exposure to an elevated CO_2 concentration should stimulate nitrogen absorption to maintain a faster growth rate (Kruse *et al.*, 2002). However, more efficient photosynthetic carbon uptake at elevated CO_2 concentration in many cases stimulates plant growth without increasing nitrogen uptake by roots (Conroy and Hocking, 1993; Taub and Wang, 2008). Moreover, in the long-term exposure, an increase in the level of CO_2 leads to a decrease in the nitrogen content in plant tissues (Xu *et al.*, 2007; Zhou and Shangguan, 2009) and causes nitrogen limitation (Stitt and Krapp, 1999), which is accompanied by an increase in carbon / nitrogen ratio of C_3 plants (Cotrufo *et al.*, 1998; Kimball *et al.*, 2002). This decrease in the nitrogen content may be due to a relative increase in the leaf area or plant biomass, or a change in the nitrogen distribution in the whole plant. On the other hand, these changes can occur due to the dilution effect (Taub and Wang, 2008, Duval *et al.*, 2012), which is due to a faster growth of biomass in comparison with the uptake of nutrients with an unchanged amount of nitrogen per plant at different concentrations of CO_2 (Wang *et al.*, 2013).

Theoretically, to increase the nitrogen use efficiency of photosynthesis at elevated CO_2 concentration, it is necessary to reduce nitrogen allocation in Rubisco. However, in the long-term exposure to CO_2 enrichment, a decrease in the amount of Rubisco protein is most likely due to a decrease in the total nitrogen leaf content, rather than the redistribution of nitrogen to other components. This means that at elevated CO_2 concentration there is no optimization of the distribution of nitrogen, and the sometimes observed increase in nitrogen use efficiency in one leaf does not necessarily lead to an increase in the accumulation of biomass of the whole plant (Makino *et al.*, 2000*b*).

Thus, since nitrogen-use efficiencies were not significantly dependent on the effect of elevated CO_2 concentration, the predominant mechanism of down-regulation of photosynthesis under these conditions could be considered a decrease in leaf nitrogen content per unit area (N_{area}) (Ellsworth *et al.*, 2004). In accordance with this, the decrease in the maximum carboxylation rate (Vcmax) at elevated CO_2 concentration is largely explained by the changes in N_{area}, because nitrogen use efficiency was not significantly affected by elevated CO_2 concentration exposure (Curtis *et al.*, 2000; Miyazawa *et al.*, 2011). According to Miyazawa *et al.* (2011), N_{area} is monitored regardless of the thickness of the leaf and may be under systemic regulation through CO_2 concentration signals from mature leaves controlling the division of mesophyll cells.

Nitrogen leaf content and N_{area} correlate with the anatomical properties of the leaves (Niinemets, 1999, 2007*a*, *b*), possibly depending on the light intensity gradient. For many species growing at elevated CO_2 concentration, an increase in the thickness of the leaves was recorded (Thomas and Harvey, 1983; Radoglou and Jarvis, 1990; Kürschner *et al.*, 1998; Sims *et al.*, 1998*a*; Lin *et al.*, 2001), which may be is associated with the expansion of mesophyll cells (Radoglou and Jarvis, 1990), as well as an increase in the density of stomata per unit area, although there are conflicting data on this matter (Woodward *et al.*, 2002).

Reducing the nitrogen content in leaves during photosynthetic acclimation to elevated CO_2 concentration (Smart *et al.*, 1998; Bloom *et al.*, 2010; Zhang *et al.*, 2013; Bloom *et al.*, 2014) may be due to a variety of reasons. This can be rapid plant growth (Stitt and Krapp, 1999, Pleijel and Uddling, 2012), a decrease in nitrogen uptake due to a transpiration lowering, inhibition of nitrogen assimilation (Geiger *et al.*, 1999; Matt *et al.*, 2001; Bloom *et al.*, 2002; Bloom, 2015), an increase in nitrogen losses due to root exudation and leaf gas emissions (Smart *et al.*, 1998; Haase *et al.*, 2007; Taub and Wang, 2008; Johansson *et al.*, 2009), nitrogen dilution as a result of the accumulation of carbon-rich compounds (Stitt and Krapp, 1999; Taub and Wang, 2008), nitrogen demand reduction due to own-regulation (Stitt and Krapp, 1999; Leakey *et al.*, 2009*a*), a change nitrogen partitioning from sources to sinks (Nie *et al.*, 1995), sink limitation in the case of accumulation of carbohydrates at elevated CO_2 concentration, leading to inhibition of photosynthetic genes (Long *et al.*, 2004; Aranjuelo *et al.*, 2011*a*), ontogenic drifts (Nie *et al.*, 1995; Taub and Wang, 2008), and nitrogen signaling due to a change in the nitrogen status (Stitt and Krapp, 1999; Geiger *et al.*, 1999; Matt *et al.*, 2001).

9.5.1.5 Metabolic Changes

In addition to reducing the photosynthetic rate and the content of nitrogen compounds, under the conditions of elevated CO_2 concentration, there is also a general down-regulation of the expression of photosynthetic genes and, in particular, down-regulation of the CO_2 fixation and up-regulation genes for ribulose-1,5-diphosphate regeneration and carbohydrate metabolism (Fukayama *et al.*, 2009, 2011). There was also a decrease in the level of transcripts of nitrogen assimilation enzymes (Vicente *et al.*, 2015), accompanied by a decrease in the amino acids content, as well as the induction of genes associated with signaling, protection and functions in the use and distribution of carbon (Miyazaki *et al.*, 2004; Markelz *et al.*, 2014*a*; Markelz *et al.*, 2014*b*; Li *et al.*, 2006; Li *et al.*, 2008; Kaplan *et al.*, 2012; Watanabe *et al.*, 2014). There is evidence of an increase in the rate of dark respiration of plants and the transcripts content of respiratory proteins at elevated CO_2 concentration (Ainsworth *et al.*, 2006; Leakey *et al.*, 2009*b*; Li *et al.*, 2013), although this results are quite contradictory (Wang and Curtis, 2002; Davey *et al.*, 2004; Li *et al.*, 2013;

Azcón-Bieto and Osmond, 1983; Rubio-Asensio *et al.*, 2015). Some researchers note a decrease in respiration rate in leaves of high nitrate grown plants at elevated CO_2 concentration (Kaplan *et al.*, 2012, Rubio-Asensio *et al.*, 2015, Vicente *et al.*, 2015).

The response of plants to elevated CO_2 concentration depends on the nitrate plant nutrition (Vicente *et al.*, 2016). Under conditions of elevated CO_2 concentration, plants growing at a low level of nitrate in the soil accumulated more carbohydrates in leaves and roots (Stitt and Krapp 1999, Long *et al.*, 2004, Pérez *et al.*, 2005) than plants growing at high levels nitrate in soil (Geiger *et al.*, 1999; Vicente *et al.*, 2015). These data correlated with the increase transcript levels for carbohydrate synthesis enzymes (starch synthase, cytosolic fructose 1,6-bisphosphatase, sucrose-phosphate synthase and fructosyl transferase) and with reduce transcripts for carbohydrate degradation enzymes (sucrose synthase, cell wall invertase and fructan exohydrolases) (Li *et al.*, 2006, Fukayama *et al.*, 2009, Fukayama *et al.*, 2011, Kaplan *et al.*, 2012). In high nitrate grown plants, some carbohydrate metabolism genes were repressed at elevated CO_2 concentration. This indicates that the accumulation rate of carbohydrates exceeded the possibility of their use in plant leaves, and that nitrogen assimilation and fixation of CO_2 in this case were poorly coordinated (Vicente *et al.*, 2016).

Thus, photosynthetic acclimation induced by elevated CO_2 concentration in low nitrate grown plants is accompanied by an increase in biomass and carbohydrate content. At the same time, there is a decrease in the amount of organic nitrogen per unit area of the leaf, amino acids and transcripts for nitrogen metabolism, Rubisco and nitrate reductase activities, organic acids and intermediates of the Benson-Bassham-Calvin cycle, while nitrate uptake is unchanged. In high nitrate grown plants at elevated CO_2 concentration, photosynthesis and the amount of organic nitrogen per unit leaf area remained unchanged, but induction of transcripts of nitrogen metabolism occurred (Vicente *et al.*, 2016).

It should also be borne in mind that Rubisco is not the only enzyme that limits the rate of photosynthesis. Other Benson-Bassham-Calvin cycle enzymes, such as aldolase, sedogeptulose-1,7-bisphosphatase, and transketolase, are also capable of controlling photosynthesis (Stitt *et al.*, 2010; Parry *et al.*, 2011; Henkes *et al.*, 2001; Raines, 2003; Lefebvre *et al.*, 2005; Tamoi *et al.*, 2006; Uematsu, *et al.*, 2012). In addition, photosynthetic down-regulation may be due to nitrate reductase activity, the inhibition of which reduces nitrate assimilation, and as a result, affects the content of nitrogen compounds in the plant (Bloom *et al.*, 2002, 2010).

9.5.1.5.1 Nitrate Assimilation

An increase in the air CO_2 concentration at first glance should have led to an increase in the number of carbon skeletons, energy and reducing power available for ammonium assimilation through the glutamine synthetase / glutamate synthase cycle (Forde and Lea, 2007). The uptake of nitrate and ammonium by the roots is controlled by several specific transporters (Lanquar *et al.*, 2009, Laugier *et al.*, 2012), whose activity ultimately determines the size of the pool of free amino acids in the cells. A decrease in glutamine pool can provide a signal that is up-regulated the nitrate uptake, when nitrogen plant status is inadequate (Miller *et al.*, 2008). Nevertheless, the dilution effect of internal nitrogen at elevated CO_2 concentration does not contribute to an increase in the roots uptake power of nitrogen-compounds. This can be due to a decrease in the glutamine concentration in the roots and shoots of plants at elevated CO_2 concentration may not be enough to stimulate the nitrogen uptake (Wang *et al.*, 2013). On the other hand, according to Bloom *et al.* (2010), the lack of increased nitrogen uptake at elevated CO_2 concentration is associated with a decrease in the ability to nitrate assimilation, and not with a direct effect on nitrogen uptake.

At elevated CO_2 concentration, both maximum and actual nitrate reductase activity, and nitrate concentration in the shoot and root tissues decrease (Wang *et al.*, 2013). Reduction of nitrate levels can promote the degradation of transcripts and nitrate reductase protein (Galangau *et al.*, 1988), and lead to repression of nitrate transporter and nitrate reductase (Stitt and Krapp, 1999). However, there are conflicting data on nitrate assimilation at elevated CO_2 concentration. In some cases, inhibition of nitrate reduction was observed (Bloom *et al.*, 2002, 2010; Searles and Bloom, 2003), in other cases, nitrate reductase activity was suppressed insignificantly (Sicher, 2001) or even increased (Robredo *et al.*, 2011). Perhaps the change in nitrate reductase activity depends on the type of tissue. Under ambient CO_2 concentration conditions, the majority of nitrate reductase activity is localized in the aerial part of the plant (Wang *et al.* (2013). At elevated CO_2 concentration, most of the total plant nitrate reductase is present in the roots due to a relatively greater decrease in nitrate reductase activity in shoots compared to roots (Wang *et al.*, 2013; Kruse *et al.*, 2002, 2003*a*). Inhibition of nitrate assimilation in shoots at elevated CO_2 concentration can also be caused by a decrease in the availability of NADH and ferredoxin (Bloom *et al.*, 2010). In addition, according to Bloom *et al.* (2012), inhibition of nitrate assimilation at elevated CO_2 concentration occurs in C_3, but not in C_4 plant species.

The relationship between carbon and nitrogen metabolism can be reflected in the model according to which the accumulation of starch includes the down-regulation of the enzymes involved in the internal amino acid provision (Ezquer *et al.*, 2010*b*; Yamakawa and Hakata, 2010; Li *et al.*, 2011). The flow of carbon skeletons through the tricarboxylic acid cycle has been described as conditioning the assimilation of nitrate and ammonium into amino acids (Ezquer *et al.*, 2010*a*). According to Aranjuelo *et al.* (2013), sucrose-starch balance is one of the important factors in the responsiveness of photosynthetic machinery to elevated CO_2 concentration. Moreover, the limitation of nitrogen assimilation is induced by photosynthetic down-regulation, since available ferredoxin is more destined to the synthesis of starch than for nitrogen assimilation. The deficit of ATP and NADH also can limit high energy-demanding processes, such as nitrate assimilation (Aranjuelo *et al.*, 2013).

Under ambient CO_2 concentration conditions, glutamine synthetase participates in the assimilation and regulation of tissue and apoplastic concentrations of ammonium (Mattsson *et al.*, 1997). At elevated CO_2 concentration, glutamine synthetase activity varies little both in shoots and in roots (Wang *et al.*, 2013). This suggests that glutamine synthetase may not be the key factor regulating the level of ammonium at elevated CO_2 concentration (Wang *et al.*, 2013).

Glutamate dehydrogenase can participate in ammonium assimilation, but mainly catalyzes oxidative deamination of glutamate, thus providing carbon skeletons for respiration and oxidative phosphorylation (Labboun *et al.*, 2009). At elevated CO_2 concentration, the aminating activity glutamate dehydrogenase (NADH-GDH) remains unchanged, but increases considerably the deamination activity (NAD-GDH) in shoots and roots (Wang *et al.*, 2013). That may be appropriate to meet the energy needs for nitrate assimilation in competition with the assimilation of CO_2, increasing at elevated CO_2 concentration (Bloom *et al.*, 2010).

9.5.1.5.2 Photorespiration

Elevated CO_2 concentration decreases levels of intermediates and transcripts for the photorespiration pathway in low nitrate and high nitrate grown plants (Vicente *et al.*, 2016). At elevated CO_2 concentration, the content of serine and glycine in leaves of plants decreases (Vicente *et al.*, 2016; Geiger *et al.*, 1999). Simultaneously, the expression of photorespiration-related genes, for example, glycolate oxidase, is suppressed (Ainsworth *et al.*, 2006; Li *et al.*, 2008; Vicente *et al.*, 2015). This indicates a low level of Rubisco oxygenation and inhibition of photorespiration at elevated CO_2 concentration (Vicente *et al.*, 2016). The increase in CO_2 concentration leads to a decrease in the amount of the transcription factor (DOF1) regulating several linked genes associated with ammonium assimilation (Kumar *et al.*, 2009) and, possibly, plays a significant role in photosynthetic acclimatization to elevated CO_2 concentration (Fukayama *et al.*, 2009).

In short-term experiments with C_3 plants, an increase in the atmospheric CO_2 concentration leads to an increase in the substomatal CO_2 concentration. The greater CO_2 availability in the active site of Rubisco (Poorter, 1993) contributes to the stimulation of net carbon fixation and a decrease in the rate of photorespiration (Leegood *et al.*, 1995; Long *et al.*, 2006). The photorespiratory pathway requires additional ATP for the re-entry of the glycerate, resulting in the change redox situation in the chloroplast stroma. Increasing ATP / NADPH ratio results in an increase in the amount of carbon passing through the regenerative part of the Benson-Bassham-Calvin cycle related to PGA reduction. Therefore, one should expect that photorespiratory activity should lead to more reducable conditions in the stroma.

However, in plants exposed to elevated CO_2, more oxidized values were determined for all parameters indicative of the redox situation in the electron transport chain of the chloroplast stroma (fluorescence of chlorophyll, P700 reduction, activation of NADP-malate dehydrogenase than at ambient CO_2 concentration. In NADP-malate dehydrogenase-under-expressing plants, under ambient conditions the redox state of ferredoxin increases, but decreases with eCO_2.

The number of electrons necessary for fixing CO_2 or photorespiration is often excessive (Stitt, 1986; Backhausen *et al.*, 1994). To prevent over-reduction and associated oxidative damage, alternate electron acceptors (poising mechanisms) are present in the chloroplast stroma, such as the ferredoxin-dependent cyclic transport of electrons around the photosystem I (Bendall and Manasse, 1995), reduction O_2 (Polle, 1996) and the malate valve (Scheibe, 1987; Fridlyand *et al.*, 1998), which do not affect CO_2-assimilation or reduction nitrite (Backhausen *et al.*, 1994). NADP-malate dehydrogenase and malate valve keep the stromal ferredoxin pool in an oxidized state (Faske *et al.*, 1997; Backhausen *et al.*, 1998) and stabilize the redox state in chloroplasts in cases of sudden changes in the number of excess electrons.

Thus, inhibition of the photorespiration by elevated CO_2 concentration can be an alternative mechanism that helps to reduce the number of excess electrons in photosynthetic electron transport chain (Backhausen and Scheibe, 1999).

9.5.2 *Low CO_2 Concentration in the Atmosphere*

The carbon-deficiency that occurs when C_3 plants are grown at low CO_2 concentration results in a decrease in plant productivity due to inhibition of growth and a decrease in biomass accumulation (Ward, 2005; Li *et al.*, 2014). Low CO_2 concentration also influences the plants plants development, for example, delaying the onset of the flowering period (Ward and Strain, 1997, Li *et al.*, 2014), which, however, can contribute to the conservation of plant productivity by increasing the accumulation time of more storage substances, necessary for reproduction (Sage and Coleman, 2001; Ward, 2005). As an adaptive mechanism in conditions of low CO_2 concentration, one can consider the development of leaves with a higher stomatal density per leaf area (Woodward, 1987; Lake *et al.*, 2001; Beerling *et al.*, 2001; Franks and Beerling, 2009; Doheny-Adams *et al.*, 2012; Franks *et al.*, 2012; Li *et al.*, 2014). This allows to increase the stomatal conductance of the whole plant and contributes to overcoming CO_2 limitation of photosynthesis (Franks and Beerling, 2009; Franks *et al.*, 2012). However, the response of plants to a low level of CO_2 is distinguished by a large genetic variation (Sharma *et al.*, 1979).

In conditions of low CO_2 concentration, because of the competitive reaction of Rubisco oxygenation, the photorespiratory pathway intensifies, which is a consequence also of the amplification of the expression of the photorespiratory genes (Li *et al.*, 2014), including in C_4 plants (Sage *et al.*, 2012).

In high light and low CO_2 concentration, there is a decrease in photosynthetic photon flux density due to an insufficient rate of CO_2 fixation in all the output of the light-harvesting reactions. The imbalance between CO_2 fixation and light absorption causes a decrease in the transcription of GLK proteins (GLK1 and GLK2) (Li *et al.*, 2014), directly regulating a large suite of genes involved in light-harvesting and thylakoid protein complexes, especially those of photosystem II (Waters *et al.*, 2009). Simultaneously, the synthesis of chlorophyll *b* is reduced, apparently due to the downregulated of the chlorophyllide an oxygenase gene. In chloroplasts are observed the fewer and less-stacked granas and a higher proportion of non-stacked stromal lamellae (Li *et al.*, 2014).

In our studies, 10-days wheat sprouts were exposed to a prolonged (10 days) exposure to low [CO_2] in a sealed chamber. After opening the chamber, the activity of the photosynthetic apparatus was measured for the isolation of O_2 in the Clark cell using CO_2 or nitrite as electron acceptors. At the same time, the activity of the main enzymes of the nitrogen metabolism was measured both immediately after opening the chamber and for the next 2 days to elucidate the change in enzymatic activities after the plants returned to the conditions ambient CO_2 concentration.

CO_2 acts as the main electron acceptor in photosynthetic photon flux density. In conditions of CO_2 deficiency, the excess energy of light absorbed by chlorophyll can partially be dissipated as heat (Qiu *et al.*, 2003), but it can also be redirected to restore nitrite acting as an alternative electron acceptor (Eichelmann *et al.*, 2011). On the one hand, this allows the plant to utilize the excess energy of absorbed sunlight; on the other hand, thus, the balance of carbon / nitrogen metabolism can be regulated.

At ambient CO_2 concentration, the main flux of absorbed light energy is directed to the reduction of carbon compounds, despite the equal distribution of the electron fluxes and the enzymatic ability to photoreducte nitrite and CO_2 (Eichelmann *et al.*, 2011). The dominance of CO_2 reduction is explained by the fact that with high substrate availability for photoassimilation of each molecule of nitrite in glutamate is required 8 electrons, in comparison with 4 electrons for CO_2 assimilation (Robinson, 1986). After the incubation of plants with low CO_2 concentration, all the main parameters of photosynthesis decrease when using CO_2 as the electron acceptor (see Table 9.1 and Figure 9.2). However, the same parameters remain unchanged when nitrite is used as the electron acceptor. Consequently, when the metabolism carbon processes slow down, the pathway of primary nitrogen assimilation continues to function successfully.

Table 9.1. The parameters of the approximation for the light curves of O_2 gas exchange of wheat leaves after incubation of plants under ambient (aCO_2) and low ($-CO_2$) CO_2 concentration using CO_2 and NO_2^- as electron acceptors from photosynthetic electron transport chain

Conditions for growing plants	aCO_2		$-CO_2$	
The electron acceptors	**CO_2**	**NO_2^-**	**CO_2**	**NO_2^-**
The rate of gas exchange on the plateau light-curve, nmol O_2 mg^{-1} FW s^{-1}	173	137	118	127
O_2 evolution in the dark, nmol O_2 mg^{-1} FW s^{-1}	-96.3	-51.8	-37.2	-56.3
Quantum efficiency of photosynthesis	0.363	0.284	0.228	0.287
Light compensation point, μmol E m^{-2} s^{-1}	266	182	163	196
Light intensity when saturation light-curve, μmol E m^{-2} s^{-1}	743	656	681	638

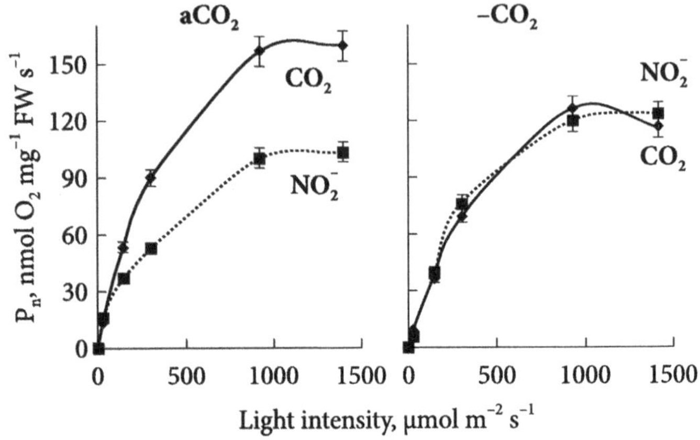

Fig. 9.2. Light curves of O_2 gas exchange of wheat leaves after incubation of plants under ambient (aCO_2) and low ($-CO_2$) CO_2 concentration using CO_2 and NO_2^- as electron acceptors from photosynthetic electron transport chain
Note: P_n – net photosynthetic rate

These data confirm measurements of the enzymatic activity of nitrite reductase, the reaction rate of which changes little after incubation of plants at low CO_2 concentration (see Figure 9.3). A brief increase in the activity of the enzyme on the second day after the opening of the incubation chamber can be explained by the effect on the activity of nitrite reductase of the intensification of the carbon metabolism under ambient CO_2 concentration (Plaut *et al.*, 1977). But already on the 2nd day after the opening of the chamber, the activity of the enzyme returns to its original level, most likely due to the work of mechanisms that maintain the constancy of the carbon / nitrogen balance.

The ability to use light energy for the reduction of nitrite is carried out in the ferredoxin-dependent nitrite reductase reaction to form ammonium (see Figure 9.4). In this reaction, the reduced ferredoxin included in the photosynthetic electron transport chain photosystem I is used as a redox factor. Since the reduction of both nitrite and CO_2 competes for the same reductant pool with similar properties (Eichelmann *et al.*, 2011), ferredoxin forms nodal points of nitrogen and carbon metabolism, distributing photosynthesically generated electrons into various metabolic processes (Hanke *et al.*, 2011).

The response of plants to a CO_2 deficit is accompanied by a change in the activity of the main enzymes of nitrogen metabolism (see Figure 9.3). After the incubation of plants at CO_2 concentration, close to the CO_2-compensation point, the nitrate reductase activity in the leaves decreases.

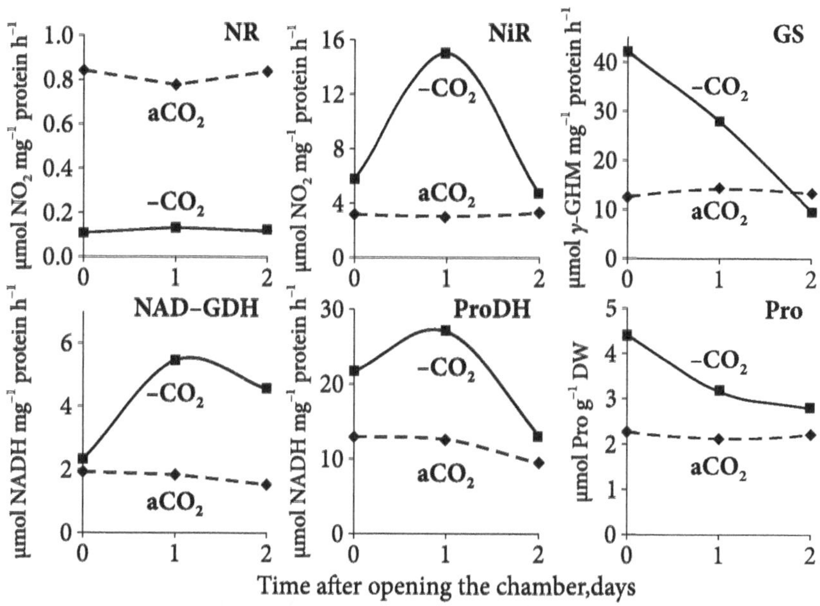

Fig. 9.3. The activity of the main enzymes of nitrogen metabolism and the content of proline in the leaves of wheat immediately after the incubation of plants at low CO_2 concentration (0 day) and during the subsequent 2 days of incubation with ambient CO_2 concentration

Note: NR - nitrate reductase, NiR- ferredoxin-dependent nitrite reductase, GS - glutamine synthetase, NAD-GDH - NAD-dependent glutamate dehydrogenase, ProDH - proline dehydrogenase, Pro - proline, γ-GHM - γ-glutamyl hydroxamate

Kaiser and Förster (1989) showed that lighting leaves in an atmosphere without CO_2 leads to inhibition of nitrate reductase activity, which again increases with the addition of CO_2. The half-time for a 50% change in enzyme activity was 30 min. This implies the presence of a mechanism for rapid inactivation / activation of nitrate reductase in leaves and the close relationship of nitrate reductase with net photosynthesis. However, according to our data, after a long-time incubation of plants in conditions of CO_2 deficiency to restore the activity of nitrate reductase to the initial level, need a time at least more than 3 days from the moment of opening the chamber.

Inhibition of nitrate reductase activity can occur at different levels of regulation. First, this regulation of expression of the enzyme genes, induced both by light and by sugars (Crawford, 1995; Jonassen *et al.*, 2009). On the other hand, the activity of nitrate reductase may decrease under

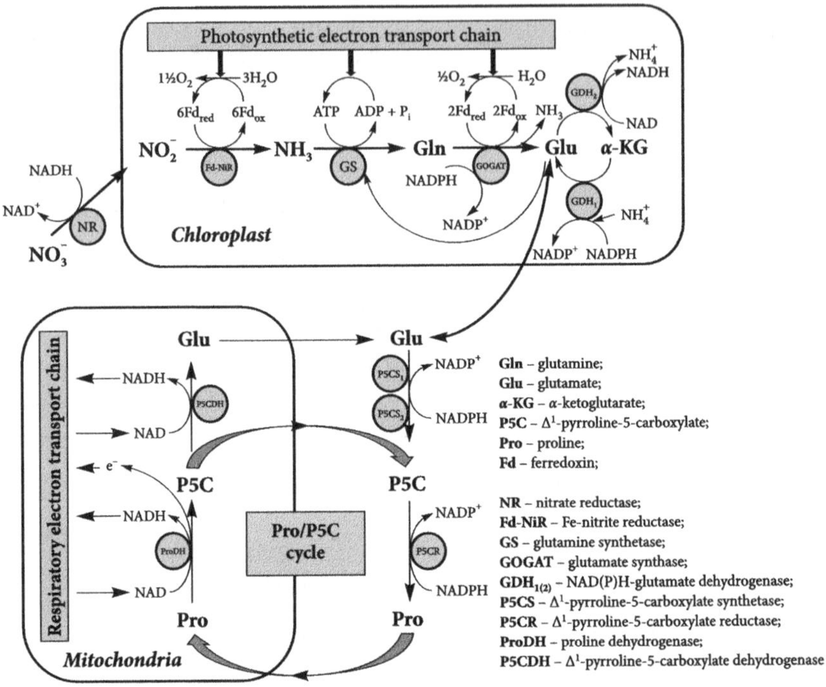

Fig. 9.4. Relationship of carbon and nitrogen metabolism and proline synthesis in plant cells

the influence of reactive oxygen species or the work of specific proteases, as well as with the combined action of these two factors enhancing each other's action (Kenis *et al.*, 1994). In our case, given the long period of recovery of enzyme activity, most likely, there is an inhibition of the expression of the enzyme in conditions of a shortage of carbohydrates, observed with CO_2 deficiency.

The activity of glutamine synthetase and proline dehydrogenase (ProDH; EC 1.5.5.2) increases, the values of which decrease to the level of control plants 3 days after the opening of the chamber, which may indicate a regulatory equalization of the imbalance of enzymatic activities caused by CO_2 deficiency. The activities of nitrite reductase and NAD-dependent glutamate dehydrogenase remain unchanged (see Figure 9.3).

Glutamine synthetase occupies a central place in the path of both primary assimilation of nitrogen and re-assimilation of ammonium produced by photorespiration (Miflin and Habash, 2002), and is the beginning of the synthesis pathway for various nitrogen-containing compounds, for example, amino acids. The activity of glutamine synthetase depends on the environment and the metabolic status of plant tissues that can fluctuate throughout the day (Stitt *et al.*, 2002).

At physiological CO_2 concentration Rubisco catalyzes two reactions in which CO_2 and O_2 compete for one active site of the enzyme with the formation of 3-phosphoglycerate and phosphoglycollate, respectively (Lorimer, 1981). With CO_2 deficiency, the value of the oxygenase function of the enzyme increases, *i.e.* the photorespiration process is intensified, as a result of which an excess amount of ammonia is formed at a rate an order of magnitude higher than when nitrate is reduced in illuminated leaves. To maintain the nitrogen balance of the leaf, it is necessary to re-fix photorespiratory ammonium, otherwise the ammonium concentration may increase to toxic levels (Osmond, 1981). Photorespiratory ammonium, formed in mitochondria of leaf cells, is re-fixed in chloroplasts using the glutamine synthetase / glutamate synthase system, but not glutamate dehydrogenase (Osmond, 1981; Wallsgrove *et al.*, 1983). At the same time, the consumption of additional energy required for ammonia-recirculation in photorespiration has relatively little effect on the stoichiometry of ATP / NADPH and the photosynthetic electron transport chain rate (Osmond, 1981). Since chloroplast glutamine synthetase is the main site of reassimilation photorespiratory ammonium (Wallsgrove *et al.*, 1983), the enhancement of its activity in the case of CO_2 deficiency under the conditions of our experiment seems quite justified (see Figure 9.3).

Along with the above enzymes, a key role in maintaining the carbon / nitrogen balance is also played by NAD-dependent glutamate dehydrogenase, which carries out the deamination of amino acids. Glutamate dehydrogenase forms the shunt to return the carbon from the amino acids back to the reactions of carbon metabolism and the tri-carboxylic acid cycle (Miflin and Habash, 2002). A glutamate dehydrogenase shunt can be particularly important when the plant is limited in carbon. There is evidence that the functioning of glutamate dehydrogenase is not directed towards assimilation of ammonium (Magalhaes *et al.*, 1990). It is more likely that glutamate dehydrogenase is a stress-responsive protein that may reflect an additional / alternative route to the glutamine synthetase / glutamate synthase pathway for ammonia assimilation under intracellular hyperammonia conditions (Skopelitis *et al.*, 2006).

After the incubation of plants under conditions of CO_2 deficiency, in our experiments there was no change in the activity of NAD-dependent glutamate dehydrogenase (see Figure 9.3). As a result, there is no increase in deamination of glutamate, which in this case is used for synthesis, for example proline, whose concentration increased. In contrast, upon the opening of the incubation chamber, the activity of NAD-dependent glutamate dehydrogenase is increased and, accordingly, the breakdown of Glu, entering from both glutamine synthetase / glutamate synthase and returning to the chloroplasts from the proline / Δ^1-pyrroline-5-carboxylate cycle (Pro / P5C cycle) (Ben Rejeb *et al.*, 2014). This process coincides with the observed decrease in the concentration of proline in the leaves on the third day after the opening of the incubation chamber.

With CO_2 deficiency, an increase in proline concentration in leaves occurs (see Figure 9.3). It should be especially emphasized that the value of the accumulation of proline did not depend on the time of incubation of plants at low CO_2 concentration (7, 10 or 14 days, data not shown). Therefore, in this case it is necessary to speak not about the accumulation of proline, but about maintaining the pool of its concentration, when the synthesis is compensated by the utilization. This statement is supported by the observed increase in proline dehydrogenase activity.

In plants, proline metabolism is based on the existence of a proline / Δ^1-pyrroline-5-carboxylate cycle, the functioning and significance of which is described in detail by Miller *et al.* (2009). The accumulation of proline depends both on activation of its biosynthesis, and inhibition of degradation (Ben Rejeb *et al.*, 2014). Biosynthesis proline occurs in the cytosol of plant cells, but under stressful conditions, its synthesis is possible in chloroplasts (Székely *et al.*, 2008). With osmotic stress, proline is synthesized mainly from glutamate (Ben Rejeb *et al.*, 2014). On this path, bifunctional Δ^1-pyrroline-5-carboxylate synthetase (EC 2.7.2.11/1.2.1.41) reduced glutamate to glutamyl-5-semialdehyde, which spontaneously converts to Δ^1-pyrroline-5-carboxylate followed by reduction to proline using Δ^1-pyrroline-5-carboxylate reductase (EC 1.2.1.88). The degradation of proline occurs in the mitochondria as a result of successive reactions of proline dehydrogenase and Δ^1-pyrroline-5-carboxylate dehydrogenase. The key points that limit the rate of biosynthesis and degradation of proline are the activities of Δ^1-pyrroline-5-carboxylate synthetase and proline dehydrogenase, respectively (Ben Rejeb *et al.*, 2014). The synthesis of proline and its use as the final electron acceptor of photosynthetic electron transport chain can reduce the use of O_2 as an electron acceptor in photosynthesis and thus reduce the production of reactive oxygen species and provide some protection against photoinhibition under unfavorable conditions (Hare *et al.*, 1998).

The metabolism of proline is directly related to the NAD(P)H / $NAD(P)^+$ redox balance, and proline itself acts as a redox transporter (Sharma *et al.*, 2011; Giberti *et al.*, 2014; Ben Rejeb *et al.*, 2014). Therefore, an increase in the synthesis of proline under stress can be part of a mechanism that supports the redox cell potential at a level acceptable for metabolism (Hare *et al.*, 1998). Closure of stomata during osmotic stress limits carbon uptake, which subsequently leads to a decrease in NADPH consumption in the Benson–Bassham–Calvin cycle. The use of NAD(P)H for the synthesis of proline during stress may increase the $NAD(P)^+$ / NAD(P)H ratio in the cell. Since it is shown that under stress conditions proline can be synthesized from glutamate in chloroplasts (Székely *et al.*, 2008), the regulation of the $NAD(P)^+$ / NAD(P)H ratio can form a link between the biosynthesis of proline and the reduction of $NAD(P)^+$ with the participation of two dehydrogenases from oxidative pentose phosphate pathway, controlling the conversion of glucose-6-phosphate to ribulose-5-phosphate (Shetty, 2004). In this case, the generation of CO_2 in the synthesis of ribulose-5-phosphate may allow to continue the carbon reduction process at a deficit of exogenous CO_2, and NAD(P)H can be used in the biosynthesis of proline, preventing photoinhibition and the production of excess reactive oxygen species in the chloroplast (Sharma *et al.*, 2011; Ben Rejeb *et al.*, 2014).

On the other hand, in non-photosynthetic plastids, the oxidation of glucose-6-phosphate through oxidative pentose phosphate pathway provides electrons for ferredoxin-dependent nitrite reductase and glutamate synthase (Oji *et al.* 1985, Bowsher *et al.*, 1989, 1992). In this case, the ferredoxin-$NADP^+$ oxidoreductase mediates the ferredoxin reduce through NADPH coming from the oxidative pentose phosphate pathway. Thus, ferredoxin and ferredoxin-$NADP^+$ oxidoreductase play a central role in the allocation of redox equivalents in both types of plastids (Sakakibara, 2003).

Thus, the carbon and nitrogen metabolism are closely interrelated (see Figure 9.4). With CO_2 deficiency, the activation of nitrogen metabolism can lead to an imbalance of the system, for the prevention of which nitrate reductase is primarily inhibited in order to reduce the amount of substrate for the light-dependent utilization of nitrite. The utilization of the already absorbed nitrate occurs via the glutamate-pathway of the synthesis of proline, which in this case acts as a by-product of the "removal" of the excess energy of absorbed photons during photosynthesis, as well as the storage of excess nitrogen in metabolically neutral substances of the proline-cycle. That is, the formation of proline under stress has no purposeful character, but serves as a means of detoxifying cells from excess nitrogen, which is consistent with the report on the possibility of synthesis of proline from glutamate in chloroplasts (Székely *et al.*, 2008). In stress-free conditions, the proline-cycle works as an "idle engine", but under stress, the cycle is broken by inhibition of proline dehydrogenase and the result is observed accumulation of proline. In this case, it is changing the purpose and functions of a given amino acid. These functions depend on the balance of spatial and temporal regulation of the synthesis and catabolism of proline, the binding or release of the reducing agent and energy in the right place and time to meet the needs of the plant, which does not obey the "more-is-better" theory when accumulating proline (Sharma *et al.*, 2011).

9.5.3 Water Stress

Drought is one of the major abiotic stresses that adversely affect the growth, development and productivity of plants (Wang and Gao, 2003; Sicher *et al.*, 2012). Adaptation of plants to drought occurs with the participation of many physiological and biochemical mechanisms covering photosynthesis and carbon-nitrogen metabolism (Robredo *et al.*, 2011). In drought conditions, a number of physiological and biochemical processes are changing, including protein synthesis (Hsiao, 1973, Bradford and Hsiao, 1982), up- and down-regulation of some genes (Bray, 2002), which may be part of the overall adaptive strategy of the plant under water deficit. Some of the up-regulated genes are involved in such processes as signaling, redox homeostasis, chaperone activities, and maintaining the water balance.

9.5.3.1 Osmotic Homeostasis

The drought disrupts the osmotic homeostasis of plant cells, which leads to osmotic stress, and, ultimately, can lead to the death of plants as a result of dehydration of the body (Basu *et al.*, 2007). Plant response to changes in relative water content in leaves depends on the plant species (Hsiao, 1973; Kaiser, 1987; Lawlor, 2002*b*). In some species, the potential photosynthetic rate does not change until the relative water content decreases by 20–30% (Kaiser, 1987; Heitholt *et al.*, 1991; Cornic, 2000; Cornic and Fresneau, 2002); in others the potential photosynthetic rate decreases linearly with a decrease in relative water content, showing progressive metabolic changes, probably related to the limitation of ATP synthesis (Lawlor, 2002*b*). Strong dehydration of mesophyll cells under severe drought can lead to metabolic changes in the leaves (Sharkey and Seeman, 1989) and to enhanced peroxidation of cell membranes (Zhang and Kirham, 1994), as well as damage to chloroplasts and photosynthetic protein functions (Kaiser, 1982; Srivali and Renu, 1998), as a result of which photosynthetic capacity and the net photosynthetic rate is decreased (Kaiser, 1987; Chaves *et al.*, 2003; Marques da Silva and Celeste, 2004; Souza *et al.*, 2004).

To overcome the effects of osmotic stress in plants, compatible solutes are synthesized and accumulated that promote additional water uptake from the environment (Sibout and Guerrier, 1998; Patakas *et al.*, 2002; Basu *et al.*, 2007; Upadhyaya *et al.*, 2013). However, some metabolites may accumulate as a result of inhibiting the pathway for their utilization in drought conditions (Muller *et al.*, 2011). Since most compatible solutes are carbon- and nitrogen-containing compounds, for example carbohydrates, amino acids and proteins (Wang *et al.*, 2012), a change in carbon-nitrogen metabolism is key to plant resistance to drought (Commichau *et al.*, 2006; Basu *et al.*, 2007; Cánchez-Rodríguez *et al.*, 2011; McDowell, 2011; Liu *et al.*, 2014).

The accumulation of certain substances under drought conditions depends on the plant species, but mainly increases the synthesis of nitrogen-containing compounds, especially proline (Gupta *et al.*, 2014). On the contrary, the content of soluble sugars and total nitrogen in leaves is usually reduced (Sinclair *et al.*, 2000; Llorens *et al.*, 2003). Nevertheless, in some plants in drought conditions an increase in the activity of sucrose hydrolysis enzymes, sucrose synthase (EC 2.4.1.13) and invertase, is observed, which may be the reason for the accumulation of hexoses (Praxedes *et al.*, 2006) that participate in the osmotic regulation of plant cells (Yang *et al.*, 2001). The increase in amylase activity during drought also leads to the accumulation of glucose as a consequence of hydrolysis of starch (Zeeman *et al.*, 2004). However, reports of changes of sucrose synthase and invertase activities (González *et al.*, 1998, Praxedes *et al.*, 2006), as well as sucrose-phosphate synthase, a key enzyme for sucrose synthesis (Praxedes *et al.*, 2006; Basu *et al.*, 2007), are rather contradictory. According to Liu *et al.* (2015), under severe drought increases the activity of amylase, which leads to an acceleration of degradation of starch and the accumulation of soluble sugars, bat the sucrose content does not change due to an increase in activities of invertase, sucrose synthase and sucrose-phosphate synthase. At the same time, the production of ammonium increases due to the accelerated hydrolysis of proteins, and the activation of the glutamine synthetase / glutamate synthase cycle stimulates ammonium assimilation, which results in the accumulation of amino acids.

9.5.3.2 *Variability of Plants' Response to Drought*

In conditions of water scarcity, the nitrogen content, net photosynthetic rate, the chlorophyll fluorescence (Delgado *et al.*, 1994, McDonald and Paulsen, 1997), the content of drought induced metabolites (Alves and

Setter, 2004) and the activity of key enzymes carbon and nitrogen metabolism significantly vary depending on the interspecific plant differences (Li M.C. *et al.*, 2009; Sánchez-Rodríguez *et al.*, 2011), the intensity of stress and its duration (Xu and Zhou, 2006*a*, *b*, 2007*a*, *b*), the age of the leaf (Pommel *et al.*, 2006; Xu *et al.*, 2008; Li M.C. *et al.*, 2009) and the plant organ (Thomas and Hilker, 2000; Gouia *et al.*, 2003).

In young leaves, drought does not have a significant effect on the carbohydrate content due to the balanced regulation of the corresponding enzymes. At the same time, nitrate reduction and ammonium assimilation is accelerated to maintain plant growth (Liu *et al.*, 2014). In old leaves, on the contrary, the carbohydrate content is reduced by spurring the activity of the main enzymes of C-metabolism, while nitrate reduction and ammonium assimilation increase only under severe drought. In the roots, carbon metabolism and nitrate reduction remain stable under mild drought, but under severe drought is enhanced ammonium assimilation. Regulation of the metabolic balance of a whole plant under mild drought can be achieved by transferring carbohydrates from old leaves to roots. At the same time, the roots could allocate more nitrogen nutrients for young leaves and less for old leaves. The distinct plant organs, with their unlike physiological strategies, ensure the adaptation of plants to drought of varying intensity for regular growth at the level of the whole plant (Liu *et al.*, 2014).

9.5.3.3 Reactive Oxygen Species

Drought-induced inhibition of photosynthetic carbon fixation during drought due to stomatal restriction (Cornic, 1994; Golding and Johnson, 2003; Zhang *et al.*, 2009) and non-stomatal limitation (Ortiz-Lopez *et al.*, 1991; Jia *et al.*, 2008) may lead to excess light energy (Smirnoff and Colombe, 1988) and the formation of reactive oxygen species (Price and Hendry, 1991, Sgherri *et al.*, 1993; Upadhyaya *et al.*, 2013), which not only cause oxidative damage to the photosynthetic apparatus (Munné-Bosch and Alegre, 2002; Krieger-Liszkay *et al.*, 2008), but also inhibit the reduction cycle (Nishiyama *et al.*, 2001, 2005; Takahashi *et al.*, 2009). In addition, light-induced inactivation of the Mn cluster of water oxidation can occur in parallel with the singlet oxygen-dependent pathway (Tyystjärvi, 2008; Vass, 2011). Drought stress can also cause up-regulation of components involved in Fd-dependent cyclic electron flow that is an important mechanism for protecting photosystem I and photosystem II (Munné-Bosch *et al.*, 2005; Lehtimaki *et al.*, 2010; Huang *et al.*, 2012).

9.5.3.4 *Metabolic Changes*

Drought adversely affects the accumulation of plant biomass, net photosynthetic rate, stomatal conductance, the maximal efficiency of photosystem II photochemistry, the actual quantum yield, and the photochemical quenching (Lu and Zhang, 1999; Chaves *et al.*, 2003; Souza *et al.*, 2004; Xu and Zhou, 2005*a*; Xu and Zhou, 2006*b*). Inhibition of photosynthesis, one of the key carbon metabolic processes, leads to depletion of energy and sugar, which reduces production (Cuellar-Ortiz *et al.*, 2008) and the growth of plants that depend on changes in carbon-nitrogen metabolism (Chaves *et al.*, 2003; Raven *et al.*, 2004).

The restriction of the rate of photosynthetic assimilation of CO_2, with drought, results in a lower rate of nitrogen assimilation and vice versa (Foyer *et al.*, 1998; González *et al.*, 1998; Heitholt *et al.*, 1991; Xu and Zhou, 2006*a*, *b*). Drought inhibits the activity enzymes of nitrogen metabolism, which leads to a change in the content of free amino acids, for example, the accumulation of proline, taking part in osmoregulation of cells (González *et al.*, 1998; Xu and Zhou, 2006*a*, *b*; Robredo *et al.*, 2011). Reduced Rubisco activity and net photosynthetic rate also correlate with lower nitrogen concentrations in leaves (Llorens *et al.*, 2003). On the other hand, a decrease in the nitrogen content (Sinclair *et al.*, 2000, Llorens *et al.*, 2003, Xu and Zhou, 2006*a*) can be closely related to the drought-induced leaf senescence (Evans, 1983; Kichey *et al.*, 2006; Xu And Zhou, 2006*b*; Pommel *et al.*, 2006), accompanied by damage to the components of the photosynthetic apparatus (Kołodziejek *et al.*, 2003). The drought-induced senescence process can cause a decrease in the activity of the key enzymes of assimilation of nitrogen and the enhancement of lipid peroxidation (Xu and Zhou, 2006b), especially in the last stages of plant growth (Evans, 1983; Kichey *et al.*, 2006). Nitrogen-limiting and senescence can enhance the effects of drought, depending on the plant species (Chaves *et al.*, 2003; Pommel *et al.*, 2006).

Dehydration of leaves can be accompanied by an increase in the glycolate pathway. As a signal molecule, may be the low CO_2 concentration in the leaves due to the closure of stomata in the light. Therefore, it should be borne in mind that the metabolic changes observed during a drought may also be a consequence of an increase in photorespiration (Cornic, 1994).

9.5.3.5 *Stomata Conductivity and Rubisco Activity*

The decrease in the photosynthetic rate in conditions of water stress can be due to 'stomatal' or 'non-stomatal' limitation (Flexas and Medrano,

2002; Lawlor and Cornic, 2002) (or 'diffusive' and 'metabolic') (Grassi and Magnani, 2005; Galmés *et al.*, 2007). The stomatal limitation is due to the resistance of diffusion of CO_2 from the atmosphere to the intercellular space of the leaves, while nonstomatal limitation has metabolic constraints. There is evidence that the value of mesophyll conductance to CO_2 is not constant, but may decrease in drought conditions (Flexas *et al.*, 2008; Warren, 2008). The decrease in photosynthesis, induce of light and mild drought, occurs mainly due to a decrease in the stomatal and mesophyll conductance to CO_2 (Chaves *et al.*, 2002; Flexas and Medrano, 2002). However, there is little information on the processes limiting photosynthesis under severe drought, as well as the limitations associated with the restoration of photosynthesis of severely stressed plants after the resumption of irrigation (Galmés *et al.*, 2007; Gallé *et al.*, 2009). Nevertheless, according to Xu and Zhou (2006*b*), the effect of severe drought leads to an increase in C_i, which was due to a decrease in net photosynthetic rate, maximal efficiency of photosystem II photochemistry (F_v / F_m), and the actual quantum yield. This may indicate the fact that the stomatal and non-stomatal limitation of photosynthesis can occur simultaneously.

When drought occurs there is a decrease in Rubisco activity, which may be due to a decrease in the availability of CO_2 caused by low stomatal and mesophyll conductance (Galmés *et al.*, 2011), rather than changing the relative water content (Parry *et al.*, 2002; Flexas *et al.*, 2006*b*). This effect may be due to the indirect effect of stomatal conductance to CO_2 in the leaves, which in turn depends on the rate of assimilation of CO_2 using Rubisco (Flexas *et al.*, 2004). Under drought, because of the stomatal closure, the CO_2 concentration in the leaves decreases, resulting in carboxylation of Rubisco becomes the main metabolic process limiting the speed of photosynthesis (Flexas *et al.*, 2004; Grassi and Magnani, 2005), although drought may also worsen the synthesis of chloroplast ATP and ribulose-1,5-diphosphate regeneration (Tezara *et al.*, 1999).

Rubisco activity depends on the activation of the enzyme in the reaction carbamylation lysyl residue in its active site with CO_2, Rubisco concentration, and concentrations of tight-binding inhibitors (Roy and Andrews, 2000). There is an inverse correlation between the carbamylation state and the Rubisco concentration (Quick *et al.*, 1991; Ruuska *et al.*, 1998; Eichelmann *et al.*, 2009). Under the conditions of water stress, Rubisco concentration decreases (Holaday *et al.*, 1992), but is still present in sufficient excess to not limit photosynthesis (Pickersgill, 1986; Warren and Adams, 2004). Drought also increases the content of tight-binding inhibitors in Rubisco (Parry *et al.*, 2002). However, the change in Rubisco activity under water stress mainly depends on carbamylation and activation state. Under water stress, stomatal and mesophyll conductance and, consequently, CO_2 concentration in leaves decrease, which leads to a decrease in carbamination of Rubisco. Deactivation of Rubisco by loss of carbamate complex in the active center at low intercellular CO_2 concentration was observed in some species, but not in others (Perchorowicz and Jensen, 1983; von Caemmerer and Edmondson, 1986; Sage *et al.*, 1990). Thus, the change in Rubisco activity during a drought is due to the CO_2 starvation and subsequent de-carbamylation of the Rubisco catalytic sites.

Initial activity and active Rubisco showed a threshold relationship with *in vivo* estimated C_c, which corresponds to the theoretical values obtained *in vitro* studies (Galmés *et al.*, 2011). When intercellular CO_2 concentration approximately coincides with the threshold CO_2 concentration at the Rubisco catalytic site, a decrease in maximum rate of carboxylation (V_{cmax}) *in vivo* (Galmés *et al.*, 2007) and Rubisco activity *in vitro* is induced (Flexas *et al.*, 2006*b*). The threshold level of CO_2 concentration at the Rubisco catalytic site causing Rubisco deactivation depends on the characteristics of the leaves and related to the maximum attained for each species under non-stressing conditions. Those species adapted to low CO_2 concentration at the Rubisco catalytic site were more capable of maintaining active Rubisco as drought stress intensified (Galmés *et al.*, 2011).

9.5.3.6 Enzymes of Nitrogen Metabolism

The activity of the main enzymes carbon and nitrogen of metabolism varies under drought conditions (Aranjuelo *et al.*, 2011*b*; Kang *et al.*, 2011), depending on the plant species (Sánchez-Rodríguez *et al.*, 2011). For example, drought adversely affects glutamine synthetase activity (Sibout and Guerrier, 1998; Venekamp, 1989), which may be due to ATP deficiency under water stress conditions. Glutamate dehydrogenase is often considered as an adaptive enzyme, the action of which depends on the source of the carbon skeleton (Lam *et al.*, 1996; Stitt *et al.*, 2002), so glutamate dehydrogenase activity can both be inhibited (Xu and Zhou, 2006*a*) and stimulated (Venekamp, 1989) under water stress.

Xu and Zhou (2006*a*) showed that with severe drought the activity of enzymes of nitrogen anabolism decreases markedly, but the activity of nitrogen catabolism enzymes (asparaginase and endopeptidase) increases, which is accompanied by a decrease in the total nitrogen content in leaves. The main reason for the change in the activity of these key enzymes was initially considered the reduction of the hydration of macromolecules under water deficiency and the increase of the regulatory molecules content of the ammonium assimilation pathways (Bourgeais-Chaillou *et al.*, 1992; Sibout and Guerrier, 1998). However, it was later shown that a change in enzyme activity is not a direct consequence of a decrease in the chemical potential of water in leaves (Fresneau *et al.*, 2007).

9.5.3.6.1 Nitrate Reductase

The nitrate reductase activity (Bardzik *et al.*, 1971, Hsiao, 1973, Shanner and Boyer, 1976; Plaut, 1974; Somers *et al.*, 1983; Mahan *et al.*, 1998; Larsson *et al.*, 1989; Ferrario-Méry *et al.*, 1998; Foyer *et al.*, 1998; Mahan *et al.*, 1998; Baki *et al.*, 2000; Garg *et al.*, 2001; Correia *et al.*, 2005; Xu and Zhou, 2006*a*), and consequently the nitrate reduction (Hsiao, 1973, Kaiser, 1987, Xu and Zhou, 2006*a*) rapidly decreases under water stress conditions. Reduction of nitrate reductase activity can occur as a result of inhibition of enzyme synthesis (Morilla *et al.*, 1973; Plaut, 1974; Kenis and Campbell, 1989). On the other hand, nitrate reductase is easily hydrolyzed by proteases (Kenis and Trippi, 1986, Kenis *et al.*, 1994), the synthesis of which is induced by oxidative stress (Yamaya *et al.*, 1980; Hamano *et al.*, 1983). The ability of nitrate reductase to lose NADH: nitrate reductase activity while retaining other additional activities such as FMNH: nitrate reductase and NADH: cytochrome *c* reductase may be interpreted as an indirect evidence of partial hydrolysis of nitrate reductase (Brown *et al.*, 1981, Hamano *et al.*, 1984, Solomonson *et al.*, 1986). Under conditions of water stress, the decrease in nitrate reductase activity under the action of proteases can be enhanced in the presence of reactive oxygen species, since an increase in the content of free radical scavengers prevents the inactivation of the enzyme from drought (Kenis *et al.*, 1994).

There is a linear relationship between the rate of net CO_2 assimilation and the nitrate reductase activity, showing that there is a close interaction between carbon and nitrogen metabolism (Ferrario-Méry *et al.*, 1998; Foyer *et al.*, 1998; Fresneau *et al.*, 2007). There is also a correlation between a decrease in the nitrate reduction and a low photosynthetic activity (Kaiser and Förster, 1989, Kaiser and Brendle-Behnisch, 1991), which may be a consequence of a decrease in intercellular CO_2 concentration after the drought induced stomata closure (Kaiser and Förster, 1989; Kaiser and Brendel-Benisch, 1991). Possibly, a low CO_2 concentration level modulates the activity of nitrate reductase through the ATP / AMP ratio (Kaiser and Spill, 1991), which is associated with the phosphorylated /

dephosphorylated state of the protein (Huber *et al.*, 1994). However, in a study by Fresneau with co-authors (Fresneau *et al.*, 2007) this statement is questioned.

The decrease in leaf intercellular CO_2 concentration due to drought may be only part of the signal that triggers the nitrate reductase activity reduction mechanism, since restoration of nitrate reductase activity in stressed leaves with elevated CO_2 concentration or after re-watering of the plants occurs only partially. Moreover, after the addition of abscisic acid or after the rapid dehydration of non-stressed leaves, the decrease in nitrate reductase activity was comparable to that after treatment with low CO_2 concentration (Fresneau *et al.*, 2007). Nitrate reductase activity can also be altered by transcription of this enzyme (Ferrario-Méry *et al.*, 1998, Foyer *et al.*, 1998), which may partly be due to a decrease in the availability of soil nitrates under drought conditions (nitrate regulation of nitrate reductase activity) (Correia *et al.*, 2005). Expression of the nitrate reductase gene is controlled by light (Deng *et al.*, 1990), nitrate (Cheng, 1986), cytokinins (Lu *et al.*, 1990) and abscisic acid (Lu J-L. *et al.*, 1992). Stability of nitrate reductase transcripts decreases with nitrate limiting (Srivastava, 1980; Galangau *et al.*, 1988; Ferrario *et al.*, 1995; Ferrario-Méry *et al.*, 1997).

9.5.3.7 Sucrose-Phosphate Synthase

The sucrose-phosphate synthase (SPS; EC 2.4.1.14) activity in drought conditions varies in a contradictory manner depending on the plant species (Praxedes *et al.*, 2006; Basu *et al.*, 2007). The maximum sucrose-phosphate synthase activity may decrease (Vasily and Sharkey, 1989, Vassey *et al.*, 1991, Castrillo, 1992, Pelleschi *et al.*, 1997; Foyer *et al.*, 1998), remain constant (Quick *et al.*, 1989, Zrenner and Stitt, 1991) or even increase (Yang *et al.*, 2002; Niedzwiedz-Siegen *et al.*, 2004; Fresneau *et al.*, 2007). Moreover, the change in maximum sucrose-phosphate synthase activity probably was not associated with relative water content or with a decrease in the water potential of the leaves (Yang *et al.*, 2002; Niedzwiedz-Siegien *et al.*, 2004; Fresneau *et al.*, 2007). As in the case of nitrate reductase, the mechanism controlling the sucrose-phosphate synthase activity depends on the phosphorylated / dephosphorylated of protein state (Pathre *et al.*, 2000).

The effect of water stress on sucrose-phosphate synthase activity may be due to inhibition of photosynthesis due to closure of stomata (Vassey *et al.*, 1991). It was shown that maximum sucrose-phosphate synthase activity decreases with dehydration of leaves at ambient CO_2 concentration, but is restored to its original values when leaf transfer to elevated CO_2 concentration. On the other hand, the incubation of non-stressed leaves at low CO_2 concentration results in a decrease in maximum sucrose-phosphate synthase activity, which, however, is completely restored after incubation with elevated CO_2 concentration. After re-watering plants, the sucrose-phosphate synthase activity can be fully restored (Vassey and Sharkey, 1989, Zrenner and Stitt, 1991) or, conversely, remain at the same level (Fresneau *et al.*, 2007). Vassey *et al.* (1991) showed that maximum sucrose-phosphate synthase activity was sensitive to an atmospheric CO_2 molar ratio. However, according to Fresneau *et al.* (2007), sucrose-phosphate synthase activity could be up-regulated by a high CO_2 molar ratio in dehydrated leaves; it appears that this enzyme is not sensitive to a decrease in CO_2 molar ratio. In some cases, sucrose-phosphate synthase activity increased under limiting substrate conditions (Zrenner and Stitt, 1991; Quick *et al.*, 1989; Foyer *et al.*, 1998). At the same time, Foyer *et al.* (1998) showed that under conditions of drought the sucrose-phosphate synthase mRNA pool slightly decreases.

The change in sucrose-phosphate synthase activity can be induced by abscisic acid synthesized during a drought (Yang *et al.*, 2002; Fresneau *et al.*, 2007). The exogenous abscisic acid fed through the transpiration stream can also increase the activity of sucrose-phosphate synthase (Fresneau *et al.*, 2007).

Changes in sucrose-phosphate synthase activity during drought are usually accompanied by an increase in the content of sucrose (Zrenner and Stitt, 1991; Pelleschi *et al.*, 1997; Ferrario-Méry *et al.*, 1998; Yang *et al.*, 2002; Niedzwiedz-Siegen *et al.*, 2004; Correia *et al.*, 2005), hexose (Keller and Ludlow, 1993; Rodriguez *et al.*, 1993) or fructans in leaves (Kerepesi *et al.*, 2002) that can function as osmoprotectors.

Modulation of sucrose-phosphate synthase activity can perform a compensatory function associated with a change in the size of the carbohydrate pool due to a decreasing photosynthetic rate during a drought (Kerepesi *et al.*, 2002; Goggin and Setter, 2004; Fresneau *et al.*, 2007). Thus, it is possible to explain the species specificity of the change in SPS activity, since different species have different osmoprotectors.

9.5.3.8 Increased Plant Resistance to Drought by Nitrogen Supply

The adverse effects of drought on plant growth can be mitigated by the addition of exogenous nitrogen. With an increase in leaf nitrogen content, the concentration of malondialdehyde decreases, which can contribute to an increase in plant drought resistance by preventing damage to cell membranes and enhancing osmoregulation (Saneoka *et al.*, 2004). On the other hand, nitrogen deficiency can increase the susceptibility of plants to photoinhibition (Lu and Zhang, 2000). Xu and Zhou, (2006*b*) suggested that induced by severe drought a decrease in the photochemical capacity of photosystem II is associated with an increase in lipid peroxidation with nitrogen-deficiency plants.

Low nitrogen addition can increase net photosynthetic rate in stressful conditions of drought (Ashraf and Foolad, 2007; Wu *et al.*, 2008*b*; Nikiema *et al.*, 2012; Yang *et al.*, 2012). However, an additional amount of nitrogen does not always play a positive role in the physiological responses of plants under drought conditions (Otoo *et al.*, 1989; DaMatta *et al.*, 2002) or has no effect on leaf gas exchange (Chen *et al.*, 2005). Excess nitrogen during drought can cause a decrease in photosynthesis by rapid closure of stomata (Morgan, 1986).

Most of the leaf nitrogen is associated with Rubisco in chloroplasts (Makino and Osmond, 1991), so it is not surprising that the addition of nitrogen improves net photosynthetic rate and plant growth (Mo *et al.*, 2008; Yang *et al.*, 2012). Since, in this case, the carbon content decreases insignificantly, an additional amount of nitrogen causes a decrease in the carbon / nitrogen ratio and nitrogen use efficiency (Yang *et al.*, 2012), which can lead to a decrease in metabolite levels and serve as a signal for gene expression (Martin *et al.*, 2002). Between water use efficiency and nitrogen use efficiency, apparently, there exists an entire trade-off, *i.e.* negative correlation between instantaneous water use efficiency (net photosynthetic rate / transpiration rate) and photosynthetic nitrogen use efficiency (Field *et al.*, 1983; Patterson *et al.*, 1997), when increasing the water use efficiency requires a decrease in the nitrogen use efficiency (Reich *et al.*, 1989; Yang *et al.*, 2012).

9.5.4 Salt Stress

High salinity is responsible for reducing plant productivity around the world (Skriver and Mundy, 1990). Reduction of yield arises due to the restriction of growth and development of plants through excessive accumulation of ions, deficiency of water and nutrients (Munns, 2002). To overcome salinity induced water stress, plants adopt different strategies, with a wide range of biochemical, physiological and morphological adaptations (Zhu, 2001). Morphological adaptation includes modification of plant growth and a change in the distribution of assimilates in the roots for the effective use of nutrients in the soil (Zhang and Forde, 1998). The physiological strategy is a higher selectivity for potassium over sodium

(Wilson *et al.*, 2000), an increase in the efficacy of potassium use (Messedi *et al.*, 2004) and the synthesis of organic low molecular weight osmolytes for osmotic protection (Sakamoto *et al.*, 2002). These osmolytes are sugars, polyols, amino acids, tertiary and quaternary ammonium, and tertiary sulphonium compounds (Rhodes and Hanson, 1993).

The elucidation of the mechanisms of plant adaptation to high salinity can promote understanding of the interaction between the carbon and nitrogen metabolism (Popova *et al.*, 2002), since the assimilation of inorganic nitrogen depends on the distribution of carbon compounds derived from carbon metabolism.

9.5.4.1 *Assimilation of Nitrogen in Salinity Conditions*

Unfavorable environmental conditions, including soil salinization, seriously affect nitrogen assimilation in plants. The activity of many plant enzymes is inhibited by strong salt stress (Flowers *et al.*, 1977; Singh and Dubey, 1994). Among nitrogen assimilative enzymes, glutamine synthetase is the most tolerant to sodium chloride, more glutamine synthetase 2 than glutamine synthetase 1 (Kumar and Dubey, 1999). Under conditions of salt stress, glutamine synthetase activity may increase (Misra and Dwivedi, 1990, Chaillou and Guerrier, 1992, Singh and Dubey, 1994), or decrease (Miranda-Ham and Loyola-Vargas, 1988; Chaillou and Guerrier, 1992), depending on the plant species, the growth stage and the type of tissue to be examined.

The main way of synthesis of glutamate is its production in the glutamine synthetase / glutamate synthase cycle. NADH-glutamate dehydrogenase can also catalyze the incorporation of ammonium into glutamate, but the alternative glutamate dehydrogenase pathway is likely function only under adverse conditions (Ferrario-Méry *et al.*, 2002; Stitt *et al.*, 2002; Lu *et al.*, 2005).

At low salinity glutamine synthetase activity is markedly increased. However, at high salinity, glutamine synthetase activity decreases, but NADH-glutamate dehydrogenase activity increases and significant amounts of proline accumulate. Nevertheless, the amount of glutamate increases with both types of salinity. This suggests that glutamine synthetase-catalysis may be the main route of glutamate synthesis at low salinity, while at high salinity glutamate is synthesized predominantly by NADH-glutamate dehydrogenase (Popova *et al.*, 2002; Wang *et al.*, 2007).

The activity of glutamine synthetase is significantly influenced by salinity-induced water stress (Miranda-Ham and Loyola-Vargas, 1988). *In vitro* experiments have shown that osmotic stress caused by polyethylene glycol reduces glutamine synthetase activity by reducing the affinity of the enzyme to its substrate (Kumar and Dubey, 1999). As protectants for glutamine synthetase can act some osmolytes.

With salt stress in plants, the content of ammonium increases due to an increase in photorespiration, since the flow of ammonium from the reduction of nitrate decreases as a result of inhibition of nitrate reductase (Wang *et al.*, 2007). Ammonium can act as a signal responsible for the induction of NADH-glutamate dehydrogenase (Ferrario-Méry *et al.*, 2002) and an increase in the activity of the enzyme (Wang *et al.*, 2007). Accumulation of ammonium also indicates a decrease in the nitrogen assimilation under NaCl stress, which can lead to a decrease in biomass accumulation and soluble protein content.

9.5.4.2 Isocitrate Dehydrogenase and Ferredoxin-Dependent Glutamate Synthase

Carbon skeletons in the form of α-ketoglutarate, required as the primary acceptor for ammonium assimilation and synthesis of glutamate, are formed through oxidative decarboxylation of isocitrate by isocitrate dehydrogenase (ICDH; EC 1.1.1.41/EC 1.1.1.42), which acts as a binding enzyme of carbon and nitrogen metabolism (Henson *et al.* 1986; Chen *et al.*, 1988; Fieuw *et al.*, 1995).

Under conditions of salt stress, the activity of NADP-isocitrate dehydrogenase is regulated due to *de novo* synthesis. With short-time NaCl-stress, the enzyme activity rises, while the long-time stress leads to a decrease in the activity of the enzyme. However, in the vascular tissue there is no increase in the isocitrate dehydrogenase content, which indicates the lack of stimulation of the transfer of nitrogen compounds in the vascular tissue in conditions of salinity (Popova *et al.*, 2002).

In contrast to NADP-isocitrate dehydrogenase, ferredoxin-dependent glutamate synthase activity is up-regulated during the first hours of salt stress (Berteli *et al.*, 1995), but then decreases in the leaves of salt-adapted plants (Lutts *et al.*, 1999; Popova *et al.*, 2002). Apparently, the expression and activity of NADP-isocitrate dehydrogenase and ferredoxin-dependent glutamate synthase, at a high salinity are regulated differently, which indicates an independent regulation of carbon and nitrogen metabolism in salt-stressed plants. However, despite the decrease in the level of transcription and ferredoxin-dependent glutamate synthase activity under long-time salt stress, proline content increases (Popova *et al.*, 2002). These data indicate that ferredoxin-dependent glutamate synthase is involved in the synthesis of proline only during the first hours of salt stress.

9.5.4.3 *Proline Accumulation*

Under the salinity stress, the free proline content increases (Lutts *et al.*, 1999; Lin *et al.*, 2002). Accumulation of proline can serve as an adaptive osmoregulatory mechanism (Kumar *et al.*, 2003) to maintain steady photosynthesis and prevent proteins degradation (Rajasekaran *et al.*, 1997). Proline can also serve as an easily accessible source of nitrogen, carbon and reducing equivalents during plant restoration from stress (Hellmann *et al.*, 2000). Under conditions of salt stress, proline synthesis occurs predominantly from glutamate (Wang *et al.*, 2007). Δ^1-pyrroline-5-carboxylate reductase activity increases significantly only at high salinity, while Δ^1-pyrroline-5-carboxylate synthetase activity does not show a specific response. Therefore, at high salinity, Δ^1-pyrroline-5-carboxylate reductase can be a limiting step in the synthesis of proline from glutamate (Rout and Shaw, 1998; Wang *et al.*, 2007).

According to Popova *et al.* (2002), the salt-induced accumulation of proline is not a consequence of the stimulated biosynthesis, but is due to the inhibition of the decomposition of the amino acid because of a decrease in proline dehydrogenase expression (Verbruggen *et al.*, 1996). The accumulation of proline is also independent of the stimulated expression and activity of the NADP-isocitrate dehydrogenase enzyme (Popova *et al.*, 2002).

9.5.4.4 *Photosynthesis, Photorespiration and Reactive Oxygen Species*

Under conditions of salt stress, the closure of the stomata occurs (Harris and Outlaw, 1991, Fendina *et al.*, 1994), which can lead to a decrease in intercellular CO_2 concentration (Evans and Loreto, 2000;) and therefore photosynthetic assimilatory capacity (Powles and Osmond, 1978; Powles *et al.*, 1979; Osmond *et al.*, 1997; Wingler *et al.*, 2000; Igamberdiev *et al.*, 2001), as well as over-reduction of photosynthetic photon flux density (Osmond and Grace, 1995). Under such conditions, Rubisco acts as an oxygenase and activates photorespiration (Fendina *et al.*, 1994), which can function as a possible way to dissipate excess light energy or reducing power (Osmond and Grace, 1995; Willekens *et al.*, 1997; Brisson *et al.*, 1998). Generated by the photorespiration ammonium is reassimilated glutamine synthetase 2. The intracellular content of ammonium and sodium in stressed plants correlates well with glutamine synthetase 2 levels, overproduction of which can increase the photorespiration potential and improve the salt tolerance of plants (Hoshida *et al.*, 2000).

On the other hand, the photorespiratory pathway is closely related to the regulation of stomatal conductance. There is a balance between stomatal opening, water use efficiency and the rate of photorespiration in plants. The need to maintain CO_2 concentration in chloroplasts requires an increased opening of stomata with a corresponding decrease in water use efficiency, *i.e.* photorespiration can participate in the regulation of the intercellular / ambient CO_2 concentration (C_i / C_a) ratio (Igamberdiev *et al.*, 2004).

In conditions of salt stress, stomata are closed, resulting in a deficit of CO_2 and a decrease in photosynthetic efficiency. Because of this, there is an over-reduction in photosynthetic electron transport chain and the energy of the electrons is transmitted directly to oxygen with the generation of reactive oxygen species (Apel and Hirt, 2004). The increased intake of NADPH and ATP for the formation of phosphoglycerate and CO_2 during photorespiration can help decrease the over-reduction of photosynthetic photon flux density (Kozaki and Takeba, 1996, Hoshida *et al.*, 2000) and the reactive oxygen species generation (Asada, 1994; Gossett *et al.*, 1994; Burdon *et al.*, 1996). The use of electrons from photosynthetic photon flux density for photorespiration and the suppression of electron transfer to oxygen molecules can perform a protective function against oxidative damage to plants under conditions of salt stress (Hoshida *et al.*, 2000).

Ferredoxin may also interfere with the formation of reactive oxygen species, either by returning an excess of reducing equivalents to photosynthetic electron transport chain by cyclic electron flow (Yamamoto *et al.*, 2006), or by activating the malate valve in the cytosol of cells (Scheibe, 2004). The content and level of ferredoxin transcripts are down-regulated under stress, including drought and salinity (Thimm *et al.*, 2001; Zimmermann *et al.*, 2004; Tognetti *et al.*, 2006, 2007). Stress, apparently, affects the stroma redox status and photosynthetic electron transport chain. Thus, light induction and stress repression, could share common ways of sensing and regulating. However, overexpression of ferredoxin in mutants does not contribute to an increase in plant resistance to oxidative stress. Apparently, light and stress-dependent expression of ferredoxin are carried out through various pathways (Ceccoli *et al.*, 2011).

9.6 Conclusion

Numerous studies devoted to the study of the relationship between carbon and nitrogen metabolism showed close coordination of the carbon and nitrogen assimilation in plant leaves. The existence of a carbon / nitrogen balance implies a regulated distribution of carbon compounds between the synthesis of carbohydrates and imino acids, which has been shown in plants growing in conditions of substrate deficiency for carbon or nitrogen metabolism. Perturbations of the carbon / nitrogen balance are particularly noticeable in the plant growing in stress conditions. Control over the distribution between carbohydrates and amino acids are provided by controlling several enzyme activities, and include a variety of metabolic effectors, as well as changes in gene expression. Coordination of carbon pathways and nitrogen assimilation in plants involves concerted actions of multiple signals that cause graded molecular and physiological responses. However, the inclusion of all these regulatory mechanisms is situational in nature and depends on species specificity, type of tissues, age and stage of plant development, type and extent of stress development. The elucidation of regulators is further complicated by the fact that carbon and nitrogen assimilation occurs on a non-competitive basis, and both processes contribute to the growth of plants, the main process in the vital activity of plants. All this implies the absence of a single management center coordinating the maintenance of carbon / nitrogen balance in the plant, or the current level of research allows us to identify only some general patterns, but can not take into account the complexity of the overall physiological state of the plant at every moment of its life.

Abbreviations

Fd – ferredoxin
Fd-NiR – ferredoxin-dependent nitrite reductase
GDH – glutamate dehydrogenase
GS – glutamine synthetase
GOGAT – ferredoxin-dependent glutamate synthase
NR – nitrate reductase

References

Aalto T., Juurola E. A three-dimensional model of CO_2 transport in airspaces and mesophyll cells of a silver birch leaf, *Plant Cell Environ.*, 2002, Vol. 25, p. 1399–1409.

Ainsworth E.A., Bush D.R. Carbohydrate export from the leaf: a highly regulated process and target to enhance photosynthesis and productivity, *Plant Physiol.*, 2011, Vol. 155, p. 64–69.

Ainsworth E.A., Long S.P. What have we learned from 15 years of free-air CO_2 enrichment (FACE)? A meta-analytic review of the responses of photosynthesis, canopy, *New Phytol.*, 2005, Vol. 165, p. 351–371.

Ainsworth E.A., Rogers A. The response of photosynthesis and stomatal conductance to rising [CO_2]: mechanisms and environmental interactions, *Plant Cell Env.*, 2007, Vol. 30, p. 258–270.

Ainsworth E.A., Davey P.A., Bernacchi C.J., Dermody O.C., Heaton E.A., Moore D.J., Morgan P.B., Naidu S.L., Ra H.S.Y., Zhu X.G., Curtis P.S., Long S.P. A metaanalysis of elevated [CO_2] effects on soybean (*Glycine max*) physiology, growth and yield, *Glob. Chan. Biol.*, 2002, Vol. 8, p. 695–709.

Ainsworth E.A., Rogers A., Leakey A.D.B., Heady L.E., Gibon Y., Stitt M. and Schurr U. Does elevated atmospheric [CO_2] alter diurnal C uptake and the balance of C and N metabolites in growing and fully expanded soybean leaves? *J. Exp. Bot.*, 2007, Vol. 58, No 3, p. 579–591.

Ainsworth E.A., Rogers A., Nelson R., Long S.P. Testing the 'source–sink' hypothesis of down-regulation of photosynthesis in elevated [CO_2] in the field with single gene substitutions in *Glycine* max, *Agric. Forest Meteorol.*, 2004, Vol. 122, p. 85–94.

Ainsworth E.A., Rogers A., Vodkin L.O., Walter A., Schurr U. The effects of elevated CO_2 concentration on soybean gene expression. An analysis of growing and mature leaves, *Plant Physiol.*, 2006, Vol. 142, p. 135–147.

Alonso A., Pérez P., Martínez-Carrasco R. Growth in elevated CO_2 enhances temperature response of photosynthesis in wheat, *Physiologia Plantarum.*, 2009, Vol. 135, p. 109–120.

Alonso A., Pérez P., Morcuende R., Martinez-Carrasco R. Future CO_2 concentrations, though not warmer temperatures, enhance wheat photosynthesis temperature responses, *Physiol. Plant.*, 2008, Vol. 132, p. 102–112.

Alves A.A.C., Setter T.M. Abscisic acid accumulation and osmotic adjustment in cassava under water deficit, *Environ. Exp. Bot.*, 2004, Vol. 51, p. 259–271.

Andreeva T.F., Stroganova L.E., Stepanenko S.Y., Maevskaya S.N., Protasova N.N., Murashov I.N. Dependence of photosynthetic apparatus activity and growth processes on the light intensity and concentration of CO_2 under long-term exposure to these factors, *Russ. J. Plant. Physiol.*, 1979, Vol. 26, p. 115–1162.

Apel K., Hirt H. Reactive oxygen species: metabolism, oxidative stress, and signal transduction, *Annu. Rev. Plant Biol.*, 2004, Vol. 55, p. 373–399.

Aranjuelo I., Cabrera-Bosquet L., Morcuende R., Avice J.C., Nogués S., Araus J.L., Martínez-Carrasco R., Pérez P. Does ear C sink strength contribute to overcoming photosynthetic acclimation of wheat plants exposed to elevated CO_2? *J. Exp. Bot.*, 2011*a*, Vol. 62, No 11, p. 3957–3969.

Aranjuelo I., Irigoyen J.J., Sánchez-Díaz M., Nogués S. Carbon partitioning in N_2 fixing Medicago sativa plants exposed to different CO_2 and temperature conditions, *Funct. Plant Biol.*, 2008, Vol. 35, p. 306–317.

Aranjuelo I., Molero G., Erice G., Avice J.C., Nogués S. Plant physiology and proteomics reveals the leaf response to drought in alfalfa (*Medicago sativa L.*), *J. Exp. Bot.*, 2011*b*, Vol. 62, p. 111–123.

Aranjuelo I., Pardo A., Biel C., Savé R., Azcón-Bieto J., Nogués S. Leaf carbon management in slow-growing plants exposed to elevated CO_2, *Glob. Chan. Biol.*, 2009, Vol. 15, p. 97–109.

Aranjuelo I., Pérez P., Hernández L., Irigoyen J.J., Zita G., Martínez-Carrasco R, Sánchez-Díaz M. The response of nodulated alfalfa to water supply, temperature and elevated CO_2: photosynthetic downregulation, *Physiol. Plant.*, 2005, Vol. 123, No 3, p. 348–358.

Aranjuelo I., Sanz-Sáez Á., Jáuregui I., Irigoyen J.J., Araus J.L., Sánchez-Díaz M., Erice G. Harvest index, a parameter conditioning responsiveness of wheat plants to elevated CO_2, *J. Exp. Bot.*, 2013, Vol. 64, No 7, p. 1879–1892.

Asada K. Production and action of active oxygen species in photosynthetic tissues, *Causes of Photooxidative Stress in Plants and Amelioration of Defense Systems*, Foyer C.H., Mullineaux P.M. eds., CRC Press, Boca Raton, FL, 1994, p. 77–104.

Ashraf M., Foolad M.R. Roles of glycine betaine and proline in improving plant abiotic stress resistance, *Environ. Exp. Bot.*, 2007, Vol. 59, p. 206–216.

Aslam M., Huffaker R.C. *In vivo* nitrate reduction in roots and shoots of barley (*Hordeum vulgare L.)* seedlings in light and darkness, *Plant Physiol.*, 1982, Vol. 70, p.1009-1013.

Aslam M., Huffaker R.C., Rains D.W., Rao K.P. Influence of light and ambient carbon dioxide concentrations on nitrate assimilation by intact barley seedlines, *Plant Physiol.*, 1979, Vol. 63, p. 1205–1209.

Auerswald K., Wittmer M.H.O.M., Bai Y.F., Yang H., Taube F., Susenbeth A., Schnyder H. C_4 abundance in an Inner Mongolia grassland system is driven by temperature-moisture interaction, not grazing pressure, *Basic Appl. Ecol.*, 2012, Vol. 13, p. 67–75.

Azcón-Bieto J., Osmond C.B. Relationship between photosynthesis and respiration. The effect of carbohydrate status on the rate of CO_2 production by respiration in darkened and illuminated wheat leaves, *Plant Physiol.*, 1983, Vol. 71, p. 574–581.

Bachmann M., Huber J.L., Liao P.C., Gage D. A., Huber S.C. The inhibitor protein of phosphorylated nitrate reductase from spinach (*Spinacea oleracea*) leaves is a 14-3-3 protei, *FEBS Lett.*, 1996, Vol. 387, p. 127–131.

Backhausen J.E., Scheibe R. Adaptation of tobacco plants to elevated CO_2: influence of leaf age on changes in physiology, redox states and NADP-malate dehydrogenase activity, *J. Exp. Bot.*, 1999, Vol. 50, No 334, p. 665–675.

Backhausen J.E., Emmerlich A., Holtgrefe S., Horton P., Nast G., Rogers J.J.M., Müller-Röber B., Scheibe R. Isolated intact chloroplasts and whole leaves from transgenic potato plants with altered levels of chloroplast NADP-MDH: interactions between photosynthetic electron transport and malate metabolism, *Planta*, 1998, Vol. 207, p. 105–114.

Backhausen J.E., Kitzmann C., Scheibe R. Competition between electron acceptors in photosynthesis: regulation of the malate valve during CO_2 fixation and nitrite reduction, *Photosynth. Res.*, 1994, Vol. 42, p. 75–86.

Badger M.R., von Caemmerer S., Ruuska S., Nakano H. Electron flow to oxygen in higher plants and algae: rate and control of direct photoreduction (Mehler reaction) and Rubisco oxygenase, *Philosophical Transactions of the Royal Society of London*, Series B, 2000, Vol. 355, p. 1433–1446.

Baki G.K.A-E., Siefritz F., Man H-M., Weiner H., Kaldenhoff R., Kaiser W.M. Nitrate reductase in Zea mays L. under salinity, *Plant Cell Environ.*, 2000, Vol. 23, p. 515–521.

Bao A., Zhao Z., Ding G., Shi L., Xu F., Cai H. The stable level of glutamine synthetase 2 plays an important role in rice growth and in carbon-nitrogen metabolic balance, *Int. J. Mol. Sci.*, 2015, Vol. 16, p. 12713–12736.

Bardzik J.M., Marsh H.V., Havis J.R. Effects of water stress on the activities of three enzymes in maize seedlings, *Plant Physiol.*, 1971, Vol. 47, p. 828–831.

Basu P.S., Ali M., Chaturvedi S.K. Osmotic adjustment increases water uptake, remobilization of assimilates and maintains photosynthesis in chickpea under drought, *Ind. J. Exp. Biol.*, 2007, Vol. 45, p. 261–267.

Becker T.W., Perrot-Rechenmann C., Suzuki A., Hirel B. Subcellular and immunocytochemical localization of the enzymes involved in ammonia assimilation in mesophyll and bundle-sheath cells of maize leaves, *Planta,* 1993, Vol. 191, p. 129–136.

Beckmann K., Dzuibany C., Biehler K., Fock H., Hell R., Mige A., Becker T.W. Photosynthesis and fluorescence quenching, and the mRNA levels of plastidic glutamine synthetase or of mitochondrial serine hydroxymethyltransferase (SHMT) in the leaves of the wild-type and of the SHMT-deficient *stm* mutant of Arabidopsis thaliana in relation to the rate of photorespiration, *Planta,* 1997, Vol. 202, p. 379–386.

Beerling D.J., Osborne C.P., Chaloner W.G. Evolution of leaf-form in land plants linked to atmospheric CO_2 decline in the Late Palaeozoic era, *Nature,* 2001, Vol. 410, p. 352–354.

Beevers L., Hageman R.H. The role of light in nitrate metabolism in higher plants, *Photophysiology,* Giese A. ed., Academic Press, New York , 1972, C 7, p. 85–113.

Ben Rejeb K., Abdelly C., Savouré A. How reactive oxygen species and proline face stress together, *Plant Physiol.Bioch.,* 2014, Vol. 80, p. 278–284.

Bendall D.S., Manasse R.S. Cyclic photophosphorylation and electron transport, *Biochim. Biophys. Acta,* 1995, Vol. 1229, p. 23–38.

Bernacchi C.J., Cloeman J.S., Bazzaz F.A., McConnaughay K.D.M. Biomass allocation in old-field annual species grown in elevated CO_2 environments: no evidence for optimal partitioning, *Global Change Biol.,* 2000, Vol. 6, p. 855–863.

Bernacchi C.J., Portis A.R., Nakano H., von Caemmerer S., Long S.P. Temperature response of mesophyll conductance. Implications for the determination of Rubisco enzyme kinetics and for limitations to photosynthesis *in vivo, Plant Physiol.,* 2002, Vol. 130, p. 1992–1998.

Bernard S.M., Moller A.L., Dionisio G., Kichey T., Jahn T.P., Dubois F., Baudo M., Lopes M.S., Tercé-Laforgue T., Foyer C.H., Parry M., Forde B.G., Araus J.L., Hirel B., Schjoerring J.K., Habash D.Z. Gene expression, cellular localisation and function of glutamine synthetase isozymes in wheat (*Triticum aestivum L.*), *Plant Mol. Biol.,* 2008, Vol. 67, p. 89–105.

Berteli F., Corrales E., Ariza M.J., Pliego F., Valpuesta V. Salt stress increases ferredoxin-dependent glutamate synthase activity and protein level in the leaves of tomato, *Physiol. Plant.,* 1995, Vol. 93, p. 259–264.

Besford R.T., Ludwig L.J., Withers A.C. The greenhouse effect: acclimation of tomato plants growing in high CO_2, photosynthesis and ribulosebisphosphate carboxylase protein, *J. Exp. Bot.*, 1990, Vol. 40, p. 925–931.

Blackwell R.D., Murray A.J.S., Lea, P.J. Inhibition of photosynthesis in barley with decreased levels of chloroplastic glutamine synthetase activity, *J. Exp. Bot.*, 1987, Vol. 38, p. 1799–1809.

Blackwell R.D., Murray A.J.S., Lea P.J., Joy K.W. Photorespiratory amino donors, sucrose synthesis and the induction of CO_2 fixation in barley deficient in glutamine synthetase and / or glutamate synthase, *J. Exp. Bot.*, 1988, Vol. 39, p. 845–858.

Bloom A.J., Chapin F.S., Mooney H.A. Resource limitations in plants-an economic analogy, *Annu. Rev. Ecol. Syst.*, 1985, Vol. 16, p. 363–392.

Bloom A.J. Photorespiration and nitrate assimilation: a major intersection between plant carbon and nitrogen, *Photosynth. Res.*, 2015, Vol. 123, p. 117–128.

Bloom A.J., Burger M., Kimball B.A, Pinter P.J., Jr. Nitrate assimilation is inhibited by elevated CO_2 in field-grown wheat, *Nat. Clim. Change*, 2014, Vol. 4, p. 477–480.

Bloom A.J., Burger M., Rubio-Asensio J.S, Cousins A.B. Carbon dioxide enrichment inhibits nitrate assimilation in wheat and *Arabidopsis*, *Science*, 2010, Vol. 328, p. 899–903.

Bloom A.J., Rubio-Asensio J.S., Randall L., Rachmilevitch S., Cousins A.B., Carlisle E.A. CO_2 enrichment inhibits shoot nitrate assimilation in C_3 but not C_4 plants and slows growth under nitrate in C_3 plants, *Ecology*, 2012, Vol. 93, p. 355–367.

Bloom A.J., Smart D.R., Nguyen D.T, Searles P.S. Nitrogen assimilation and growth of wheat under elevated carbon dioxide, *Proc. Natl. Acad. Sci. USA*, 2002, Vol. 99, p. 1730–1735.

Bourgeais-Chaillou P., Perez-Alfocea F., Guerrier G. Compared effects of NO_3, NH_4 and NO_3:NH_4 sources on growth and physiological responses of soybean exposed to NaCl stress, *J. Exp. Bot.*, 1992, Vol. 43, p. 1225–1233.

Bowes G. Facing the inevitable: plants and increasing atmospheric CO_2. *Annu. Rev. Plant Physiol. Plant Mol. Biol.*, 1993, Vol. 44, p. 309–332.

Bowsher C.G., Boulton E.L., Rose J., Nayagam S., Emes M.J. Reductant for glutamate synthase is generated by the oxidative pentose phosphate pathway in non-photosynthetic root plastids, *Plant J.*, 1992, Vol. 2, p. 893–898.

Bowsher C.G., Hucklesby D.P., Emes M.J. Nitrite reduction and carbohydrate metabolism in plastids purified from roots of *Pisum sativum L*, *Planta*, 1989, Vol. 177, p. 359–366.

Bradford K.J., Hsiao T.C. Physiological responses to moderate water stress, *Encyclopedia of Plant Physiology New Series. Physiological Plant Ecology II*, Lange O.R., Nobel P.S., Osmond C.B., Ziegler H. eds., Springer-Verlag, Berlin, 1982, Vol. 12B, p. 264–324.

Bray E.A. Classification of genes differentially expressed during water-deficit stress in Arabidopsis thaliana: an analysis using microarray and differential expression data, *Ann. Bot.*, 2002, Vol. 89, p. 803–811.

Brisson L.F., Zelitch I., Havi E.A. Manipulation of catalase levels produces altered photosynthesis in transgenic tobacco plants, *Plant Physiol.*, 1998, Vol. 116, p. 259–269.

Brown J., Small I.S., Wray J.L. Age-dependent conversion of nitrate reductase to cytochrome c reductase species in barley leaf extracts, *Phytochem.*, 1981, Vol. 20, p. 389–398.

Brueck H., Senbayram M. Low nitrogen supply decreases water-use efficiency of oriental tobacco, *J. Plant Nutr. Soil Sci.*, 2009, Vol. 172, p. 216–223.

Burdon R.H., O'Kane D., Fadzillah N., Gill V., Boyd P.A., Finch R.R. Oxidative stress and responses in *Arabidopsis thaliana* and *Oryza sativa* subjected to chilling and salinity stress, *Biochem. Soc. Trans.*, 1996, Vol. 24, p. 468–472.

Bush D.R. Amino acid transport, *Plant Amino Acids. Biochemistry and Biotechnology*, Singh B.K. ed., Inc, American Cyanamid Company, Marcel Dekker, Princeton, New Jersey, New York, Basel, Hong Kong, 1999, p. 305–318.

Cabello P., de la Haba P., Maldonado J.M. Isoforms of glutamine synthetase in cotyledons, leaves and roots of sunflower plants, *J. Plant Physiol.*, 1991, Vol. 137, p. 378–380.

Cai H., Xiao J., Zhang Q., Lian X. Co-suppressed glutamine synthetase 2 gene modifies nitrogen metabolism and plant growth in rice, *Chin. Sci. Bull.*, 2010, Vol. 55, p. 823–833.

Calza R., Huttner E., Vincerntz M., Rouze P., Galangau F., Vaucheret H., Cheral I., Meyer C., Kronenberger J., Caboche M. Cloning of DNA fragments complementary to tobacco nitrate reductase mRNA and encoding epitopes common to the nitrate reductase from higher plants, *Mol. Gen. Genet.*, 1987, Vol. 209, p. 552–562.

Campbell W.H. Structure and regulation of nitrate reductase in higher plants, *Molecular and Genetic Aspects of Nitrate Assimilation*, Wray J.L., Kinghorn J.R. eds., Oxford Sci. Publ., Oxford, U.K., 1989, p. 125–154.

Campbell W.H., Kinghorn J.R. Functional domains of assimilatory nitrate reductases and nitrite reductases, *TIBS*, 1990, Vol. 1990, p. 315–319.

Carlisle E., Myers S., Raboy V., Bloom A. The effects of inorganic nitrogen form and CO_2 concentration on wheat yield and nutrient accumulation and distribution, *Front. Plant Sci.*, 2012, Vol. 3, p. 1–13.

Carvalho J.D.C., Madgwick P.J., Powers S.J., Keys A.J., Lea P.J., Parry M.A.J. An engineered pathway for glyoxylate metabolism in tobacco plants aimed to avoid the release of ammonia in photorespiration. *BMC Biotechnology*, 2011, Vol. 11, p. 111.

Castrillo M. Sucrose metabolism in bean plants under water deficit, *J. Exp. Bot.*, 1992, Vol. 43, p. 1557–1561.

Ceccoli R.D., Blanco N.E., Medina M., Carrillo N. Stress response of transgenic tobacco plants expressing a cyanobacterial ferredoxin in chloroplasts, *Plant Mol. Biol.*, 2011, Vol. 76, p. 535–544.

Cen Y-P., Turpin D.H., Layzell D.B. Whole-plant gas exchange and reductive biosynthesis in white lupin, *Plant Physiol.*, 2001, Vol. 126, p. 1555–1565.

Centritto M., Loreto F., Chartzoulakis K. The use of low [CO_2] to estimate diffusional and non-diffusional limitations of photosynthetic capacity of salt-stressed olive saplings, *Plant Cell Environ.*, 2003, Vol. 26, p. 585–594.

Ceulemans R., Taylor G., Bosac C., Wilkins D., Besford R.T. Photosynthetic acclimation to elevated CO_2 in poplar grown in glasshouse cabinets or in open top chambers depends on duration of exposure, *J. Exp. Bot.*, 1997, Vol. 48, p. 1681–1689.

Chaillou P.B., Guerrier G. Salt-responses in *lycopersicon esculentum* calli and whole plants, *J. Plant Physiol.*, 1992, Vol. 140, No 4, p. 494–501.

Chaves M.M., Maroco J.P., Pereira J.S. Understanding plant responses to drought – from genes to the whole plant, *Funct. Plant Biol.*, 2003, Vol. 30, p. 239–264.

Chaves M.M., Pereira J.S., Maroco J., Rodrigues M.L., Ricardo C.P.P., Osório M.L., Clávalo I., Faria T., Pinheiro C. How plants cope with water stress in the field. Photosynthesis and growth, *Ann. Bot.*, 2002, Vol. 89, p. 907–916.

Chen S.P., Bai Y.F., Zhang L.X., Han X.G. Comparing physiological responses of two dominant grass species to nitrogen addition in Xilin River Basin of China, *Environ. Exp. Bot.*, 2005, Vol. 53, p. 65–75.

Chen R., Le Marerchal P., Vidal J., Jacquot J.P., Gadal P. Purification and comparative properties of the cytosolic isocitrate dehydrogenases (NADP) from pea (*Pisum sativum*) roots and green leaves, *Eur. J. Biochem.*, 1988, Vol. 175, p. 565–572.

Chen Y-Q., Yi F., Cai M., Luo J-X. Effects of amino acids, nitrate, and ammonium on the growth and taxol production in cell cultures of Taxus yunnanensis, *Plant Growth Regul.*, 2003, Vol. 44, p. 265–268.

Cheng C.L. Cloning and nitrate induction of nitrate reductase mRNA, *Proc. Natl. Acad. Sci. USA*, 1986, Vol. 83, p. 6835–6828.

Cheng C-L., Acedo G.N., Cristinsin M., Conkling M.A. Sucrose mimics the light induction of Arabidopsis nitrate reductase gene transcription, *Proc. Natl. Acad. Sci. USA*, 1992, Vol. 89, p. 1861–1864.

Cheng L.L., Fuchigami L.H. Rubisco activation state decreases with increasing nitrogen content in apple leaves, *J. Exp. Bot.*, 2000, Vol. 51, p. 1687–1694.

Chu Z., Peng K., Zhang L., Ma J.S., Orienter H. Construction and characterization of a normalized whole-life-cycle cDNA library of rice, *Chinese Sci. Bul.*, 2003, Vol. 48, p. 229–235.

Cock J.M., Brock I.W., Watson A.T., Swarup R., Morby A.P., Cullimore J.V. Regulation of glutamine synthetase genes in leaves of Phaseolus vulgaris, *Plant Mol. Biol.*, 1991, Vol. 17, p. 761–771.

Commichau F.M., Forchhammer K., Stülke J. Regulatory links between carbon and nitrogen metabolism, *Curr. Opin. Microbiol.*, 2006, Vol. 9, No 2, p. 167–172.

Conroy J., Hocking P. Nitrogen nutrition of C_3 plants at elevated atmospheric CO_2 concentrations, *Physiol. Plant.*, 1993, Vol. 89, p. 570–576.

Cornic G. Drought stress and high light effects on leaf photosynthesis, *Photoinhibition of Photosynthesis*, Baker N., Bowyer J.S. eds., Bios Scientific Publishers, Oxford, 1994, p. 297–313.

Cornic G. Drought stress inhibits photosynthesis by decreasing stomatal aperture – not by affecting ATP synthesis, *Trends Plant Sci.*, 2000, Vol. 5, p. 187–188.

Cornic G., Fresneau C. Photosynthetic carbon reduction and carbon oxidation cycles are the main sinks for photosystem II activity during a mild drought, *Ann. Bot.*, 2002, Vol. 89, p. 887–894.

Correia M.J., Fonseca F., Azedo-Silva J., Dias C., David M.M., Barrote I., Osório M.L., Osório J. Effects of water deficit on the activity of nitrate reductase and content of sugars, nitrate and free amino acids in the leaves and roots of sunflower and white lupin plants growing under two nutrient supply regimes, Physiol. Plant., 2005, Vol. 124, p. 61–70.

Correll C.C., Batie J.C., Ballou D.P., Ludwig M.L. Phtalate dioxygenase reductase: a modular structure for electron transfer from pyridine nucleotides to [2Fe-2S], *Science,* 1992, Vol. 258, p. 1604–1610.

Correll C.C., Ludwig M.L., Bruns C.M., Karplus P.A. Structural prototypes for an extended family of flavoprotein reductases: comparison of phtalate dioxygenase with ferredoxin reductase and ferredoxin, *Protein Sci.,* 1993, Vol. 2, p. 2112–2133.

Coruzzi G.M., Zhou L. Carbon and nitrogen sensing and signaling in plants: Emerging "matrix effects", *Curr. Opin. Plant Biol.,* 2001, Vol. 4, p. 247–253.

Coschigano K.T., Melo-Oliveira R., Lim J., Coruzzi G.M. Arabidopsis gls mutants and distinct Fd-GOGAT genes: implication for photorespiration and primary nitrogen assimilation, *Plant Cell,* 1998, Vol. 10, p. 741–752.

Cotrufo M.F., Ineson P., Scott A. Elevated CO_2 reduces the nitrogen concentration of plant tissues, *Global Change Biol.,* 1998, Vol. 4, p. 43–54.

Crawford N.M. Nitrate nutrient and signal for plant growth, *Plant Cell,* 1995, Vol. 7, p. 859–868.

Cren M., Hirel B. Glutamine synthetase in higher plants: regulation of gene and protein expression from the organ to cell, *Plant Cell Physiol.*, 1999, Vol. 40, p. 1187–1193.

Cuellar-Ortiz S.M., Arrieta-Montiel M.D.L.P., Acosta-Gallegos J., Covarrubias A.A. Relationship between carbohydrate partitioning and drought resistance in common bean, Plant Cell Environ., 2008, Vol. 31, No 10, p. 1399–1409.

Curtis P.S., Vogel C.S., Wang X., Pregitzer K.S., Zak D.R., Lussenhop J., Kubiske M., Teeri J.A. Gas exchange, leaf nitrogen, and growth efficiency of Populus tremuloides in a CO_2-enriched atmosphere, *Ecol. Appl.,* 2000, Vol. 10, p. 3–17.

Dai Z., Ku M.S.B., Edwards G.E. C_4 photosynthesis: the CO_2-concentrating mechanism and photorespiration, *Plant Physiol.,* 1993, Vol. 103, p. 83–90.

Dalling M.J., Tolbert N.E., Hageman R.H. Intraeellular location of nitrate reductase and nitrite reductase, I. Spinach and tobacco leaves, *BBA*, 1972, Vol. 283, p. 505–512.

DaMatta F.M., Loos R.A., Silva E.A., Loureiro M.E., Ducatti C. Effects of soil water deficit and nitrogen nutrition on water relations and photosynthesis of pot-grown Coffea canephora Pierre, *Trees*, 2002, Vol. 16, p. 555–558.

Davey P.A., Hunt S., Hymus G.J., DeLucia E.H., Drake B.G., Karnosky D.F., Long S.P. Respiratory oxygen uptake is not decreased by an instantaneous elevation of [CO_2], but is increased with long-term growth in the field at elevated [CO_2], *Plant Physiol.*, 2004, Vol. 134, p. 520–527.

Delgado E., Mitchell R.A.C., Parry M.A.J., Driscoll S.P., Mitchell V.J., Lawlor D.W. Interacting effects of CO_2 concentration, temperature and nitrogen supply on photosynthesis and composition of winter wheat leaves, *Plant Cell Environ.*, 1994, Vol. 17, p. 1205–1213.

Deng M-D., Moureaux T., Leydecker M-T., Caboche M. Nitrate reductase expression is under the control of a circadian rhythm and is light inducible in Nicotiana tabacum leaves, *Planta*, 1990, Vol. 180, p. 257–261.

Deng Y.-W., Zhang Y-D., Chen Y., Wang S., Tang D-M., Huang D-F. Isolation and characterization of a GS2 gene in melon (*Cucumis melo L.*) and its expression patterns under the fertilization of different forms of N, *Mol. Biotechnol.*, 2010, Vol. 44, p. 51–60.

Doheny-Adams T., Hunt L., Franks P.J., Beerling D.J., Gray J.E. Genetic manipulation of stomatal density influences stomatal size, plant growth and tolerance to restricted water supply across a growth carbon dioxide gradient, *Philosophical Transactions of the Royal Society B*, 2012, Vol. 367, p. 547–555.

Drake B.G., Gonzàlez-Meler M.A., Long S.P. More efficient plants: a consequence of rising atmospheric CO2? *Annu. Rev. Plant Physiol. Plant Mol. Biol.*, 1997, Vol. 48, p. 609–639.

Eaton-Rye J., Govidjee. Electron transfer through the quinone acceptor complex of photosystem II after one or two actinic flashes in bicarbonate-depleted spinach thylakoid membranes, *Biochem. Biophys. Acta*, 1988, Vol. 935, p. 248–257.

Eckes P., Schmitt P., Daub W., Wegenmayer F. Overproduction of alfalfa glutamine synthetase in transgenic tobacco, *Mol. Gen. Genet.*, 1989, Vol. 217, p. 263–268.

Edwards J.W., Coruzzi G.M. Photorespiration and light act in concert to regulate the expression of the nuclear gene for chloroplast glutamine synthetase, Plant Cell, 1989, Vol. 1, p. 241–248.

Eichelmann H., Laisk A. Ribulose-1,5-bisphosphate carboxylase / oxygenase content, assimilatory charge, and mesophyll conductance in leaves, *Plant Physiol.*, 1999, Vol. 119, p. 179–189.

Eichelmann H., Oja V., Peterson R.B., Laisk A. The rate of nitrite reduction in leaves as indicated by O_2 and CO_2 exchange during photosynthesis, *J. Exp. Bot.*, 2011, Vol. 62, No 6, p. 2205–2215.

Eichelmann H., Talts E., Oja V., Padu E., Laisk A. Rubisco in planta kcat is regulated with photosynthetic electron transport, *J. Exp. Bot.*, 2009, Vol. 60, p. 4077–4088.

Ellsworth D.S., Reich P.B., Naumberg E.S., Koch G.W., Kubiske M.E., Smith S.D. Photosynthesis, carboxylation and leaf nitrogen responses of 16 species to elevated pCO_2 across four free-air CO_2 enrichment experiments in forest, grassland and desert, *Global Change Biol.*, 2004, Vol. 10, p. 2121–2138.

Epron D., Godard D., Cornic G., Genty B. Limitation of net CO_2 assimilation rate by internal resistance to CO_2 transfer in the leaves of two tree species (*Fagus sylvatica L.* and *Castanea sativa Mill.*), *Plant Cell Env.*, 1995, Vol. 18, p. 43–51.

Ethier G.J., Livingston N.J. On the need to incorporate sensitivity to CO_2 transfer conductance into the Farquhar–von Caemmerer–Berry leaf photosynthesis model, *Plant Cell Env.*, 2004, Vol. 27, p. 137–153.

Evans J.R. Nitrogen and photosynthesis in the flag leaf of wheat (*Triticum aestivum L.*), *Plant Physiol.*, 1983, Vol. 72, p. 297–302.

Evans J.R. Photosynthesis and nitrogen relationship in leaves of C_3 plants, *Oecologia*, 1989, Vol. 78, p. 9–19.

Evans J.R., Loreto F. Acquisition and diffusion of CO_2 in higher plant leaves, *Photosynthesis: Physiology and Metabolism*, Leegood R.C., Sharkey T.D., von Caemmerer S. eds., Kluwer Academic Publishers, Dordrecht, Netherlands, 2000, p. 321–351.

Evans J.R., Kaldenhoff R., Genty B., Terashima I. Resistances along the CO_2 diffusion pathway inside leaves, *J. Exp. Bot.*, 2009, Vol. 60, p. 2235–2248.

Evans J.R., Sharkey T.D., Berry J.A., Farquhar G.D. Carbon isotope discrimination measured concurrently with gas exchange to investigate CO_2 diffusion in leaves of higher plants, *Aust. J. Plant Physiol.*, 1986, Vol. 13, p. 281–292.

Evans J.R., von Caemmerer S., Setchell B.A., Hudson G.S. The relationship between CO_2 transfer conductance and leaf anatomy in transgenic tobacco with a reduced content of Rubisco, *Aust. J. Plant Physiol.*, 1994, Vol. 21, p. 475–495.

Ezquer I., Li J., Ovecka M., Baroja-Fernández E., Muñoz F.J., Montero M., de Cerio J.D., Hidalgo M., Sesma M.T., Bahaji A., Etxeberria E., Pozueta-Romero J. A suggested model for potato MIVOISAP involving functions of central carbohydrate and amino acid metabolism, as well as actin cytoskeleton and endocytosis, *Plant Signal. Behav.*, 2010*a*, Vol. 5, p. 1638–1641.

Ezquer I., Li J., Ovecka M., Baroja-Fernández E., Muñoz F.J., Montero M., de Cerio J.D., Hidalgo M., Sesma M.T., Bahaji A., Etxeberria E., Pozueta-Romero J. Microbial volatile emissions promote accumulation of exceptionally high levels of starch in leaves in mono- and dicotyledonous plants, *Plant Cell Physiol.*, 2010*b*, Vol. 51, p. 1674–1693.

Farquhar G.D., Firth P.M., Wetselaar R., Weir B. On the gaseous exchange of ammonia between leaves and the environment: Determination of the ammonia compensation point, *Plant Physiol.*, 1980*a*, Vol. 66, p. 710–714.

Farquhar G.D., von Caemmerer S., Berry J.A. A biochemical model of photosynthetic CO_2 assimilation in leaves of C_3 species, *Planta*, 1980*b*, Vol. 149, p. 78–90.

Faske M., Backhausen J.E., Sendker M., Singer-Bayerle M., Scheibe R., von Schaewen A. Transgenic tobacco plants expressing pea chloroplast *Nmdh* cDNA in sense and antisense orientation: effects on NADP-malate dehydrogenase level, stability of transformants, and plant growth, *Plant Physiol.*, 1997, Vol. 115, p. 705–715.

Fendina I.S., Tsonev T.D., Guleva E.I. ABA as a modulator of the response of *Pisum sativum* to salt stress, *J. Plant Physiol.*, 1994, Vol. 143, p. 245–249.

Feraud M., Masclaux-Daubresse C., Ferrario-Mérry S., Pageau K., Lelandais M., Ziegler C., Leboeuf E., Jouglet T., Viret L., Spampinato A., Paganelli V., Hammouda M.B., Suzuki A. Expression of a ferredoxin-dependent glutamate synthase gene in mesophyll and vascular cells and functions of the enzyme in ammonium assimilation in *Nicotiana tabacum* (L.), *Planta*, 2005, Vol. 222, p. 667–677.

Ferrario S., Valadier M-H., Morot-Gaudry J-F., Foyer C.H. Effects of constitutive expression of nitrate reductase in transgenic *Nicotiana plumbaginifolia L.* in response to varying nitrogen supply, *Planta*, 1995, Vol. 196, p. 288–294.

Ferrario-Méry S., Hodges M., Hirel B., Foyer C.H. Photorespiration-dependent increases in phosphoenol pyruvate carboxylase, isocitrate dehydrogenase and glutamate dehydrogenase in transformed tobacco plants deficient in ferredoxin-dependent glutamine–a-ketoglutarate aminotransferase, *Planta,* 2002, Vol. 214, p. 877–886.

Ferrario-Méry S., Masclaux C., Suzuki A., Valadier M-H., Hirel B., Foyer C.H. Glutamine and α-ketoglutarate are metabolite signals involved in nitrate reductase gene transcription in untransformed and transformed tobacco plants deficient in ferredoxin-glutamine-α-ketoglutarate aminotransferase, *Planta,* 2001, Vol. 213, p. 265–271.

Ferrario-Méry S., Suzuki A., Kunz C., Valadier M.H., Roux Y., Hirel B., Foyer C.H. Modulation of amino acid metabolism in transformed tobacco plants deficient in Fd-GOGAT, *Plant Soil,* 2000, Vol. 221, p. 67–79.

Ferrario-Méry S., Thibaud M-C., Betsche T., Valadier M-H., Foyer C.H. Modulation of carbon and nitrogen metabolism, and of nitrate reductase, in untransformed and transformed Nicotiana plumbaginifolia during CO_2 enrichment of plants grown in pots and in hydroponic culture. *Planta,* 1997, Vol. 202, p. 510–521.

Ferrario-Méry S., Valadier M-H., Foyer C.H. Overexpression of nitrate reductase in tobacco delays drought-induced decreases in nitrate reductase activity and mRNA, *Plant Physiol.,* 1998, Vol. 117, p. 293–302.

Field C., Merino J., Mooney H.A. Compromises between water-use efficiency and nitrogen-use efficiency in five species of California evergreens, *Oecologia,* 1983, Vol. 60, p. 384–389.

Field C.B., Mooney H.A. The photosynthesis-nitrogen relationship in wild plants, *On the Economy of Plant Form and Function,* Givinish T.A. ed., Cambridge University Press, London, 1986, p. 25–55.

Fieuw S., Müller-Röber B., Gálvez S., Willmitzer L. Cloning and expression analysis of the cytosolic NADP($^+$)-dependent isocitrate dehydrogenase from potato. Implications for nitrogen metabolism, Plant *Physiol.,* 1995, Vol. 107, p. 905–913.

Finnemann J., Schjoerring J.K. Translocation of NH_4^+ in oilseed rape plants in relation to glutamine synthetase isogene expression and activity, *Physiol. Plant.,* 1999, Vol. 105, p. 469–477.

Finnemann J., Schjoerring J.K. 2000. Post-translational regulation of cytosolic glutamine synthetase by reversible phosphorylation and 14-3-3 protein interaction, *Plant J.,* 2000, Vol. 24, p. 171–181.

Flexas J., Medrano H. Drought-inhibition of photosynthesis in C_3 plants: stomatal and non-stomatal limitations revisited, *Ann. Bot.*, 2002, Vol. 89, p. 183–189.

Flexas J., Bota J., Galmés J., Medrano H., Ribas-Carbó M. Keeping a positive carbon balance under adverse conditions: responses of photosynthesis and respiration to water stress, *Physiol Plant,* 2006*a*, Vol. 127, p. 343–352.

Flexas J., Bota J., Loreto F., Cornic G., Sharkey T.D. Diffusive and metabolic limitations to photosynthesis under drought and salinity in C_3 plants, *Plant Biol.*, 2004, Vol. 6, p. 269–279.

Flexas J., Díaz-Espejo A., Galmés J., Kaldenhoff R., Medrano H., Ribas-Carbo M. Rapid variations of mesophyll conductance in response to changes in CO_2 concentration around leaves, Plant Cell Environ., 2007, Vol. 30, p. 1284–1298.

Flexas J., Ribas-Carbó M., Bota J., Galmés J., Henkle M., Martínez-Cañellas S., Medrano H. Decreased Rubisco activity during water stress is induced by stomatal closure, not by decreased relative water content, *New Phytol.*, 2006*b*, Vol. 172, p. 73–82.

Flexas J., Ribas-Carbó M., Diaz-Espejo A., Galmés J., Medrano H. Mesophyll conductance to CO_2: current knowledge and future prospects, *Plant Cell Environ.*, 2008, Vol. 31, p. 602–621.

Flowers T.J., Troke P.F., Yeo A.R. The mechanism of salt tolerance in halophytes, *Annu. Rev. Plant Physiol.*, 1977, Vol. 28, p. 89–121.

Forde B.G., Lea P.J. Glutamate in plants: metabolism, regulation, and signaling, *J. Exp. Bot.,* 2007, Vol. 58, p. 2339–2358.

Foyer C.H., Parry M., Noctor G. Markers and signals associated with nitrogen assimilation in higher plants, *J. Exp. Bot.,* 2003, Vol. 54, p. 585–593.

Foyer C.H., Valadier M-H., Migge A., Becker T.W. Droughtinduced effects on nitrate reductase activity and mRNA on the coordination of nitrogen and carbon metabolism in maize leaves, *Plant Physiol.*, 1998, Vol. 117, p. 283–292.

Franks P.J., Beerling D.J. Maximum leaf conductance driven by CO_2 effects on stomatal size and density over geologic time, *Proc. Natl. Acad. Sci. USA,* 2009, Vol. 106, p. 10343–10347.

Franks P.J., Leitch I.J., Ruszala E.M., Hetherington A.M., Beerling D.J. Physiological framework for adaptation of stomata to CO_2 from glacial to future concentrations. *Philosophical Transactions of the Royal Society B,* 2012, Vol. 367, p. 537–546.

Fresneau C., Ghashghaie J., Cornic G. Drought effect on nitrate reductase and sucrose-phosphate synthase activities in wheat (*Triticum durum L.*): role of leaf internal CO_2, *J. Exp. Bot.,* 2007, Vol. 58, No 11, p. 2983–2992.

Fridlyand L.E., Backhausen J.E., Scheibe R. Flux control of the malate valve in leaf cells, *Arch. Biochem. Biophys.,* 1998, Vol. 349, p. 290–298.

Fukayama H., Fukuda T., Masumoto C., Taniguchi Y., Sakai H., Cheng W., Hasegawa T., Miyao M. Rice plant response to long term CO_2 enrichment: gene expression profiling, *Plant Sci.,* 2009, Vol. 177, p. 203–210.

Fukayama H., Sugino M., Fukuda T., Masumoto C., Taniguchi Y., Okada M., Sameshima R., Hatanaka T., Misoo S., Hasegawa T., Miyao M. Gene expression profiling of rice grown in free air CO_2 enrichment (FACE) and elevated soil temperature, *Field Crop. Res.,* 2011, Vol. 121, No 1, p. 195–199.

Gadaleta A., Nigro D., Giancaspro A., Blanco A. The glutamine synthetase (GS2) gene is in relation to grain protein content of durum wheat, *Funct. Integr. Genomics,* 2011, Vol. 11, p. 665–670.

Galangau F., Danielvedele F., Moureaux T., Dorbe M.F., Leydecker M.T., Caboche M. Expression of leaf nitrate reductase genes from tomato and tobacco in relation to light-dark regimes and nitrate supply, *Plant Physiol.,* 1988, Vol. 88, p. 383–388.

Gallé A., Flórez-Sarasa I., Tomás M., Pou A., Medrano H., Ribas-Carbó M., Flexas J. The role of mesophyll conductance during water stress and recovery in tobacco (*Nicotiana sylvestris*): acclimation or limitation, *J. Exp. Bot.,* 2009, Vol. 60, p. 2379–2390.

Galmés J., Medrano H., Flexas J. Photosynthetic limitations in response to water stress and recovery in Mediterranean plants with different growth forms, *New Phytol.,* 2007, Vol. 175, p. 81–93.

Galmés J., Ribas-Carbó M., Medrano H., Flexas J. Rubisco activity in Mediterranean species is regulated by the chloroplastic CO2 concentration under water stress. *J. Exp. Bot.,* 2011, Vol. 62, No 2, p. 653–665.

García-Calderón M., Chiurazzi M., Espuny M.R., Márquez A.J. Photorespiratory metabolism and nodule function: Behavior of Lotus japonicas mutants deficient in plastid glutamine synthetase, *Mol. Plant. Microbe Interact.,* 2012, Vol. 25, p. 211–219.

Garg B.K., Kathju S., Burman U. Influence of water stress on water relation, photosynthetic parameters and nitrogen metabolism of moth bean genotype, *Biol. Plant.,* 2001, Vol. 44, p. 289–292.

Geiger M., Haake V., Ludewig F., Sonnewald U., Stitt M. The nitrate and ammonium nitrate supply have a major influence on the response of photosynthesis, carbon metabolism, nitrogen metabolism and growth to elevated carbon dioxide in tobacco, *Plant Cell Environ.*, 1999, Vol. 22, p. 1177–1199.

Geiger M., Walch-Liu P., Engels C., Harnecker J., Schulze E.D., Ludewig F., Sonnewald U., Scheible W.R., Stitt M. Enhanced carbon dioxide leads to a modified diurnal rhythm of nitrate reductase activity in older plants, and a large stimulation of nitrate reductase activity and higher levels of amino acids in young tobacco plants, *Plant Cell Environ.*, 1998, Vol. 21, p. 253–268.

Ghisi R., Trentin A.R., Masi A., Ferretti M. Carbon and nitrogen metabolism in barley plants exposed to UV-B radiation, *Physiol. Plant.*, 2002, Vol. 116, No 2, p. 200–205.

Giberti S., Funck D., Forlani G. Δ^1-pyrroline-5-carboxylate reductase from Arabidopsis thaliana: stimulation or inhibition by chloride ions and feedback regulation by proline depend on whether NADPH or NADH acts as co-substrate, *New. Phytol.*, 2014, Vol. 202, p. 911–919.

Gillon J.S., Yakir D. Internal conductance to CO_2 diffusion and (COO)-O-18 discrimination in C_3 leaves. *Plant Physiol.*, 2000, Vol. 123, p. 201–213.

Givan C.V., Joy K.W., Kleczkowski L.A. A decade of photorespiratory nitrogen cycling, *Trends Biochem. Sci.*, 1988, Vol. 13, p. 433–437.

Goggin D.E., Setter T.L. Fructosyltransferase activity and fructan accumulation during development in wheat exposed to terminal drought, *Funct. Plant Biol.*, 2004, Vol. 31, p. 11–21.

Golding A.J., Johnson G.N. Down-regulation of linear and activation of cyclic electron transport during drought, *Planta,* 2003, Vol. 218, p. 107–114.

González E.M., Aparicio-Tejo P.M., Gordon A.J., Minchin F.R., Royuela M., Arrese-Igor C. Water-deficit effects on carbon and nitrogen metabolism of pea nodules, *J. Exp. Bot.*, 1998, Vol. 49, No 327, p. 1705–1714.

Gossett D.R., Millholland E.P., Lucas M.C. Antioxidant response to salt stress in salt-tolerant and salt-sensitive cultivars of cotton, *Crop Sci.*, 1994, Vol. 34, p. 706–714.

Gouia H., Suzuki A., Brulfert J., Ghorbal M.H. Effects of cadmium on the co-ordination of nitrogen and carbon metabolism in bean seedlings, *J. Plant Physiol.*, 2003, Vol. 160, No 4, p. 367–376.

Grassi G., Magnani F. Stomatal, mesophyll conductance and biochemical limitations to photosynthesis as affected by drought and leaf ontogeny in ash and oak trees, *Plant Cell Environ.,* 2005, Vol. 28, p. 834–849.

Guerrero M. G., Vega J. M., Losada M. The assimilatory nitrate-reducing system and its regulation, *Anu. Rev. Plant Physiol.,* 1981, Vol. 32, p. 169–204.

Gupta N., Thind S.K., Bains N.S. Glycine betaine application modifies biochemical attributes of osmotic adjustment in drought stressed wheat, *Plant Growth Regul.*, 2014, Vol. 72, p. 221–228.

Gutiérrez D., Gutiérrez E., Pérez P., Morcuende R., Verdejo A.L., Martínez-Carrasco R. Acclimation to future atmospheric CO_2 levels increases photochemical efficiency and mitigates photochemistry inhibition by warm temperatures in wheat under field chambers, *Physiol. Plant.*, 2009, Vol. 137, p. 86–100.

Gutiérrez D., Morcuende R., Del Pozo A., Martínez-Carrasco R., Pérez P. Involvement of nitrogen and cytokinins in photosynthetic acclimation to elevated CO_2 of spring wheat, *J. Plant Physiol.*, 2013, Vol. 170, p. 1337–1343.

Gutiérrez R.A., Lejay L., Dean A., Chiaromonte F., Shasha D.E., Coruzzi G.M. Qualitative network models and genome wide expression data define carbon / nitrogen-responsive molecular machines in *Arabidopsis, Genome Biol.*, 2007, Vol. 8, R7.

Haase S., Neumann G., Kania A., Kuzyakov Y., Römheld V., Kandeler E. Elevation of atmospheric CO_2 and N-nutritional status modify nodulation, nodule-carbon supply, and root exudation of *Phaseolus vulgaris L., Soil Biol. Biochem.*, 2007, Vol. 39, p. 2208–2221.

Hamano T., Oji Y., Mitsuhashi Y., Matsuki Y., Okamoto S. Action of thiol proteinase on nitrate reductase in leaves of *Hordeum disticum L.*, *Plant Cell Physiol.*, 1984, Vol. 25, p. 1469–1475.

Hamano T., Oji Y., Okamoto S., Mitsuhashi Y., Matsuki Y. A nitrate reductase inactivating factor from barley (*Hordeum disticum L.*) leaves, *Plant Cell Physiol.*, 1983, Vol. 24, p. 1337–1341.

Hanke G.T., Satomi Y., Shinmura K., Takao T., Hase T. A screen for potential ferredoxin electron transfer partners uncovers new, redox dependent interactions, *BBA,* 2011, Vol. 1814, p. 366–374.

Hare P.D., Cress W.A., Van Staden J. Dissecting the roles of osmolyte accumulation during stress, *Plant Cell Environ.*, 1998, Vol. 21, p. 535–553.

Harris M.J., Outlaw W.H. Rapid adjustment of guard-cell abscisic acid levels to current leaf-water status. *Plant Physiol.*, 1991, Vol. 95, p. 171–173.

Hausler R.E., Blackwell R.D., Lea P.J., Leegood R.C. Control of photosynthesis in barley leaves with reduced activities of glutamine synthetase or glutamate synthase. I. Plant characteristics and changes in nitrate, ammonium and amino acids, *Planta,* 1994*a*, Vol. 194, p. 406–417.

Hausler R.E., Lea P.J., Leegood R.C. Control of photosynthesis in barley leaves with reduced activities of glutamine synhetase or glutamate synthase. II. Control of electron transport and CO_2 assimilation, *Planta,* 1994*b*, Vol. 194, p. 418–435.

Hayakawa T., Kamachi K., Oikawa M., Ojima K., Yamaya T. Response of glutamine synthetase and glutamate synthase isoforms to nitrogen sources in rice cell cultures, *Plant Cell Physiol.*, 1990, Vol. 31, p. 1071–1077.

Heitholt J.J., Johnson R.C., Ferris D.M. Stomatal limitation to carbon dioxide assimilation in nitrogen and drought-stressed wheat, *Crop Sci.*, 1991, Vol. 31, p. 135–139.

Hellmann H., Funk D., Rentsch D., Frommer W.B. Hypersensitivity of an arabidopsis sugar signaling mutant toward exogenous proline application, *Plant Physiol.*, 2000, Vol. 122, p. 357–68.

Henkes S., Sonnewald U., Badur R., Flachmann R., Stitt M. A small decrease of plastid transketolase activity in antisense tobacco transformants has dramatic effects on photosynthesis and phenylpropanoid metabolism, *Plant Cell,* 2001, Vol. 13, p. 535–551.

Henson C.A., Duke S.H., Collins M. Characterization of $NADP^+$-isocitrate dehydrogenase from the host plant cytosol of lucerne (*Medicago sativa*) root nodules, *Physiol. Plant.*, 1986, Vol. 67, p. 538–544.

Hewitt E.J. Assimilatory nitrate-nitrite reduction, *Annu. Rev. Plant Physiol.*, 1975, Vol. 26, p. 73–100.

Hiei Y., Ohta S., Komari T. Efficient transformation of rice (*Oryza sativa* L.) mediated by *Agrobacterium* and sequence analysis of the boundaries of the T-DNA, *Plant J.*, 1994, Vol. 6, No 2, p. 271–282.

Hilliard N.P., Hirasawa M., Knaff D.B., Shaw R.W. A reexamination of the properties of spinach nitrite reductase: Protein and siroheme content heterogeneity in purified preparations, *Arch. Biochem. Biophys.*, 1991, Vol. 291, No 1, p. 195–199.

Hirasawa M., Tollin G., Salamon Z., Knaff D.B. Transient kinetic and oxidation-reduction studies of spinach ferredoxin: nitrite oxidoreductase. *BBA,* 1994, Vol. 1185, p. 336–345.

Hirel B., Gadal P. Glutamine synthetase in rice: A comparative study of the enzymes from roots and leaves, *Plant Physiol.*, 1980, Vol. 66, p. 619–623.

Hirel B., Lea P.J. Ammonium assimilation, *Plant Nitrogen*, Lea P.J., Morof-Gaudry J.F. eds., Springer-Verlag, Berlin, 2001, p. 79–99.

Hirel B., Marsolier M-C., Hoarau A., Hoarau J., Brangeon J., Schafer R., Verma D.P.S. Forcing expression of a soybean root glutamine synthetase gene in tobacco leaves induces a native gene encoding the cytosolic enzyme, *Plant Mol. Biol.*, 1992, Vol. 20, p. 207–218.

Hirel B., Miao G.H., Verma D.P.S. Metabolic and developmental control of glutamine synthetase genes in legume and non-legume plants, *Control of Plant Gene Expression,* Verma D.P.S. ed., CRC Press, Boca Raton, 1993, p. 443–458.

Hodges M. Enzyme redundancy and the importance of 2-oxoglutarate in plant ammonium assimilation, *J. Exp. Bot.*, 2002, Vol. 53, No 370, p. 905–916.

Hoff T., Truong H-N., Caboche M. The use of mutants and transgenic plants to study nitrate assimilation, *Plant Cell Environ.*, 1994, Vol. 17, p. 489–506.

Högy P., Zörb C., Langenkämper G., Betsche T., Fangmeier A. Atmospheric CO_2 enrichment changes the wheat grain proteome, *J. Cer. Sci.*, 2009, Vol. 50, p. 248–254.

Holaday A.S., Ritchie S.W., Nguyen H.T. Effect of water deficit on gas exchange parameters and ribulose 1,5-bisphosphate carboxylase activation in wheat, *Env. Exp. Bot.*, 1992, Vol. 32, p. 403–409.

Hoshida H., Tanaka Y., Hibino T., Hayashi Y., Tanaka A., Takabe T., Takabe T. Enhanced tolerance to salt stress in transgenic rice that overexpresses chloroplast glutamine synthetase, *Plant Mol. Biol.*, 2000, Vol. 43, p. 103–111.

Hsiao T.C. Plant responses to water stress, *Annu. Rev. Plant Physiol.*, 1973, Vol. 24, p. 519–570.

Huang W., Yang S-J., Zhang S-B., Zhang J-L., Cao K-F. Cyclic electron flow plays an important role in photoprotection for the resurrection plant Paraboea rufescens under drought stress, *Planta,* 2012, Vol. 235, p. 819–828.

Huber J.L., Redinbaugh M.G., Huber S.C., Campbel W.H. Regulation of maize leaf nitrate reductase activity involves both gene expression and protein phosphorylation, *Plant Physiol.*, 1994, Vol. 106, p. 1667–1674.

Huber S.C., Bachmann M., Huber J.L. Post-translational regulation of nitrate reductase activity: a role for Ca_2^+ and 14-3-3 proteins, *Trends Plant Sci.*, 1996, Vol. 1, p. 432–438.

Hungate B.A., Dukes J.S., Shaw M.R., Luo Y., Field C.B. Nitrogen and climate change, *Science*, 2003, Vol. 320, p. 1512–1513.

Huppe H.C., Turpin D.H. Integration of carbon and nitrogen metabolism in plant and algal cells, *Annu. Rev. Plant Physiol. Plant Mol. Biol.*, 1994, Vol. 45, p. 577–607.

Husic D.W., Husic H.D., Tolbert N.E. The oxidative photosynthetic carbon cycle or C_2 cycle, *CRC Crit. Rev. Plant Sci.*, 1987, Vol. 5, p. 45–100.

Husted S., Mattsson M., Mollers C., Wallbraun M., Schjoerring J.K. Photorespiratory NH_4^+ production in leaves of wild-type and glutamine synthetase 2 antisense oilseed rape, *Plant Physiol.*, 2002, Vol. 130, No 2, p. 989–998.

Ida S., Mikami B. Purification and characterization of assimilatory nitrate reductase from the cyanobacterium plectonema boryanum, *Plant Cell Physiol.*, 1983, Vol. 24, No 4, p. 649–658.

Igamberdiev A.U., Bykova N.V., Lea P.J., Gardeström P. The role of photorespiration in redox and energy balance of photosynthetic plant cells: a study with a barley mutant deficient in glycine decarboxylase, *Physiol. Plant.*, 2001, Vol. 111, p. 427–438.

Igamberdiev A.U., Mikkelsen T.N., Ambus P., Bauwe H., Lea P.J., Gardeström P. Photorespiration contributes to stomatal regulation and carbon isotope fractionation: a study with barley, potato and Arabidopsis plants deficient in glycine decarboxylase, *Photosynth. Res.*, 2004, Vol. 81, p. 139–152.

Ireland R.J., Lea P.J. The enzymes of glutamine, glutamate, asparagine, and aspartate metabolism, *Plant Amino Acids. Biochemistry and Biotechnology*, Singh B.K. ed., Marcel Dekker, Inc., New York, Basel, Hong Kong, 1999, p. 49–109.

Ishiyama K., Inoue E., Tabuchi M., Yamaya T., Takahashi H. Biochemical background and compartmentalized functions of cytosolic glutamine synthetase for active ammonium assimilation in rice roots, *Plant Cell Physiol.*, 2004, Vol. 45, p. 1640–1647.

Jach M.E., Laureysens I., Ceulemans R. Above- and belowground production of young scots pine (*Pinus sylvestris L.*) trees after three years of growth in the field under elevated CO_2, *Ann. Bot.*, 2000, Vol. 85, p. 789–798

Jang J.C., Sheen J. Sugar sensing in higher plants. *Plant Cell,* 1994, Vol. 6, p. 1665–1679.

Jeong M.L., Jiang H.Y., Chen H.S., Tsai C.J., Harding S.A. Metabolic profiling of the sink-to-source transition in developing leaves of quaking aspen, *Plant Physiol.*, 2004, Vol. 136, p. 3364–3375.

Jia H., Oguchi R., Hope A.B., Barber J., Chow W.S. Differential effects of severe water stress on linear and cyclic electron fluxes through Photosystem I in spinach leaf discs in CO_2-enriched air, *Planta,* 2008, Vol. 228, p. 803–812.

Jin S.H., Hong J., Li X.Q., Jiang D.A. Antisense inhibition of Rubisco activase increases Rubisco content and alters the proportion of Rubisco activase in stroma and thylakoids in chloroplasts of rice leaves, *Ann. Bot.*, 2006, Vol. 97, p. 739–744.

Johansson E.M., Fransson P.M.A., Finlay R.D., van Hees P.A.W. Quantitative analysis of soluble exudates produced by ectomycorrhizal roots as a response to ambient and elevated CO_2, *Soil Biol. Biochem.*, 2009, Vol. 41, p. 1111–1116.

Jonassen E.M., Sandsmark B.A.A., Lillo C. Unique status of NIA2 in nitrate assimilation: NIA2 expression is promoted by HY5 / HYH and inhibited by PIF4, *Plant Signal Behav.*, 2009, Vol. 4, p. 1084–1086.

Jones T., Tucker D.E., Ort D.R. Chilling delays circadian pattern of sucrose phosphate synthase and nitrate reductase activity in tomato, *Plant Physiol.*, 1998, Vol. 118, p. 149–158.

Joy K.W. Ammonia, glutamine and asparagine: A carbonnitrogen interface, *Can. J. Bot.*, 1988, Vol. 66, p. 2103–2109.

Kaiser W.M. Correlation between changes in photosynthetic activity and changes in total protoplast volume in leaf tissue from hygro-, meso-, and xerophytes under osmotic stress, *Planta,* 1982, Vol. 154, p. 538–545.

Kaiser W.M. Effects of water deficit on photosynthetic capacity, *Phsyiol. Plant.*, 1987, Vol. 71, p. 142–149.

Kaiser W.M., Brendle-Behnisch E. Rapid modulation of spinach leaf nitrate reductase activity by photosynthesis. I. Modulation *in vivo* by CO_2 availability, *Plant Physiol.*, 1991, Vol. 96, p. 363–367.

Kaiser W.M., Förster J. Low CO_2 prevents nitrate reduction in leaves, *Plant Physiol.*, 1989, Vol. 91, p. 970–974.

Kaiser W.M., Spill D. Rapid modulation of spinach leaf nitrate reductase activity by photosynthesis. II. *In vitro* modulation by ATP and AMP, *Plant Physiol.*, 1991, Vol. 96, p. 368–375.

Kamachi K., Yamaya T., Maie T., Ojima K. A role for glutamine synthetase in the remobilization of leaf nitrogen during natural senescence in rice leaves, *Plant Physiol.*, 1991, Vol. 96, p. 411–417.

Kang Y., Han Y.H., Torres-Jerez I., Wang M.Y., Tang Y.H., Monteros M., Udvardi M. System responses to long-term drought and re-watering of two contrasting alfalfa varieties, *Plant J.*, 2011, Vol. 68, p. 871–889.

Karplus P.A., Daniels M.J., Herriott J.R. Atomic structure of ferredoxin-$NADP^+$ reductase: prototype for a structurally novel flavoenzyme family, *Science*, 1991, Vol. 251, p. 62–66.

Kaschuk G., Hungria M., Leffelaar P.A., Giller K.E., Kuyper T.W. Differences in photosynthetic behaviour and leaf senescence of soybean (*Glycine max* [L.] Merrill) dependent on N_2 fixation or nitrate supply, *Plant Biol.*, 2010, Vol. 12, p. 60–69.

Keegstra K., Cline K. Protein import and routing system of chloroplasts, *Plant Cell*, 1999, Vol. 11, p. 557–570.

Keller F., Ludlow M.M. Carbohydrate metabolism in drought-stressed leaves of pigeonpea (*Cajanus cajan*), *J. Exp. Bot.*, 1993, Vol. 44, p. 1351–1357.

Kendall A.C., Wallsgrove R.M., Hall N.P., Turner J.C., Lea P.J. Carbon and nitrogen metabolism in barley (*Hordeum vulgare* L.) mutants lacking ferredoxin-dependent glutamate synthase, *Planta*, 1986, Vol. 168, p. 316–323.

Kenis J.D., Campbell W.H. Oxygen inhibition of nitrate reductase biosynthesis in detached corn leaves via inhibition of total soluble protein synthesis, *Plant Physiol.*, 1989, Vol. 91, p. 883–888.

Kenis J.D., Trippi V.S. Regulation of nitrate reductase in detached oat leaves by light and oxygen, *Physiol. Plant.*, 1986, Vol. 68, p. 387–390.

Kenis J.D., Rouby M.B., Edelman M.O., Silvente S.T. Inhibition of nitrate reductase by water stress and oxygen in detached oat leaves: A possible mechanism of action, *J. Plant Physiol.*, 1994, Vol. 144, p. 735–739.

Kerepesi I., Bányai-Stevanovits E., Galiba G. Fructans in wheat under stress conditions, *Acta Biologica Szegediensis*, 2002, Vol. 46, p. 101–102.

Kichey T., Heumez E., Pocholle D., Pageau K., Vanacker H., Dubois F., Le Gouis J., Hirel B. Combined agronomic and physiological aspects of nitrogen management in wheat highlight a central role for glutamine synthetase, *New Phytol.*, 2006, Vol. 169, p. 265–278.

Kimball B.A., Kobayashi K., Bindi M. Responses of agricultural crops to free-air CO_2 enrichment, *Advances in Agronomy,* 2002, Vol. 77, p. 293–368.

Kleinhofs A., Warner R.L. Advances in nitrate assimilation, *The Biochemistry of Plants,* Miflin B.J., Lea P.J. eds., Academic Press, San Diego, 1990, Vol. 16, p. 89–120.

Klimov V.V., Allachverdiev S.I., Feyziev Y.M., Baranov S.V. Bicarbonate requirements for the donor side of photosystem II, *Febs Lett.,* 1995, Vol. 363, p. 252–255.

Knaff D.B., Hirasawa M. Ferredoxin-dependent chloroplast enzymes, *BBA,* 1991, Vol. 1056, p. 93–125.

Koch K.E. Carbohydrate-modulated gene expression in plants, *Annu. Rev. Plant Physiol. Plant Mol. Biol.,* 1996, Vol. 47, p. 509–540.

Kołodziejek I., Kozioł J., Wałeza M., Mostowska A. Ultrastructure of mesophyll cells and pigment content in senescing leaves of maize and barley, *J Plant Growth Regul.,* 2003, Vol. 22, p. 217–227.

Kozaki A., Takeba G. Photorespiration protects C_3 plants from photooxidation, *Nature,* 1996, Vol. 384, p. 557–560.

Krapp A. Plant nitrogen assimilation and its regulation: a complex puzzle with missing pieces, *Curr. Opin. Plant Biol.,* 2015, Vol. 25, p. 115–122.

Krieger-Liszkay A., Fufezan C., Trebst A. Single oxygen production in photosystem II and related protection mechanism, *Photosynth. Res.,* 2008, Vol. 98, p. 551–564.

Kruse J., Hetzger I., Hansch R., Mendel R.R., Walch-Liu P., Engels C., Rennenberg H. Elevated pCO_2 favours nitrate reduction in the roots of wild-type tobacco (*Nicotiana tabacum* cv. Gat.) and significantly alters N-metabolism in transformants lacking functional nitrate reductase in the roots, *J. Exp. Bot.,* 2002, Vol. 53, p. 2351–2367.

Kruse J., Hetzger I., Hansch R.H., Mendel R.R., Rennenberg H. Elevated pCO_2 affects C and N metabolism in wild type and transgenic tobacco exhibiting altered C / N balance in metabolite analysis, *Plant Biol.,* 2003*b*, Vol. 5, p. 540–549.

Kruse J., Hetzger I., Mai C., Polle A., Rennenberg H. Elevated pCO_2 affects N-metabolism of young poplar plants (*Populus tremula* × *P. alba*) differently at deficient and sufficient N-supply, *New Phytol.,* 2003*a*, Vol. 157, p. 65–81.

Kumagai E., Araki T., Hamaoka N., Ueno O. Ammonia emission from rice leaves in relation to photorespiration and genotypic differences in glutamine synthetase activity, *Ann Bot.,* 2011, Vol. 108, p. 1381–1386.

Kumar R., Taware R., Gaur V., Guru S., Kumar A. Influence of nitrogen on the expression of TaDof1 transcription factor in wheat and its relationship with photosynthetic and ammonium assimilating efficiency, *Mol. Biol. Rep.*, 2009, Vol. 36, p. 2209–2220.

Kumar R.G., Dubey R.S. Glutamine Synthetase isoforms from rice seedlings: effects of stress on enzyme activity and the protective roles of osmolytes, *J. Plant Physiol.*, 1999, Vol. 155, p. 118–121.

Kumar S.G, Reddy A.M., Sudhakar C. NaCl effects on proline metabolism in two high yielding genotypes of mulberry (*Morus alba L.*) with contrasting salt tolerance, *Plant Sci.*, 2003, Vol. 165, p. 1245–1251.

Kürschner W.M., Stulen I., Wagner F., Kuiper P.J.C. Comparison of palaeobotanical observations with experimental data on the leaf anatomy of durmast oak [*Quercus petraea* (Fagaceae)] in response to environmental change, *Ann. Bot.*, 1998, Vol. 81, p. 657–664.

Labboun S., Terce-Laforgue T., Roscher A., Bedu M., Restivo F.M., Velanis C.N., Skopelitis D.S., Moschou P.N., Moshou P.N., Roubelakis-Angelakis K.A., Suzuki A., Hirel B. Resolving the role of plant glutamate dehydrogenase. I. *In vivo* real time nuclear magnetic resonance spectroscopy experiments, *Plant Cell Physiol.*, 2009, Vol. 50, p. 1761–1773.

Lahners K., Kramer V., Back E., Privalle L., Rothstein S. Molecular cloning of complementary DNA encoding maize nitrite reductase: molecular analysis and nitrate induction, *Plant Physiol.*, 1988, Vol. 88, p. 741–746.

Laisk A., Eichelmann H., Oja V. Leaf C_3 photosynthesis in silico: integrated carbon / nitrogen metabolism, *Photosynthesis in Silico: Understanding Complexity from Molecules to Ecosystems,* Laisk A., Nedbal L., Govindjee eds., Springer Science+Business Media BV, Netherlands, 2009, p. 295–322.

Laisk A., Eichelmann H., Oja V., Talts E., Scheibe R. Rates and roles of cyclic and alternative electron flow in potato leaves, *Plant Cell Physiol.*, 2007, Vol. 48, p. 1575–1588.

Lake J.A., Quick W.P., Beerling D.J., Woodward F.I. Plant development: signals from mature to new leaves, *Nature,* 2001, Vol. 411, p. 154.

Lam H-M., Coschigano K.T., Oliveira I.C., Melo-Oliveira R., Coruzzi G.M. The molecular-genetics of nitrogen assimilation into amino acids in higher plants, *Annu. Rev. Plant Physiol. Plant Mol. Biol.*, 1996, Vol. 47, p. 569–593.

Lam H-M., Coshigano K., Schultz C., Melo-Oliveira R., Tjaden G., Oliveira I., Ngai N, Hsieh M-H., Coruzzi G.M. Use of Arabidopsis mutants and genes to study amide amino acid biosynthesis, *Plant Cell,* 1995, Vol. 7, p. 887–898.

Lancien M., Martin M., Hsieh M-H., Leustek T., Goodman H., Coruzzi G.M. Arabidopsis glt-T mutant defines a role for NADH-GOGAT in the non-photorespiratory ammonium assimilatory pathway, *Plant J.*, 2002, Vol. 29, p. 347–358.

Lanquar V., Loque D., Hormann F., Yuan L.X., Bohner A., Engelsberger W.R., Lalonde S., Schulze W.X., von Wiren N., Frommer W.B. Feedback inhibition of ammonium uptake by a phospho-dependent allosteric mechanism in Arabidopsis, *Plant Cell,* 2009, Vol. 21, p. 3610–3622.

Larios B., Agüera E., Cabello P., Maldonado J. M., de la Haba P. The rate of CO_2 assimilation controls the expression and activity of glutamine synthetase through sugar formation in sunflower (*Helianthus annuus L.*) leaves, *J. Exp. Bot.,* 2004, Vol. 55, No 394, p. 69–75.

Larios B., Agüera E., de la Haba P., Pérez-Vicente R., Maldonado J.M. A short-term exposure of cucumber plants to rising atmospheric CO_2 increases leaf carbohydrate content and enhances nitrate reductase expression and activity, *Planta,* 2001, Vol. 212, p. 305–312.

Larsson M., Larsson C-M., Whitford P.N., Clarkson D.T. Influence of osmotic stress on nitrate reductase activity in wheat (*Triticum aestivum L.*) and the role of abscisic acid, *J. Exp. Bot.,* 1989, Vol. 40, p. 1265–1271.

Laugier E., Bouguyon E., Mauries A., Tillard P., Gojon A., Lejay L. Regulation of high-affinity nitrate uptake in roots of *Arabidopsis* depends predominantly on posttranscriptional control of the NRT2.1 / NAR2.1 transport system, *Plant Physiol.,* 2012, Vol. 158, p. 1067–1078.

Lawlor D.W. Carbon and nitrogen assimilation in relation to yield: mechanisms are the key to understanding production systems, *J. Exp. Bot.*, 2002*a*, Vol. 53, No 370, p. 773–787.

Lawlor D.W. Limitation to photosynthesis in water-stressed leaves: stomata vs. metabolism and role of ATP, *Ann. Bot.*, 2002*b*, Vol. 89, p. 871–885.

Lawlor D.W., Cornic G. Photosynthetic carbon and assimilation and associated metabolism in relation to water deficits in higher plants, *Plant Cell Environ.*, 2002, Vol. 25, p. 275–294.

Lea P.J. Primary nitrogen metabolism, *Plant Biochemistry,* Dey P.M., Harborne J.B. eds., Academic Press, San Diego, 1997, p. 273–313.

Lea P.J., Blackwell R.B., Joy K.W. Ammonia assimilation in higher plants. *Nitrogen Metabolism of Plants,* Mengel K., Pilbeam D.J. eds., Oxford Scientific Publishers, New York, 1992, p. 153–186.

Lea P.J., Robinson S.A., Stewart G.R. The enzymology and metabolism of glutamine, glutamate and asparagine, *The Biochemistry of Plants*, Miflin B.J., Lea P.J. eds., Academic Press, New York, Vol. XVI, 1990, p. 121–159.

Lea U.S., Leydecker M-T., Quilleré I., Meyer, C., Lillo C. Posttranslational regulation of nitrate reductase strongly affects the levels of free amino acids and nitrate, whereas transcriptional regulation has only minor influence, *Plant Physiol.*, 2006, Vol. 140, p. 1085–1094.

Leakey A.D.B., Ainsworth E.A., Bernacchi C.J., Rogers A., Long S.P., Ort D.R. Elevated CO_2 effects on plant carbon, nitrogen, and water relations: six important lessons from FACE, *J. Exp. Bot.*, 2009*a*, Vol. 60, p. 2859–2876.

Leakey A.D.B., Xu F., Gillespie K.M., McGrath J.M., Ainsworth E.A., Ort D.R. Genomic basis for stimulated respiration by plants growing under elevated carbon dioxide, *Proc. Natl. Acad. Sci. USA, 2009b*, Vol. 106, p. 3597–3602.

Lefebvre S., Lawson T., Zakhleniuk O.V., Lloyd J.C., Raines C.A. Increased sedoheptulose-1,7-bisphosphatase activity in transgenic tobacco plants stimulates photosynthesis and growth from an early stage in development, *Plant Physiol.,* 2005, Vol. 138, p. 451–460.

Lehtimaki N., Lintala M., Allahverdiyeva Y., Aro E.M., Mulo P. Drought stress-induced upregulation of components involved in ferredoxin-dependent cyclic electron transfer, *J. Plant. Physiol.*, 2010, Vol. 167, p. 1018–1022.

Lekshmy S., Jain V., Khetarpal S., Pandey R. Inhibition of nitrate uptake and assimilation in wheat seedlings grown under elevated CO_2, *Ind. J. Plant Physiol.*, 2013, Vol. 18, p. 23–29.

Lewis J.D., Lucash M., Olszyk D., Tingey D.T. Seasonal patterns of photosynthesis in Douglas fir seedlings during the third and fourth year of exposure to elevated CO_2 and temperature, *Plant Cell Environ.*, 2001, Vol. 24, p. 539–548.

Lewis J.D., Wang X.Z., Griffin K.L., Tissue D.T. Effects of age and ontogeny on photosynthetic responses of a determinate annual plant to elevated CO_2 concentrations, *Plant Cell Environ.*, 2002, Vol. 25, p. 359–368.

Li J., Ezquer I., Bahaji A., Montero M., Ovecka M., Baroja-Fernández E., Muñoz F.J., Mérida Á., Almagro G., Hidalgo M., Sesma M.T., Pozueta-Romero J. Microbial volatile-induced accumulation of exceptionally high levels of starch in Arabidopsis leaves is a process involving NTRC and starch synthase classes III and IV, *MPMI,* 2011, Vol. 24, No 10, p. 1165–1178.

Li M.C., Kong G.Q., Zhu J.J. Vertical and leaf-age-related variations of nonstructural carbohydrates in two alpine timberline species, southeastern Tibetan Plateau, *J. For. Res.*, 2009, Vol. 14, No 4, p. 229–235.

Li M-G., Villemur R., Hussey P.J., Silflow C.D., Gantt J.S., Snustad D.P. Differential expression of six glutamine synthetase genes in Zea mays, *Plant Mol. Biol.*, 1993, Vol. 23, p. 401–407.

Li P., Ainsworth E.A., Leakey A.D.B., Ulanov A., Lozovaya V., Ort D.R., Bohnert H. Arabidopsis transcript and metabolite profiles: ecotype-specific responses to open-air elevated CO_2, *Plant Cell Environ.*, 2008, Vol. 31, p. 1673–1687.

Li P., Sioson A., Mane S., Ulanov A., Grothaus G., Heath L Murali T., Bohnert H., Grene R. Response diversity of Arabidopsis thaliana ecotypes in elevated [CO_2] in the field, *Plant Mol. Biol.*, 2006, Vol. 62, p. 593–609.

Li X., Zhang G., Sun B., Zhang S., Zhang Y., Liao Y., Zhou Y., Xia X., Shi K., Yu J. Stimulated leaf dark respiration in tomato in an elevated carbon dioxide atmosphere, *Sci. Rep.*, 2013, Vol. 3, p. 3433.

Li Y., Gao Y., Xu X., Shen Q., Guo S. Light-saturated photosynthetic rate in high-nitrogen rice (Oryza sativa L.) leaves is related to chloroplastic CO_2 concentration, *J. Exp. Bot.*, 2009, Vol. 60, No 8, p. 2351–2360.

Li Y., Xu J., Haq N.U., Zhang H., Zhu X-G. Was low CO_2 a driving force of C_4 evolution: *Arabidopsis* responses to long-term low CO_2 stress, *J. Exp. Bot.*, 2014, Vol. 65, No 13, p. 3657–3667.

Lightfoot D.A., Green N.K., Cullimore J.V. The chloroplast-located glutamine synthetase of *Phaseolus vulgaris* L.: nucleotide sequence, expression in different organs and uptake into isolated chloroplast, *Plant Mol. Biol*, 1988, Vol. 11, p. 191–202.

Lillo C. Signalling cascades integrating light-enhanced nitrate metabolism, *Biochem. J.*, 2008, Vol. 415, p. 11–19.

Lillo C., Smith L.H., Nimmo H.G., Wilkins M.B. Regulation of nitrate reductase and phosphoenolpyruvate carboxylase activities in barley leaf protoplasts, *Planta,* 1996, Vol. 200, p. 181–185.

Lin C.C., Hsu Y.T., Kao C.H. The effect of NaCl on proline accumulation in rice leaves, *Plant Growth Regul.*, 2002, Vol. 36, p. 275–85.

Lin J., Jach M.E., Ceulemans R. Stomatal density and needle anatomy of Scots pine (*Pinus sylvestris*) are affected by elevated CO_2, *New Phytol.*, 2001, Vol. 150, p. 665–674.

Liu C., Wang Y., Pan K., Jin Y., Liang J., Li W., Zhang L. Photosynthetic carbon and nitrogen metabolism and the relationship between their metabolites and lipid peroxidation in dwarf bamboo (Fargesia rufa Yi) during drought and subsequent recovery. *Trees,* 2015, Vol. 29, p. 1633–1647.

Liu C.G., Wang Y.J., Pan K.W., Zhu T.T., Li W., Zhang L. Carbon and nitrogen metabolism in leaves and roots of dwarf bamboo (*Fargesia denudata Yi*) subjected to drought for two consecutive years during sprouting period, *J Plant Growth Regul.*, 2014, Vol. 33, p. 243–255.

Liu J.X., Zhang D.Q., Zhou G.Y., Duan H.L. Changes of leaf nutrient traits and photosynthesis of four tree species: effects of elevated CO_2, N fertilization and canopy positions, *J. Plant Ecol.*, 2012, Vol. 5, p. 376–390.

Llorens L., Peñelas J., Estiarte M. Ecophysiological responses of two Mediterranean shrubs, Erica multiflora and Globularia alypum, to experimentally drier and warmer conditions, *Physiol. Plant.*, 2003, Vol. 119, p. 231–243.

Long S.P., Ainsworth E.A., Rogers A., Ort D.R. Rising atmospheric carbon dioxide: plants FACE the future, *Annu. Rev. Plant Biol.*, 2004, Vol. 55, p. 591–628.

Lorimer G.H. The carboxylation and oxygenation of ribulose 1,5-bisphosphate. The primary events in photosynthesis and photorespiration, *Annu. Rev. Plant Physiol.*, 1981, Vol. 32, p. 349–383.

Lu B., Yuan Y., Zhang C., Ou J., Zhou W., Lin Q. Modulation of key enzymes involved in ammonium assimilation and carbon metabolism by low temperature in rice (*Oryza sativa L.*) roots, *Plant Sci.,* 2005, Vol. 169, p. 295–302.

Lu C., Zhang J. Effects of water stress on photosystem II photochemistry and its thermostability in wheat plants, *J. Exp. Bot.*, 1999, Vol. 50, p. 1139–1206.

Lu C., Zhang J. Photosynthetic CO_2 assimilation, chlorophyll fluorescence and photoinhibition as affected by nitrogen deficiency in maize plants, *Plant Sci.*, 2000, Vol. 151, p. 135–143.

Lu G., Campbel W., Lindqvist Y., Schneider G. Crystallization and preliminary crystallographic studies of the fad domain of corn nadh: nitrate reductase, *J. Mol. Biol.*, 1992, Vol. 224, p. 277–279.

Lu G., Campbell W.H., Schneider G., Lindqvist Y. Crystal structure of the FAD-containing fragment of corn nitrate reductase at 2.5 A resolution: relationship to other flavoprotein reductases, *Structure,* 1994, Vol. 2, No 9, p. 809–821.

Lu J-L., Ertl J.R., Chen C-M. Cytokinin enhancement of the light induction of nitrate reductase transcript levels in etiolated barley leaves, *Plant Mol. Biol.*, 1990, Vol. 14, p. 585–594.

Lu J-L., Ertl J.R., Chen C-M. Transcriptional regulation of nitrate reductase mRNA levels by cytokinin-abscisic acid interactions in etiolated barely leaves, *Plant Physiol.*, 1992, Vol. 98, p. 1255–1260.

Luo Y., Field C.B., Mooney H.A. Predicting responses of photosynthesis and root fraction elevated [CO_2]: interactions among carbon, nitrogen, and growth, *Plant Cell Environ.*, 1994, Vol. 17, p. 1195–1204.

Luo Y., Su B., Currie W.S., Dukes J.S., Finzi A., Hartwig U., Hungate B., Mcmurtrie R.E., Oren R., Parton W.J., Pataki D.E., Shaw M.R., Zak D.R., Field C.B. Progressive nitrogen limitation of ecosystem responses to rising atmospheric carbon dioxide, *Bioscience,* 2004, Vol. 54, p. 731–739.

Lutts S., Majerus V., Kinet J.M. NaCl effects on proline metabolism in rice (*Oryza sativa*) seedlings, *Physiol. Plant.*, 1999, Vol. 105, p. 450–458.

MacKintosh C., Douglas P., Lillo C. Identification of a protein that inhibits the phosphorylated form of nitrate reductase from spinach (*Spinacia oleracea*) leaves, *Plant Physiol.*, 1995, Vol. 107, p. 451–457.

MacKintosh C., Meek S.E.M. Regulation of plant NR activity by reversible phosphorylation, 14-3-3 proteins and proteolysis, *Cell. Mol. Life Sci.*, 2001, Vol. 58, p. 205–214.

Magalhaes J.R., Ju G.C., Rich P.J., Rhodes D. Kinetics of Nitrogen-15-labeled ammonium ion assimilation in *Zea mays*, Preliminary studies with a glutamate dehydrogenase GDH1 null mutant, *Plant Physiol.*, 1990, Vol. 94, p. 647–656.

Mahan J.R., Oliver M.J., Sherman T.D. Nitrate reductase activity during desiccation and rehydration of the desiccationtolerant moss *Tortula ruralis, Environ. Exp. Bot.*, 1998, Vol. 39, p. 67–76.

Makino A., Osmond B. Effects of nitrogen nutrition on nitrogen partitioning between chloroplasts and mitochondria in pea and wheat, *Plant Physiol.*, 1991, Vol. 96, p. 355–362.

Makino A., Harada M., Kaneko K., Mae T., Shimada T., Yamamoto N. Whole-plant growth and N allocation in transgenic rice plants with decreased content of ribulose-1,5-bisphosphate carboxylase under different CO_2 partial pressures, *Aust. J. Plant Physiol.*, 2000*a*, Vol. 27, p. 1–12.

Makino A., Harada M., Sato T., Nakano H., Mae T. Growth and N allocation in rice plants under CO_2 enrichment, *Plant Physiol.,* 1997*a*, Vol. 115, p. 199–203.

Makino A., Nakano H., Mae T., Shimada T., Yamamoto N. Photosynthesis, plant growth and N allocation in transgenic rice plants with decreased Rubisco under CO_2 enrichment, *J. Exp. Bot.*, 2000*b*, Vol. 51, GMP Special Issue, p. 383–389.

Makino A., Sakuma H., Sudo E., Mae T. Differences between maize and rice in N-use efficiency for photosynthesis and protein allocation, *Plant Cell Physiol.*, 2003, Vol. 44, p. 952–956.

Makino A., Shimada T., Takumi S., Kaneko K., Matsuoka M., Shimamoto K., Nakano H., Miyao-Tokutomi M., Mae T., Yamamoto N. Does decrease in ribulose-1,5-bisphosphate carboxylase by antisense rbcS lead to a higher N-use efficiency of photosynthesis under conditions of saturating CO_2 and light in rice plants? *Plant Physiol.*, 1997*b*, Vol. 114, p. 483–491.

Mann A.F., Fenten P.A., Stewart G.R. Identification of two forms of glutamine synthetase in barley (*Hordeum vulgare L.*), *Biochem. Biophys. Res. Commun.*, 1979, Vol. 88, p. 515–521.

Manter D.K., Kerrigan J. A / C_i curve analysis across a range of woody plant species: influence of regression analysis parameters and mesophyll conductance, *J. Exp. Bot.*, 2004, Vol. 55, p. 2581–2588.

Markelz R.J.C., Lai L.X., Vosseler L.N., Leakey A.D.B. Transcriptional reprogramming and stimulation of leaf respiration by elevated CO_2 concentration is diminished, but not eliminated, under limiting nitrogen supply, *Plant Cell Environ.*, 2014*a*, Vol. 37, p. 886–898.

Markelz R.J.C., Vosseller L.N., Leakey A.D.B. Developmental stage specificity of transcriptional, biochemical and CO_2 efflux responses of leaf dark respiration to growth of Arabidopsis thaliana at elevated [CO_2], *Plant Cell Environ.*, 2014*b*, Vol. 37, p. 2542–2552.

Marques da Silva J., Celeste A.M. Photosynthesis in the water-stressed C_4 grass Setaria sphacelata is mainly limited by stomata with both rapidly and slowly imposed water deficits, *Physiol. Plant.*, 2004, Vol. 121, p. 409–420.

Martin T., Oswald O., Graham I.A. Arabidopsis seedling growth, storage lipid mobilization and photosynthetic gene expression are regulated by carbon: nitrogen availability, *Plant Physiol.*, 2002, Vol. 128, p. 472–481.

Martínez-Carrasco R. Diurnal changes of Rubisco in response to elevated CO_2, temperature and nitrogen in wheat grown under temperature gradient tunnels, *Env. Exp. Bot.*, 2005, Vol. 53, p. 13–27.

Masclaux C., Valadier M-H., Brugiere N., Morot-Gaudry J-F., Hirel B. Characterization in the source / sink in tobacco (*Nicotiana tabacum L.*)

shoots in relation to nitrogen management and leaf senescence, *Planta,* 2000, Vol. 211, p. 510–518.

Matt P., Geiger M., Walch-Liu P., Engels C., Krapp A., Stitt M. The diurnal changes of nitrogen metabolism in leaves of nitrate-replete tobacco are driven by an imbalance between nitrate reductase activity and the rate of nitrate uptake and ammonium and glutamine metabolism during the first part of the light period, *Plant Cell Environ.*, 2001, Vol. 24, p. 177–190.

Mattsson M., Schjoerring J.K. Ammonia exchange between plants and the atmosphere: Effects of ammonium supply to the roots, dark-induced senescence and reduced GS activity, *Plant Nutrition – for Sustainable Food Production and Environment,* Ando T. et al. eds., Kluwer Academic Publishers, 1997, p. 827–831.

Mattsson M., Schjoerring J.K. Dynamic and steady-state responses of inorganic nitrogen pools and NH_3 exchange in leaves of *Lolium perenne* and *Bromus erectus* to changes in root nitrogen supply, *Plant Physiol.,* 2002, Vol. 128, p. 742–750.

Mattsson M., Hausler R.E., Leegood R.C., Lea P.J., Schjoerring J.K. Leaf-atmosphere NH_3 exchange in barley mutants with reduced activities of glutamine synthetase, *Plant Physiol.,* 1997, Vol. 114, p. 1307–1312.

Mattsson M., Husted S., Schjoerring J.K. Influence of nitrogen nutrition and metabolism on ammonia volatilization in plants, *Nutr. Cycl. Agroecosyst.*, 1998, Vol. 51, p. 35–40.

May T., Soll J. Chloroplast precursor protein translocon, *FEBS Lett.*, 1999, Vol. 452, p. 53–56.

McDonald G.K., Paulsen G.M. High temperature effects on photosynthesis and water relations of grain legumes, *Plant Soil.*, 1997, Vol. 196, p. 47–58.

McDowell N.G. Mechanisms linking drought, hydraulics, carbon metabolism, and vegetation mortality, *Plant Physiol.*, 2011, Vol. 155, No 3, p. 1051–1059.

McGrath R.B., Coruzzi G.M. A gene network controlling glutamine and asparagine biosynthesis in plants, *Plant J.*, 1991, Vol. 1, p. 275–280.

McNally S.F., Hirel B., Gadal P., Mann A.F., Stewart G.R. Glutamine synthetase from higher plants: evidence for a specific isoform content related to their possible physiological role and their compartmentation within the leaf, *Plant Physiol.*, 1983, Vol. 72, p. 22–25.

Medlyn B.E. The optimal allocation of nitrogen within the C_3 photosynthetic system at elevated CO_2. *Aust. J. Plant Physiol.*, 1996, Vol. 23, p. 593–603.

Melo P.M., Lima L.M., Santos I.M., Carvalho H.G., Cullimore J.V. Expression of the plastid-located glutamine synthetase of *Medicago truncatula*. Accumulation of the precursor in root nodules reveals an *in vivo* control at the level of protein import into plastids, *Plant Physiol.*, 2003, Vol. 132, p. 390–399.

Messedi D., Labidi N., Grignon C., Abdelly C. Limits imposed by salt to the growth of the halophyte *Sesuvium portulacastrum*, *J. Plant Nutr. Soil Sci.*, 2004, Vol. 167, p. 720–725.

Miflin B.J., Habash D.Z. The role of glutamine synthetase and glutamate dehydrogenase in nitrogen assimilation and possibilities for improvement in the nitrogen utilization of crops, *J. Exp. Bot.*, 2002, Vol. 53, No 370, p. 979–987.

Miflin B.J., Lea P.J. The pathway of nitrogen assimilation in plants. *Phytochemistry*, 1976, Vol. 15, p. 873–885.

Miflin B.J., Lea P.J. Ammonia assimilation, *The Biochemistry of Plants*, Miflin B.J. ed., Academic Press, New York, NY, USA, Vol. 5, 1980, p. 169–202.

Migge A., Carrayol E., Hirel B., Becker T.W. Leaf-specific overexpression of plastidic glutamine synthetase stimulates the growth of transgenic tobacco seedlings, *Planta*, 2000, Vol. 210, p. 252–260.

Migge A., Carrayol E., Kunz C., Hirel B., Fock H., Becker T. The expression of tobacco genes encoding plastidic glutamine synthetase or ferredoxin-dependent glutamate synthase does not depend on the rate of nitrate reduction, and is unaffected by suppression of photorespiration, *J. Exp. Bot.*, 1997, Vol. 48, p. 1175–1184.

Mikami B., Ida S. Purification and properties of ferredoxin-nitrate reductase from the cyanobacterium *Plectonema bor yanum*, *BBA*, 1984, Vol. 791, p. 294–304.

Miller A., Tsai C-H., Hemphill D., Endres M., Rodermel S., Spalding M. Elevated CO_2 effects during leaf ontogeny: a new perspective on acclimation, *Plant Physiol.*, 1997, Vol. 115, p. 1195–1200.

Miller A.J., Fan X.R., Shen Q.R., Smith S.J. Amino acids and nitrate as signals for the regulation of nitrogen acquisition, *J. Exp. Bot.*, 2008, Vol. 59, p. 111–119.

Miller G., Honig A., Stein H., Suzuki N., Mittler R., Zilberstein A. Unraveling Δ^1-pyrroline-5-carboxylate-proline cycle in plants by uncoupled expression of proline oxidation enzymes, *J. Biol. Chem.*, 2009, Vol. 284, p. 26482–26492.

Miranda-Ham M.L., Loyola-Vargas V.M. Ammonia assimilation in canavalia ensiformis plants under water and salt stress, *Plant Cell Physiol.*, 1988, Vol. 29, No 5, p. 743–753.

Misra N., Dwivedi U.N. Nitrogen assimilation in germinating *Phaseolus aureus* seeds under saline stress, *J. Plant Physiol.*, 1990, Vol. 135, No 6, p. 719–724.

Mitchell R.A.C., Theobald J.C., Parry M.A.J., Lawlor D.W. Is there scope for improving balance between RuBP-regeneration and carboxylation capacities in wheat at elevated CO_2? *J. Exp. Bot.,* 2000, Vol. 51, GMP Special Issue, p. 391–397.

Miyazaki S., Fredricksen M., Hollis K.C., Poroyko, V., Shepley D., Galbraith D.W., Long S.P., Bohnert H.J. Transcript expression profiles of Arabidopsis thaliana grown under controlled conditions and open-air elevated concentrations of CO_2 and of O_3, *Field Crop. Res.*, 2004, Vol. 90, p. 47–59.

Miyazawa S-I., Warren C.R., Turpin D.H., Livingston N.J. Determination of the site of CO_2 sensing in poplar: is the area-based N content and anatomy of new leaves determined by their immediate CO_2 environment or by the CO_2 environment of mature leaves? *J. Exp. Bot.,* 2011, Vol. 62, No 8, p. 2787–2796.

Mo J.M., Li D.J., Gundersen P. Seedling growth response of two tropical tree species to nitrogen deposition in southern China, *Eur. J. For. Res.*, 2008, Vol. 127, p. 275–283.

Moore B.D., Cheng S-H., Sims D., Seemann J.R. The biochemical and molecular basis for photosynthetic acclimation to elevated atmospheric CO_2, *Plant Cell Environ.*, 1999, Vol. 22, p. 567–582.

Moorhead G., Douglas P., Morrice N., Scarabel M., Aitken A., Mackintosh C. Phosphorylated nitrate reductase from spinach leaves is inhibited by 14-3-3 proteins and activated by fusicoccin, *Curr. Biol.*, 1996, Vol. 6, p. 1104–1113.

Morcuende R., Krapp A., Hurry V., Stitt M. Sucrose feeding leads to increased rates of nitrate assimilation, increased rates of α-oxoglutarate synthesis, and increased synthesis of a wide spectrum of amino acids in tobacco leaves, *Planta,* 1998, Vol. 206, p. 394–409.

Morgan J.A. The effects of N nutrition on the water relations and gas exchange characteristics of wheat (*Triticum aestivum L.*), *Plant Physiol.*, 1986, Vol. 80, p. 52–58.

Morilla C.A., Boyer J. S., Hageman R.H. Nitrate reductase activity and polyribosomal content of corn (*Zea mays L.*) having low water potentials, *Plant Physiol.*, 1973, Vol. 51, p. 817–824.

Muller B., Pantin F., Génard M., Turc O., Freixes S., Piques M., Gibon Y. Water deficits uncouple growth from photosynthesis, increase C content, and modify the relationships between C and growth in sink organs, *J. Exp. Bot.*, 2011, Vol. 62, p. 1715–1729.

Munné-Bosch S. and Alegre L. Plant aging increases oxidative stress in chloroplasts, *Planta,* 2002, Vol. 214, p. 608–615.

Munné-Bosch S., Shikanai T., Asada K. Enhanced ferredoxin-dependent cyclic electron flow around photosystem I and α-tocopherol quinone accumulation in water-stressed ndhB-inactivated tobacco mutants, *Planta,* 2005, Vol. 222, p. 502–511.

Munns R. Comparative physiology of salt and water stress, *Plant Cell Environ.*, 2002, Vol. 25, p. 239–250.

Nakano H., Makino A., Mae T. The effect of elevated partial pressures of CO_2 on the relationship between photosynthetic capacity and N content in rice leaves, *Plant Physiol.*, 1997, Vol. 115, p. 191–198.

Naumburg E., Housman D.C., Huxman T., Charlet T.N., Loik M.E., Smith S.D. Photosynthetic responses of Mojave Desert shrubs to free-air CO_2 enrichment are greatest during wet years, *Glob. Chan. Biol.*, 2004, Vol. 8, p. 276–285.

Neales T.F., Nichols A.O. Growth responses of young wheat plants to a range of ambient CO_2 levels, *Aust. J. Plant Physiol.*, 1978, Vol. 5, p. 560–563.

Neame P.J., Barber M. J. Conserved domains in molybdenum hydroxylases- the amino acid sequence of chicken hepatic sulfite oxidase, *J. Biol. Chem.*, 1989, Vol. 264, p. 20894–20901.

Newingham B.A., Vanier C.H., Charlet T.N., Ogle K., Smith S.D., Nowak R.S. No cumulative effect of 10 years of elevated CO_2 on perennial plant biomass components in the Mojave Desert, *Glob. Chan. Biol.*, 2013, Vol. 19, p. 2168–2181.

Neyra C.A., Hageman R.H. Dependence of nitrite reduction on electron transport in chloroplasts, *Plant Physiol.*, 1974, Vol. 54, p. 480–483.

Nie G.Y., Hendrix D.L., Webber A.N., Kimball B.A., Long S.P. Increased accumulation of carbohydrates and decreased photosynthetic gene transcript levels in wheat grown at an elevated CO_2 concentration in the field, *Plant Physiol.*, 1995, Vol. 108, p. 975–983.

Niedzwiedz-Siegen I., Bogatek-Leszczynska R., Côme D., Corbineau F. Effects of drying rate on dehydration sensitivity of excised wheat seedling shoots as related to sucrose metabolism and antioxidant enzyme activities, *Plant Sci.*, 2004, Vol. 167, p. 879–888.

Nielsen T.H., Stitt M. Tobacco transformants with strongly decreased expression of pyrophosphate:fructose-6-phosphate expression in the base of their sink leaves contain much higher levels of fructose-2,6-bisphosphate but exhibit no major changes in fluxes, *Planta,* 2001, Vol. 214, p. 106–116.

Niinemets Ü. Components of leaf dry mass per area – thickness and density – alter leaf photosynthetic capacity in reverse directions in woody plants, *New Phytol.*, 1999, Vol. 144, p. 35–47.

Niinemets Ü. Photosynthesis and resource distribution through plant canopies, *Plant Cell Environ.*, 2007*a*, Vol. 30, p. 1052–1071.

Niinemets Ü. Plasticity in mesophyll volume fraction modulates light-acclimation in needle photosynthesis in two pines, *Tree Physiol.*, 2007*b*, Vol. 27, p. 1137–1151.

Nikiema P., Nzokou P., Rothstein D. Effects of groundcover management on soil properties, tree physiology, foliar chemistry and growth in a newly established Fraser fir (*Abies fraseri* [Pursh] Poir) plantation in Michigan, United States of America, *New Forests,* 2012, Vol. 43, No 2, p. 213–230.

Nishiyama Y., Allakhverdiev S.I., Murata N. Inhibition of the repair of photosystem II by oxidative stress in cyanobacteria, *Photosynth. Res.*, 2005, Vol. 84, p. 1–7.

Nishiyama Y., Yamamoto H., Allakhverdiev S.I., Inaba M., Yokota A., Murata N. Oxidative stress inhibits the repair of photodamage to the photosynthetic machinery, *EMBO J.*, 2001, Vol. 20, p. 5587–5594.

Niu S.L., Jiang G.M., Li Y.G., Gao L.M., Liu M.Z. Diurnal gas exchange and superior resources use efficiency of typical C_4 species in Hunshandak Sandland, China, *Photosynthetica,* 2003, Vol. 41, p. 221–226.

Noctor G., Foyer C.H. A re-evaluation of the ATP: NADPH budget during C_3 photosynthesis: a contribution from nitrate assimilation and its associated respiratory activity? *J. Exp. Bot.*, 1998, Vol. 49, p. 1895–1908.

Novitskaya L., Trevanion S.J., Driscoll S., Foyer C.H., Noctor G. How does photorespiration modulate leaf amino acid contents? A dual approach through modelling and metabolite analysis, *Plant Cell Environ.*, 2002, Vol. 25, p. 821–835.

Nunes-Nesi A., Fernie A.R., Stitt M. Metabolic and signaling aspects underpinning the regulation of plant carbon nitrogen interactions, *Mol. Plant,* 2010, Vol. 3, p. 973–996.

Oaks A. Evidence for deamination by glutamate dehydrogenase in higher plants: reply, *Can. J. Bot.*, 1995, Vol. 73, p. 1116–1117.

Ogren W.L. Photorespiration: pathways, regulation and modification, *Annu. Rev. Plant Physiol.*, 1984, Vol. 35, p. 415–442.

Oji Y., Watanabe M., Wakiuchi N., Okamoto S. Nitrite reduction in barley root plastids: dependence on NADPH coupled with glucose-6-phosphate and 6-phosphogluconate dehydrogenase, and possible involvement of an electron carrier and a diaphorase, *Planta,* 1985, Vol. 165, p. 85–99.

Oliveira I.C., Coruzzi G.M. Carbon and amino acids reciprocally modulate the expression of glutamine synthetase in Arabidopsis, *Plant Physiol.*, 1999, Vol. 121, p. 301–309.

Oliveira I.C., Brears T., Knight T.J., Clark A., Coruzzi G.M. Overexpression of cytosolic glutamine synthetase: Relation to nitrogen, light and photorespiration, *Plant Physiol.*, 2002, Vol. 129, p. 1170–1180.

Ono K., Terashima I., Watanabe A. Interaction between nitrogen deficit of a plant and nitrogen content in the old leaves, *Plant Cell Physiol.*, 1996, Vol. 37, p. 1083–1089.

Onoda Y., Hikosaka K., Hirose T. Seasonal change in the balance between capacities of RuBP carboxylation and RuBP regeneration affects CO_2 response of photosynthesis in *Polygonum cuspidatum*, *J. Exp. Bot.*, 2005, Vol. 56, No 412, p. 755–763.

Ookawa T., Naruoka Y., Sayama A., Hirasawa T. Cytokinin effects on ribulose-1,5-bisphosphate carboxylase / oxygenase and nitrogen partitioning in rice during ripening, *Crop Sci.*, 2004, Vol. 44, p. 2107–2115.

Oren R., Ellsworth D.S., Johnsen K.H., Phillips N., Ewers B.E., Maier C., Schäfer K.V.R., McGarthy H., Hendrey G., McNulty S.G., Katul G.G. Soil fertility limits carbon sequestration by forest ecosystems in a CO_2-enriched atmosphere, *Nature,* 2001, Vol. 411, p. 469–472.

Ortiz-Lopez A., Ort D.R., Boyer J.S. Photophosphorylation in attached leaves of Helianthus annuus at low water potentials, *Plant Physiol.*, 1991, Vol. 96, p. 1018–1025.

Osmond C.B. Photorespiration and photoinhibition. Some implications for the energetics of photosynthesis, *BBA,* 1981, Vol. 639, p. 77–98.

Osmond C.B., Grace S.C. Perspectives on photoinhibition and photorespiration in the field: quintessential inefficiencies of the light and dark reactions of photosynthesis? *J. Exp. Bot.*, 1995, Vol. 46, p. 1351–1362.

Osmond C.B., Badger M.R., Maxwell K., Björkman O., Leegood R.C. Too many photons: photorespiration, photoinhibition and photooxidation, *Trends Plant Sci.*, 1997, Vol. 2, p. 119–121.

Otoo E., Ishii R., Kumura A. Interaction of nitrogen addition and soil water stress on photosynthesis and transpiration in rice, *Jpn. J. Crop Sci.*, 1989, Vol. 58, No 3, p. 424–429.

Pal M., Karthikeyapandian V., Jain V., Srivastava A.C., Raj A., Sengupta U.K. Biomass production and nutritional levels of berseem (*Trifolium alexandrium*) grown under elevated CO_2, *Agr. Ecosyst. Environ.*, 2004, Vol. 101, p. 31–38.

Pant B.D., Musialak-Lange M., Nuc P., May P., Buhtz A., Kehr J., Walther D., Scheible W.R. Identification of nutrient responsive Arabidopsis and rapeseed microRNAs by comprehensive real-time PCR profiling and small RNA sequencing, *Plant Physiol.*, 2009, Vol. 150, p. 1541–1555.

Parkhurst D.F., Mott K.A. Intercellular diffusion limits to CO_2 uptake in leaves, *Plant Physiol.*, 1990, Vol. 94, p. 1024–1032.

Parry M.A.J., Andralojc P.J., Khan S., Lea P.J., Keys A.J. Rubisco activity: effects of drought stress. *Ann. Bot.*, 2002, Vol. 89, p. 833–839.

Parry M.A.J., Reynolds M., Salvucci M.E., Raines C., Andralojc P.J., Zhu X., Price G.D., Condon A.G., Furbank R.T. Raising yield potential of wheat. II. Increasing photosynthetic capacity and efficiency, *J. Exp. Bot.*, 2011, Vol. 62, p. 453–467.

Patakas A., Nikolaou N., Zioziou E., Radoglou K., Noitsakis B. The role of organic solute and ion accumulation in osmotic adjustment in drought-stressed grapevines, *Plant Sci.*, 2002, Vol. 163, p. 361–367.

Pathre U.V., Sinha A.K., Ranade S.A., Shirke P.A. Activation of sucrose-phosphate synthase from Prosopis juliflora in light: effects of protein kinase and protein phosphatase inhibitors, *Physiol. Plant.*, 2000, Vol. 108, p. 249–254.

Patterson T.B., Guy R.D., Dang Q.L. Whole-plant nitrogen- and water-relations traits, and their associated trade-offs, in adjacent muskeg and upland boreal spruce species, *Oecologia*, 1997, Vol. 110, p. 160–168.

Peeters K.M.U., Van Laere A.J. Ammonium and amino acid metabolism in excised leaves of wheat *(Triticum aestivum)* senescing in the dark, *Physiol. Plant.*, 1992, Vol. 84, p. 243–249.

Pelleschi S., Rocher J-P., Prioul J-L. Effect of water restriction on carbohydrate metabolism and photosynthesis in mature maize leaves, *Plant Cell Environ.*, 1997, Vol. 20, p. 493–503.

Perchorowicz J.T., Jensen R.G. Photosynthesis and activation of ribulose bisphosphate carboxylase in wheat seedlings. Regulation by CO_2 and O_2, *Plant Physiol.*, 1983, Vol. 71, p. 955–960.

Pereira S., Carvalho H., Sunkel C., Salema R. Immunocytolocalization of glutamine synthetase in mesophyll and phloem of leaves of *Solanum tuberosum L., Protoplasma,* 1992, Vol. 167, p. 66–73.

Pérez P., Alonso A., Zita G., Morcuende R., Martínez-Carrasco R. Down-regulation of Rubisco activity under combined increases of CO_2 and temperature minimized by changes in Rubisco kcat in wheat, *Plant Growth Regul.,* 2011, Vol. 65, p. 439–447.

Pérez P., Morcuende R., Martín del Molino I.M., Martínez-Carrasco R. Diurnal changes of Rubisco in response to elevated CO_2, temperature and nitrogen in wheat grown under temperature gradient tunnels, *Environ. Exp. Bot.*, 2005, Vol. 53, p. 13–27.

Pérez P., Zita G., Morcuende R., Martínez-Carrasco R. Elevated CO_2 and temperature differentially affect photosynthesis and resource allocation in flag and penultimate leaves of wheat, *Photosynthetica,* 2007, Vol. 45, p. 9–17.

Pickersgill R.P. An upper limit to the active site concentration of ribulose bisphosphate carboxylase in chloroplasts, *Biochem. J.*, 1986, Vol. 236, p. 311.

Plaut Z. Nitrate reductase activity of wheat seedlings during exposure to and recovery from water stress and salinity, *Physiol. Plant.*, 1974, Vol. 30, p. 212–217.

Plaut Z., Lendzian K., Bassham J.A. Nitrite reduction in reconstituted and whole spinach chloroplasts during carbon dioxide reduction, *Plant Physiol.*, 1977, Vol. 59, p. 184–188.

Pleijel H., Uddling J. Yield vs. quality trade-offs for wheat in response to carbon dioxide and ozone, *Global Change Biol.*, 2012, Vol. 18, p. 596–605.

Polle A. Mehler reaction: friend or foe in photosynthesis? *Botanica Acta,* 1996, Vol. 109, p. 84–89.

Pommel B., Gallais A., Coque M., Quilleré I., Hirel B., Prioul J.L., Andrieu B., Floriot M. Carbon and nitrogen allocation and grain filling in three maize hybrids differing in leaf senescence, *Eur. J. Agron.*, 2006, Vol. 24, No 3, p. 203–211.

Poorter H., Evans J.R. Photosynthetic nitrogen-use efficiency of species that differ inherently in specific leaf area, *Oecologia,* 1998, Vol. 116, p. 26–37.

Poorter H., Niklas K.J., Reich P.B., Oleksyn J., Poot P., Mommer L. Biomass allocation to leaves, stems and roots: meta-analyses of interspecific variation and environmental control, *New Phytol.*, 2012, Vol. 193, p. 30–50.

Popova O.V., Ismailov S.F., Popova T.N., Dietz K-J., Golldack D. Salt-induced expression of NADP-dependent isocitrate dehydrogenase and ferredoxin-dependent glutamate synthase in *Mesembryanthemum crystallinum*, *Planta,* 2002, Vol. 215, p. 906–913.

Pouteau S., Cherel I., Vaucheret H., Caboche M. Nitrate reductase mRNA regulation in Nicotiana plumbaginifolia nitrate reductase-deficient mutants, *Plant Cell,* 1989, Vol. 1, p. 1111–1120.

Powles S.B., Osmond C.B. Inhibition of the capacity and efficiency of photosynthesis in bean leaflets illuminated in a CO_2-free atmosphere at low O_2: a possible role for photorespiration, *Aust. J. Plant Physiol.*, 1978, Vol. 5, p. 619–629.

Powles S.B., Osmond C.B., Thorne S.W. Photoinhibition of intact attached leaves of C_3 plants illuminated in the absence of both CO_2 and of photorespiration, *Plant Physiol.*, 1979, Vol. 64, p. 982–988.

Praxedes S.C., DaMatta F.M., Loureiro M.E.G., Ferrão M.A., Cordeiro A.T. Effects of long-term soil drought on photosynthesis and carbohydrate metabolism in mature robusta coffee (*Coffea canephora Pierre* var. kouillou) leaves, *Environ. Exp. Bot.*, 2006, Vol. 56, No 3, p. 263–273.

Price A.H., Hendry G.A.F. Iron-catalysed oxygen radical formation and its possible contribution to drought damage in nine native grasses and three cereals, *Plant Cell Environ.*, 1991, Vol. 14, p. 477–484.

Prior S.A., Rogers H.H., Mullins G.L., Runion G.B. The effects of elevated atmospheric CO_2 and soil P placement on cotton root deployment, *Plant Soil,* 2003, Vol. 255, p. 179–187.

Provan F., Lillo C. Photosynthetic Post-Translational Activation of Nitrate Reductase, *J. Plant Physiol.*, 1999, Vol. 154, p. 605–609.

Qiu N., Lu Q., Lu C. Photosynthesis, photosystem II efficiency and the xanthophyll cycle in the salt-adapted halophyte *Atriplex centralasiatica*, *New Phytol.*, 2003, Vol. 159, p. 479–486.

Quick P., Siegl G., Neuhaus E., Feil R., Stitt M. Short-term water stress leads to a stimulation of sucrose synthesis by activating sucrose-phosphate synthase, *Planta,* 1989, Vol. 177, p. 535–546.

Quick W.P., Schurr U., Cheibe R., Schulze E-D., Rodermel S.R., Bogorad L., Stitt M. Decrased ribulose-1,5-bisphosphate carboxylase-oxygenase in transgenic tobacco transformed with 'antisense' rbcS.I.

Impact on photosynthesis in ambient growth conditions, *Planta,* 1991, Vol. 183, p. 542–554.

Rachmilevitch S., Cousins A.B., Bloom A.J. Nitrate assimilation in plant shoots depends on photorespiration, *Proc. Natl. Acad. Sci. USA,* 2004, Vol. 101, p. 11506–11510.

Radoglou K.M., Jarvis P.G. Effects of CO_2 enrichment on four poplar clones. I. Growth and leaf anatomy, *Ann. Bot.*, 1990, Vol. 65, p. 617–626.

Raines C.A. The Calvin cycle revisited, *Photosynth. Res.,* 2003, Vol. 75, p. 1–10.

Rajasekaran L.R., Kriedemann P.E., Aspinall D., Paleg L.G. Physiological significance of proline and glycinebetaine maintaining photosynthesis during NaCl stress in wheat, *Photosynthetica,* 1997, Vol. 34, p. 357–366.

Raven J.A., Handley L.L., Andrews M. Global aspects of C / N interactions determining plant-environment interactions, *J. Exp. Bot.*, 2004, Vol. 55, p. 11–25.

Ray D., Sheshshayee M.S., Mukhopadhyay K., Bindumadhava H., Prasad T.G., Kumar M.U. High nitrogen use efficiency in rice genotypes is associated with higher net photosynthetic rate at lower Rubisco content, *Biol. Plant.*, 2003, Vol. 46, p. 251–256.

Redinbaugh M.G., Campbell W.H. Quaternary structure and composition of squash NADH: nitrate reductase, *J. Biol. Chem.,* 1985, Vol. 260, p. 3380–3385.

Reich P.B., Kloeppel B.D., Ellsworth D.S. Different photosynthesis-nitrogen relations in deciduous hardwood and evergreen coniferous tree species, *Oceologia,* 1995, Vol. 104, p. 24–30.

Reich P.B., Walters M.B., Tabone T.J. Response of Ulmus americana seedlings to varying nitrogen and water status. 2. Water- and nitrogen-use efficiency in photosynthesis, *Tree Physiol.*, 1989, Vol. 5, p. 173–184.

Rhodes D. Hanson A.D. Quaternary ammonium and tertiary sulfonium compounds in higher plants, *Annu. Rev. Plant Physiol.,* 1993, Vol. 44, p. 357–384.

Riedel J., Tischner R., MaÈck G. The chloroplastic glutamine synthetase (GS2) of tobacco is phosphorylated and associated with 14-3-3 proteins inside the chloroplast, *Planta,* 2001, Vol. 213, p. 396–401.

Robinson J.M. Carbon dioxide and nitrite photoassimilatory processes do not intercompete for reducing equivalents in spinach and soybean leaf chloroplasts, *Plant Physiol.*, 1986, Vol. 80, p. 676–684.

Robredo A., Pérez-López U., Miranda-Apodaca J., Lacuesta M., Mena-Petite A., Muñoz-Rueda A. Elevated CO_2 reduces the drought effect on nitrogen metabolism in barley plants during drought and subsequent recovery, *Environ. Exp. Bot.*, 2011, Vol. 71, No 3, p. 399–408.

Rodriguez J.L., Chaves M.M., Wendler R., David M.M., Quick W.P., Leegood R.C., Stitt M., Pereira J.S. Osmotic adjustment in water-stressed grapevine leaves in relation to carbon assimilation, *Aust. J. Plant Physiol.*, 1993, Vol. 20, p. 309–321.

Rogers A., Humphries S.W. A mechanistic evaluation of photosynthetic acclimation at elevated CO_2, *Glob. Chan. Biol.*, 2000, Vol. 6, p. 1005–1011.

Rogers A., Ainsworth E.A. The response of foliar carbohydrates to elevated carbon dioxide concentration, *Managed Ecosystems and CO_2: Case Studies, Processes and Perspectives*, Nösberger J. ed., Springer Verlag, Berlin, 2006, p. 293–308.

Rogers A., Ainsworth E.A., Leakey A.D.B. Will elevated carbon dioxide concentration amplify the benefits of nitrogen fixation in legumes? *Plant Physiol.*, 2009, Vol. 151, p. 1009–1016.

Rogers A., Fischer B.U., Bryant J., Frehner M., Blum H., Raines C.A., Long S.P. Acclimation of photosynthesis to elevated CO_2 under low-nitrogen nutrition is affected by the capacity for assimilate utilization. Perennial ryegrass under free-air CO_2 enrichment, *Plant Physiol.*, 1998, Vol. 118, p. 683–689.

Romanova A.K. Mechanisms of biosynthesis and genetic engineering of ribulose-bisphosphate carboxylase / oxygenases, *Biological Chemistry Reviews*, 1997, Vol. 37, p. 211–260. [in Russsian]

Romanova A.K. Physiological and biochemical aspects and molecular mechanisms of plant adaptation to the elevated concentration of atmospheric CO_2, *Russ. J. Plant. Physiol.*, 2005, Vol. 52, No 1, p. 112–126.

Rout N.P., Shaw B.P. Salinity tolerance in aquatic macrophytes: probable role of proline, the enzymes involved in its synthesis and C_4 type of metabolism, *Plant Sci.*, 1998, Vol. 136, p. 121–130.

Rouze P., Caboche M. Nitrate reduction in plants: molecular approaches to function and regulation, *Inducible Plant Proteins: Their Biochemistry and Molecular Biology*, Wray J.L. ed., Cambridge University Press, Cambridge, 1992, p. 45–77.

Rowland-Bamford A.I., Baker J.T., Allen L.H., Bowers G. Acclimation of rice to changing atmospheric carbon dioxide concentration, *Plant Cell Environ.*, 1991, Vol. 14, p. 577–583.

Roy H., Andrews T.J. Rubisco: assembly and mechanism, *Photosynthesis: Physiology and Mechanism,* Leegood R.C., Sharkey T.D., von Caemmerer S. eds., Kluwer Academic Publishers, Dordrecht, Netherlands, 2000, p. 53–83.

Rubio-Asensio J.S., Rachmilevitch S., Bloom A.J. Responses of Arabidopsis and wheat to rising CO_2 depend on nitrogen source and nighttime CO_2 levels, *Plant Physiol.*, 2015, Vol. 168, p. 156–163.

Ruuska S., Andrews T.J., Badger M.R., Hudson G.S., Laisk A., Price G.D., von Caemmerer S. The interplay between limiting processes in C_3 photosynthesis studied by rapid-response gas exchange using transgenic tobacco impaired in photosynthesis, *Aust. J. Plant Physiol.*, 1998, Vol. 25, p. 859–870.

Ruuska S.A., Badger M.R., Andrews T.J., von Caemmerer S. Photosynthetic electron sinks in in transgenic tobacco with reduced amounts of Rubisco: little evidence for significant Mehler reaction, *J. Exp. Bot.*, 2000, Vol. 51, p. 357–368.

Sage R.F. A model describing the regulation of ribulose-1,5-bisphosphate carboxylase, electron transport, and triose phosphate use in response to light intensity and CO_2 in C_3 plants, *Plant Physiol.*, 1990, Vol. 94, p. 1728–1734.

Sage R.F. Acclimation of photosynthesis to increasing atmospheric CO_2: the gas exchange perspective, *Photosynth. Res.*, 1994, Vol. 39, p. 351–368.

Sage R.F., Coleman J.R. Effects of low atmospheric CO_2 on plants: more than a thing of the past, *Trends Plant Sci.*, 2001, Vol. 6, p. 18–24.

Sage R.F., Pearcy R.W. The nitrogen use efficiency of C_3 and C_4 plants. Leaf nitrogen effects on the gas exchange characteristics of *Chenopodium album* (L.) and *Amaranthus retroflexus* (L.), *Plant Physiol.*, 1987, Vol. 84, p. 959–963.

Sage R.F., Sage T.L., Kocacinar F. Photorespiration and evolution of C_4 photosynthesis. *Annu. Rev. Plant Biol.*, 2012, Vol. 63, p. 19–47.

Sage R.F., Sharkey T.D., Seemann J.R. Regulation of Ribulose-1,5-bisphosphate carboxylase activity in response to light intensity and CO_2 in the C_3 annuals *Chenopodium album L.* and *Phaseolus vulgaris L*, *Plant Physiol.*, 1990, Vol. 94, p. 1735–1742.

Sakakibara H. Differential response of genes for ferredoxin and ferredoxin: $NADP^+$ oxidoreductase to nitrate and light in maize leaves, *J. Plant Physiol.*, 2003, Vol. 160, p. 65–70.

Sakamoto W., Tamura T., Hanba-Tomita Y., Sodmergen and Murata M. The *VAR1* locus of *Arabidopsis* encodes a chloroplastic FtsH and is responsible for leaf variegation in the mutant alleles. *Genes to Cells,* 2002, Vol. 7, p. 769–780.

Sánchez-Rodríguez E., Rubio-Wilhelmi M.M., Rios J.J., Blasco B., Rosales M.A., Melgarejo R., Romero L., Ruiz J.M. Ammonia production and assimilation: its importance as a tolerance mechanism during moderate water deficit in tomato plants, *J. Plant Physiol.*, 2011, Vol. 168, p. 816–823.

Saneoka H., Moghaieb R.E.A., Premachandra G.S., Fujita K. Nitrogen nutrition and water stress effects on cell membrane stability and leaf water relations in Agrostis palustris Huds, *Environ. Exp. Bot.*, 2004, Vol. 52, p. 131–138.

Scheibe R. $NADP^+$-malate dehydrogenase in C_3-plants: regulation and role of a light-activated enzyme, *Physiol. Plant.*, 1987, Vol. 71, p. 393–400.

Scheibe R. Malate valves to balance cellular energy supply, *Physiol. Plant.*, 2004, Vol. 120, p. 21–26.

Searles P.S., Bloom A.J. Nitrate photo-assimilation in tomato leaves under short-term exposure to elevated carbon dioxide and low oxygen, *Plant Cell Environ.*, 2003, Vol. 26, p. 1247–1255.

Seneweera S., Makino A., Hirotsu N., Norton R., Suzuki Y. New insight into photosynthetic acclimation to elevated CO_2: the role of leaf nitrogen and ribulose-1,5-bisphosphate carboxylase / oxygenase content in rice leaves. *Environ. Exp. Bot.*, 2011, Vol. 71, p. 128–136.

Sgherri C.L.M., Pinzino C., Navari-Izzo F. Chemical changes and O_2^{-1} production membranes under water stress, *Physiol. Plant.*, 1993, Vol. 87, p. 211–216.

Shanner D.L., Boyer J.S. Nitrate reductase activity in maize (*Zea mays L.*) leaves. II. Regulation by nitrate flux at low leaf water potential, *Plant Physiol.*, 1976, Vol. 58, p. 505-509.

Sharkey T.D. Photosynthesis in intact leaves of C_3 plants: physics, physiology and rate limitations, *Bot. Rev.*, 1985, Vol. 51, p. 53–105.

Sharkey T.D., Seemann J.R. Mild water stress effects on carbon-reduction-cycle intermediates, ribulose bisphosphate carboxylase activity and spatial homogeneity of photosynthesis in intact leaves, Plant Physiol., 1989, Vol. 89, p. 1060–1065.

Sharma R.K., Griffing B., Scholl R.L. Variations among races of *Arabidopsis thaliana* (L.) Heynh for survival in limited carbon dioxide, *Theor. App.Gen.*, 1979, Vol. 54, p. 11–15.

Sharma S., Villamor J.G., Verslues P.E. Essential role of tissue-specific proline synthesis and catabolism in growth and redox balance at low water potentiall essential role of tissue-specific proline synthesis and catabolism in growth and redox balance at low water potential, *Plant Physiol.*, 2011, Vol. 157, p. 292–304.

Sheehy Skeffington M.J., Jeffrey D.W. Growth performance of an inland population of Plantago maritima in response to nitrogen and salinity, *Plant Ecol.*, 1985, Vol. 61, p. 265–272.

Sheen J. Metabolic repression of transcription in higher plants, *Plant Cell*, 1990, Vol. 2, p. 1027–1038.

Shetty K. Role of proline-linked pentose phosphate pathway in biosynthesis of plant phenolics for functional food and environmental applications: a review, *Process Biochem.*, 2004, Vol. 39, p. 789–803.

Sibout R., Guerrier G. Solute incompatibility with glutamine synthetase in water-stressed (*Populus nigra*), *Environ. Exp. Bot.*, 1998, Vol. 40, No 2, p. 173–178.

Sicher R.C. Responses of nitrogen metabolism in N-sufficient barley primary leaves to plant growth in elevated atmospheric carbon dioxide, *Photosynth. Res.*, 2001, Vol. 68, p. 193–201.

Sicher R.C., Timlin D., Bailey B. Responses of growth and primary metabolism of water-stressed barley roots to rehydration. *J. Plant Physiol.*, 2012, Vol. 169, No 7, p. 686–695.

Siebke K., Ghannoum O., Conroy J.P., Badger M.R., von Caemmerer S. Photosynthetic oxygen exchange in C_4 grasses: the role of oxygen as electron acceptor, *Plant Cell Environ.*, 2003, Vol. 26, p. 1963–1972.

Simpson R.J., Dalling M.J. Nitrogen redistribution during grain growth in wheat *(Triticum aestivum* L.) III. Enzymology and Transport of amino acids from senescing flag leaves, *Planta,* 1981, Vol. 151, p. 447–456.

Sims D.A., Seemann J.R., Luo Y. Elevated CO_2 concentration has independent effects on expansion rates and thickness of soybean leaves across light and nitrogen gradients, *J. Exp. Bot.,* 1998, Vol. 49, p. 583–591.

Sinclair T.R., Pinter P.J., Kimball B.A., Adamsen F.J., LaMorte R.L., Wall G.W., Hunsaker D.J., Adam N., Brook T.J., Garcia R.L., Thompson T., Leavitt S., Mattias A. Leaf nitrogen concentration of wheat subjected to elevated [CO_2] and either water or N deficits, *Agric. Ecosyst. Environ.*, 2000, Vol. 79, p. 53–60.

Singh A.K., Dubey R.S. Salinity induced stimulation of glutamine synthetase activity in rice seedlings differing in salt tolerance, *Plant Physiol. Biochem. (India),* 1994, Vol. 21, p. 84–90.

Singsaas E.L., Ort D.R., DeLucia E.H. Elevated CO_2 effects on mesophyll conductance and its consequences for interpreting photosynthetic physiology, *Plant Cell Environ.*, 2003, Vol. 27, p. 41–50.

Sivasankar S., Oaks A. Nitrate assimilation in higher plants: The effects of metabolites and light, *Plant Physiol. Biochem.*, 1996, Vol. 34, p. 609–620.

Skopelitis D.S., Paranychianakis N.V., Paschalidis K.A., Pliakonis E.D., Delis I.D., Yakoumakis D.I., Kouvarakis A., Papadakis A.K., Stephanou E.G., Roubelakis-Angelakis K.A. Abiotic stress generates ROS that signal expression of anionic glutamate dehydrogenases to form glutamate for proline synthesis in tobacco and grapevine, *Plant Cell,* 2006, Vol. 18, p. 2767–2781.

Skriver K., Mundy J. Gene expression in response to abscisic acid and osmotic stress, *Plant Cell,* 1990, Vol. 2, p. 503–512.

Smart, D.R., Ritchie K., Bloom A.J., Bugbee B.B. Nitrogen balance for wheat canopies (*Triticum aestivum* cv. Veery 10) grown under elevated and ambient CO_2 concentrations, *Plant Cell Environ.*, 1998, Vol. 21, p. 753–763.

Smirnoff N., Colombe S.V. Drought influences the activity of enzymes of the chloroplast hydrogen peroxide scavenging system, *J. Exp. Bot.*, 1988, Vol. 39, p. 1097–1188.

Solomonson L.P., Barber M.J. Assimilatory nitrate reductase, functional properties and regulation, *Annu. Rev. Plant Physiol. Plant Mol. Biol.*, 1990, Vol. 41, p. 225–253.

Solomonson L.P., Barber M.J., Robbins A.P. Functional domains of assimilatory NADH: nitrite reductase from Chiarella, *J. BioI. Chern.*, 1986, Vol. 261, p. 11290–11294.

Somers D.A., Kuo T.M., Kleinhofs A., Warner R.L., Oaks A. Synthesis and degradation of barley nitrate reductase, *Plant Physiol.*, 1983, Vol. 72, p. 949–952.

Somerville C.R., Ogren W.L. Inhibition of photosynthesis in Arabidopsis mutants lacking leaf glutamate synthase activity, *Nature,* 1980, Vol. 286, p. 257–259.

Souza R.P., Machado E.C., Silva J.A.B., Lagôa A.M.M.A., Silveir J.A.G. Photosynthetic gas exchange, chlorophyll fluorescence and some associated metabolic changes in cowpea (*Vigna unguiculata*) during water stress and recovery, *Environ. Exp. Bot.*, 2004, Vol. 51, p. 45–56.

Spreitzer R.J., Salvucci M.E. Rubisco: structure, regulatory interactions, and possibilities for a better enzyme. *Annu. Rev. Plant Biol.*, 2002, Vol. 53, p. 449–479.

Srivali B., Renu K.C. Drought-induced enhancement of protease activity during monocarpic senescence in wheat, *Curr. Sci.*, 1998, Vol. 75, p. 1174–1176.

Srivastava H.S. Regulation of nitrate reductase activity in higher plants, *Phytochemistry*, 1980, Vol. 19, p. 725–733.

Stewart G.R., Mann A.F., Fentem P.A. Enzymes of glutamate formation: glutamate dehydrogenase, glutamine synthetase and glutamate synthase, *The Biochemistry of Plants: Amino Acids and Derivatives*, Miflin B.J. ed., Academic Press, New York, 1980, p. 271–327.

Stitt M. Limitation of photosynthesis by carbon metabolism. 1. Evidence for excess electron transport capacity in leaves carrying out photosynthesis in saturating light and CO_2, *Plant Physiol.*, 1986, Vol. 81, p. 1115–1122.

Stitt M. Rising CO_2 levels and their potential significance for carbon flow in photosynthetic cells, *Plant Cell Environ.*, 1991, Vol. 14, p. 741–762.

Stitt M., Krapp A. The interaction between elevated carbon dioxide and nitrogen nutrition: the physiological and molecular background, *Plant Cell Environ.*, 1999, Vol. 22, p. 583–621.

Stitt M., Schulze D. Does Rubisco control the rate of photosynthesis and plant growth? An exercise in molecular ecophysiology, *Plant Cell Environ.*, 1994*a*, Vol. 17, p. 465–487.

Stitt M., Schulze E.D. Plant growth, storage, and resource allocation: from flux control in a metabolic chain to the whole plant level, *Flux Control in Biological Systems*, Schulze E.D. ed., Academic Press Inc., San Diego, USA, 1994*b*, p. 57–118.

Stitt M., Lunn J., Usadel B. Arabidopsis and primary photosynthetic metabolism – more than the icing on the cake, *Plant J.*, 2010, Vol. 61, p. 1067–1091.

Stitt M., Müller C., Matt P., Gibon Y., Carillo P., Morcuende R., Scheible W-R., Krapp A. Steps towards an integrative view of nitrogen metabolism, *J. Exp. Bot.*, 2002, Vol. 53, p. 959–970.

Suzuki A., Knaff D.B. Glutamate synthase: structural, metchistic and regulatory properties, and role in the amino acid metabolis, *Photosynth. Res.*, 2005, Vol. 83, p. 191–217.

Syvertsen J.P., Lloyd J., McConchie C., Kriedemann P.E., Farquhar G.D. On the relationship between leaf anatomy and CO_2 diffusion through the mesophyll of hypostomatous leaves, *Plant Cell Environ.*, 1995, Vol. 18, p. 149–157.

Székely G., Abrahám E., Cséplo A., Rigó G., Zsigmond L., Csiszár J., Ayaydin F., Strizhov N., Jásik J., Schmelzer E., Koncz C., Szabados L. Duplicated P5CS genes of Arabidopsis play distinct roles in stress regulation and developmental control of proline biosynthesis, *Plant J.*, 2008, Vol. 53, p. 11–28.

Tabuchi M., Sugiyama K., Ishiyama K., Inoue E., Sato T., Takahashi H., Yamaya T. Severe reduction in growth rate and grain filling of rice mutants lacking OsGS1;1, a cytosolic glutamine synthetase1;1, *Plant J.*, 2005, Vol. 42, p. 641–651.

Takahashi S., Milward S.E., Fan D.Y., Chow W.S., Badger M.R. How does cyclic electron flow alleviate photoinhibition in Arabidopsis? *Plant Physiol.*, 2009, Vol. 149, p. 1560–1567.

Takashima T., Hikosaka K., Hirose T. Photosynthesis or persistence: nitrogen allocation in leaves of evergreen and deciduous Quercus species, *Plant Cell Environ.*, 2004, Vol. 27, p. 1047–1054.

Tamoi M., Nagaoka M., Miyagawa Y., Shigeoka S. Contribution of fructose-1,6-bisphosphatase and sedoheptulose-1,7-bisphosphatase to the photosynthetic rate and carbon flow in the Calvin cycle in transgenic plants, *Plant Cell Physiol.*, 2006, Vol. 47, p. 380–390.

Taub D.R., Wang X. Why are nitrogen concentrations in plant tissues lower under elevated CO2? A critical examination of the hypotheses, *J. Integr. Plant Biol.*, 2008, Vol. 50, p. 1365–1374.

Temple S.J., Knight T.J., Unkefer P.J., Sengupta-Gopalan C. Modulation of glutamine synthetase gene expression in tobacco by the introduction of an alfalfa glutamine synthetase gene in sense and antisense orientation: molecular and biochemical analysis, *Mol. Gen. Genet.*, 1993, Vol. 236, p. 315–325.

Temple S.J., Kunjibettu S., Roche D., Sengupta-Gopalan C. Total glutamine synthetase activity during soybean nodule development is controlled at the level of transcription and holoprotein turnover, *Plant Physiol.*, 1996, Vol. 112, p. 1723–1733.

Temple S.J., Vance C.P., Gantt J.S. Glutamate synthase and nitrogen assimilation, *Trends Plant Sci.*, 1998, Vol. 3, p. 51–56.

Terashima I., Ono K. Effects of $HgCl_2$ on CO_2 dependence of leaf photosynthesis: evidence indicating involvement of aquaporins in CO_2 diffusion across the plasma membrane, *Plant Cell Physiol.*, 2002, Vol. 43, p. 70–78.

Teskey R.O. Combined effects of elevated CO_2 and air temperature on carbon assimilation of *Pinus taeda* trees, *Plant Cell Environ.*, 1997, Vol. 20, p. 373–380.

Tezara W., Mitchell V.J., Driscoll S.P., Lawlor D.W. Water stress inhibits plant photosynthesis by decreasing coupling factor and ATP, *Nature*, 1999, Vol. 401, p. 914–917.

Thimm O., Essigmann B., Kloska S., Altmann T., Buckhout T.J. Response of Arabidopsis to iron deficiency stress as revealed by microarray analysis, *Plant Physiol.*, 2001, Vol. 127, p. 1030–1043.

Thomas F.M., Hilker C. Nitrate reduction in leaves and roots of young pedunculate oaks (*Quercus robur*) growing on different nitrate concentrations, *Environ. Exp. Bot.*, 2000, Vol. 43, No 1, p. 19–32.

Thomas J.F., Harvey C.N. Leaf anatomy of four species grown under continuous CO_2 enrichment, *Bot. Gaz.*, 1983, Vol. 14, p. 303–309.

Thomas R.B., Strain B.R. Root restriction as a factor in photosynthetic acclimation of cotton seedlings grown in elevated carbon dioxide, *Plant Physiol.*, 1991, Vol. 96, p. 627–634.

Thum K.E., Shasha D.E., Lejay L.V., Coruzzi G.M. Light- and carbon-signaling pathways. Modeling circuits of interactions, *Plant Physiol.*, 2003, Vol. 132, p. 440–452.

Tingey S.V., Tsai F-Y., Edwards J W, Walker E.L., Coruzzi G.M. Chloroplast and cytosolic glutamine synthetase are encoded by homologous nuclear genes which are differentially expressed *in vivo*, *J. Biol. Chem.*, 1988, Vol. 163, p. 9651–9657.

Tobin A.K., Yamaya T. Cellular compartmentation of ammonium assimilation in rice and barley, *J. Exp. Bot.*, 2001, Vol. 53, p. 591–604.

Tognetti V.B., Palatnik J.F., Fillat M.F., Melzer M., Hajirezaei M.R., Valle E.M., Carrillo N. Functional replacement of ferredoxin by a cyanobacterial flavodoxin in tobacco confers broad-range stress tolerance, *Plant Cell*, 2006, Vol. 18, p. 2035–2050.

Tognetti V.B., Zurbriggen M.D., Morandi E.N., Fillat M.F., Valle E.M., Hajirezaei M.R., Carrillo N. Enhanced plant tolerance to iron starvation by functional substitution of chloroplast ferredoxin with a bacterial flavodoxin, *Proc. Natl. Acad. Sci. USA*, 2007, Vol. 104, p. 11495–11500.

Trepp G.B., van de Mortel M., Yoshioka H., Miller S.S., Samac D.A., Gantt J.S., Vance C.P. NADH-glutamate synthase in alfalfa root nodules. Genetic regulation and cellular expression, *Plant Physiol.*, 1999, Vol. 119, p. 817–828.

Tyystjärvi E. Photoinhibition of photosystem II and photodamage of the oxygen evolving manganese cluster, *Coord. Chem. Rev.*, 2008, Vol. 252, p. 361–376.

Uddling J., Gelang-Alfredsson J., Karlsson P.E., Selldén G., Pleijel H. Source–sink balance of wheat determines responsiveness of grain production to increased [CO_2] and water supply, *Agr. Ecosys. Environ.*, 2008, Vol. 127, p. 215–222.

Uematsu K., Suzuki N., Iwamae T., Inui M., Yukawa H. Increased fructose 1,6-bisphosphate aldolase in plastids enhances growth and photosynthesis of tobacco plants, *J. Exp. Bot.*, 2012, Vol. 44, p. 318–324.

Upadhyaya H., Sahoo L., Panda S.K. Molecular physiology of osmotic stress in plants, *Molecular Stress Physiology of Plants*, Rout G.R., Das A.B. eds., Springer, New York, 2013, p. 179–192.

Van Oosten J-J., Beford R.T. Some relationships between the gas exchange, biochemistry and molecular biology of photosynthesis during leaf development of tomato plants after transfer to different carbon dioxide concentrations, *Plant Cell Environ.*, 1995, Vol. 18, p. 1253–1266.

Vass I. Role of charge recombination processes in photodamage and photoprotection of the photosystem II complex, *Physiol. Plant.*, 2011, Vol. 142, p. 6–16.

Vassey T.L., Quick W.P., Sharkey T.D., Stitt M. Water stress, carbon dioxide and light effects on sucrose-phosphate synthase activity in Phaseolus vulgaris, *Physiol. Plant.*, 1991, Vol. 81, p. 37–44.

Vassey T.L., Sharkey T.D. Mild water stress of Phaseolus vulgaris plants leads to reduced starch synthesis and extractable sucrose-phosphate synthase activity, *Plant Physiol.*, 1989, Vol. 89, p. 1066–1070.

Venekamp J.H. Regulation of cytosol acidity in plants under conditions of drought, *Physiol. Plant.*, 1989, Vol. 76, p. 112–117.

Verbruggen N., Hua X.J., May M., Montagu M.V. Environmental and developmental signals modulate proline homeostasis: evidence for a negative transcriptional regulator, *Proc. Natl. Acad. Sci. USA*, 1996, Vol. 93, p. 8787–8791.

Vicente R., Pérez P., Martínez-Carrasco R., Feil R., Lunn J.E., Watanabe M., Arrivault S., Stitt M., Hoefgen R., Morcuende R. Metabolic and transcriptional analysis of durum wheat responses to elevated CO_2 at low and high nitrate supply, *Plant Cell Physiol.*, 2016, Vol. 57, No 10, p. 2133–2146.

Vicente R., Pérez P., Martínez-Carrasco R., Usadel B., Kostadinova S., Morcuende R. Quantitative RT–PCR platform to measure transcript levels of C and N metabolism-related genes in durum wheat: transcript profiles in elevated [CO_2] and high temperature at different nitrogen supplies, *Plant Cell Physiol.*, 2015, Vol. 56, p. 1556–1573.

Vidal E.A., Gutiérrez R.A. A systems view of nitrogen nutrient and metabolite responses in Arabidopsis, *Curr. Opin. Plant Biol.*, 2008, Vol. 11, p. 521–529.

Vidal E.A., Araus V., Lu C., Parry G., Green P.J., Coruzzi G.M., Gutiérrez R.A. Nitrate-responsive miR393 / AFB3 regulatory module controls root system architecture in Arabidopsis thaliana, *Proc. Natl. Acad. Sci. USA*, 2010, Vol. 107, p. 4477–4482.

Vincent R., Fraisier V., Chaillou S., Limani M., Deleens E., Phillipson B., Douet C., Boutin J-P., Hirel B. Overexpression of a soybean gene encoding cytosolic glutamine synthetase in shoots of transgenic Lotus corniculatus plants triggers changes in ammonium assimilation and plant development, *Planta*, 1997, Vol. 201, p. 424–433.

Vincentz M., Moureaux T., Leydecker M-T., Vaucheret H., Caboche M. Regulation of nitrate and nitrite reductase expression in Nicotiana plumbaginifolia leaves by nitrogen and carbon metabolites, *Plant J.*, 1993, Vol. 3, p. 315–324.

Vogan P.J., Sage R.F. Effects of low atmospheric CO_2 and elevated temperature during growth on the gas exchange responses of C_3, C_3–C_4 intermediate, and C_4 species from three evolutionary lineages of C_4 photosynthesis, *Oecologia*, 2012, Vol. 169, p. 341–352.

von Caemmerer S., Edmondson D.L. The relationship between steady-state gas exchange, *in vivo* ribulose bisphosphate carboxylase activity and some carbon reduction cycle intermediates in *Raphanus sativus*, *Aust. J. Plant Physiol.*, 1986, Vol. 13, p. 669–688.

von Caemmerer S., Evans J.R. Determination of the average partial pressure of CO_2 in chloroplasts from leaves of several C_3 plants, *Aust. J. Plant Physiol.*, 1991, Vol. 18, p. 287–305.

Walch-Liu P., Filleur S., Gan Y., Forde B.G. Signaling mechanisms integrating root and shoot response to changes in the nitrogen supply, *Photosynth. Res.*, 2005, Vol. 83, p. 239–250.

Walcroft A.S., Whitehead D., Silvester W.B., Kelliher F.M. The response of photosynthetic model parameters to temperature and nitrogen concentration in *Pinus radiata* D. Don, *Plant Cell Environ.*, 1997, Vol. 20, p. 1338–1348.

Wallsgrove R.M., Keys A.J., Eea P.J., Mieein B.J. Photosynthesis, photorespiration and nitrogen metabolism, *Plant Cell Environ.*, 1983, Vol. 6, No 4, p. 301–309.

Wallsgrove R.M., Lea P.J., Miflin B.J. Distribution of the enzymes of nitrogen assimilation within the pea leaf cell, *Plant Physiol.*, 1979, Vol. 63, p. 232–236.

Wallsgrove R.M., Turner J.C., Hall N.P., Kendall A.C., Bright S.W.J. Barley mutants lacking chloroplast glutamine synthetase – Biochemical and genetic analysis, *Plant Physiol.*, 1987, Vol. 83, p. 155–158.

Wang H., Zhang M.S., Guo R., Shi D.C., Liu B., Lin X.Y., Yang C.W. Effects of salt stress on ion balance and nitrogen metabolism of old and young leaves in rice (*Oryza sativa L.*), *BMC Plant Biol.*, 2012, Vol. 12, No 1, p. 1–11.

Wang L., Pedas P., Eriksson D., Schjoerring J. K. Elevated atmospheric CO_2 decreases the ammonia compensation point of barley plants, *J. Exp. Bot.*, 2013, Vol. 64, No 10, p. 2713–2724.

Wang P.S., Ali M., Chaturvedi S.K. Osmotic adjustment increases water uptake, remobilization of assimilates and maintains photosynthesis in chickpea under drought, *Indian J. Exp. Biol.*, 2007, Vol. 45, No 3, p. 261–267.

Wang R.Z., Gao Q. Climate-driven changes in shoot density and shoot biomass in Leymus chinensis (Poaceae) on the Northeast China Transect (NECT), *Global Ecol. Biogeogr.*, 2003, Vol. 12, p. 249–259.

Wang X., Curtis P. A meta-analytical test of elevated CO_2 effects on plant respiration, *Plant Ecol.*, 2002, Vol. 161, p. 251–261.

Ward J.K. Evolution and growth of plants in a low CO_2 world, *A History of Atmospheric CO_2 and Its Effects on Plants, Animals, and Ecosystems*, Ehleringer J.R., Cerling T.E., Dearing M.D. eds., Springer, New York, 2005, p. 232–257.

Ward J.K., Strain B.R. Effects of low and elevated CO_2 partial pressure on growth and reproduction of *Arabidopsis thaliana* from different elevations, *Plant Cell Environ.*, 1997, Vol. 20, p. 254–260.

Warren C.R. The photosynthetic limitation posed by internal conductance to CO_2 movement is increased by nutrient supply, *J. Exp. Bot.*, 2004, Vol. 55, No 406, p. 2313–2321.

Warren C.R. Stand aside stomata, another actor deserves centre stage: the forgotten role of the internal conductance to CO_2 transfer, *J. Exp. Bot.*, 2008, Vol. 59, p. 1475–1487.

Warren C.R., Adams M.A. Evergreen trees do not maximize photosynthesis, *Trends Plant Sci.*, 2004, Vol. 9, p. 270–274.

Warren C.R., Dreyer E., Adams M.A. Photosynthesis-Rubisco relationships in foliage of Pinus sylvestris in response to nitrogen supply and the proposed role of Rubisco and amino acids as nitrogen stores, *Trees*, 2003, Vol. 17, p. 359–366.

Watanabe C.K., Sato S., Yanagisawa S., Uesono Y., Terashima I., Noguchi K. Effects of elevated CO_2 on levels of primary metabolites and transcripts of genes encoding respiratory enzymes and their diurnal patterns in Arabidopsis thaliana: possible relationships with respiratory rates, *Plant Cell Physiol.*, 2014, Vol. 55, p. 341–357.

Waters M.T., Wang P., Korkaric M., Capper R.G., Saunders N.J., Langdale J.A. GLK transcription factors coordinate expression of the photosynthetic apparatus in *Arabidopsis*, *Plant Cell*, 2009, Vol. 21, p. 1109–1128.

Willekens H., Chamnogpol S., Schraudner M., Langebartels C., Van Montagu M., Inzé D., Van Camp W. Catalase is a sink for H_2O_2 and is indispensable for stress defence in C_3 plants, *EMBO J.*, 1997, Vol. 16, p. 4806–4816.

Wilson C., Lesch S.M., Grieve C.M. Growth stage modulates salinity tolerance of New Zealand spinach (*Tetragonia tetragonioides*, Pall.) and Red Orach (*Atriplex hortensis* L.), *Ann. Bot.*, 2000, Vol. 85, p. 501–509.

Wingler A., Lea P.J., Quick W.P., Leegood R.C. Photorespiration, metabolic pathways and their role in stress protection, *Phil. Trans. R. Soc. London B*, 2000, Vol. 355, p. 1517–1529.

Wong I. E. Elevated atmospheric partial pressure of CO_2 and plant growth, *Photosynth. Res.*, 1990, Vol. 23, p. 171–180.

Woodrow I.E., Berry J.A. Enzymatic regulation of photosynthetic CO_2 fixation in C_3 plants, *Annu. Rev. Plant Physiol. Plant Mol. Biol.*, 1988, Vol. 39, p. 533–594.

Woodward F.I. Stomatal numbers are sensitive to increases in CO_2 from pre-industrial levels, *Nature*, 1987, Vol. 327, p. 617–618.

Woodward F.I., Lake J.A., Quick W.P. Stomatal development and CO_2: ecological consequences, *New Phytol.*, 2002, Vol. 153, p. 477–484.

Wu F., Bao W., Li F., Wu N. Effects of drought stress and N supply on the growth, biomass partitioning and water-use efficiency of *Sophora davidii* seedlings, *Environ. Exp. Bot.*, 2008*a*, Vol. 63, p. 248–255.

Wu F.Z., Bao W.K., Li F.L., Wu N. Effects of water stress and nitrogen addition on leaf gas exchange and fluorescence parameters of *Sophora davidii* seedlings, *Photosynthetica*, 2008*b*, Vol. 46, p. 40–48.

Xu Z.Z., Zhou G.S. Effects of water stress and nocturnal temperature on carbon allocation in the perennial grass, *Leymus chinensis*, *Physiol. Plant.*, 2005*a*, Vol. 123, p. 272–280.

Xu Z.Z., Zhou G.S. Effects of water stress and high nocturnal temperature on photosynthesis and nitrogen level of a perennial grass *Leymus chinensis*, *Plant Soil*, 2005*b*, Vol. 269, p. 131–139.

Xu Z.Z., Zhou G.S. Nitrogen metabolism and photosynthesis in *Leymus chinensis* in response to long-term soil drought, *J. Plant Growth. Regul.*, 2006*a*, Vol. 25, No 3, p. 252–266.

Xu Z.Z., Zhou G.S. Combined effects of water stress and high temperature on photosynthesis, nitrogen metabolism and lipid peroxidation of a perennial grass *Leymus chinensis*, *Planta*, 2006*b*, Vol. 224, No 5, p. 1080–1090.

Xu Z.Z., Zhou G.S. Photosynthetic recovery of a perennial grass *Leymus chinensis* after different periods of soil drought, *Plant Prod. Sci.*, 2007*a*, Vol. 10, No 3, p. 277–285.

Xu Z.Z., Zhou G.S., Wang Y.H. Combined effects of elevated CO_2 and soil drought on carbon and nitrogen allocation of the desert shrub Caragana intermedia, *Plant Soil*, 2007*b*, Vol. 301, p. 87–97.

Xu Z.Z., Zhou G.S., Wang Y.L., Han G.X., Li Y.J. Changes in chlorophyll fluorescence in maize plants with imposed rapid dehydration at different leaf ages, *J. Plant Growth. Regul.*, 2008, Vol. 27, No 1, p. 83–92.

Yamakawa H., Hakata M. Atlas of rice grain filling-related metabolism under high temperature: joint analysis of metabolome and transcriptome demonstrated inhibition of starch accumulation and induction of amino acid accumulation, *Plant Cell Physiol.*, 2010, Vol. 51, p. 795–809.

Yamamoto H., Kato H., Shinzaki Y., Horiguchi S., Shikanai T., Hase T., Endo T., Nishioka M., Makino A., Tomizawa K., Miyake C. Ferredoxin limits cyclic electron flow around PSI (CEF-PSI) in higher plants-stimulation of CEF-PSI enhances non-photochemical quenching of Chl fluorescence in transplastomic tobacco, *Plant Cell Physiol.*, 2006, Vol. 47, p. 1355–1371.

Yamaya T., Oaks A., Boesel I.L. Characteristics of nitrate reductase-inactivating proteins obtained from corn roots and rice cell cultures, *Plant Physiol.*, 1980, Vol. 65, p. 141–145.

Yanagisawa S. Transcription factors involved in controlling the expression of nitrate reductase genes in higher plants, *Plant Sci.*, 2014, Vol. 229, p. 167–171.

Yang J., Zhang J., Wang Z., Zhu Q., Lijun L. Abscisic acid and cytokinins in the root exudates and leaves and their relationship to senescence and remobilisation of carbon reserves in rice subjected to water stress during grain filling, *Planta*, 2002, Vol. 215, p. 645–652.

Yang J.C., Zhang J.H., Wang Z.Q., Zhu Q.S. Activities of starch hydrolytic enzymes and sucrose-phosphate synthase in the stems of rice subjected to water stress during grain filling, *J. Exp. Bot.*, 2001, Vol. 52, No 364), p. 2169–2179.

Yang Y., Guo J., Wang G., Yang L., Yang Y. Effects of drought and nitrogen addition on photosynthetic characteristics and resource allocation of *Abies fabri* seedlings in eastern Tibetan Plateau, *New Forests,* 2012, Vol 43, p. 505–518.

Zeeman S.C., Smith S.M., Smith A.M. The breakdown of starch in leaves, *New Phytol.*, 2004, Vol. 163, No 2, p. 247–261.

Zhang H., Forde B.G. An *Arabidopsis* MADS box gene that controls nutrient-induced changes in root architecture, *Science,* 1998, Vol. 279, p. 407–409.

Zhang J.L., Meng L.Z., Cao K.F. Sustained diurnal photosynthetic depression in uppermost-canopy leaves of four dipterocarp species in the rainy and dry seasons: does photorespiration play a role in photoprotection? *Tree Physiol.*, 2009, Vol. 29, p. 217–228.

Zhang J.X., Kirham M.B. Drought stress-induced changes in activities of superoxide dismutase, catalase, and peroxidase in wheat species, *Plant Cell Physiol.*, 1994, Vol. 35, p. 785–791.

Zhang Q.F. Strategies for developing green super rice, *Proc. Natl. Acad. Sci. USA,* 2007, Vol. 104, p. 16402–16409.

Zhang X.C., Yu X.F., Ma Y.F. Effect of nitrogen application and elevated CO_2 on photosynthetic gas exchange and electron transport in wheat leaves, *Photosynthetica,* 2013, Vol. 51, p. 593–602.

Zhang Y.H., Zhang G., Liu L.Y., Zhao K., Wu L.S., Hu C.X., Di H.J. The role of calcium in regulating alginate-derived oligosaccharides in nitrogen metabolism of *Brassica campestris L.* var. Tsen et Lee, *J. Plant Growth. Regul.*, 2011, Vol. 64, No 2, p. 193–202.

Zheng Z. Carbon and nitrogen nutrient balance signaling in plants, *Plant Signal. Behav.*, 2009, Vol. 4, p. 584–591.

Zhou Z.C., Shangguan Z.P. Effects of elevated CO_2 concentration on the biomasses and nitrogen concentrations in the organs of Sainfoin (*Onobrychis viciaefolia* Scop.), *Agr. Sci. China,* 2009, Vol. 8, p. 424–430.

Zhu J.K. Plant salt tolerance, *Trends Plant Sci.,* 2001, Vol. 6, p. 66–71.

Ziegler C., Feraud M., Jouglet T., Viret L., Spampinato A., Paganelli V., Hammouda M.B., Suzuki A. Regulation of promoter activity of ferredoxin-dependent glutamate synthase, *Plant Physiol. Bioch.*, 2003, Vol. 41, p. 649–655.

Zimmermann P., Hirsch-Hoffmann M., Hennig L., Gruissem W. GENEVESTIGATOR. Arabidopsis microarray database and analysis toolbox, *Plant Physiol.*, 2004, Vol. 136, p. 2621–2632.

Ziska L.H., Morris C.F., Goins E.W. Quantitative and qualitative evaluation of selected wheat varieties released since 1903 to increasing atmospheric carbon dioxide: can yield sensitivity to carbon dioxide be a factor in wheat performance? *Global Change Biol.,* 2004, Vol. 10, p. 1810–1819.

Zrenner R., Stitt M. Comparison of the effect of rapidly and gradually developing water-stress on carbohydrate metabolism in spinach leaves, *Plant Cell Environ.*, 1991, Vol. 14, p. 939–946.

10
Physiological Parameters of *Fucus vesiculosus* and *Fucus serratus* in the Barents Sea during a Tidal Cycle

Inna V. Ryzhik[1,2,*], Anatoly A. Kosobryukhov[3,*], Evgeniya F. Markovskaya[4], Mikhail V. Makarov[1]

[1]*Murmansk Marine Biological Institute, Kola Scientific Center, Russian Academy of Sciences, Murmansk 183010, Russia*
[2]*Murmansk Arctic State University, Murmansk 183038, Russia*
[3]*Institute of Basic Biological Problems, Russian Academy of Sciences, Pushchino, Moscow Region 142290, Russia*
[4]*Petrozavodsk State University, Petrozavodsk 185910, Russia*

Abstract

A comparative study of the photosynthetic rate, catalase activity, photosynthetic pigments content and water content in thalli of *Fucus vesiculosus* L. and *Fucus serratus* L., during a tidal cycle, was carried out. The aforementioned species grow in various horizons of the Barents Sea littoral, so, during the study period, at low tide, bladder wrack was unsubmerged for a time length of six hours, the duration of toothed wrack being exposed to air amounted to two hours. The photosynthetic pigments content in both species was identical and did not change over the course of the tidal cycle. The photosynthetic rate measured by the CO_2 uptake in *F. serratus* was twice as high as in *F. vesiculosus*. As for the photosynthetic rate dynamic pattern at low tide, both species demonstrated three increase peaks (at the onset and halfway through the low tide, and at the onset of high tide), differing in specific values and duration of the draining period of the studied algae. The analysis of carbon-dioxide curves

[*] *Corresponding authors*: alaria@yandex.ru; kosobr@rambler.ru

of the CO_2 gas exchange in bladder wrack thalli during a tidal cycle performed with the model by Farquhar *et al.* (1980), revealed that photosynthesis was limited by the rate of light and dark reactions. The increase in thiobarbituric acid reactive substances content and catalase activity in *Fucus vesiculosus* thalli is indicative of reversible oxidative stress occurring at low tide.

Keywords: *Fucus vesiculosus, Fucus serratus,* Tidal cycle, Draining, Photosynthesis, Photosynthetic pigments, Lipid peroxidation, Catalase activity, Water content

10.1 Introduction

The causes of the different vertical distribution of various algal species throughout the littoral have interested scientists ever since the 19th century. The first person to point out the draining duration as a key factor for such distribution of these plants was Gottfried Berthold, a German scientist, who, in 1882, distinguished five horizons in the littoral zone (Berthold, 1882, cited by Zaneveld, 1937).

Macroalgae growing in the littoral zone are exposed to a drastic change in environmental conditions four times a day, which is associated with the tidal pattern. At low tide, when the water environment gives way to a ground-air one, they are exposed to various negative factors: draining, variations in temperature, light and salinity, *etc.* During this period, macroalgae can demonstrate a decrease in the water content in the thallus (Kamnev, 1989; Bisson and Kirst, 1995; Schagerl and Moostl, 2011), as well as the rate of photosynthetic and respiratory gas exchange (Quadir *et al.*, 1979; Dring and Brown, 1982; Andreev *et al.*, 2012) and the growth rate. The adaptation of algae to periodic draining is often accounted for by the accumulation of moisture-retentive substances (Wiltens *et al.*, 1978; Quadir, *et al.*, 1979, Kamnev, 1989; Davison and Pearson, 1996; Kuznetsov *et al.*, 2006). The adaptation to changes in the intensity of illumination during the tidal cycle in fucoids is largely determined by the change in the activity of processes at the level of light-dependent reactions of photosynthesis (Andreev *et al.*, 2012).

Genus *Fucus* is well represented worldwide. In some seas (the Barents Sea, the White Sea, *etc.*) they can grow in the upper sublittoral. In the Barents Sea, with a proper substrate, they only inhabit the littoral zone. In this case, *Fucus vesiculosus* usually inhabits the upper horizon of the littoral, with *F. serratus* growing in the lower horizon. Consequently, *F. vesiculosus* is exposed to air for a longer period of time, whereas *F. serratus*, at dead tides, may not be drained at all. The differences in the hydrological regime of the habitat of the aforementioned species determine the differences in the regulating mechanisms of physiological processes with changes in environmental factors. Besides, in our opinion, special attention should be given to *F. vesiculosus*, which, on account of its growing in the upper horizon of the littoral, has to survive in more extreme environmental conditions. A comparison of parameters of similar species growing at high latitudes with identical conditions (except for the hydrological regime) has never been implemented before.

In recent decades, to study reactions of the photosynthetic apparatus (PAR) to various external factors, researchers mostly applied the fluorimetric method. Among its notable advantages we can mention ease of technical implementation and the unambiguous biophysical interpretation of the defined parameters. However, using this method, one can only assess the PAR efficiency without identifying the mechanisms of a whole range of photosynthetic processes, such as photoadaptation to excessive light and changes in PAR when exposed to environmental factors.

The objective of the present study was a comparative analysis of the PAR functional state in two *Fucus* species (*F. vesiculosus and F. serratus)* during a tidal cycle *in vivo* and the identification of adaptation mechanisms of *F. vesiculosus* to an extended period of draining.

10.2 Materials and Methods

The study was conducted from July to August 2013 and 2014 on the premises of the Seasonal Biological Station "Dalniye Zelentsy" of the Murmansk Marine Biological Institute, Karelian Research Center of the Russian Academy of Sciences, *in situ* in the littoral zone of the Zelenetskaya Bay (69.07 N, 36.04 E).

The objects of the study were even-aged plants (4-5 dichotomous branches) of bladder wrack (*F. vesiculosus* Linnaeus, 1753; Phaeophyceae: Fucales), growing in the upper horizon of the littoral, and toothed wrack (*F. serratus* Linnaeus, 1753; Phaeophyceae: Fucales), growing in the lower horizon. During the experiments, the duration of draining (a period of low tide) for *F. vesiculosus* amounted to approximately 6 hours, for *F. serratus* it was about 2 hours.

The thalli of *Fucus* species are highly-branched, grow densely, with a projective degree of substrate coverage up to 100%. At low tide, an algal layer of 20 cm is formed on the littoral, and particular parts of the thalli may come to be on the surface of the layer ("the shrub surface") and dry up to a large extent. Inside the layer ("the shrub depth"), there are thallus parts or whole thalli not subjected to drying up.

For the purposes of the study, the apical parts of thalli were used. The sample collection was carried out during the tidal cycle in the daytime, at daylight illumination and at an air temperature of 20–25 °C. First, samples were collected from plants that had stayed underwater for over six hours. Then, as the water was receding at low tide, samples were collected every 30 to 60 minutes from plants that were emerging on the surface of the algal layer and were prone to severe drying. After the onset of the next high tide, when the plants were submerged in water again, samples were taken every 10–15 minutes for 2 hours. The frequency of sampling was determined by the tool capacity and the speed of the initial processing of samples. During the experiment, for the assessment of the relationship between the water content in plants and the photosynthetic rate, samples were collected both from the surface of the layer (dried up), and from the inner part of the layer (moist).

The photosynthetic rate was measured by means of a portable infrared gas analyzer LCPro+ (ADC BioScientific Ltd., Great Britain). The apical part of the thallus was placed inside the assimilative leaf chamber of the gas analyzer, and when determining the stationary values, the rate of CO_2 gas exchange was measured in the setting of atmospheric CO_2, to which Rubisco has the highest affinity. The response of photosynthetic rate to internal CO_2 concentration were determined by means of altering the CO_2 concentration in the air supplied to the assimilation chamber of the gas analyzer within the range of 0–1600 μmol CO_2 mol^{-1} with saturating light intensity. The levels of CO_2 concentration supplied to the leaf chamber were preset using the gas analyzer microprocessor and changed sequentially: 50, 100, 200, 400, 800, 1200, 1600 μmol CO_2 mol^{-1}. The analysis of the carbon dioxide curves was performed using the Farquhar model (Farquhar *et al.*, 1980) with some modifications (Caemmerer and Farquhar., 1982; Harley and Sharkey, 1991) and with the software Photosynthesis Assistant Ver. 1.1.2 (Parsons and Ogston, 1999). According to the authors' equations, the model makes it possible to determine the highest rate of carboxylation (Vc_{max}), the light-saturated electron transport rate (J_{max}), the rate of triose phosphate utilization and some other parameters.

The content of photosynthetic pigments (chlorophylls *a* and *c* ($c_1 + c_2$), and carotenoids (*β*-carotin, violanxanthin, fucoxanthin)) was measured using the method of paper chromatography. The qualitative and quantitative composition of pigments was studied using modified methods (Sapozhnikov, 1964; Li, 1978; Maslova *et al.*, 1986).

The water content in the thalli was calculated by the ratio of dry / wet weight of thallus cuttings (n = 20). The dry weight of the cuttings was measured using scales (VLTE-210) to an accuracy of 0.001 g after drying in the drying oven for 24 hours (T = 105 °C).

The activity of catalase and the level of lipid peroxidation were measured at low tide only in *F. vesiculosus*, since this species is exposed to drying for a much longer period of time. The level of lipid peroxidation, which serves as an oxidative stress marker in algal tissues, was assessed by the accumulation of thiobarbituric acid reactive substances (Esterbauer and Cheesman, 1990; Olenichenko *et al.*, 2008). The activity of the catalase was assessed using the modified spectrometric method (Korolyuk,

1988). The measurements were taken with the spectrophotometer JENWAY 6305 UV/VIS.

All the measurements were performed in biological and analytical triplicate. For statistical data processing, the software applications Statistica 6.0 and MS Excel 2007 were used. The study validity was assessed with the use of Student's *t*-test, where $P = 0.95$. The mean values and the confidence interval are depicted in the charts below.

10.3 Results

10.3.1 Water Content in Algal Thalli

The water content in the thalli of both species of the studied macroalgae at high tide amounted to approximately 80%.

The duration of *F. vesiculosus* being exposed to drying at low tide amounted to 6 hours, for *F. serratus* the duration was 2 hours. During this period of time, the water content in *F. vesiculosus* dropped down to 30%, in *F. serratus* – to 60%, besides, in *F. serratus* significant differences were only observed between the extreme values (the water content in plants staying underwater and in plants exposed to prolonged draining and drying).

With the onset of high tide, the water content in the tissues of macroalgae was restored back to the initial level (Figures 10.1 and 10.2).

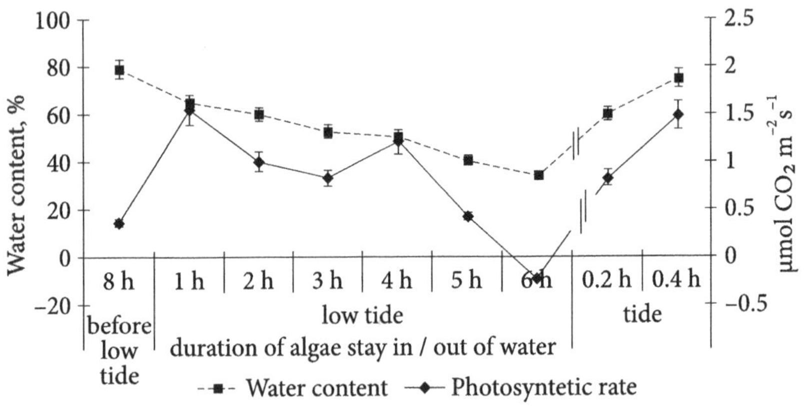

Fig. 10.1. Dynamics of the photosynthetic rate and water content in the thalli of *Fucus vesiculosus* during a tidal cycle

Note: On the abscissa axis, "8 h before low tide" is the beginning of measurements after the 8-hours tide; "1 h, 2 h, 3 h … low tide" are the measurement points at low tide; "0.2 h, 0.4 h tide" are the measurement points during the new tide

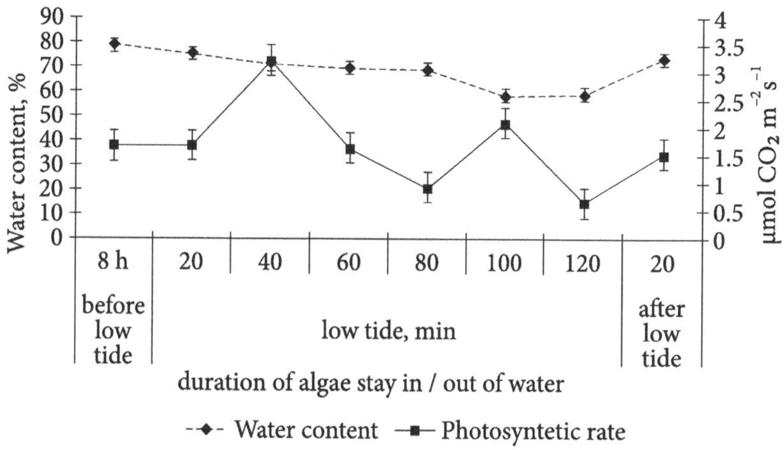

Fig. 10.2. Dynamics of the photosynthetic rate and water content in the thalli of *Fucus serratus* during a tidal cycle

Note: On the abscissa axis, "8 h before low tide" is the beginning of measurements after the 8-hours tide; "20 min, 40 min, 60 min ... low tide" are the measurement points at low tide; "20 min after low tide" is the measurement point after low tide

10.3.2 The Rate of Photosynthesis

For both species during a tidal cycle (the end of high tide – low tide – the onset of high tide), a trimodal curve of photosynthesis was achieved, with two peaks at low tide and one peak at the onset of high tide (Figures 10.1 and 10.2).

10.3.3 Photosynthetic Pigments: Content and Proportion

In the thalli of plants of both species, the photosynthetic pigments content was identical. No significant variations in their content during a tidal cycle were observed (Figure 10.3).

Table 10.1. Chlorophyll *a* / carotenoids (Chl *a* / Car) ratio and the relative size of the light-harvesting complex (xanthosome), assessed by the Chl *a* / (Chl *c* + fucoxanthin) ratio of *Fucus vesiculosus* and *Fucus serratus* thalli during a tidal cycle

Ratios	***F. vesiculosus***		***F. serratus***	
	high tide	low tide	high tide	low tide
Chl *a* / Car	1.92	2.02	2.02	1.92
Chl *a* / (Chl *c* + fucoxanthin)	4.03	3.98	4.90	4.92

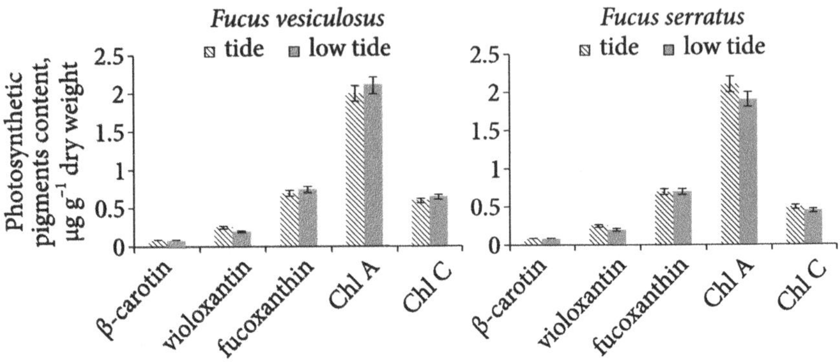

Fig. 10.3. Photosynthetic pigments content in *Fucus vesiculosus* and *Fucus serratus* thalli during a tidal cycle

Furthermore, no significant alterations in the chlorophylls / carotenoids ratio at high tide and at low tide, and likewise no changes in the size of the light-harvesting complex (xanthosome), assessed by the ratio of chlorophyll *a* to the sum of chlorophyll *c* and fucoxanthin (Makarov, 2012), were observed (Table 10.1).

10.3.4 Dependence of the Photosynthetic Rate on the Water Content in the Thallus

To assess the correlation between the rate of photosynthesis and the water content, we conducted a study of the thalli of *Fucus vesiculosus* selected from the surface and from the depth of the algal layer at low tide, and, correspondingly, exposed to a different degree of drying: during low tide, in thalli taken from the surface the water content dropped twofold, whereas inside the layer the decrease amounted to approximately 10% (Figure 10.4). Yet, the rate of photosynthesis decreases in both cases: in the thalli from the surface it drops down to negative values, in the thalli from the depth of the algal layer the resulting values were very low, yet still positive. This is indicative of the fact that water content in a thallus is not a key factor of photosynthetic rate decrease in *F. vesiculosus* at low tide.

10.3.5 Potential Rate of Photosynthesis of Fucus vesiculosus

The possibilities of the device used in the study enable us to obtain carbon-dioxide dependence of photosynthesis at maximum illumination, *i.e.* assess the potential possibilities of the CO_2 gas exchange. This part of the study was performed for *Fucus vesiculosus* only, which is exposed to drying for a long time and loses a considerable amount of water during this period.

The assessment of the photosynthetic rate in algae was made during three-time segments of the tidal cycle: plants emerging at low tide for a long time, plants during the first hours of submersion at high tide and plants staying underwater for a long time (Table 10.2). The comparative analysis of the carbon-dioxide curves was carried out using the Farquhar model.

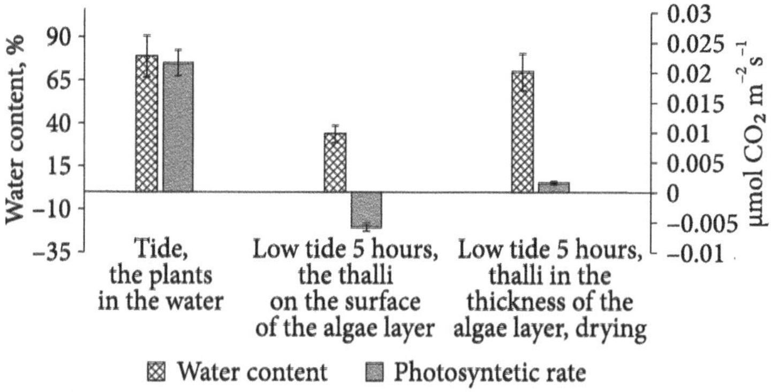

Fig. 10.4. Water content and photosynthetic rate of *Fucus vesiculosus* depending on the location of thalli during low tide

Table 10.2. Approximation parameters of carbon dioxide gas exchange curves in *Fucus vesiculosus* using the Farquhar *et al.* (1980) model

Parameters	Conditions			
	5 hours without water	10 min in the water, a depth of 10 cm	30 min in the water, a depth of 30–40 cm	over 10 hours in the water, a depth of 40 cm
Maximum CO_2 absorption rate, μmol CO_2 m^{-2} s^{-1}	2.2 ± 0.28	4.07 ± 0.11	4.67 ± 0.94	1.78 ± 0.46
Rate of CO_2 dark respiration, μmol CO_2 m^{-2} s^{-1}	0.23 ± 0.03	0.24 ± 0.04	0.49 ± 0.08	0.67 ± 0.16
Maximum ribulose bisphosphate carboxylation rate, μmol CO_2 m^{-2} s^{-1}	0.81 ± 0.1	4.65 ± 0.35	3.57 ± 0.95	3.07 ± 0.74
Carboxylation efficiency, μmol CO_2 m^{-2} s^{-1} $pascal^{-1}$	0.022 ± 0.011	0.032 ± 0.016	0.050 ± 0.013	0.064 ± 0.013
Light-saturated electron transport rate, μmol m^{-2} s^{-1}	2.1 ± 0.4	13.1 ± 1.1	16.6 ± 1.5	7.73 ± 2.6
Triose phosphate utilization rate, μmol m^{-2} s^{-1}	0.93 ± 0.12	1.19 ± 0.11	1.46 ± 0.14	0.80 ± 0.20
Carbon dioxide compensation point, μmol CO_2 m^{-2} s^{-1}	177 ± 5	204 ± 14	114 ± 12	217 ± 14

The highest possible values of the photosynthetic rate in algal thalli that were drained and exposed to drying for a long time were similar to those that were submerged at high tide. However, they were considerably higher than the actual values: at low tide, the gas exchange was negative, at high tide; the photosynthetic rate was three times lower. The calculation of the ratio of light / dark reactions of photosynthesis (J_{max} / Vc_{max}) showed that in plants exposed to air for a long time (during low tide) and in plants staying underwater (over 5 hours at high tide), the J_{max} / Vc_{max} index was the same: 2.4 and 2.5 respectively. The similar correlations of light and dark reactions of photosynthesis result in a similarity in the photosynthetic rate values.

10.3.6 *Lipid Peroxidation and Catalase Activities in Fucus vesiculosus*

The drastic change in the environmental conditions at low tide, resulting in the changes of water content and photosynthetic rate in *Fucus vesiculosus* suggests the action of oxidative stress in algal cells. Indeed, the analysis of thiobarbituric acid reactive substances content in the tissues proved a significant, almost threefold increase in these compounds in the course of the first two hours of draining. However, three hours later, their content dropped down to the initial level which remained unchanged till high tide (Figure 10.5).

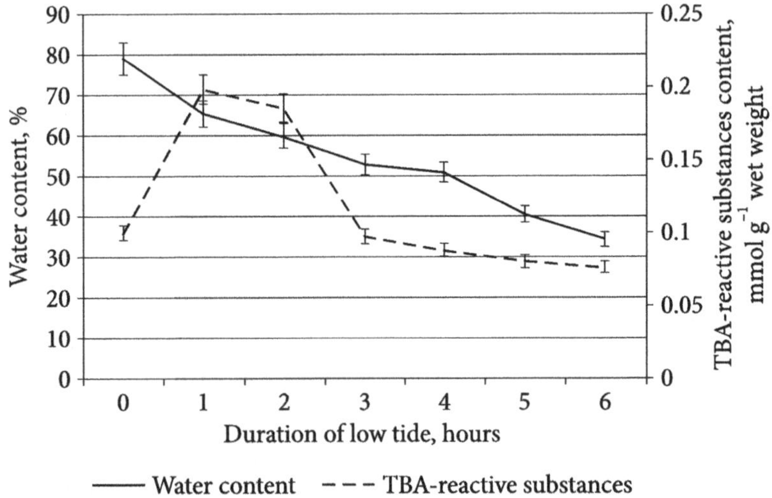

Fig. 10.5. Water and thiobarbituric acid (TBA) reactive substances content in thalli of *Fucus vesiculosus* during low tide

These findings are indicative of oxidative stress occurring during the first hours of low tide, which is associated with the disruption of the membrane structure on account of lipid peroxidation. Such changes are reversible, and by the end of low tide, a restoration process can be observed, which can also be catalyzed by the ferment catalase. The activity of this ferment is noted during the whole period of plants' draining at low tide, and the highest activity can be observed during the 4th hour of low tide (Figure 10.6).

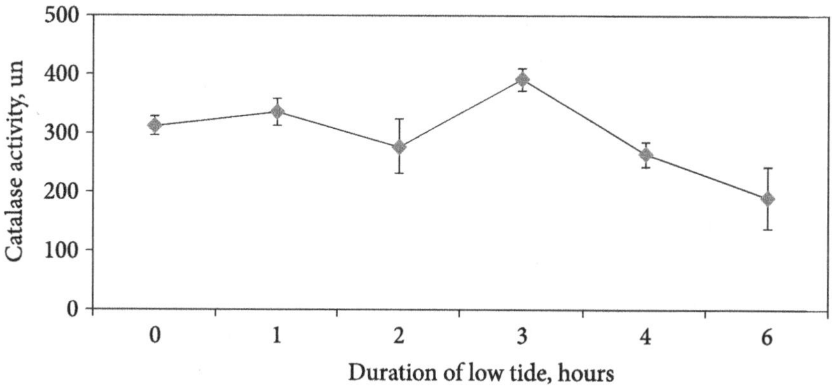

Fig. 10.6. Catalase activity in the thalli of *Fucus vesiculosus* during low tide

The measurement of catalase activity over the course of several tidal cycles showed that in all the cases the rate of the curve remained the same. A certain shift of the activity peak of the ferment either way was determined by the weather conditions (temperature, cloudiness, wind velocity, humidity): on hotter days, periods of increases and decreases in the ferment activity occurred earlier.

10.4 Discussion

On the Barents Sea coast, in the vicinity of the Biological Station "Zelenetskaya Bay", Murmansk Marine Biological Institute, during a tidal cycle, the draining duration of the upper littoral zone amounts to 6–8 hours, for the lower littoral the figure is 2-3 hours. For the implementation of the present work, two species of genus Fucus were selected, which can grow under a wide range of various conditions in the littoral. In the setting of Subarctic, ecotope preferences are observed: *F. vesiculosus* grows in the upper littoral zone and is exposed to more prolonged draining, whereas *F. serratus* inhabits the lower littoral and is less prone to draining.

The comparative analysis of both species showed that in *Fucus serratus* the rate of photosynthesis is two to three times higher than in *Fucus vesiculosus*. However, the dynamic pattern of the rate of photosynthesis in both species was the same – a trimodal curve with two peaks at low tide and one peak at the onset of high tide. The differences between the species manifest themselves in the values and times of photosynthetic peaks. Since *F. serratus* is exposed to draining at low tide for a shorter period of time, in this species the aforementioned pattern occurs within a shorter timeframe. Another similar feature in both species was a decrease in photosynthesis after the second peak towards the end of low tide, but in *F. vesiculosus* it dropped down to negative values of gas exchange, whereas in *F. serratus* it only went down to low, but still positive values. It's also worth noting that the rates of photosynthesis at the onset and at the end of high tide in *Fucus serratus* were the same, whereas for *Fucus vesiculosus* they were different (higher at the end of high tide, than at the beginning of it). The negative values of gas exchange in the thalli of *F. vesiculosus* suggest a considerable inhibition of photosynthesis in this species.

The analysis of photosynthetic pigments content and proportions revealed a similarity of the photosynthetic apparatus of both species. In the course of a tidal cycle, no changes in the photosynthetic pigments content and proportions or in the xanthosome size were observed. These findings suggest that the revealed changes in the rate of photosynthesis are due to the change in the rate of photosynthetic reaction, and are associated with the functional (not structural) rearrangement of the photosynthetic apparatus. If this were not the case, with the structural rearrangement (an enlargement of the lamellar system, an increase in the number of chloroplasts, an increased size and structure of xanthosomes, *etc.*, besides the changes in chloroplast position), changes in the content and proportions of various pigments would also be observed.

Further calculations based on the obtained data showed that in plants at high tide there was three- or fourfold intensification of such processes as electron transport rate, maximum carboxylation rate and efficiency and the rate of CO_2 dark respiration. At the same time, the rate of triose phosphate utilization and the carbon dioxide compensation point had close values. The rate of photosynthesis was already twice as high in as little as 10 minutes, and then after 30 minutes of being underwater at the onset of high tide. This transition into an aquatic environment, which is natural for algae, is immediately (within 10 minutes) accompanied by an increase in the maximum carboxylation rate and electron transport rate, and after 30 minutes other processes also intensified (the rate of triose phosphate utilization and the rate of CO_2 dark respiration). These findings suggest that when passing into an aquatic environment, other carboxylation mechanisms get involved, which are not registered with such method of photosynthesis rate measurement.

The main difference in the living conditions of the studied species is the duration of draining and drying at low tide, which can be the cause of the observed differences. In our work we used a method that only registers the absorption of atmospheric carbon in the form of CO_2, which carboxylates with the involvement of Rubisco. However, in the water environment, the carbon dioxide content is too small, and the main source of carbon for aquatic plants, including algae, is bicarbonate, which is transformed into an accessible form by means of the ferment carboxylase. This means that, in the studied species, the differences in the rate of photosynthesis are associated with the different accessibility of substrates for carboxylation: in *F. vesiculosus,* the lower photosynthetic values are predetermined by the longer period of draining and drying, and the rate of photosynthesis is limited by the lack of bicarbonate.

The possibility of switching between the carboxylation mechanisms and the ability of littoral algae to use atmospheric CO_2 as a source of inorganic carbon for photosynthesis instead of bicarbonate in the water was first discovered in the middle of the 20th century by a pioneer of research in this field, R. G. S. Bidwell (Bidwell, 1958). Although, in some works, a decrease in photosynthesis of littoral algae at low tide was pointed out (Bidwell and Craigie, 1963; Williams and Dethier, 2005), yet, it can be associated with the fact that such studies mostly focused on the prolonged exposure of algae to air. In the present study, in line with studies by other authors (Quadir *et al.*, 1979; Gao *et al.*, 1999), an increase in the photosynthetic rate of littoral algae during the first phase of low tide, with short-term draining, has been revealed. The differences in the degree of response of different algal species to the exposure to air (being outside the water environment) can be associated with the horizon of their growth in the littoral zone (and, correspondingly, the period of being in the air environment during low tide) and the different abilities to switch from one carboxylation mechanism to another (Kremer and Schmitz, 1973). Besides, in *F. vesiculosus* an ability to store carbon in the tissue was also demonstrated, like in CAM-plants, thus this species is capable of performing photosynthetic processes for a certain length of time without an external carbon source (Kawamitsu and Boyer, 1999).

It can be assumed that the resilience of algae to draining is predetermined by their ability to use atmospheric carbon dioxide (for a prolonged period of time). Yet, most published works mainly focus on the problem of hypovolia (a decrease in water content) in thalli. For example, it is known from literary sources that *Fucus vesiculosus* is capable of withstanding a longer period of draining and drying and then quickly restoring the lost moisture: these plants can lose up to 91% of water (Kanwisher, 1957) and remain in the air environment for up to 32 hours without losing viability (Kawamitsu *et al.*, 2000; Andreev *et al.*, 2012).

Our studies have also demonstrated that *F. vesiculosus* can live in the hydric environment for about a month, quickly reactivating its photosynthetic activity as soon as it gets into the water (Ryzhik and Makarov, unpublished data). As exemplified in fucoids of the White Sea, it was also shown that littoral species lose moisture more slowly than sublittoral ones, and *Fucus serratus* has a low physiological plasticity (Andreev *et al.*, 2012).

A number of mechanisms preventing the dehydration of algae at low tide have also been described: the exudation of polysaccharides (*e.g.*, fucoidan) on the thallus surface (Lobban *et al.*, 1985), the accumulation of substances with osmotic properties (*e.g.*, hydrin-like proteins) in the cells (Wiltens, 1975; Quadir *et al.*, 1979; Li *et al.*, 1998), changes in the concentration of ions inside the cells (Bisson and Kirst, 1995), a decrease in the specific surface area (Haylov *et al.*, 1992; Ryzhik, 2005). On the other hand, studies conducted on various algal species have demonstrated that the specific horizon of the littoral zone they inhabit (that is how high or low in the littoral they grow) does not depend on the mechanisms regulating the rate of moisture loss by the thallus (Dorgelo, 1976; Schonbeck and Norton, 1979; Ji and Tanaka, 2002). This can imply that the aforementioned mechanisms mainly provide conditions for maintaining the carbon dioxide metabolism in the aquatic environment and while being drained.

The conducted analysis showed that both species share a similar rate of photosynthesis pattern in the course of a tidal cycle: an increase in the rate of photosynthesis during the initial phase of high tide and low tide and a considerable inhibition of this process during a prolonged period of draining and drying of the plants at low tide. The revealed regularities suggest that this species adapts to its habitat located in the intertidal zone, adjusting for the local conditions of the low tide duration. *Fucus* species are aquatic plants, so the duration of low tide for them is more crucial than that of high tide. In support of this reasoning, we can provide similar data on pigments content and closer photosynthetic values at high tide (*F. serratus* – 2.0 μmol CO_2 m^{-2} s^{-1} and *F. vesiculosus* – 1.5 μmol CO_2 m^{-2} s^{-1}). The focus of the studies carried out on *Fucus vesiculosus* has allowed us to analyze the response of the species to a broader range of environmental changes. Plants remaining in the air environment for a long time at low tide had lower functional photosynthetic parameters (model-calculated), than plants submerged in water. During a long period of being underwater, algae demonstrate an increase in dark respiration, a high rate and efficiency of the carboxylation process, intensified electron-transport activities and even an increase in the carbon dioxide compensation point.

At the onset of high tide, the rate of photosynthesis in *F. vesiculosus* increased almost twofold and this process was accompanied by an electron-transport rate increase in the chloroplast electron-transport chain, and, consequently, an increase in the regeneration rate of the CO_2 acceptor – ribulose bisphosphate. The electron transport rate within the first 10-30 minutes increased six to eight times. The maximum carboxylation rate (Rubisco activities) also increased almost 6 times, which happened to be comparable to the increased rate of electron transport. However, the significant increase in the electron transport rate and the activities of the carboxylation reaction did not lead to a similar rise in the photosynthetic rate. At the onset of high tide, the J_{max} / Vc_{max} parameter increased to 4.6 in plan staying under water for 30 minutes, suggesting a possible limitation of the photosynthetic rate by the activity of Rubisco. As a result, at the onset of high tide, the activity of Rubisco, albeit higher compared to the low tide conditions, never manifests itself through a similar increase in the photosynthetic rate due to the absence of any other source of carbon (bicarbonate).

The obtained data on the photosynthetic activity of the dark and light phase of photosynthesis in *Fucus vesiculosus* revealed a reversible disruption of the photosynthetic activity at low tide, which could have been caused by an increase in illumination and temperature and the developing oxidative stress.

The revealed similar pattern of photosynthetic variations during a tidal cycle in both species can be accounted for by similar mechanisms. An increase in the photosynthesis in algae during the first hours of low tide was reported, when plants get exposed to high illumination, which increases fivefold and more (like in *F. vesiculosus)*, since in the littoral zone of the Barents Sea over 80% of photosynthetically active radiation is entrapped by a meter-deep water layer (Makarov *et al.*, 2010). On the other hand, the rate of photosynthesis of different species of littoral algae during the first hours of low tide can either increase or decrease significantly, and according to researchers, it is highly unlikely that this effect can be associated with an increase in illumination alone (Johnson *et al.*, 1974;

Quadir *et al.*, 1979; Wiltens *et al.*, 1978, Oquist and Fork, 1982; Hanelt, 1998; Gan-Lin and Shuang-Lin, 2008).

Changes in the photosynthetic rate can be associated with an enhanced electron transport rate and increased content of reactive oxygen intermediate, as well as a further reactivation of defense responses during the photochemical stage of photosynthesis (an increase in thermal dissipation, fluorescence, *etc.*) (Kolupaev, 2007; Collén and Davison, 1999; Yoshinobu *et al.*, 2000; Heber *et al.*, 2014). The obtained data on the accumulation of lipid peroxidation products in *F. vesiculosus*, which occurred during the early stage of draining, also suggest an increase in reactive oxygen intermediate (Figure 10.5). The consequent decrease in reactive oxygen intermediate is associated with the activation of antioxidant systems, the active functioning of which can ensure and maintain the second peak of photosynthetic activity (Figure 10.6). An increase in the catalase content and / or activity during the first peak is accounted for by the accumulation of reactive oxygen intermediate, and the further decrease in its activity is associated with a decrease in the concentration of hydrogen peroxide (Radyukina, 2013) and with the involvement of other systems of antioxidant defense. Along with catalase, other enzymatic components of antioxidant defense known to be present in algae can also get involved.

The decrease in the rate of photosynthesis in *Fucus vesiculosus* down to negative values after the exposure of thalli to the air environment for an extended period of time can indicate either a limited ability of fixing atmospheric CO_2, or a limited capacity of antioxidant systems, which ensure the activity of the photosynthetic apparatus. In support of the first assumption we can provide our findings that indicate an ability to quickly reactivate the photosynthetic apparatus of algae after the onset of high tide and also the data on the continued activity of the photosynthetic apparatus in plants exposed to prolonged dehydration (the findings of the present study concerning the measurement of the activity of the photosynthetic apparatus. in plants sampled from inside the algal layer at low tide, as well as literature data) (Bidwell and Craigie, 1963; Williams and Dethier, 2005; Quadir *et al.*, 1979; Schagerl and Moostl, 2011).

The fact that the studied species grow in different littoral horizons may account for the application of different mechanisms of resistance to draining, which can manifest themselves either in the form of hydrolability (an ability to lose a considerable amount of water at low tide and then quickly restore its content with the onset of high tide), as is evidenced by *Fucus vesiculosus*, or in the form of hydrostability – in the case of *Fucus serratus*. Moreover, their adaptation to specific habitat conditions is obviously connected with a different degree of activation of protective antioxidant systems and carboxylation mechanisms, depending on the time of exposure to the water or air environment. The duration of the effective work of all defense mechanisms is likely to be in correlation with the mean duration of the draining period at low tide in the local growth conditions of the species (Flores-Molina *et al.*, 2014). For the confirmation of the assumptions made, additional research needs to be carried out.

The prolonged periodic exposure of littoral algae to the air environment at low tide has its positive effect as well. Quite a big number of marine species, both animals and plants, are incapable of sustaining prolonged draining and drying. And although the habitat conditions of *Fucus* species are far from optimal, their existence in the littoral zone proves to be advantageous, protecting them from epiphytes and herbivores (Makarov *et al.*, 2013; Ryzhik *et al.*, 2014).

Abbreviations

[CO_2] – carbon dioxide concentration
Rubisco – ribulose bisphosphate carboxylase / oxygenase
TBA – thiobarbituric acid

References

Andreev V.P, Maslov Y.I., Sorokoletova Y.F. Functional peculiarities of the photosynthetic apparatus in three *Fucus* species in the setting of the White Sea. The impact of dehydration, *Physiology of Plants,* 2012, Vol. 59, No. 2, p. 244–250.

Berthold G.D.W. *Über die Vertheilung der Algen im Golf von Neapel nebst einem Verzeichnis der bisher daselbst beobachteten Arten*, 1882.

Bidwell R. G. S. Photosynthesis and metabolism of marine algae. II. A survey of rates and products of photosynthesis in $C^{14}O_2$, *Can. J. Bot.*, 1958, Vol. 36, p. 337–349.

Bidwell R.G.S., Craigie J.S. A note on the greatly reduced ability of *Fucus vesiculosus* to absorb or evolve CO_2 when not submerged, *Can. J. Bot.*, 1963, Vol. 41, No. 2, p.179–182.

Bisson M.A., Kirst G.O. Osmotic acclimation and turgor pressure regulation in algae, *Naturwissenschaften,* 1995, Vol. 82, p. 461–471.

Caemmerer S., Farquhar G.D. Some relationships between the biochemistry of photosynthesis and the gas exchange rates of leaves, *Planta*, 1982, Vol. 153, p. 376–387.

Collén J., Davison I.R. Stress tolerance and reactive oxygen metabolism in the intertidal red seaweeds *Mastocarpus stellatus* and *Chondrus crispus*, *Plant, Cell Environ.*, 1999, Vol. 22, p. 1143–1151.

Davison I.R., Pearson G.A. Stress tolerance in intertidal seaweeds, *J. Phycol.*, 1996, Vol. 32, p. 197–211.

Dorgelo J. Intertidal fucoid zonation and desiccation, *Hydrobiol. Bull.*, 1976, Vol. 10, No 2, p. 115–122.

Dring M.J., Brown F.A. Photosynthesis of intertidal brown algae during and after periods of emersion: a renewed search for physiological causes of zonation, *Mar. Ecol. Prog. Ser.*, 1982, Vol. 8, p. 301–308.

Esterbauer H., Cheesman U. Determination of aldehydic lipid peroxidation products: malonaldehyde and 4-hydroxymalonaldehyde, *Methods in Enzymology*, San Diego, 1990, Vol. 186, p. 302-310.

Farquhar G.D., Caemmerer S., Berry J.A. A biochemical model of photosynthetic CO_2 assimilation in leaves of C_3 plants, *Planta*, 1980, Vol. 149, No. 1, p. 78–90.

Flores-Molina M.R., Thomas D., Lovazzano C., Nunez A., Zapata J., Kumar M., J.A. Correa, Contreras-Porcia A.C. Desiccation stress in intertidal seaweeds: Effects on morphology, antioxidant responses and photosynthetic performance, *Aquatic Botany*, 2014, Vol. 113, p. 90–99.

Gan-Lin G., Shuang-Lin D. Effects of desiccation on the growth and photosynthetic rate of four intertidal macroalgae from different vertical locations, *Trans. Oceanol. Limnol.*, 2008-04.

Gao K., Ji Y., Aruga Y. Relationship of CO_2 concentrations to photosynthesis of intertidal macroalgae during emersion, *Sixteenth International Seaweed Symposium*, Springer Netherlands, 1999, p. 355–359.

Hanelt D. Capability of dynamic photoinhibition in Arctic macroalgae is related to their depth distribution, *Mar. Biol.*, 1998, Vol. 131, p. 361–369.

Harley P.C., Sharkey T.D. An improved model of C_3 photosynthesis at high CO_2: Reversed O_2 sensitivity explained by lack of glycerate reentry into the chloroplast, *Photosynth. Res.*, 1991, Vol. 27, p. 169–178.

Haylov K.M., Prazukin A.V., Kovardakov S.A., Rygalov V.Y. *Functional Morphology of Marine Multicellular Algae*, Naukova Dumka, Kiev, 1992, 280 p.

Heber W., Lange O.L., Shuvalov V.A. The accumulation and dissipation of light energy by plants as complementary processes involved in the support of plants existence, *Problems of Regulation in Biological Systems*, Rubin A.B. ed., Moscow–Izhevsk, RDS "Regular and Chaotic Mechanics", 2006, 480 p.

Ji Y., Tanaka J. Effect of desiccation on the photosynthesis of seaweeds from the intertidal zone in Honshu, *Japan. Phycol. Res.*, 2002, Vol. 50, p. 145–153.

Johnson W.S., Gigon A., Gulmon S.L., Mooney H.A. Comparative photosynthetic capacities of intertidal algae under exposed and submerged conditions, *Ecology*, 1974, Vol. 55, No. 2, p. 450–453.

Kamnev A.N. The *Structure and Functions of Brown Algae*, Moscow, MSU Publ. House, 1989, 200 p.

Kanwisher J. Freezing and drying in intertidal algae, *Biol. Bull.*, 1957, Vol. 113, No. 2, p. 275–285.

Kawamitsu Y., Boyer J.S. Photosynthesis and carbon storage between tides in a brown alga, *Fucus vesiculosus, Mar. Biol.*, 1999, Vol.133, No 2, p. 361–369.

Kawamitsu Y., Driscoll T., Boye J.S. Photosynthesis during desiccation in an intertidal alga and a land plant, *Plant Cell Physiol.*, 2000, Vol. 41, No. 3, p. 344–353.

Kolupaev Y.Y. Reactive oxygen intermediate in plants in the presence of stress factors: formation and possible functions, *Bulletin of Kharkiv National Agrarian University, Biology Series*, 2007, Vol. 3, No 12, p. 6–26.

Korolyuk M.A. The method of assessment of catalase activity, *Laboratory Science,* 1988, No. 1.

Kuznetsov V.V., Radyukina N.L., Shevyakova N.I. Polyamines under stress: biological role, metabolism and regulation, *Russ. J. Plant Physiol.*, 2006, Vol. 53, No. 5, p. 658–683.

Kremer B.P., Schmitz K. CO_2-fixierung und stofftransport in benthischen marinen algen. IV. Zur ^{14}C assimilation einiger litoraler braunalgen in submersen und emersen zustand, *Z. Pflanzenphysiol.*, 1973, Vol. 68, p. 357.

Li R., Brawley S.H., Close T.J. Proteins immunologically related to dehydrins in fucoid algae, *J. Phycol.*, 1998, Vol. 34, p. 642–650.

Li B.D. Ecological aspects of photosynthesis in marine plants, Publ. House of Dalnevostochnyi Research Centre AS USSR, Vladivostok, 1978, p. 38–54.

Lobban C.S., Harrison P.J., Duncan M.J. *The Physiological Ecology of Seaweeds*, Cambridge University Press, Cambridge, 1985, 242 p.

Makarov M.V. Adaptation of the light-harvesting complex of the Barents Sea brown seaweed *Fucus vesiculosus* L. to light conditions, *Dokl. Biol. Sci.*, 2012, Vol. 442, No. 1, p. 58–61.

Makarov M.V., Ryzhik I.V., Voskoboinikov G.M. The effect of *Fucus vesiculosus* L. (Phaeophyceae) depth of vegetation in the Barents sea (Russia) on its morphophysiological parameters, *Int. J. Algae,* 2013, Vol. 15, No. 1, p. 77–90.

Makarov M.V., Ryzhik I.V., Voskoboinikov G.M., Matishov G.G. The relationship between the depth of growth and morphological-physiological parameters of *Fucus vesiculosus* L. of the Barents Sea, *Doklady Biological Sciences,* 2010, Vol. 430, No 3. p. 427–429.

Maslova T.G., Popova I.A., Popova O.F. A critical assessment of the spectrophotometric method for quantitative measurement of carotenoids, *Soviet Plant Physiol.,* 1986, Vol. 33, No. 3, p. 615–619.

Olenichenko N.A, Gorodkova Y.S., Zagoskina N.V. The relationship between exogenous phenol compounds and lipid peroxidation in wheat plants, *Agricultural biology,* 2008, No 3. p. 58–61.

Oquist G., Fork D.C. Effects of desiccation on the excitation energy distribution from phycoerythrin to the two photosystems in the red alga *Porphyra perforate, Physiol. Plant.,* 1982, Vol. 56, p. 56–62.

Parsons R., Ogston S. *Photosynthesis Assistant: Tools for Analysis of Photosynthesis,* Version 1.1.2, Dundee Sci., Dundee, UK, 1999.

Sapozhnikov D.I. ed. *Pigments of Green Plants Plastids and Methods of Their Study,* Nauka, Moscow-Leningrad, 1964, 120 p. [in Russian]

Ryzhik I.V., Makarov M.V., Voskoboinikov G.M. The physiological state of intertidal brown seaweeds *Fucus serratus* Linnaeus, 1753 and *Fucus distichus* Linnaeus, 1767 cultivated on a biofiltration system in the Barents Sea, *Russ. J. Mar. Biol.,* 2014, Vol. 40, No. 2, p. 119–124.

Quadir P., J. Harrison, DeWreede R.E. The effects of emergence and submergence on the photosynthesis and respiration of marine macrophytes, *Phycologia,* 1979, Vol. 18, No. 1, p. 83–88.

Radyukina N.L. The functioning of the antioxidant system of wild-growing plants species under the short-term impact of stress factors, *Doctoral Thesis in Biological Sciences,* Moscow, 2015, 200 p.

Ryzhik I.V. Morphological and functional peculiarities of industrially cultivated algal species from different biotopes of the Barents Sea, *Published Summary of Ph.D. Thesis in Biological Sciences,* Murmansk, 2005.

Schagerl M., Moostl M. Drought stress, rain and recovery of the intertidal seaweed *Fucus spiralis, Mar. Biol.,* 2011, Vol. 158, p. 2471–2479.

Schonbeck, M.W., Norton, T.A. An investigation of drought avoidance in intertidal fucoid seaweeds, *Bot. Mar.*, 1979, Vol. 22. p. 133–44.

Williams S.N., Dethier M.N. High and dry: variation in net photosynthesis of the intertidal seaweed *Fucus gardneri*, *Ecology*, 2005, Vol. 86, p. 2373–2379.

Wiltens J.S. Ultrastructural and chlorophyll fluorescence responses on intertidal marine algae to desiccation, M.Sc. Thesis, Simon Fraser University, B.C., 1975.

Wiltens J., Schreiber U., Vidaver W. Chlorophyll fluorescence induction: an indicator of photosynthetic activity in marine algae undergoing desiccation, *Can. J. Bot.*, 1978, Vol. 56, p. 2787–2794.

Yoshinobu K., Driscoll T., Boyer J.S. Photosynthesis during Desiccation in an Intertidal Alga and a Land Plant, *Plant Cell Physiol.*, 2000, Vol. 41, No 3. p. 344–353.

Zaneveld J.S. The littoral zonation of some *Fucaceae* in relation to desiccation, *J. Ecol.*, 1937, Vol. 25, No. 2, p. 431–468.

History and Biography – Tribute

11
Benson's Protocol[Φ]

Arthur M. Nonomura[1,*], Karl Y. Biel[2,3,*], Irina R. Fomina[2,3,*], Wai-Ki "Frankie" Lam[4], Daniel P. Brummel[5], Allison Lauria[4], Michael S. McBride[1]

[1]*Carbon Reactions of Photosynthesis Sector, BRANDT iHammer, 479 Village Park Drive, Powell, Ohio 43065, USA*
[2]*Institute of Basic Biological Problems, Russian Academy of Sciences, Pushchino, Moscow Region 142290, Russia*
[3]*Biosphere Systems International Foundation, Tucson, Arizona 85755, USA*
[4]*BRANDT Consolidated, 3654 S. Willow Ave., Fresno, California 93725, USA*
[5]*Department of Agronomy of Iowa State University, 2104 Agronomy Hall, Ames, Iowa 50011-1010, USA*

"Those adventures began a delightful series of scientific and cultural exchanges which have enriched both our lives and our Science."– Andrew A. Benson
Philosophy of the Tracer Method, *Radioisotopes*, 1977, Vol. 26, No. 5, pp. 348–356

Abstract
Andrew A. Benson was the first to recognize the building blocks for the synthesis of organic compounds that satisfy the energetic needs and demands of life – whether autotrophic, mixotrophic, or heterotrophic. Bio-

[Φ] We have published Benson's biography in different journals and books (see, for example, Biel, Fomina *et al.*, 2014; Biel and Fomina, 2015; Fomina and Biel, 2014, 2016) because we sincerely believe in widely disseminating his scientific contributions. Here we present Benson's Protocol with original research that upholds Benson's contributions to agriculture.

**Corresponding authors*: Art.N@BRANDTiHammer.co, karlbiel@hotmail.com; irafomi@rambler.ru chemical optimization and ecological evolutionary stability of these fundamental metabolic mechanisms unfolded elegantly resulting in Benson's Protocol that has been applied for decades to lead science to the elucidation of most of the metabolic pathways known at this time. The carbon reactions of photosynthesis work together with other energy and oxidative reactions and secondary biosynthetic processes and, thereby, ensure maintenance of the biosphere. Protection of life processes is by an array of agents that neutralize ever-present reactive species, as well as, repair the damage they cause. Final discoveries with Benson combined with robust crop investigations by colleagues have resulted in the recognition of a most demonstrable modulation of glycoconjugation for introduction to agriculture of a new class of plant growth regulator, BRANDT iH026a. Indeed, we present investigations of iH026a in the field that show enhanced qualities and quantities of crops. Finally, we celebrate what would have been Benson's 101st birthday.

Keywords: Autoradiography, Carbon reactions of photosynthesis, Chromatographic analysis, Glycoconjugation, Lectin, Radioisotope, Reductive pentose phosphate cycle

11.1 Introduction

As a 30-year-old scientist in 1947, Andrew A. Benson had the entire world's supply of concentrated radioactive carbon dioxide, $^{14}CO_2$, synthesized from the cyclotron of the Old Radiation Lab, University of California, Berkeley. Benson boldly fed this $^{14}CO_2$ to photosynthetic organisms to track formation of photosynthates (Calvin and Benson, 1948). One of the greatest achievements in biochemistry of the 20th century, elucidation of the cyclic sequence of reactions of carbon fixation, ensued; humanity gaining a key to understanding the process by which terrestrial life flourished. Indeed, the life sciences began a new stage in the study of metabolism that has been described in many reviews (Benson, 2002*a*, 2002*b*; Adrianov *et al.*, 2007; Buchanan *et al.*, 2007; Titlyanov and Titlyanova, 2007; Lichtenthaler *et al.*, 2008, 2015*a*, 2015*b*; Govindjee, 2010; Buchanan, 2012; Buchanan and Wong, 2013; Fomina and Biel, 2014, 2016; Buchanan and Douce, 2015; Biel and Fomina, 2015; and others). Benson's Protocol (Nonomura, Lorimer *et al.*, 2016; Nonomura, Holtz *et al.*, 2017) combines the use of definitive tracers and chemical separation of the products of metabolism. Its significance for biology and science in general should be fully recognized because Benson, in laying the groundwork for the elucidation of the Benson–Bassham–Calvin cycle from 1946–1954, had, as a consequence, founded his eponymous protocol that was to be applied, thereafter, to unveil most of the biochemical pathways known today (Nonomura, Holtz *et al.*, 2017). For example, Benson's Protocol was recently applied to symbiotic dinoflagellates to elucidate metabolic pathways on release of photosynthate (Gates *et al.*, 1995; Biel *et al.*, 2007, and others). In the course of 75 years of scientific research, Benson collaborated with hundreds of scientists to apply at least 15 different isotopes of different elements to reveal structures and metabolic pathways in animals, plants, microbes, cell membranes and ecosystems. When Benson fed and tracked various radioisotopes in plants and animals, he and his colleagues discovered the paths of sulfocarbohydrates, sulfolipids, phospholipids, and arsinolipids; the paths of photosynthates fed to animals; the path to waxes; the path from waxes into the marine food chain; the path to membranes; and the path to glycoregulation. Each steppingstone laid down by Benson paved the path to ever-more discoveries of higher order.

Benson's Protocol has been charted (Nonomura, Holtz *et al.*, 2017) and here we supplement details of his initial separation of the various metabolites. In the 1940s, Benson's innovations, particularly of chromatographic mobile phases, made 2-dimensional (2D) paper chromato-

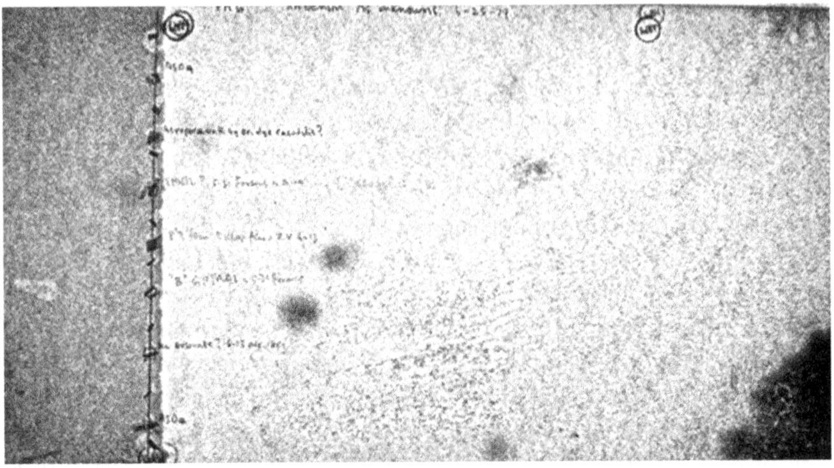

Chromatogram 11.1. Benson explored how metabolites could be isolated. He experimented with various radiochemical preparations and, in this case, Benson directly viewed separation of metabolites of the arsenic radioisotope, ^{74}As, by paper chromatography in 1973. Later, on 25 June 1979, Benson signed the paper chromatogram after recording notes

graphy the metabolite separation tool of choice. At the onset, exploratory isolations of radiolabel metabolites were run, such as exemplified by Benson's signed paper chromatography experiment using radioactive arsenic, ^{74}As, displayed in Chromatogram 11.1. These chromatograms were analyzed to determine the suitability of mobile phases and treatments. The next step was to undertake 2D paper chromatography with the best preparations and radiolabels, but the question remained of what exactly were the unknowns that were being isolated. The answer was found in chromatographic keys to autoradiograms, such as shown in Chromatogram 11.2, charting the position of each and every known standard run on 2D paper chromatograms. The positions of the known chromatographic spots were compared to unknowns. This took years of chromatographic runs to achieve highly reproducible standards. The metabolites had to be tracked with absolute certainty and this method was developed by feeding photosynthetic organisms with a cyclotron-synthesized radioactive carbon that had a long life, ^{14}C. This combination of autoradiography and Benson's keys elucidated the R_f for each of the metabolites of $^{14}CO_2$, for the most part, using two different mobile phases that Benson had developed, Phenol-Water and Butanol-Propionic Acid-Water. The key exhibited in Figure 11.2 is exceptional because it also designated variable movements according to pH of origin – indicated by acidic origin,

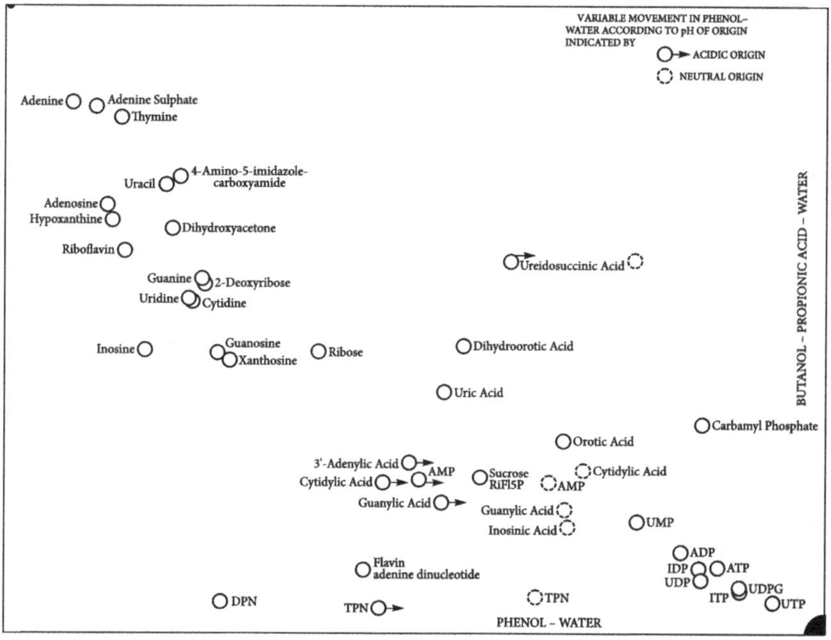

Chromatogram 11.2. Founding Benson's Protocol: Keys to autoradiograms charted the position of each and every known standard run on 2-dimensional paper chromatograms. This took many state of the art chromatographic runs to achieve highly reproducible standards. The positions of chromatographic spots were then compared to each unknown. Thereafter, numerous experiments that revealed corresponding chromatographic spots of ^{14}C-metabolites established Benson's Protocol. This key mapped the R_f for each of the metabolites using two different mobile phases, Phenol-Water and Butanol-Propionic Acid-Water and also designated variable movement according to pH of origin indicated by acidic origin, circle; or neutral origin, broken circle

circle; or neutral origin, broken circle. Benson focused on the group of compounds in the lower right corner of his 2D autoradiography because, since 1949, he had discovered those compounds appeared within the first seconds of exposure to $^{14}CO_2$ (Benson *et al.*, 1950). Examples of Benson's meticulously hand-labeled autoradiograms are exhibited in Chromatograms 11.3, 11.4, 11.5 and 11.6.

Autoradiography was neither quick nor easy. Each 2D paper chromatographic run consumed at least two days, and sometimes more. Typically, X-ray film required a week and longer exposure times. Indeed, the fact that Benson had in the course of his 76-year scientific lifetime produced over 1000 autoradiograms is miraculous. Here, we summarize how Benson and his colleagues first described the photosynthetic reductive pentose phosphate cycle; and then we add original research investigations that conclusively show benefits of reversible glycoconjugation to agriculture (see also: Biel, Fomina *et al.*, 2014; Biel and Fomina, 2015; Fomina and Biel, 2016).

11.2 Benson–Bassham–Calvin and Lectin Cycles

American scientists, Andrew A. Benson, James A. Bassham, and Melvin Calvin (Photo 11.1) have been recorded as the first to recognize the mechanism of photosynthetic carbon metabolism. Their polished and sophisticated scientific inquiries and profound insights combined to describe the chemical reactions that fix atmospheric CO_2 in the cells of all photosynthetic organisms. To be true, the chemical reactions came together in a relatively short period by 1954 (Calvin and Benson, 1948; Benson and Calvin, 1950*a*, 1950*b*; Benson *et al.*, 1950; Benson, 1952; Bassham, Benson *et al.*, 1954) and the cycle was named in honor of them.

Photo 11.1. Discoverers of the path of carbon in photosynthesis (from left to right): James A. Bassham, Andrew A. Benson, and Melvin Calvin (Benson presented this photograph as a gift to Karl Biel on 9 September 1988, and it may be seen also in the following: Biel, Fomina *et al.*, 2014; Biel and Fomina, 2015; Govindjee *et al.*, 2016)

The Benson–Bassham–Calvin cycle (also known as C_3 photosynthesis) was developed mainly based on experiments with unialgal cultures of either *Scenedesmus* or *Chlorella*. Benson cultivated large volumes of algae under fluorescent illumination in defined water nutrient media and, yet, he was primarily a fine atomic chemist. Prerequisite to working with radioactivity, Benson assembled and operated sophisticated instrumentation, one of which is shown in Photo 11.3, Benson's electrometer flow-through detector, circa 1950, for measuring background radiation and $^{14}CO_2$-gas flow.

By the earliest applications of Benson's Protocol, it was proven that primary CO_2-assimilation occurs in the carboxylation reaction of ribulose-1,5-bisphosphate (RuBP) and the first stable products are two molecules of 3-phosphoglyceric acid (3-PGA), then reduced to 3-phosphoglyceric aldehyde and dihydroxyacetone phosphate (DOAP) (Figure 11.1). These triose phosphates yielded hexose phosphate esters that were then used for the synthesis of glucose, sucrose and various other photosynthates. Regeneration of RuBP provided interconversion of phosphoric sugar esters (PES) in the sequence of the transketolase and aldolase reactions that lead ultimately to the formation of ribulose-5-phosphate. Then ribulose-5-phosphate is phosphorylated with the expenditure of ATP.

The diagram of Figure 11.1 presently links the lectin cycle (Nonomura and Benson, 2014) with the Benson–Bassham–Calvin cycle (adapted from Bassham *et al.*, 1954; Benson, 2002*a*) by feeding photosynthates, such as sucrose, into it. Sucrose, glucose and methyl-*ß*-D-glucopyranoside (MeG) are naturally present in vacuoles and they chemically compete for binding sites in lectins. In nature, MeG binds reversibly, and therefore, when the concentration of sucrose is reduced, for example, by respiratory metabolism, then MeG wins, binding to lectin while displacing sucrose. During crop stress, deconjugated sucrose may be utilized toward protection or tolerance to environmental extremes, such as, low light intensity and midday depression. Deconjugated sucrose also may be transported to enhance flavor of fruit and increase productivity. On the other hand, when sugars build in concentration by the supply of fresh photosynthates from the Benson–Bassham–Calvin cycle, then they may outcompete MeG to bind lectins while displacing MeG.

During the reduction of CO_2 under illumination, plants produce organic and amino acids along with PES, such as glycolate, glycine, glycerate, malate, alanine, aspartate, and serine. The glycolate (GOGAT) pathway is physiologically associated with photorespiration and origi-

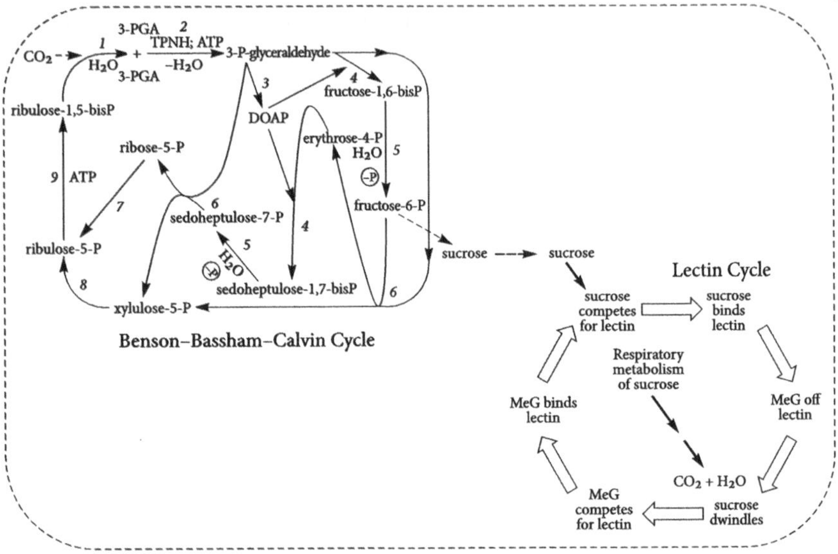

Fig. 11.1. The Benson–Bassham–Calvin cycle, top, (adapted from Bassham *et al.*, 1954; Benson, 2002*a*) feeds photosynthates such as sucrose into the plant lectin cycle, bottom. Sucrose and MeG are naturally present in vacuoles and may chemically compete for binding sites in lectins. In nature, MeG binds reversibly, and therefore, when sucrose dwindles, for example by respiration, then MeG binds to lectin while displacing sucrose. On the other hand, when sucrose builds up concentration from, *e.g.*, fresh supplies from the Benson–Bassham–Calvin cycle, then sucrose may outcompete MeG for lectin binding sites.
Note: BisP – bisphosphate, P – phosphate, TPNH – reduced triphosphopyridine nucleotide, 1 – carboxydismutase system, 2 – triosephosphate dehydrogenase, 3 – phosphotriose isomerase, 4 – aldolase, 5 – phosphatase, 6 – transketolase, 7 – phosphopentoisomerase, 8 – phosphoketopentose epimerase, 9 – phosphopentokinase; MeG – methyl-*ß*-*D*-glucopyranoside

nates from 5-carbon intermediates of the Benson–Bassham–Calvin cycle (Zelitch, 1973, 1972; Tolbert, 1981, 1997; and others). Formation of alanine and C_4 acids occurs by conversion of 3-PGA, leading out of the reductive pentose phosphate cycle. Analogous research was carried out by scientists in the Soviet Union (Russia) on species of higher plant families confirming the existence of this cycle in all photosynthetic eukaryotes (Doman, 1952, Nichiporovich, 1953, Zalensky *et al.*, 1955; Doman *et al.*, 1958; and others). The reductive pentose phosphate cycle is inherent to eukaryotic plants, prokaryotic cyanobacteria, and to most prokaryotic chemoautotrophic bacteria and purple phototrophic bacteria (Kondratieva, 1963, 1996; Bassham and Buchanan, 1982).

11.3 Types of Photosynthetic Carbon Metabolism in Prokaryotes and Eukaryotes

After discovery of the reductive pentose phosphate cycle, many different paths of photosynthetic carbon fixation and reduction were discovered and described, thanks to Benson's Protocol, in both prokaryotes and eukaryotes, such as the 3-hydroxypropionate bicycle (Herter *et al.*, 2002, Zarzycki *et al.*, 2009); reducing citrate or Arnon–Buchanan cycle (Evans *et al.*, 1966; Buchanan and Sirevåg, 1977; Buchanan and Arnon, 1990), Crassulacean acid metabolism (CAM) (Osmond and Allaway, 1974; Kluge, 1976; Osmond, 1978; Winter and Troughton, 1978; and others), C_4 or cooperative photosynthesis (Hartt *et al.*, 1954; Nezgovorova, 1956; Karpilov, 1960, 1969; Kortschak *et al.*, 1965; Hatch and Slack, 1966; and others); C_3 / C_4 photosynthesis (Monson *et al.*, 1984; and others), and C_4-CAM photosynthesis (Bil' and Gedemov, 1980; Bil' *et al.*, 1981, 1983; Shomer-Ilan *et al.*, 1981; Nishio and Ting, 1987; Guralnick *et al.*, 2002).

All of these variations of prokaryotic and eukaryotic phototrophic pathways contribute to the total photosynthetic productivity of aquatic and terrestrial landscapes and are currently referred collectively as the carbon reactions of photosynthesis (Buchanan 2016).

11.4 Regulation of Photosynthates

Targeted regulation of photosynthetic carbon metabolism in plants and algae (Benson and Muscatine, 1974; Benson *et al.*, 1978, 2009; Blanquet *et al.*, 1979; Benson and Nonomura, 1992; Nonomura and Benson, 1992, 2013; Biel *et al.*, 2010; Nonomura *et al.*, 2011) has resulted from the first recognition of a role for the plant lectin cycle by Nonomura and Benson (2014). Lectins possess "lock and key" protein conformations that allow them to selectively and reversibly bind photosynthates. Specific substrates bind with greater affinities than glucose and sucrose and displace them from lectins, freeing the energy sources to growth points. Enhanced content of soluble sugars after treatment of Curcubitaceae, Vitaceae, and Rosaceae with modulators of glycolconjugation in the field is entirely consistent with this concept (Nonomura, Pedersen *et al.*, 2018).

Therefore, adaptation of this novel technology has broad implications that can lead to crop improvement by means of regulation of the carbon reactions of photosynthesis.

11.5 The Origin and Development of the Carbon Reactions of Photosynthesis

The simplest sequence of evolutionary origin and development of the carbon reactions of photosynthesis are likely reflected in the diversity of autotrophic carbon cycles (Marakushev and Belonogova, 2011) in modern anaerobic, chemo- and phototrophic prokaryotes; whereas, higher plants exhibit complex, cooperative interactions of different organelles, cells, and tissues involved in the process of photosynthesis (Biel *et al.*, 2014).

In general, the strategy of adaptation of carbon metabolism in the evolution of photosynthesis may be summarized as follows: Anoxygenic phototrophs and chemotrophs brought to life, the opportunity to replenish organic compounds consumed by heterotrophs for nutriment. Pigments and the mechanisms of the photosynthetic electron transport chain enabled these primitive systems to use water as an electron donor. This cardinal step in the evolutionary history of photosynthesis could be performed only with the simultaneous development of aerobic respiration and increasing stability of cellular structures to the adverse effects of oxygen. The use of oxygen as an electron acceptor by the oxidative electron transport chain allowed primitive oxygenic photosynthetic organisms to greatly intensify recycling of reduced nicotinamide adenine dinucleotide and break down organic compounds $(CH_2O)_n$ to CO_2 and H_2O in the central link of the respiratory metabolism, *i.e.*, in the Krebs (TCA) cycle. In turn, an increase of the oxygen concentration in the environment and the decrease of its reducing properties intensified the important accumulation of photosynthetic products.

Populating landmasses with floral species created an urgent need for their adaptation to the deficit of previously inexhaustible substrates in the aqueous medium – water and bicarbonate – and to the extensive solar radiation in unprotected terrestrial habitats having no water column to absorb harmful wavelengths. In terrestrial environments, especially in higher plants, this caused a great number of structural and functional adaptations throughout the plant body. In photosynthesis, the cooperation between carbon pathways among different organelles and differentiated tissues were deployed to optimize the flow of metabolites from light and carbon reactions of photosynthesis. As a consequence of multifunctional adaptations, the reductive pentose phosphate carbon cycle has not changed its fundamental biochemical nature in the course of hundreds of millions of years. Plants attended to and protected the cycle, creating the necessary structural and functional improvements, *i.e.*, the development of biochemical and structural add-ins around it. For example, C_4 and CAM plants are not deprived of benefits of the cycle.

Thus, the Benson–Bassham–Calvin cycle was, and remains for now, the central depot for the production of basic blocks of organic material.

Today, we may see that photosynthetic organisms exhibit wide ecological diversity, while cycling in energy equivalents of the photosynthetic electron transport chain is the main industry for fixing atmospheric carbon. The flow of atmospheric, as well as, internal CO_2 released during the oxidative processes of respiration to the sites of RuBP-carboxylation can be diverse and complex, using special biochemical mechanisms developed in the course of evolution.

Indeed, the primary mechanisms of CO_2 delivery are:

1) CO_2 – Carbonic anhydrase concentrating mechanism, which is inherent to the majority of aquatic algal and a number of terrestrial phototrophs;

2) Dark fixation of carbon in the β-carboxylation reactions of phosphoenolpyruvate and its accumulation in the form of malic acid in cell vacuoles in CAM plants;

3) C_4 pumps of malate and aspartate groups of C_4 plants, and multiple combinations thereof; and

4) Respiration – in which CO_2 and water are end products, both of which may be recycled into photosynthesis.

An important task for these biochemical superstructures is an uninterrupted carbon sequestration from atmospheric or aqueous phases and its timely delivery to sites of RuBP carboxylation in the reductive pentose phosphate cycle. However, these biochemical superstructures also perform some other necessary functions. For example, the C_4 pump in the mesophyll cells of C_4 plants are a source of many organic and amino acids (see *e.g.*, Biel, 1993) and even have the ability to create the primary product of photorespiration – glycolate (see ref. in Biel *et al.*, 2014; Fomina and Biel, 2014) – in addition to its primary task of capture and delivery of CO_2 to the reductive pentose phosphate cycle. The mesophyll cells of C_4 plants can also protect the Benson–Bassham–Calvin cycle; their location around the bundle sheath cells reducing exposure to photodegradation and reactive oxygen species.

11.6 The Next Steps

In the field of photosynthetic biochemistry, there have been ongoing efforts to improve the efficiency of the carbon reactions of photosynthesis.

To date, the most successful result in this direction may be the aforementioned link to regulation of the carbon reactions of photosynthesis, the first recognition of a specific modulation of a glycoconjugate pathway (Nonomura and Benson, 2014), for which Benson was recognized by Brandt iHammer for "*Conferring the Greatest Benefit on Mankind*" (Buchanan and Douce 2015).

Replicated trials of iH026a were conducted on *Antirhinum majus* L., *Coffea arabica* L., *Eustoma grandiflorum* (Raf.) Shinn., *Plumeria obtusa* L., *Ranunculus* L., *Phalaenopsis* Blume, *Ipomoea batatus* L., *Allium* L., *Gossypium* L., *Festuca* L., *Brassica olera* L., *Brassica oleracea* L. *sabellica* L., and *Citrullus lanatus* (Thunb.) Matsum. & Nakai var. *lanatus*. Initially, we observed a reduction of midday depression and, as a consequence of treatments with iH026a, confirmed enhancements of qualities and quantities of leaf expansion, foliar biomass, stem diameter, fruiting and flowering as compared to controls.

11.6.1 Materials and Methods

Abbreviations: αMN – α-mannoside; Bu – bushel; Kg – kilogram; L ha^{-1} – Liter per hectare; lb – pound; PGR – plant growth regulator; 95CI – 95% confidence interval; UTC – untreated nutrient control

Experiments with farm cultivars were conducted from 2013 through 2017 in field research plots in California-like agriculture conditions, USA. BRANDT iH026a PGR product formulation contained the αMN-active ingredient obtained from BRANDT iHammer; United States Patent numbers 8093182; 9072304. Control plots were in the same fields as treated plots, therefore, all plots were cultivated according to the same growers' standards except for applications of iH026a; thus, controls without the experimental treatments are abbreviated, UTC. Treatments were applied in effective dose ranges of 0.3 to 1.2L iH026a ha^{-1} applied at 1 to 4-week intervals. Plastic trays were filled with aggregate-free media, BM, Berger, 121, 1er rang, Saint-Modeste, QC Canada G0L 3W0. Plastic containers were filled with high drainage media such as Super Soil® Palm and Cactus Mix, Rod McLellan Company, Marysville OH 43041. Measurements of leaf and stem dimensions were made with a NAPA Professional 1500 Series digital caliper, calibrated daily to 0.01 mm resolution. Dry weights of individual leaves were taken with a 4-place Mettler digital balance. Data analyses compared mean values, SD and SE for different treatments (*t*-test, significant at $p \leq 0.05$), and in graphs ±95CI for which mean is the center point. Software included IBM® SPSS® Statistics Premium (www.spss.com); Microsoft Excel; and ARM (Agriculture Research

Manager) 2016, Gylling Data Management, Inc., used for data management, crop production experiments and analysis of variance by field researchers.

Plumeria: Among the first indications that crops were responsive to treatments with iH026a were observations of foliar elevation in *Plumeria obtusa* var. *obtusa*; and the lanceolate leaves, as long as a 30 cm on upright branches, were effective pointers. Foliar wilt was observed nearly every afternoon due to high light saturation intensity (1700–2000 μEinstein $m^{-2}sec^{-1}$), rises in midday temperatures, and moderate humidity, approximately 30% to 70%. This daily movement of leaf elevation from high +cm to low -cm were visually discernible. A decrease of wilt may be measured according to the positive change of elevation of foliage, especially during lengthy periods of midday depression lasting three or more hours in UTC that received the same liquid volume of fertilizers. Controlled replicate trials were undertaken on plumeria cultivar Ruby Star, for experimentation in a walled setting that minimized wind in Orange County, California USA. Plumeria plants were potted in 4 L plastic containers with high drainage Super Soil® media. Healthy plants were allowed at least a week to acclimate to environmental conditions of direct sunlight and were further observed for consistency of diurnal changes in leaf elevations as air temperatures rose from 25–35 °C, with daily evening fertigation of grower's standard plant nutrients from Brandt 12-26-26 Micronutrient, 5 gm 3.78 L^{-1} water. The baseline elevations of leaves in each flush of growth were measured and marked on walls and meter sticks with pushpins and adhesive arrows and compared against elevations (mm) of the same leaves hours after treatment. At the starting time of the experiments, 25 ml aqueous volume with grower's standard nutrients were applied to soil directly above roots of each control plant; and treatments were applied in the same manner with 25 ml of 2% iH026a.

Snapdragon: Field confirmations of the efficacy of iH026a were conducted with *Antirhinum majus* cv. Maryland White snapdragon from spring through summer 2015, in irrigated nurseries of a temperate climate zone at Oxnard, Ventura County, CA, USA (34.1975° N; 119.176111° W). Plastic 392-well plug trays and hoop houses were filled with soilless planting media. Plants received fertigation with grower's standard primary, secondary and micronutrients. Foliar treatments with 1.2 L iH026a ha^{-1} delivering approximately 1 ml $plant^{-1}$ were applied manually on the following schedule: 7 January 2015, after seated in trays; and 19 January 2015, after another twelve days. After transplanting to hoop houses on 20 February 2015, snapdragons were treated with monthly foliar sprays, as follow: 20 February 2015; 20 March 2015; and

17 April 2015. On each of these dates, one (1X) or two applications (2X) per treatment date were compared in separate populations of snapdragons to determine the effects of doubling the rate of application on flowering. The experiments were in a randomized complete block design of replicates. Treated and UTC plants were analyzed for several characteristics of growth. Juvenile snapdragons are decussate, therefore, the total length of a pair of opposite leaves from one end to the other was measured as leaf mean mm length. The width of a single expanded leaf, second from the apex, was measured as leaf mm width. Finally, stem mm diameter and flower spike cm lengths were measured.

***Coffee*:** Confirmations of the efficacy of iH026a were conducted on *Coffea arabica* coffee from spring to summer 2015. Coffee plants were cultivated under modified methods described for the aforementioned plumeria as they were potted in 20 L ceramic containers with high drainage media and were allowed a month to acclimate to environmental conditions of direct sunlight. They were further observed for consistency of leaf and stem development at 25–33 °C, with daily morning fertigation of grower's standard plant nutrients from Brandt 20-20-20 Micronutrients, 5 gm 3.78 L^{-1} water. Root treatments with 0.25% iH026a, 100 ml per container, were applied at the base of each plant. Equal volumes of stock nutrient media were applied to roots of UTC. Treatments were applied 1 April and 1 May to matched plants. Leaves were harvested, second tier from the base of stems at the same height and, thereafter, measured and dried.

***Lisianthus*:** Field confirmations of the efficacy of iH026a were conducted on *Eustoma grandiflorum* (Raf.) Shinn. double-flowered variety ABC White lisianthus during summer 2015, in the aforementioned irrigated temperate climate zone nurseries at Oxnard. Experiments were laid out in randomized complete block design with replications. Plastic 392-well plug trays and hoop houses were filled with proprietary soilless planting media and marked. Lisianthus seeds were sown on 6 June 2014. Treated and UTC received daily fertigation with grower standard primary, secondary and micronutrients. The iH026a was diluted in grower standard nutrient control solution. Foliar 1.2 L iH026a ha^{-1} was applied manually in a volume of 114 L water. Both treated and control plants were analyzed for comparisons of mean of leaf mm length and width. The following schedules of manual foliar applications and measurements were undertaken in greenhouses to compare leaf growth: Leaf length 8 July 2015 and 10 July 2015; leaf width over the course of 2 weeks on 14 July 2015, 16 July 2015, 21 July 2015, and 21 July 2015. On 30 September 2015, a week prior to the 5 October 2015 harvest of flowers, open blossoms were counted from treated and control plants.

Ranunculus: Experiments on *Ranunculus* cv. Sprinkles Series, pink garden hybrid, were conducted on iH026a and UTC during spring 2016. Ranunculus plants were potted in 1-pint plastic containers with high drainage media, Palm and Cactus Mix, Super Soil® Rod McLellan Company, Marysville OH 43041. The healthy plants were allowed at least a week to acclimate to environmental conditions of direct sunlight. Potted plants were observed for consistency of growth and selected for uniform bud development at 30°C, with daily morning and evening fertigation of grower's standard plant nutrients from Brandt 12-26-26 Micronutrient, 5 gm 3.78 L^{-1} water. Root treatments with 0.25% iH026a were 30 ml per container applied at the base of plants. Equal volumes of stock nutrient media were applied to UTC roots. Treatments were applied at first bloom in matched plants.

Orchid: Experiments on *Phalaenopsis* sp. orchid were conducted with iH026a and UTC during spring 2016. Orchid plants were potted in 1-cup plastic containers in orchid potting bark with sphagnum moss to retain moisture. Healthy plants were allowed a week to acclimate to in-house environmental conditions of diffuse sunlight. Potted orchids were observed for consistency of growth and selected for uniform bud development at 30°C, with daily morning and evening mist fertigation of grower's standard plant nutrients from Brandt 12-26-26 Micronutrient, 1 gm 3.78L^{-1} water. Root treatments with 0.25% iH026a were 30 ml per container spray-applied at the exposed root and base. UTC was given equal volumes of stock nutrient medium applied to roots.

Sweet Potato: Field experiments with iH026a and UTC were conducted on *Ipomoea batatas* sweet potato during 2014 and in 2015 in a dryland system on a farm in Roseboro, Sampson County, North Carolina USA; Latitude: 34° 52' 46.765" N, Longitude: 78° 34' 7.46" W. For the first trial, 0.6 L iH026a $acre^{-1}$ was applied as a transplant drench. In the second trial, one application was applied in-furrow at planting on 24 April 2014. Each experimental unit was 3-meter row. For 2015, one application was made at the growth stage when the plants were starting to run. The experiment was laid out in a strip block design with seven UTC and eight treatment replications; and with the appropriate dilutions as per treatment with foliar surfactant. Sweet potatoes were harvested by hand-digging each plot and yield $acre^{-1}$ was calculated from the plot yield totals recorded on 6 October 2014, and on 24 August 2015.

Onion: Experiments with iH026a and UTC were conducted on *Allium* onion during summer 2015 in Othello, Adams County, Washington, USA; Latitude: 46° 44' 16.897" N, Longitude: 119° 2' 23.423" W. Onions were direct-seeded and applications were made at 2-3 leaf growth stages for the first spray, and 30% bulb diameter for the second spray. Plot size adopted for the experiment was 1.7-row meter. Experiments were laid out in a randomized block design with 5 replicates. Treated plots were given 0.45L iH026a ha^{-1} with surfactants for proper adherence. The UTC was treated likewise, but without iH026a. Onion bulbs were collected on 3 September 2015, as fresh live weight.

Cotton: Field experiments with iH026a and UTC were conducted on *Gossypium* cotton during spring and summer 2014 on dryland farms of Roseboro, Sampson County, North Carolina, USA; Latitude: 34° 52' 46.765" N, Longitude: 78° 34' 7.46" W. One treatment was applied in-furrow with seeds at planting on 24 April 2014 at a rate of 0.75L iH026a ha^{-1}. Each experimental unit was one 2 m x 3.1 m plot with 1 m row spacing, and population was 40,000 plants $acre^{-1}$. The experiment was laid out in a randomized complete block design with five replications of a single treatment of iH026a and UTC. The spray volume was based on appropriate dilutions per treatment with organosiloxane Brandt Super Wetter surfactant. Observations of boll counts per plot were recorded on 15 September 2014. At harvest, cotton was hand-picked and lint weight yields per plot were calculated.

Grass: Field confirmations on the efficacy of iH026a as compared to UTC were conducted on *Festuca* grass microplots in controlled environment research facilities in Village Park Drive, Powell, Delaware County, Ohio, USA; Latitude: 40° 10' 3.515" N, Longitude: 83° 4' 49.252" W; during spring 2016. Grass trials were conducted on tall fescue "Trophy blend" comprising the following: 49.94% Speedway Tall Fescue; 24.78% Grande 3 Tall Fescue; 24.64% Blackwatch 2 Tall Fescue; 0.01% Other Crop; 0.63% Inert Matter; 0.00% Weed Seed / Noxious Weed; Seed tested December 2015; 90% guaranteed germination; Oregon, USA origin. Replicate plots each measuring 3X3 cm sown with 2 seeds $plot^{-1}$, NK Seed Starter; and UTC and iH026a were cultivated side-by-side in lightweight soil mix. For treatments, seeds were sprayed to coat manually with 10 ml iH026a L^{-1} or nutrient control and subsequent weekly applications followed. UTC was grown in the same environment, irrigation, light and nutrients as iH026a. In 36 replicates, counts of germinated sprouts were taken after 5 days when treated cells were approximately 75–80% germinated. From 27 replicate plots, tiller counts were taken 5 days after the first clipping.

Leafy Vegetables: Field trials were undertaken in commercial greenhouses, Nipomo, San Luis Obispo County, CA, USA, Latitude: 35° 2' 34" N, Longitude: 120° 28' 34" W in a temperate climate zone. Trials with *Brassica olera*, broccoli, were conducted from spring through summer 2015.

Plastic 392-well plug trays and hoop houses were filled with the grower's standard proprietary soilless planting media and primary, secondary and micronutrients. Foliar iH026a 480 ml acre^{-1} was applied spray to drench (sprench) manually delivering approximately 1 ml plant^{-1} on the following schedule of applications: 11 and 20 August 2015. Both treated and UTC populations were analyzed for stem thickness (mm diameter) on 1 September 2015. Measurements were made with a NAPA Professional 1500 Series 0.01 mm resolution digital caliper, calibrated daily. Each tray of plants was tagged for recording observations. Field experiments with *B. oleracea sabellica* curly kale were conducted from 27 May to 17 July 2015 on a sprinkler-irrigated farm in Modesto, Stanislaus County, California, USA; Latitude: 37° 42' 2.772" N, Longitude: 121° 9' 13.917" W. Three foliar treatments were applied: the first application on 4 June 2014 was sprayed at 6-leaf stage (43 days preharvest). The following two applications were sprayed on 18 June and 3 July 2015. Each experimental unit was one 1 m × 3.1 m plot with 1 m beds and 2 plant lines per bed. The experiment was laid out in a randomized complete block design with replicated plots treated with iH026a 300 ml acre^{-1} and UTC. The spray volume used in this study was based on 568 L acre^{-1} water with the appropriate dilutions as per treatment with Brandt organosiloxane surfactant added for proper adherence. Coverage of foliage was achieved with a carbon dioxide gas-powered backpack spray rig with a 3-nozzle boom. Rows were flagged for recording observations. Ten plants were selected from the center of each plot and all harvestable leaves were stripped, counted and weighed, and mean leaf weights were calculated. All leaves that were considered harvestable from the selected plants were at least 15 cm long and had not started to turn yellow.

Melon: Field confirmations of the efficacy of iH026a PGR were conducted on *Citrullus lanatus* watermelon in 2013 and 2017, by modified methods of Nonomura, Pedersen *et al.* (2018). Exploratory melon trials on side-by-side 15.2 m rows were initiated without replicates in 2013, in Roseboro, NC, USA (LL Corner: 34.88197°N; 78.5690556°W). Melons for 2013 trials were given foliar applications of 1.46 L(iH026a) ha^{-1}, in volumes of 468 L ha^{-1}. Treatments of fall watermelons were initiated at first bloom, followed by weekly foliar sprays for a total of four applications and harvested 23 October 2013. In these early field trials, melons that were treated with iH026a showed enhanced Brix and fresh mass yields that provided the basis for further field investigations of cantaloupes, as follow. Replicated watermelon trials were undertaken on *C. lanatus* variety Jubilee, in a drip irrigated system at 891 W. Fox, Raymondville, Texas, USA (NW Corner: 26.46503°Latitude; -97.79938°Longitude). Soil was analyzed as Racombes sandy clay loam, pH 7.4, comprised of 55% sand, 10% silt, and 35% clay. Plants were machine transplanted into beds at 1.93 m row spacing, 0.6 m spacing within row. These trials were randomized complete block design with 6 replicates in 3.9 × 15.2 m plots, flagged to identify treatments and map locations for recording observations. These trials included UTC and treatments with iH026a through drip-tape (Netafim Streamline Plus lay-flat, 30 cm emitter spacing to polyline to 4000 L nurse tank) applications of 0.75 L iH026a ha^{-1}. In 2017, two root applications of iH026a were made to each treated plot. The first treatment on 16 May 2017 was applied at first bloom. The second treatment on 19 June 2017, was applied two weeks preharvest. Fruit harvest and evaluation was conducted on 5 July 2017.

11.6.2 Results

Plumeria: In the afternoon, during periods in which all control plants showed midday depression, plants treated earlier with 2% iH026a showed leaves rising up 10–20 mm in elevation. Our statistical summary of leaf elevation measurements, Table 11.1, showed significant upward leaf mean 12.6 mm elevations on plants treated with iH026a as compared to negative mean −5.8 mm downward drop in elevations of control foliage from their respective starting points. The rise of foliage after treatment with iH026a was greater in plants treated late in the afternoon, after 3 PM; and at the same time, control foliage showed the most pronounced loss of elevation in the afternoon. In general, large older leaves showed less elevation as compared to medium leaves. Figure 11.2 summarizes the foliar elevation response of plumeria after treatment with iH026a.

Table 11.1. Treatment of plumeria with 2% iH026a resulted in a statistically significant enhancement of foliar mean mm elevations as compared to control

Plumeria	*p*	*n*	Elevation mm	SD	SE
Control		5	−5.8000	6.8	3.0
iH026a	0.001	5	12.6000	4.2	1.9

Snapdragon: Several indicators of growth including, leaf length, leaf width, stem diameter, and open flowers were measured. Foliar 1.2 L iH026a ha^{-1} significantly enhanced leaf 29 mm mean length (SD 1.10, SE 0.09), as compared to control leaf 26 mm mean length (SD 0.95, SE 0.08), Table 11.2. Significant increases during the entire course of measurements, 12, 19, 21, 22 and 23 days after first treatment are further summarized in Figure 11.3. Foliar iH026a enhanced leaf width after a single treatment and, as shown in Table 11.3, significantly enhanced stem 13.6 mm mean width (SD 1.22, SE 0.1), as compared to control stem 12.1 mm mean width (SD 0.87, SE 0.07). This enhancement of leaf width is further summarized in Figure 11.4. Foliar treatments with iH026a resulted in significant increases of growth of snapdragons as compared to control.

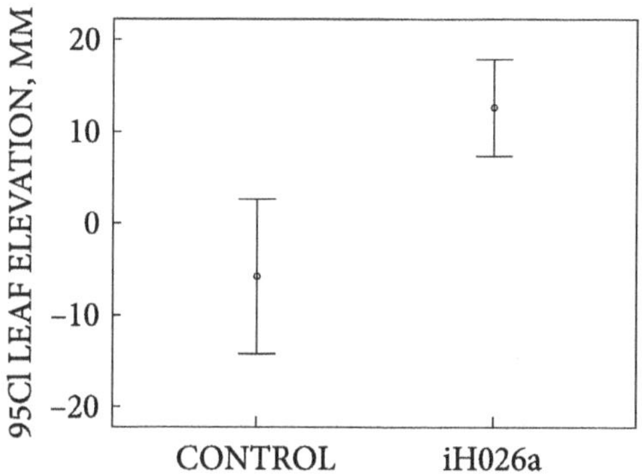

Fig. 11.2. Treatment of plumeria with iH026a resulted in a significant ($n = 5$; $p = 0.001$) increase of mean leaf 12.6 mm elevation (*right* 95CI bar; SD 4.2, SE 1.9) as compared to control −5.8 mm mean elevation (*left* 95CI bar; SD 6.8, SE 3.0). The center points of 95CI bars indicate mean leaf elevations 1 hour after treatment. Negative values are symptomatic of midday depression

Table 11.2. Treatment of snapdragons with iH026a resulted in significant overall enhancement of leaf mean mm lengths as compared to control

Leaf length	*p*	*n*	mm	SD	SE
Control		140	26	0.95	0.08
iH026a	0.000	140	29	1.10	0.09

Table 11.3. Statistical analysis of overall leaf mean widths of snapdragons after treatment with iH026a as compared to control indicated significant enhancement of foliar growth

Leaf width	*p*	*n*	mm	SD	SE
Control		140	12.1	0.87	0.07
iH026a	0.000	140	13.6	1.22	0.10

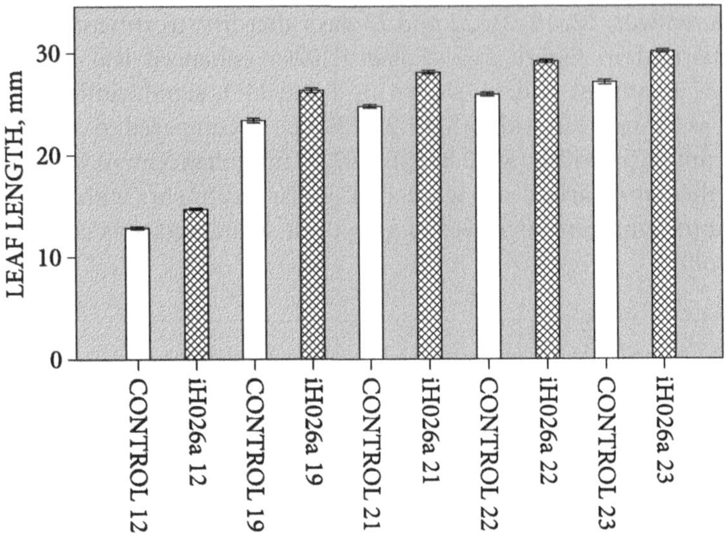

Fig. 11.3. Snapdragons given a single foliar treatment with iH026a showed significant increases in leaf length *(stipple bars)* during the entire course of measurements 12, 19, 21, 22 and 23 day after first treatment as compared to control (*clear bar*)

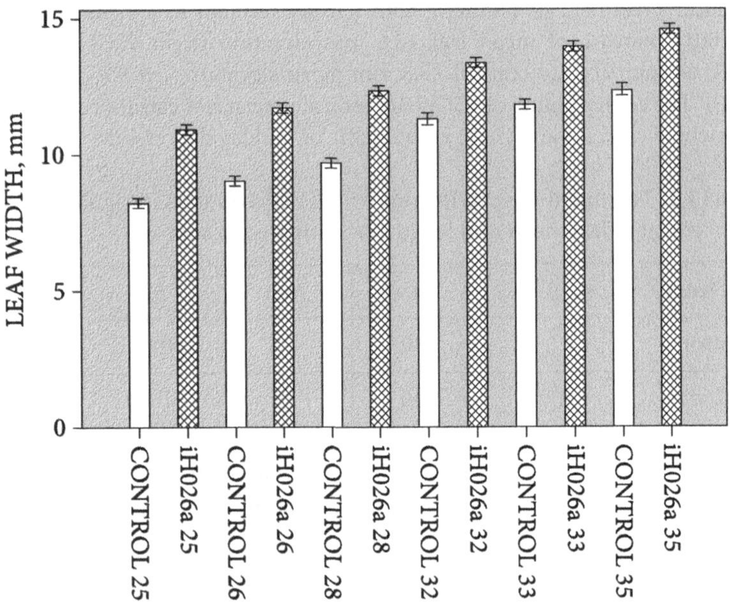

Fig. 11.4. Snapdragons given a single foliar treatment with iH026a showed a significant increase of leaf width *(stipple bars)* during the entire course of measurements 25, 26, 28, 32, 33, and 35 day after first treatment as compared to control (*clear bar*)

After treatment with foliar iH026a, stems showed statistically significant mean 8.4 mm diameter increase (SD 0.49, SE 0.05), as compared to control stem width mean 7.5 mm diameter (SD 0.51, SE 0.06). Snapdragons treated with foliar iH026a developed thicker stems than controls that may have been beneficial for supporting the weight of enlarged foliage and flower spikes of treated plants; thus, for example, greater stem width contributed to resistance against lodging and toppling over. Long-term stem enhancement in response to treatment with iH026a is presented in Table 11.4 and is shown in Figure 11.5, in which treatments increased stem width mm diameter of snapdragon over the course of two weeks after the first treatment in hoop houses.

Table 11.4. Statistical analysis of snapdragon stem mean mm diameter after treatment with iH026a as compared to control indicates significant enhancement of shoot growth

Stem Diameter	*p*	*n*	mm	SD	SE
Control		80	7.5	0.51	0.06
iH026a	0.000	80	8.4	0.49	0.06

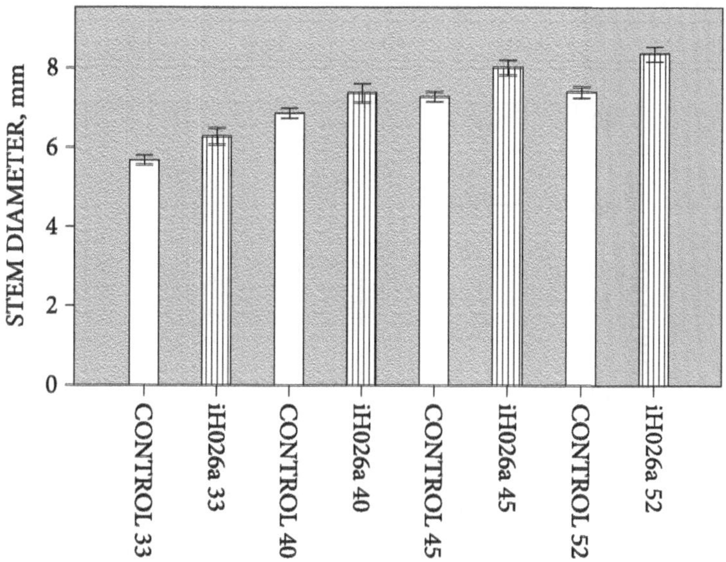

Fig. 11.5. Snapdragons given a single foliar treatment with iH026a showed significant increases in stem diameter during the entire course of measurements 33, 40, 45, and 52 day after first treatment (*crosshatch bars*) as compared to control (*clear bar*)

For the commercial grower of ornamentals, the most important outcome may be the quality and quantity of blossoms and as a result of treatment with iH026a, snapdragons grew longer spikes with more flowers than controls. The effect of two foliar applications per treatment of iH026a in Table 11.5, showed significant (2X $n = 40$; $p = 0.002$) increases of flower spike mean length, as compared to a single foliar application of 1.2 L iH026a ha^{-1}. Furthermore, after the first treatment with iH026a, a significant enhancement of flower spikes as compared to control (1X $n = 40$; $p = 0.000$) was seen. Mean spike length increase after 1X or 2X iH026a as compared to control are exhibited in Figure 11.6. Moreover, a visual comparison of exemplary control to treated trays of snapdragons in Figure 11.7 clearly showed enhancement of leaf length and width after treatment with iH026a.

Table 11.5. Statistical analysis of flower spike cm lengths after 1X or 2X passes per treatment of snapdragon with iH026a as compared to control indicated significant enhancement of blossoming

Flower Spike	*p*	*n*	cm	SD	SE
Control		40	30.7	3.5	0.6
iH026a 1X	0.000	40	33.2	1.8	0.3
iH026a 2X	0.002	40	34.4	1.5	0.2

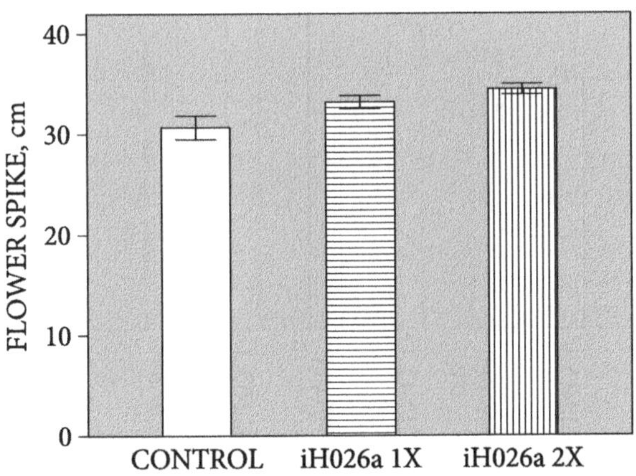

Fig. 11.6. The effects of one (1X) as compared to two (2X) foliar passes per treatment of snapdragons with iH026a (*filled bars*) showed the following: Significant increases of iH026a 1X flower spike mean 33.2 cm length (SD 1.8, SE 0.3), iH026a 2X spike mean 34.4 cm (SD 1.5, SE 0.2) as compared to each other; and also as compared to control spike mean 30.7 cm length (SD 3.5, SE 0.6) (*clear bar*)

Fig. 11.7. The effect of foliar iH026a on plug trays of snapdragons, grown side-by-side with control, was visibly discernible. These young snapdragon plants responded to treatment with iH026a with enhanced vegetative growth as compared to control. Particularly evident are larger leaf dimensions that completely covered the tray after foliar application of iH026a; control, left, showed smaller foliage and openings of incomplete canopy coverage of the tray between plants

Table 11.6. Summary of the several effects of treatment of snapdragons with iH026a as compared to control are presented

Snapdragon	Increase
Leaf length	11%
Leaf width	12%
Stem width	12%
Flower spike	20%

In summary, treatment of snapdragon with iH026a resulted in an improvement in the strength and quality of young plants as compared to control. Table 11.6 summarizes the responses of snapdragons to treatment with iH026a including larger foliage, thicker stems and longer flower spikes than controls.

Table 11.7. Statistical analyses of individual leaf mean dry gram weight indicated a significant increase of productivity in coffee after treatment with iH026a as compared to control

Coffee	*p*	*n*	gm leaf^{-1}	SD	SE
Control		21	0.30	0.01	0.003
iH026a	0.002	21	0.33	0.04	0.009

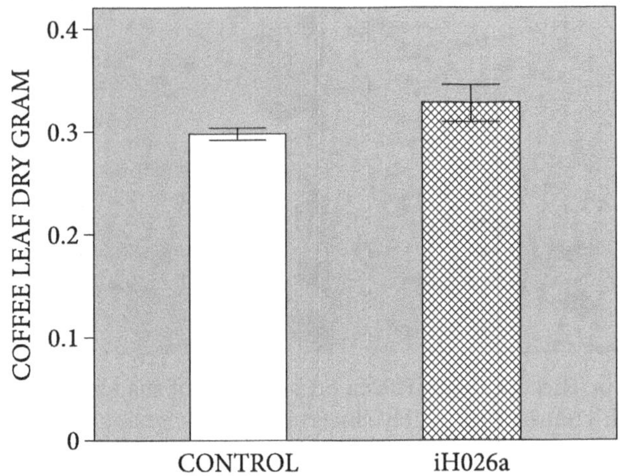

Fig. 11.8. The effect of iH026a application to coffee (*stipple bar*) showed a statistically significant (n = 21, p = 0.002) increase of leaf mean 0.33 gm dry weight (SD 0.04, SE 0.009) as compared to control leaf mean 0.30 gm dry weight (SD 0.01, SE 0.003) (*clear bar*)

Coffee: Treatments of coffee plants with iH026a resulted in significant enhancement of vegetative biomass as compared to controls. After two treatments with 0.25% iH026a, the resulting mean dry 0.33 g leaf^{-1} (SD 0.04, SE 0.009) showed a significant 10% increase as compared to the mean dry weight of control, 0.30 g leaf^{-1} (SD 0.01, SE 0.003) $n = 21$, $p = 0.002$. Coffee leaf results are statistically summarized in Table 11.7 and exhibited graphically in Figure 11.8. Vegetative responses of coffee to iH026a were similar to those of snapdragons and large leaves are attractive and functional features in coffee plants.

Lisianthus: Analysis of data from lisianthus populations of individuals treated with iH026a showed greater leaf mean length and increased leaf mean width as compared to control. In Table 11.8, lisianthus treated with foliar 1.2 L iH026a ha^{-1} showed statistically significant ($n = 84$, p - value 0.000) increases of 35 day leaf mean 27.3 mm length (SD 1.52, SE 0.17) as compared to 35 day control mean 25.2 mm length (SD 1.74 SE 0.19); as well as, enhanced 37 day leaf mean 28.5 mm length (SD 1.68, SE 0.18) as compared to 37 day control leaf mean 26.3 mm length (SD 1.73, SE 0.19); and these results are graphically summarized in Figure 11.9. Statistical summaries of Table 11.9 indicated that treatments significantly (n = 84; p = 0.000) enhanced leaf mean mm width as compared to control. Furthermore, Figure 11.10 indicates responses of foliar 1.2 L iH026a ha^{-1} on lisianthus of persistently enhanced leaf widths through the course of measurements for 2 weeks in July 2015 at 41, 43, 48, and 55 day. For example, at 41 day, lisianthus showed significantly enhanced leaf mean 11.9 mm width (SD 0.92, 0.10) after treatment with iH026a as compared to control mean 10.7 mm width (SD 1.06, SE 0.12); and later at 55 day, treatment significantly enhanced leaf mean 16.1 mm width (SD 0.88, SE 0.10), as compared to 55 day control leaf mean 14.8 mm width (SD 0.67, SE 0.07).

Table 11.8. Statistical analysis of lisianthus leaf mean mm lengths after treatment with iH026a as compared to control mean indicate significant enhancement of foliar growth at 35 day and 37 day

Leaf Length	*p*	*n*	mm	SD	SE
Control 35		84	25.2	1.74	0.19
iH026a 35	0.000	84	27.3	1.52	0.17
Control 37		84	26.3	1.73	0.19
iH026a 37	0.000	84	28.5	1.68	0.18

Table 11.9. Statistical analysis of lisianthus leaf mean mm widths after treatment with iH026a indicates significant enhancements of foliar growth as compared to control

Leaf Width	*p*	*n*	mm	SD	SE
Control 41		84	10.7	1.06	0.12
iH026a 41	0.000	84	11.9	0.92	0.10
Control 43		84	11.3	0.92	0.10
iH026a 43	0.000	84	12.6	0.81	0.09
Control 48		84	13.9	0.90	0.10
iH026a 48	0.000	84	15.4	0.76	0.08
Control 55		84	14.8	0.67	0.07
iH026a 55	0.000	84	16.1	0.88	0.10

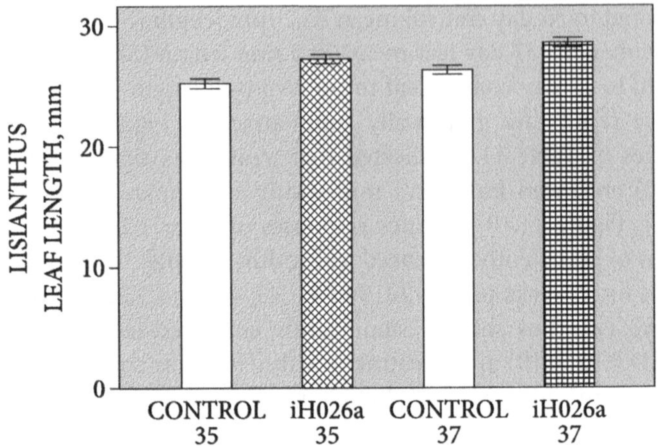

Fig. 11.9. Effects of foliar iH026a on lisianthus showed (*filled bars*) statistically significant ($n = 84$; $p = 0.000$) enhancement of 35 day leaf mean 27.3 mm length (SD 1.52, SE 0.17) as compared to 35 day control mean 25.2 mm length (SD 1.74 SE 0.19); as well as, enhanced 37 day leaf mean 28.5 mm (SD 1.68, SE 0.18) as compared to 37 day control mean 26.3 mm (SD 1.73, SE 0.19) (*clear bars*)

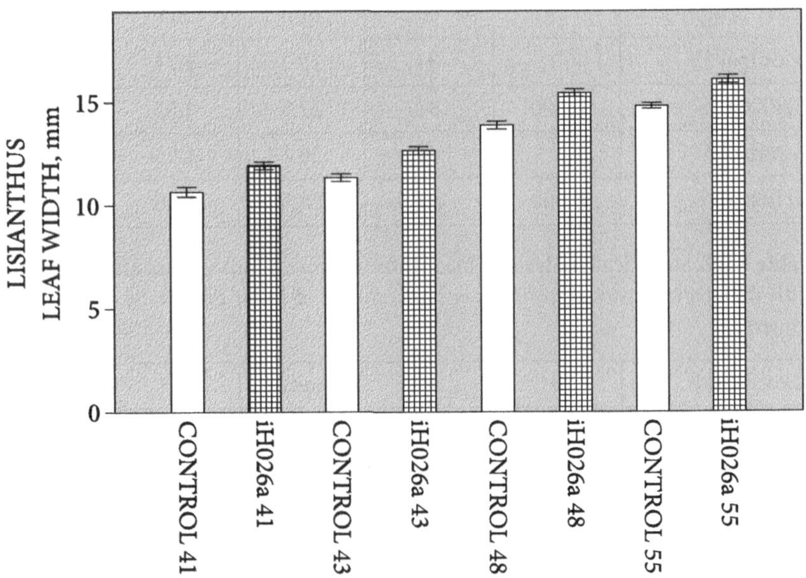

Fig. 11.10. Effects of foliar iH026a on lisianthus (*crosshatch bars*) showed, for example at 55 day, significantly ($n = 84$; $p = 0.000$) enhanced leaf mean 16.1 mm width (SD 0.88, SE 0.10) as compared to 55 day control leaf mean 14.8 mm width (SD 0.67, SE 0.07) (*clear bars*); and showed such persistently enhanced leaf widths through the course of measurements for 2 weeks in July 2015; at 41, 43, 48, and 55 day

Table 11.10. Statistical analysis of bloom mean count plot^{-1} after treatment of lisianthus with iH026a as compared to control indicates significant acceleration of flowering

Bloom count	p	n	Count plot^{-1}	SD	SE
Control		6	0.3	0.82	0.33
iH026a	0.001	6	4.3	1.63	0.67

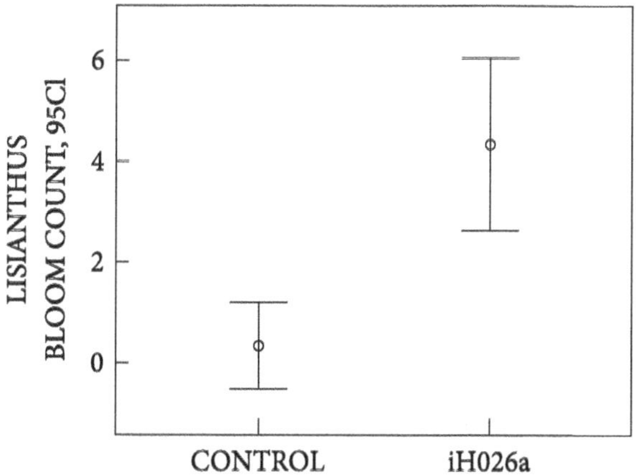

Fig. 11.11. Effects of foliar iH026a on lisianthus showed significantly ($n = 6$; $p = 0.001$) greater open flowers mean 4.3 count plot^{-1} (*right* 95CI bar; SD 1.63, SE 0.67) as compared to control mean 0.3 count plot^{-1}, (*left* 95CI bar; SD 0.82, SE 0.33). Mean is the center point of 95CI bars

Analysis of flowering data from six replicate plots treated with 1.2 L iH026a ha^{-1} as compared to controls are presented in Table 11.10, treatments with iH026a advancing harvest a week earlier than control. Treatments with foliar 1.2 L iH026a ha^{-1} on lisianthus, Figure11.11, showed significantly greater 4.3 open flower mean count plot^{-1} (SD 1.63, SE 0.67) as compared to control 0.3 mean count plot^{-1}, (SD 0.82, SE 0.33) $n = 6$; $p = 0.000$. At the time of the 30 September count, there were only two open flowers among all of the control plots, therefore, cutting flowers from the control group did not commence until later, on 12 October, after the treated population was harvested.

Ranunculus: Treatment with iH026a resulted in the opening of flower buds after a week, while control showed far fewer blossoms. The mean opened bloom count, 2 plant^{-1}, after treatment of ranunculus with iH026a was significant ($n = 6$; $p = 0.022$); as compared to the 1.2 count of opened bloom per control plant, as presented in Table 11.11.

Table 11.11. Statistical analyses of bloom mean count plant^{-1} after treatment of ranunculus with iH026a as compared to control indicates significantly greater flowering

Ranunculus blooms	*p*	*n*	Count plant^{-1}	SD	SE
Control		6	1.2	0.41	0.17
iH026a	0.02	6	2	0.63	0.26

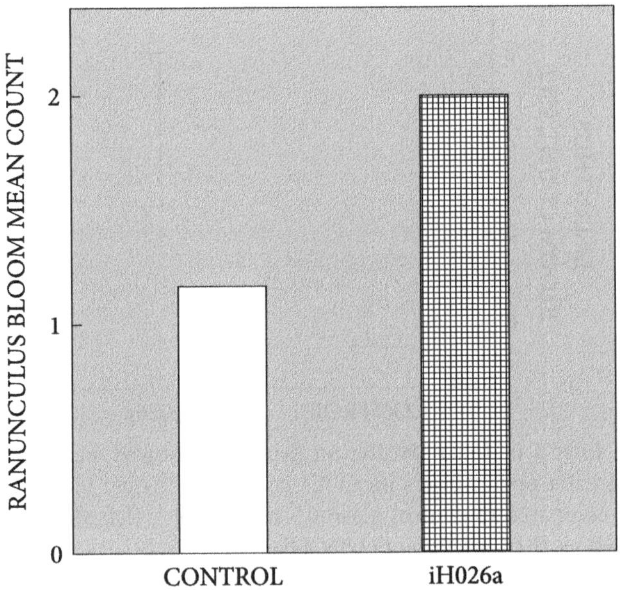

Fig. 11.12. The effect of root application of iH026a showed a significant ($n = 6$; $p = 0.022$) increase of ranunculus mean count of 2 open blooms, right, as compared to the nutrient control mean count of 1.2 open blooms, left

Figure 11.12 graphically summarizes the bloom count response of ranunculus to treatment with iH026a, indicating the same. Furthermore, in Figure 11.13, an increase in open bloom counts of iH026a-treated ranunculus is visibly discernible, right, as compared to nutrient control, left.

Orchid: After treatment of orchid roots with iH026a, buds opened earlier than control. The opened orchid bloom mean count 9 plant^{-1} after treatment with iH026a showed a significant increase ($n = 6$; $p = 0.05$) as compared to the mean count 7 open blooms plant^{-1} of control.

In Figure 11.14 this increase in open bloom counts of iH026a-treated orchids, right, as compared to nutrient control, left, was visibly discernible.

Fig. 11.13. After treatment with iH026a, right, ranunculus showed a visually distinguishable increase of open blooms as compared to control
Note: cm / inch scale, center yellow bar

Sweet Potato: The first trial with iH026a resulted in mean 907 Bu acre^{-1} yield as compared to control mean 605 Bu acre^{-1}. Furthermore, treatment with iH026a resulted in a 5% increase in graded #1 sweet potato with an ($n = 4$, $p = 0.026$) increase of sweetness from Brix 8.25% in control to Brix 8.79% in crops treated with iH026a. Statistical analysis for 2014 is presented in Table 11.12 and graphically summarized in Figure 11.15, and show that mean weight yields increased significantly with the application of iH026a as compared to controls.

Onion: Foliar treatment of onion with iH026a resulted in significantly ($n = 5$; $p = 0.004$) higher estimated mean weight yields of 108,270 lb acre^{-1} (121,354 Kg ha^{-1}) as compared to Control 100,768 lb acre^{-1} (112,945 Kg ha^{-1}); as presented in Table 11.13 and are shown in Figure 11.16.

Fig. 11.14. The effect of iH026a on higher mean 9 counts of open orchid flowers plant^{-1} was significant (n = 6; p = 0.05) as compared to 7 open flowers plant^{-1} in control. 1 cm scale

Table 11.12. Statistical analysis of sweet potato mean lb plot^{-1} after treatment with iH026a as compared to control indicates a significant increase of yield

Sweet Potato	p	n	lb plot^{-1}	SD	SE
Control		7	25	6.75	2.55
iH026a	0.001	8	33	5.01	1.77

Table 11.13. Statistical analysis of estimated mean lb acre^{-1} indicates a significant increase of onion yield after treatment with iH026a as compared to control

Onion	p	n	lb acre^{-1}	SD	SE
Control		5	100,768	2086	933
iH026a	0.004	5	108,270	3603	1611

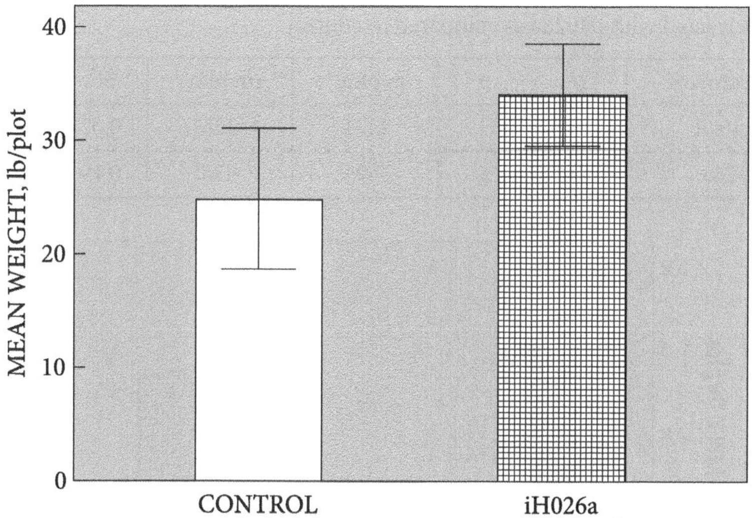

Fig. 11.15. A significant effect of iH026a on enhancement of mean 33 lb plot^{-1} yield of sweet potato as compared to control 25 lb plot^{-1} was observed (Control $n = 7$; iH026a $n = 8$; $p = 0.02$)

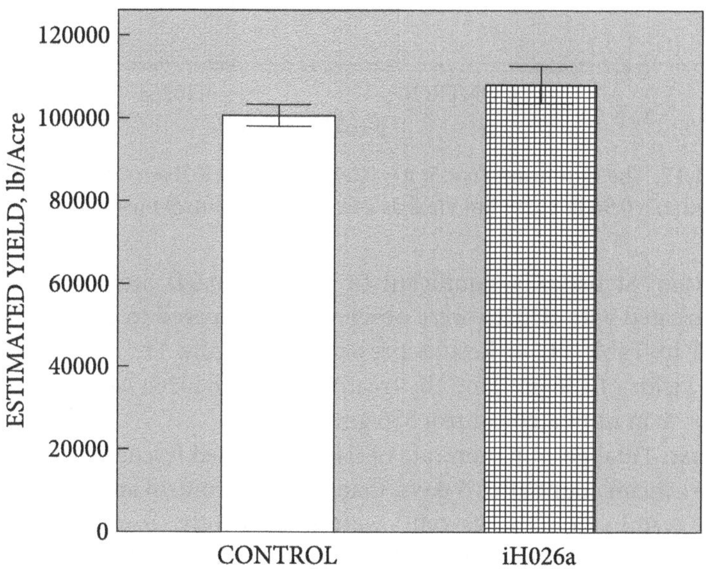

Fig. 11.16. Treatment of onion with iH026a, right, resulted in a statistically significant ($n = 5$; $p = 0.004$) enhanced weight yield of 108,270 lb acre^{-1} (121,354 Kg ha^{-1}) as compared to control 100,768 lb acre^{-1} (112,945 Kg ha^{-1}), left

Table 11.14. Data from field trials of cotton showed greater mean lint yields of plots treated with iH026a as compared to control

Treatment	*p*	*n*	lb plot^{-1}	gm plot^{-1}	SD	SE
Control		5	0.74	336	0.07	0.03
iH026a	0.04	5	0.96	436	0.19	0.08

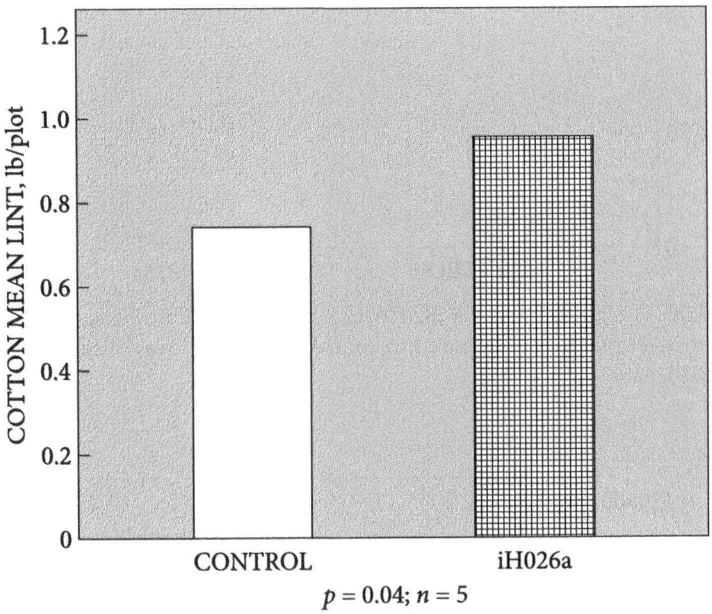

Fig. 11.17. The significant ($n = 5$; $p = 0.04$) effect of iH026a treatments on cotton resulted in a 0.96 lb plot^{-1} lint yield as compared to control mean 0.74 lb plot^{-1}

Cotton: Statistically significant ($n = 5$; $p = 0.04$) increases in cotton rows treated with iH026a were observed as compared to controls, as presented in Table 11.14. Results presented in Figure 11.17 show that lint mean yields of 436 gm plot^{-1} by treatment with iH026a were significantly greater than untreated control 336 gm plot^{-1}.

Grass: Total germination rate of iH026a-treated fescue grass was higher at 61 mean count after 5 days, than nutrient control at 13 mean count. Mean germination counts cell^{-1} were significantly ($n = 36$; $p = 0.000$) higher after treatment with iH026a as compared to control and as shown by statistical analysis in Table 11.15. Tiller counts of iH026a-treated grass were significantly ($n = 27$; $p = 0.011$) higher, showing tiller mean count 9 plot^{-1}, as compared to control at mean tiller count 7 plot^{-1}, also in Ta-

Table 11.15. Statistics from fescue grass indicated significantly accelerated germination and improved tiller count in populations treated with iH026a as compared to control

Germination	*p*	*n*	Count	SD	SE
Control		36	0.4	0.54	0.09
iH026a	0.000	36	1.7	0.95	0.16
Tiller					
Control		27	7	2.8	0.5
iH026a	0.000	27	9	3.0	0.6

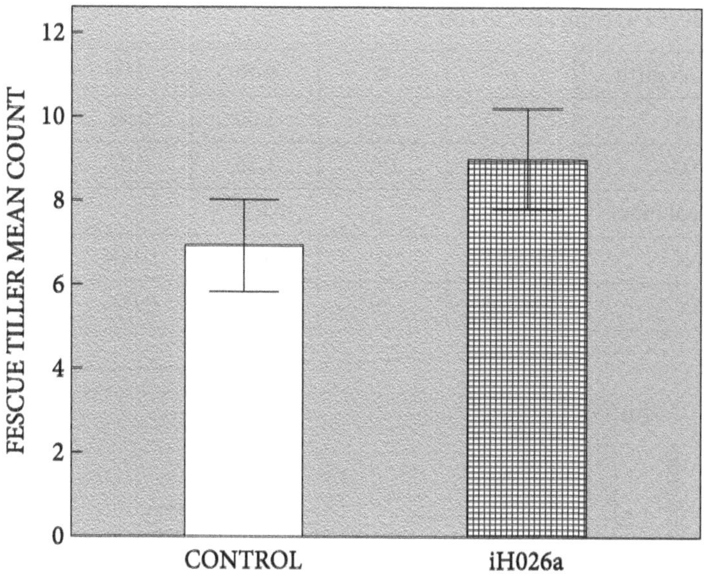

Fig. 11.18. Fescue showed significant ($n = 27$; $p = 0.011$) enhancement of tiller mean count after treatment with iH026a as compared to nutrient control

ble 11.15. Furthermore, greater tiller counts after treatment with iH026a as compared to control are shown in Figure 11.18.

Leafy Greens: Broccoli stem diameter thickness was measured on 180 plants each of iH026a-treated and control populations to determine growth in girth, a diagnosis of long-term turgidity and vigorous strength of the plant. As a result of treatment of broccoli with iH026a, enhanced growth in girth expressed as stem mean 1.55 mm diameter (SD 0.07, SE 0.01) was statistically significant ($n = 180$; $p = 0.000$) as compared to control 1.50 mm diameter (SD 0.08, SE 0.01); and these results are shown in Table 11.16 and exhibited in Figure 11.19. The enhanced productivity in broccoli was borne out by results of field applications of iH026a to another member of the same plant family, kale. As presented in Table 11.16 and shown in Figure 11.20, curly kale showed statistically significant ($n = 5$; $p = 0.016$) higher mean leaf 0.10 lb leaf^{-1} (SD 0.006, SE 0.0025) as compared to control mean 0.09 lb leaf^{-1} (SD 0.006, SE 0.0027). Furthermore, estimated mean yield of kale treated with iH026a 47,453 lb acre^{-1} (53,188 Kg ha^{-1}) was greater than that of control 38,219 lb acre^{-1} (42,838 Kg ha^{-1}).

Table 11.16. Statistics on leafy greens indicated significantly increased stem thickness in broccoli girth and improved leaf biomass of kale after treatments with iH026a as compared to control

Broccoli girth	*p*	*n*	mm	SD	SE
Control		180	1.50	0.08	0.01
iH026a	0.000	180	1.55	0.07	0.01
Kale leaf mass			lb leaf^{-1}		
Control		5	0.09	0.006	0.003
iH026a	0.016	5	0.10	0.006	0.003

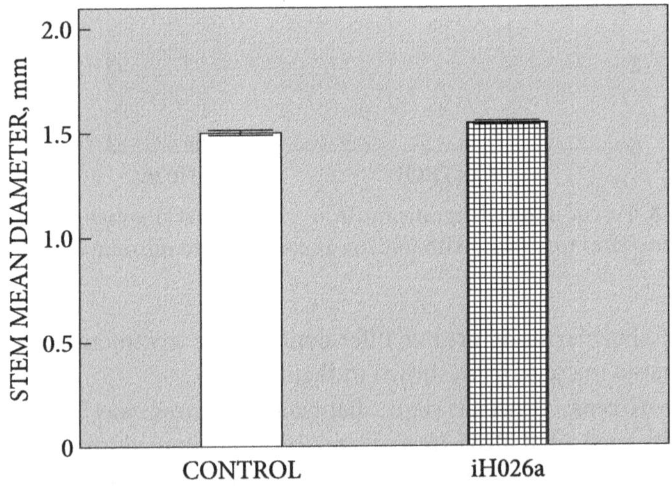

Fig. 11.19. Treatment of young broccoli with iH026a, right, resulted in significantly ($n = 180$; $p = 0.000$) enhanced growth in girth, expressed as stem mean mm diameter over control, left

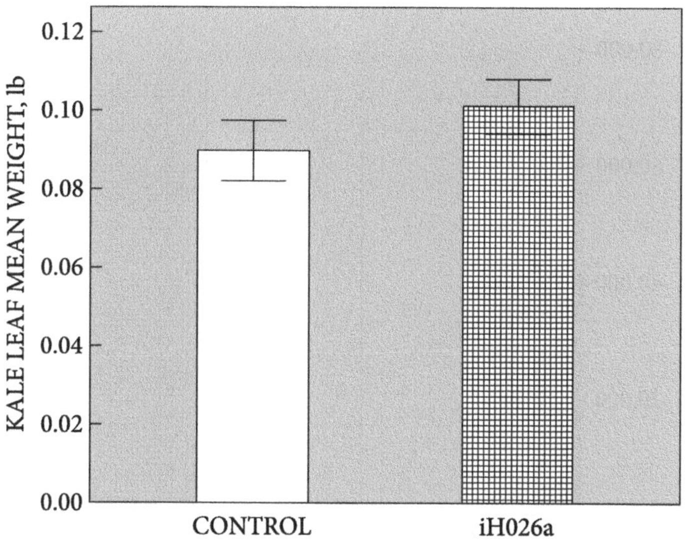

Fig. 11.20. Treatment of curly kale with iH026a in field rows, right, resulted in statistically significant ($n = 5$; $p = 0.016$) enhanced mean lb leaf^{-1} (1 lb = 454 grams) over control, left

Melon: Treatment with iH026a resulted in significantly higher mean weight yields ($n = 6$; $p = 0.016$) of 881 fruit lb plot^{-1} (988 Kg plot^{-1}) as compared to Control 769 lb plot^{-1} (862 Kg plot^{-1}) based on data provided in Table 11.17. The graphic summary of estimated yields of watermelon treated with iH026a 75.788 lb acre^{-1} (84,947 Kg ha^{-1}) as compared to control 66,136 lb acre^{-1} (74,128 Kg ha^{-1}) are exhibited in Figure 11.21.

Table 11.17. Statistical analysis of watermelon treated with iH026a showed significantly greater yields than control

Treatment	p	n	lb plot^{-1}	Kg plot^{-1}	SD	SE
Control		6	769	862	92	37
iH026a	0.016	6	881	988	102	42
			lb acre^{-1}	Kg ha^{-1}		
Control		6	66,136	74,128	7887	3217
iH026a	0.016	6	75,788	84,947	8766	3579

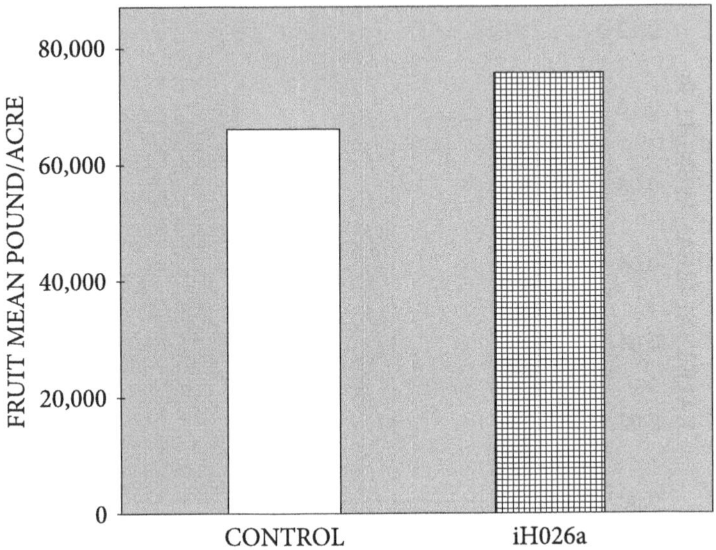

Fig. 11.21. Treatment of watermelons with iH026a in large field beds, right, resulted in an (n = 6; p = 0.016) increase of marketable 75,788 lb acre^{-1} (84,947 Kg ha^{-1}) fruit as compared to control 66,136 lb acre^{-1} (74,128 Kg ha^{-1}) fruit yield, left

11.6.3 Conclusion

Productivity and quality of ornamental crops were enhanced by our field treatments with iH026a PGR and these improvements are consistent with previous field investigations that included higher counts of cherry blossoms, increased sweetness of raisins, and greater cantaloupe harvests than control (Nonomura, Pedersen *et al.*, 2018). Moreover, plant growth regulation by iH026a may reduce crop cycling times, inputs, irrigation, and runoff, while the health of crops is reflected by enhanced strength and quality after treatment with iH026a.

In foliage, the lectin cycle may be oft repeated daily and the accumulated effects by modulation of glycoconjugation over time may be notable in elevated foliage, thicker stems, greater leaf area and biomass, more open blooms, and longer flower spikes than controls that wilt during the same season of growth. Applications of iH026a reduced wilt, providing a means for diagnosis and correction of midday depression. Midday wilt, a daily event that has been observable in agricultural crops persistently over time, may be regarded as a symptom of a response to extreme environments by entire fields of plants. The conquest of midday wilt accomplished here may be demonstrative of the linked metabolic cycles shown in Figure 11.1. Furthermore, the promotion of growth in treated plots for the long duration, up to 3 weeks at a time, may be attributable to the benefits of photosafening.

Development of glycoregulation stemming from autoradiography is rooted in the aforementioned Benson's protocol and in our treatments of crops with substrates that have potent binding specificity and binding affinities for vacuolar lectins.

We are finding clear means of adaptation in agriculture as foretold by Fomina and Biel (2016). Therefore, we shall continue to pursue further field investigations on a wide variety of crops with regard to photosynthetic and respiratory pathways that may enhance qualities and quantities of agricultural harvests for the economic benefit of growers who would fuel, clothe, and feed the world.

11.7 Felicitation

The authors of this paper honor Andrew A. Benson for his brilliance as a Scientist, outstanding trailblazing of numerous metabolic pathways, and lifelong achievements of the first order. Some of us worked side-by-side with him and we always enjoyed his friendship over many visits through several decades. We cherish the memories of his intellect, heart and soul; and herein, we celebrate what would have been his 101st birthday.

Photo 11.2. Andrew Alm Benson (*24 September 1917 – †16 January 2015)

Also, we honor Dee Benson, the light of Andrew's life for 45 years, with our profound admiration. Dee contributed the portrait, right, and original autoradiograms from personal archives. Thank you, Dee, for adding sparkle to our lives, too.

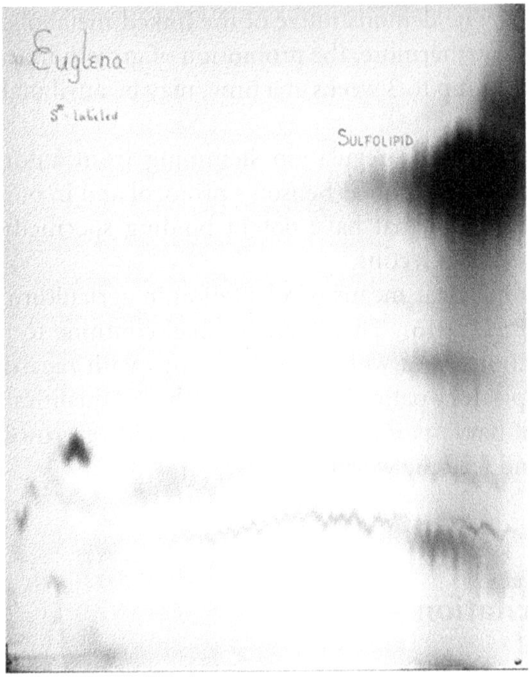

Chromatogram 11.3. Autoradiogram by Benson showing a labeled sulfolipid metabolite from *Euglena* on exposure to ^{35}S.

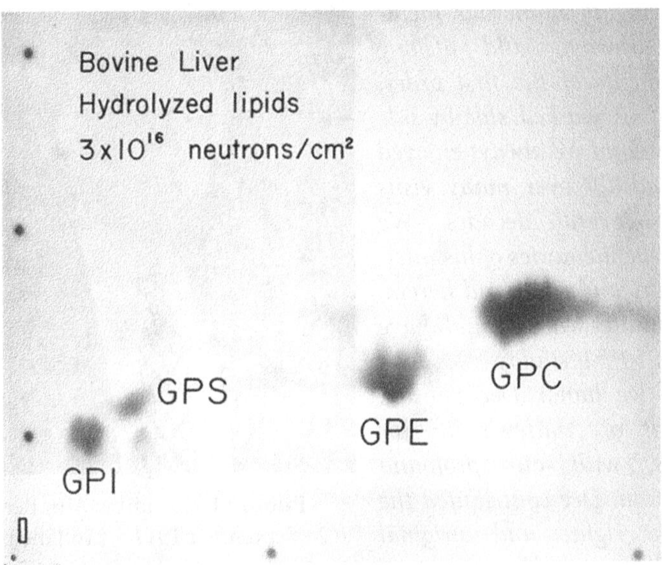

Chromatogram 11.4. Neutron activation of bovine liver by Benson resulted in GPI, GPS, GPE, and GPC

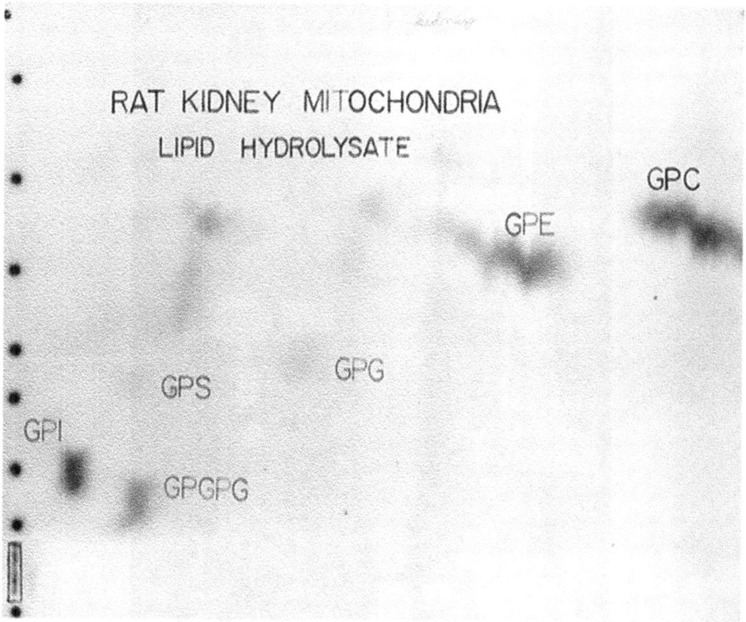

Chromatogram 11.5. Autoradiogram by Benson showing labeled metabolites in rat kidney mitochondria included the following: GPGPG, GPI, GPS, GPG, GPE, and GPC

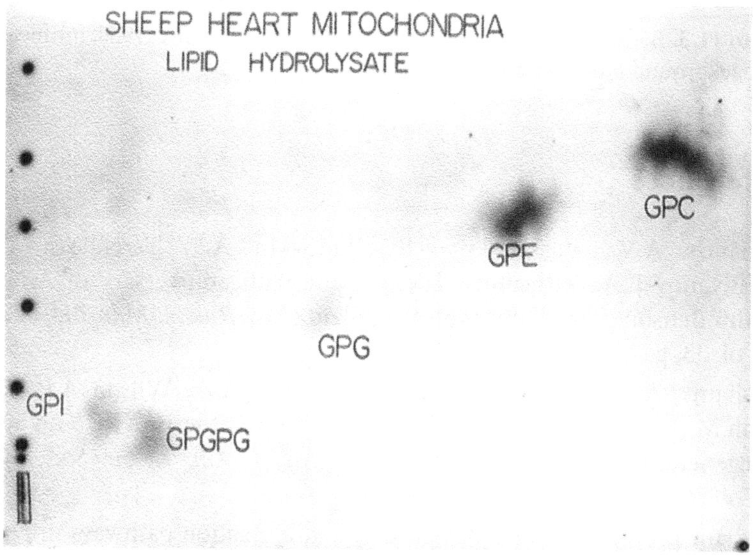

Chromatogram 11.6. Autoradiogram by Benson of sheep heart mitochondria lipid hydrolysate, the labeled metabolites included the following: GPGPG, GPI, GPG, GPE, and GPC

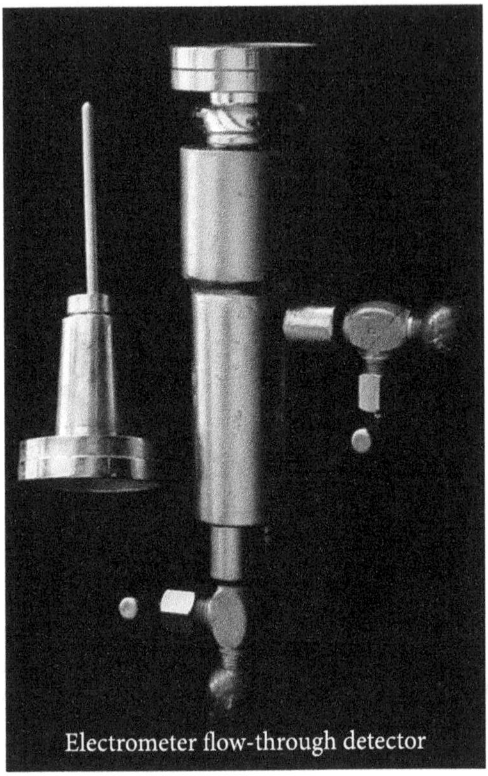

Photo 11.3. Benson's electrometer flow-through detector, circa1950, for measuring background radiation and $^{14}CO_2$-gas flow

References

Adrianov A.V., Vaskovsky V.E., Pudovkin A.I., Terekhova T.A., Titlyanov E.A., Titlyanova T.V. On the 90th anniversary of Andrew Alm Benson great scientist and excellent man, *Russ. J. Mar. Biol.*, 2007, Vol. 33, p. 343–346.

Bassham J.A., Benson A.A., Kay L.D., Harris A.Z., Wilson A.P., Calvin M. The path of carbon dioxide in photosynthesis. XXI. The cyclic regeneration of carbon dioxide acceptor, *J. Am. Chem. Soc.*, 1954, Vol. 76, p. 1760–1770.

Bassham J.A., Buchanan B.B. Carbon dioxide fixation pathways in plants and bacteria. *Photosynthesis, Vol 2: Development, Carbon Metabolism, and Plant Productivity*, Govindjee ed., Academic Press, New York, 1982, p. 141–189.

Benson A.A. Mechanism of biochemical photosynthesis, *Z. Elektrochem.*, 1952, Vol. 56, p. 848–854.

Benson A.A. Paving the Path, *Annu. Rev. Plant Biol.*, 2002*a*, Vol. 53, p. 1–25.

Benson A.A. Following the path of carbon in photosynthesis: a personal story, *Photosynth Res.*, 2002*b*, Vol. 73, p. 29–49.

Benson A.A., Bassham J.A., Calvin M., Goudate T.C., Haas U.A., Stepka W. The path of carbon in photosynthesis. 5. Paper chromatography and radioautography of the products, *J. Am. Chem. Soc.*, 1950, Vol. 72, p. 1710–1718.

Benson A.A., Calvin M. Carbon dioxide fixation by green plants, *Annu. Rev. Plant Physiol.*, 1950*a*, Vol. 1, p. 25–42.

Benson A.A., Calvin M. The path of carbon in photosynthesis. VII. Respiration and photosynthesis, *J. Exp. Bot.*, 1950*b*, Vol. 1, p. 63–68.

Benson A.A., Muscatine L. Wax in coral mucus: Energy transfer from corals to reef fishes, *Limnol. Oceanogr.*, 1974, Vol. 19, p. 810–814.

Benson A.A., Nonomura A.M. The path of carbon in photosynthesis: Methanol inhibition of glycolic acid accumulation, *Research in Photosynthesis 1, Proceedings of the IX International Congress on Photosynthesis*, Murata N., ed., Kluver, Nagoya, 1992, P-522.

Benson A.A., Nonomura A.M., Gerard V.A. The path of carbon in photosynthesis. XXV. Plant and algal growth responses to glycopyranosides, *J. Plant Nutr.*, 2009, Vol. 32, Vol. 1185–1200.

Benson A.A., Patton J.S., Abraham S. Energy exchange in coral reef ecosystems, *Atoll Res. Bull.*, 1978, Vol. 220, p. 33–54.

Biel K.Y. *Ecology of Photosynthesis*, Nauka, Moscow, 1993, 221 p. [In Russian]

Biel K.Y., Gates R.D., Muscatine L. The effect of free amino acids on the photosynthetic carbon metabolism of symbiotic dinoflagellates, *Russian J. Plant Physiol.*, 2007, Vol. 54, No. 2, p. 171–183.

Biel K.Y., Fomina I.R. Benson-Bassham-Calvin cycle contribution to the organic life on our planet, *Photosynthetica*, 2015, Vol. 53, No. 2, p. 161–167.

Biel K.Y., Fomina I.R., Nazarova G.N., Soukhovolsky V.G., Khlebopros R.G., Nishio J.N. Untangling metabolic and spatial interactions of stress tolerance in plants. 1. Patterns of carbon metabolism within leaves, *Protoplasma*, 2010, Vol. 245, p. 49–73.

Biel K.Y., Fomina I.R., Yensen N.P., Nishio J.N., Matichenkov V.V., Nazarova G.N., Soukhovolsky V.G., Khlebopros R.G. *Complex Biologi-*

cal Systems: Adaptation and Tolerance to Extreme Environments, Gorod, Pushchino-Krasnoyarsk, 2014, 343 p.

Biel K.Y., Nonomura A.M., Benson A.A., Nishio J.N. The path of carbon in photosynthesis. XXVI. Uptake and transport of methylglucopyranoside throughout plants, *J. Plant Nutr.*, 2010, Vol. 33, p. 902–913.

Bil' K.Y., Gedemov T. Structural and functional properties of assimilation apparatus in *Suaeda arcuata* Bunge (Chenopodiaceae) related to cooperative photosynthesis, *Proc. Acad. Sci. USSR*, 1980, Vol. 250, p. 9–13.

Bil' K.Y., Lyubimov V.Y., Demidova R.N., Gedemov T. Assimilation of CO_2 by plants of the family *Chenopodiaceae* with three types of autotrophic tissues in the leaves, *Soviet Plant Physiol.*, 1981, Vol. 28, p. 1119–1127. [In Russian]

Bil' K.Y., Lyubimov V.Y., Trusov M.F., Gedemov T. Atakhanov B.O. Participation of three types of autotrophic tissues in diurnal dynamics of assimilation CO_2 in some plants of Chenopodiaceae family, *Bot. Zhurnal*, 1983, Vol. 68, p. 54–61. [In Russian]

Blanquet R.S., Nevenzel J.C., Benson A.A. Acetate incorporation into the lipids of the sea anemone *Anthopleura elegantissima* and its associated zooxanthellae, *Mar. Biol.*, 1979, Vol. 54, p. 185–194.

Buchanan B.B. The carbon (formerly dark) reactions of photosynthesis, *Photosynth. Res.*, 2016, Vol. 128, p. 215–217.

Buchanan B.B., Arnon D.I. A reverse Krebs cycle in photosynthesis: consensus at last, *Photosynth. Res.*, 1990, Vol. 24, p. 47–53.

Buchanan B.B., Sirevåg R. Ribulose-l,5-diphosphate carboxylase and *Chlorobium thiosulfatophilum*, *Arch. Microbiol.*, 1977, Vol. 109, p. 15–19.

Buchanan B.B., Wong J.H. A conversation with Andrew Benson: reflections on the discovery of the Calvin–Benson cycle, *Photosynth. Res.*, 2013, Vol. 114, p. 207–214.

Buchanan B.B. A conversation with Andrew Benson. Reflections on the discovery of the Calvin–Benson cycle, *You Tube* 2012, https://www.youtube.com/watch?v=GfQQJ2vR_xEE

Buchanan B.B., Douce R. Andrew Benson honored on birthday no 97, *Photosynth. Res.*, 2015, Vol. 123, p. 115–116.

Buchanan B.B., Douce R., Lichtenthaler H.K. Andrew A Benson, *Photosynth. Res.*, 2007, Vol. 92, p. 143–144; also see 12 other papers that follow, up to p. 271.

Calvin M., Benson A.A. The path of carbon in photosynthesis, *Science*, 1948, Vol. 107, p. 476–480.

Doman N.G. A question about the nature of rapidly labeled photosynthetic products, *Dokl. Akad. Nauk. SSSR*, 1952, Vol. 84, p. 1017–1019. [In Russian]

Doman N.G., Khadzhi-Murat L.N., Demina S.E. Unities and peculiarities of carbon assimilation pathway(s) in different plant species, *Dokl. Akad. Nauk. SSSR*, 1958, Vol. 122, p. 111–113. [In Russian]

Evans M.C.W., Buchanan B.B., Arnon D.I. A new ferredoxin dependent carbon reduction cycle in a photosynthetic bacterium, *Proc. Natl. Acad. Sci. USA*, 1966, Vol. 55, p. 928–934.

Fomina I.R., Biel K.Y. Photosynthetic carbon metabolism: Strategy of adaptation, *Contemporary Problems of Photosynthesis*, Allakhverdiev S.I., Rubin A.B., Shuvalov V.A., eds., Institute of Computer Science, Moscow-Izhevsk, 2014, Vol. 2, p. 415–484. [In Russian]

Fomina I.R., Biel K.Y. Photosynthetic Carbon Metabolism: Strategy of Adaptation over Evolutionary History, *Photosynthesis: New Approaches to the Molecular, Cellular, and Organismal Levels*, Allakhverdiev S.I., ed., © Scrivener Publishing LLC, 2016, p. 233–326.

Gates, R.D., Hoegh-Guldberg O., McFall-Ngai M.J., Bil' K.Y., Muscatine L. Free amino acids exhibit anthozoan "Host factor" activity: They induce the release of photosynthate from symbiotic dinoflagellates *in vitro*, *Proc. Natl. Acad. of Sci. USA*, 1995, Vol. 92, p. 7430–7434.

Govindjee. Celebrating Andrew Alm Benson's 93rd birthday, *Photosynth. Res.*, 2010, Vol. 105, p. 201–208.

Govindjee, Bassham H., Bassham S. Remembering James Alan Bassham (1922–2012), *Photosynth. Res.*, 2016, Vol. 128, p. 3–13.

Guralnick L.J., Edwards G.E., Ku M.S.B., Hockema B., Franceschi V. Photosynthetic and anatomical characteristics in the C_4 / CAM-cycling species, *Portulaca grandiflora*, *Funct. Plant. Biol.*, 2002, Vol. 29, p. 763–773.

Hatch M.D., Slack C.R. Photosynthesis by sugar-cane leaves: a new carboxylation reaction and the pathway of sugar formation, *Biochem. J.*, 1966, Vol. 101, p. 103–111.

Hartt C.E., Kortschak H.P., Burr G.O. Photosynthesis by sugar cane fed radioactive carbon dioxide, *Proc. Hawaii Acad. Sci.*, 1953, p. 13–14.

Herter S., Fuchs G., Bacher A., Eisenreich W. A bicyclic autotrophic CO_2 fixation pathway in *Chloroflexus aurantiacus*, *J. Biol. Chem.*, 2002, Vol. 277, p. 20277–20283.

Karpilov Y.S. Distribution of ^{14}C among the photosynthetic products of maize, *Trudy Kazan Selskokhoz. Inst.*, 1960, Vol. 41, p. 15–24. [In Russian]

Karpilov Y.S. Features of the functions and structure of the photosynthetic apparatus of some plant species of tropical origin, *Trudy Mold. NII Orosh. Zemledel. Ovoshchevod.*, 1969, Vol. 11, p. 1–35. [In Russian]

Kluge M. Crassulacean acid metabolism (CAM): CO_2 and water economy, *Ecological Studies, Vol 19: Water and Plant Life: Problems and Modern Approaches*, Lange O.L., Kappen L., Schulze E-D, ed., Springer, Berlin Heidelberg, 1976, p. 313–322.

Kondratieva E.N. *Photosynthetic Bacteria, Izd. AN SSSR*, Moscow, 1963, 314 p. [In Russian]

Kondratieva E.N. *Autotrophic Prokaryotes, Publ Moscow University*, Moscow, 1996, 304 p. [In Russian]

Kortschak H.P., Hartt C.E., Burr G.O. Carbon dioxide fixation in sugarcane leaves, *Plant Physiol.*, 1965, Vol. 40, p. 209–213.

Lichtenthaler H.K., Buchanan B.B., Douce R. Honoring Andrew Benson in Paris. A tribute on his 90th birthday, *Photosynth Res.*, 2008, Vol. 96, p. 181–184.

Lichtenthaler H.K., Buchanan B.B., Douce R., Govindjee. Andrew A. Benson, 1917–2015, *Photosynth. Res.* 2015*a*, Vol. 124, p. 131–135.

Lichtenthaler H.K., Buchanan B.B., Douce R., Govindjee. Obituary: Andrew A. Benson 1917–2015, *ASPB (American Society for Plant Biology) News*, March / April 2015, 2015*b*, p. 25–26.

Marakushev S.A., Belonogova O.V. Emergence of the chemoautotrophic metabolism in hydrothermal environments and the origin of ancestral bacterial taxa, *Doklady Biochem. Biophys.*, 2011, Vol. 439, p. 161–166.

Monson R.K., Edwards G.E., Ku M.S.B. C_3-C_4 intermediate photosynthesis in plants, *Bio. Sci.*, 1984, Vol. 34, p. 563–574.

Nezgovorova L.A. To a question about the products of photosynthesis, *Soviet Plant Physiol.*, 1956, Vol. 3, p. 497–507. [In Russian]

Nishio J.N., Ting I.P. Carbon flow and metabolic specialization in the tissue layers of the Crassulacean acid metabolism plant, *Peperomia camptotricha, Plant Physiol.*, 1987, Vol. 84, p. 600–604.

Nonomura A.M., Benson A.A. The path of carbon in photosynthesis: Improved crop yields with methanol, *Proc. Natl. Acad. Sci. USA*, 1992, Vol. 89, p. 9794–9798. (XXIV in the Nobel series)

Nonomura A.M., Benson A.A. The path of carbon in photosynthesis. XXIX. Glass microbeads, *J. Plant Nutr.*, 2012, Vol. 35, p. 1896–1909.

Nonomura A.M., Benson A.A., Biel K.Y. The path of carbon in photosynthesis. XXVII. Sugar-conjugated plant growth regulators enhance general productivity, *J. Plant Nutr.*, 2011, Vol. 34, p. 653–664.

Nonomura A.M., Cullen B.A., Benson A.A. The path of carbon in photosynthesis. XXVIII. Responses of plants to polyalkylglycopyranose and polyacylglycopyranose, *Advances in Photosynthesis – Fundamental Aspects*, Part 3, Chapter 13, Najafpour M.M., ed., InTech, Rijeka, 2012, p. 259–271.

Nonomura A.M., Benson A.A. The path of carbon in photosynthesis - XXX - α-mannosides, *Photosynthesis*, Dubinsky Z, ed., InTech, Rijeka, 2013, p. 3–22.

Nonomura A.M., Benson A.A. The path of carbon in photosynthesis. XXXI. The role of lectins, J. *Plant Nutr.*, 2014, Vol. 37, p. 785–794.

Nonomura A., Lorimer G., Holtz B., Vacquier V., Biel K., Govindjee. Andrew A. Benson: personal recollections, *Photosynth. Res.*, 2016, Vol. 127, p. 369–378. DOI 10.1007/s11120-015-0186-x

Nonomura A.M., Holtz B., Biel K.Y., Cooney R., Lorimer G., Govindjee. The paths of Andrew A. Benson: a radio-autobiography, *Photosynth. Res.*, 2017, Vol. 134, p. 93–105. DOI 10.1007/s11120-017-0410-y

Nonomura A.M., Pedersen A., Brummel D.P., Loveless L., Lauria A., Haschemeyer B., McBride M.S. Brandt iH026a plant growth regulator, *Photosynthetica*, 2018, Vol. 56, p. 411–417. DOI 10.1007/11099-017-0739-7

Osmond C.B. Crassulacean acid metabolism: A curiosity in context, *Annu. Rev. Plant Physiol.*, 1978, Vol. 29, p. 379–414.

Osmond C.B., Allaway W.G. Pathway of CO_2 fixation in the CAM plant *Kalanchoe daigremontiana*. I. Patterns of $^{14}CO_2$ fixation in the light, *Aust. J. Plant Physiol.*, 1974, Vol. 1, p. 503–511.

Shomer-Ilan A., Nissenbaum A., Waisel Y. Photosynthetic pathway and ecological distribution of the Chenopodiaceae in Israel, *Oecologia*, 1981, Vol. 48, p. 244–248.

Titlyanov E.A., Titlyanova T.V. Andy Benson in Russia and at Home, *Russian J. Mar. Biol.*, 2007, Vol. 33, p. 347–348.

Tolbert N.E. Metabolic pathways in peroxisomes and glyoxysomes, *Annu. Rev. Biochem.*, 1981, Vol. 50, p.133–157.

Tolbert N.E. The C_2 oxidative photosynthetic carbon cycle, *Annu. Rev. Plant Phys.*, 1997, Vol. 48, p. 1–25.

Winter K., Troughton J.H. Carbon assimilation pathway in *Mesembryanthemum nodiflorum* under natural conditions, *Z. Pflanzenphysiol.*, 1978, Vol. 88, p. 153–162

Zalensky O.V., Semikhatova O.A., Voznesenski V.L. *The Application of ^{14}C-Methods for Research of Photosynthesis, Publ. USSR Acad. Sci.*, Moscow–Leningrad, 1955, 90 p. [in Russian]

Zarzycki J., Brecht V., Müller M., Fuchs G. Identifying the missing steps of the autotrophic 3-hydroxypropionate CO_2 fixation cycle in *Chloroflexus aurantiacus*, *Proc. Natl. Acad. Sci. USA*, 2009, Vol. 106, p. 21317–21322.

Zelitch I. The photooxidation of glyoxylate by envelope-free spinach chloroplasts and its relation to photorespiration, *Arch. Biochem. Biophys.*, 1972, Vol. 150, p. 698–707.

Zelitch I. Alternate pathways of glycolate synthesis in tobacco and maize leaves in relation to rates of photorespiration, *Plant Physiol.*, 1973, Vol. 51, p. 299–305.

12

Recollection of Yuri S. Karpilov's Scientific and Social Life

Karl Y. Biel[1,2,*], **Irina R. Fomina**[1,2,*]

[1]*Institute of Basic Biological Problems, Russian Academy of Sciences, Pushchino, Moscow Region 142290, Russia*
[2]*Biosphere Systems International Foundation, Tucson, Arizona 85755, USA*

Abstract
We are dedicating this recollection to Soviet plant physiologist Professor Yuri S. Karpilov, brilliant scientist and a wonderful person, who 41 years ago, on September 21, 1977 passed away in a tragic traffic accident in Pushchino Moscow Region, Russia. He was only 46½ years old. In this tribute, we highlight the great importance of Yuri S. Karpilov's research of photosynthetic carbon metabolism, as well as the lack of public recognition of him as one of the pioneers in the discovery and development of Cooperative photosynthesis (C_4 photosynthesis) at a time when the scientific society had already accepted the Benson–Bassham–Calvin cycle as a universal mechanism of the photosynthetic carbon fixation.

Keywords: Karpilov's protocol, Carbon metabolism, Cooperative photosynthesis

* *Corresponding authors*: karlbiel@hotmail.com; irafomi@rambler.ru

12.1 Introduction

In this tribute, we highlight the research of the Soviet scientist Y. S. Karpilov in the field of C_4 photosynthesis (1959–1978) and explain the significance of his important contributions which have not received proper recognition.

In the last century, by the early 1950s a number of fundamental ideas and technical approaches had been developed for studying the mechanism of photosynthesis in plants, which led to future studies which revealed the amazing diversity in adaptations in the mechanism. This happened because Professor Andrew A. Benson and colleagues (Benson, Bassham *et al.*, 1950; Bassham *et al.*, 1954; Benson, 2002; and others) offered to the scientific community the $H^{14}CO_3^-$ / $^{14}CO_2$-chromatographic method which provided information on the sequence of biochemical reactions during photosynthetic carbon assimilation. This development led to the discovery of the reductive pentose phosphate cycle (named by the authors Benson–Bassham–Calvin cycle) from their initial studies on *Chlorella* (see Benson, 2002). This method has been remarkably accurate and replicable, and Benson's technique for the separating and identifying products was adopted by the World community of scientists.

The biochemical scheme of the reductive pentose phosphate cycle was so convincingly confirmed by experiments with various algae and some higher plants, that the American team of scientists was recognized as authorities on the process. A high level of confidence developed in the universal uniformity of carbon path in photosynthesis which became quite entrenched. This made it more likely that any exceptional results would be dismissed.

12.2 Some Contradictory Discoveries

In 1953 the results obtained from initial studies on sugarcane in Hawaii by the Kortschak group (see Hartt *et al.*, 1953, and Benson and Maretzki, 2011 in which some proceedings of meetings were reported in a tribute they made to Hartt in the 2011 C_4 book) did not fit into the scheme of the primary CO_2 photoassimilation proposed by Benson and colleagues. Thereafter Kortschak was visiting with Benson and discussed his results; Benson became convinced of the reliability of the results of the Hawaiian researchers, and strongly recommended that they urgently publish the material (A. Benson, per. com. with K. Biel). Eventually, Kortschak and colleagues did (see Kortschak *et al.*, 1965) but, unfortunately with a significant delay.

At the same time in the USSR, the scientific centers of Moscow (Doman, 1952; Doman *et al.*, 1952, 1958; Nichiporovich, 1953, 1955, 1958; Nezgovorova, 1956, 1957), Leningrad (Zalensky, 1954, 1976; see also Magomedov, 2015), Kazan (Karpilov, 1959, 1960; Tarchevsky and Karpilov, 1963) and other towns actively began to study the photosynthetic carbon metabolism in different groups of higher plants.

Like most foreign scientists, many Russian plant biologists perfectly mastered the $H^{14}CO_3^-$ / $^{14}CO_2$-chromatographic method. They considered primary carbon photoassimilation occurs according to the scheme proposed by Benson *et al.* without thinking there may be other options. Consequently, these scientists, working with a variety of higher plants either confirmed the presence of the reductive pentose phosphate cycle, or, they were left with a serious dilemma if they were studying a species now known to be C_4 (*e.g.*, *Zea mays* whose culture was very popular in the Soviet Union in those years). Namely, in maize leaves the primary products of photosynthesis at short light exposures were identified as C_4 dicarboxylic acids, not 3-phosphoglyceric acid (3-PGA). Thus, unusual data were considered either an artifact; or, if researcher(s) were confident in their methodology, the results were filed away, "till better times". In rare cases, some progressive thoughts were made with care (see *e.g.*, Nezgovorova, 1956, 1957).

12.3 Official Statement of a Young Scientist in the USSR and His Deed

Most likely, only this can explain the fact that in the Soviet Union, among a large galaxy of physiologists and biochemists of photosynthesis, only one researcher, in 1960 publicly stated the occurrence of a C_4 phenomenon in higher plants based on results with maize. The individual was a young scientist from the Agricultural Institute of Kazan, Yuri S. Karpilov.

After a series of experiments on maize leaves, he received data, which like the data of Kortschak and colleagues on sugarcane, contradicted the generally accepted concept of the primary photosynthetic carbon assimilation, and he was not afraid to publish them (see Karpilov, 1960). In the USSR, it was the first experimental publication associated with C_4 photosynthesis.

In those years Karpilov noted that "...In maize leaves exposed to light for 15 sec in the atmosphere of $^{14}CO_2$, significant labelling was not appearing in the compound, which was classified by other researchers as (3-PGA (13)[1], "as the first product" (10)[1]..." (see Figure 12.1). "...Incubation of the leaves in a normal atmosphere after a brief exposure to $^{14}CO_2$ led to ^{14}C transfer from C_4 dicarboxylic acids into compounds of the Calvin cycle and sugars" (Karpilov, 1960).

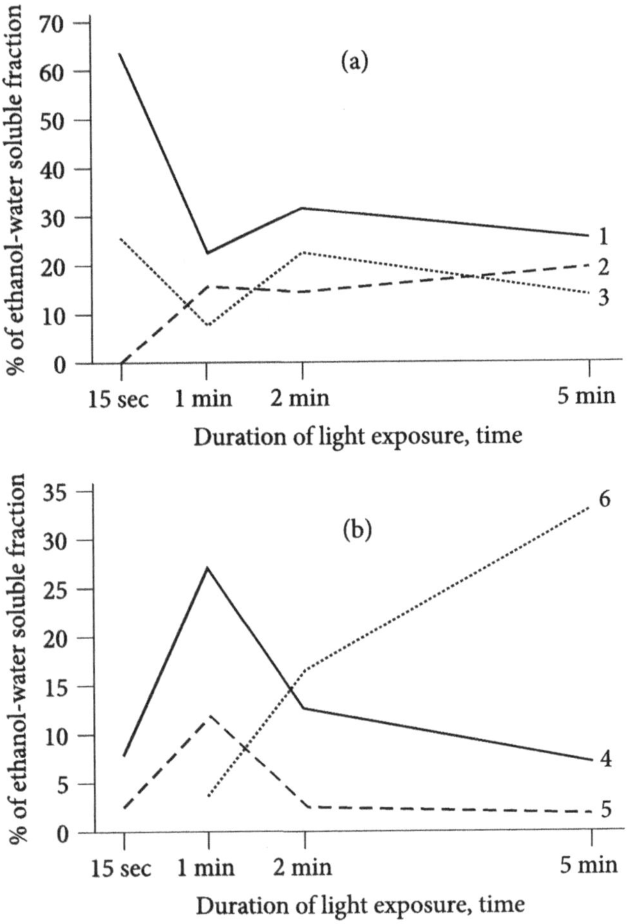

Fig. 12.1. Distribution of ^{14}C between photosynthetic products of *Zea mays* leaves after different duration of light exposure (label with $^{14}CO_2$ 15 sec, 1 min, 2 min, and 5 min): a) 1 – malate, 2 – alanine, 3 – aspartate; b) 4 – phosphoric esters of sugars, 5 – monosaccharides, 6 – sucrose (adapted from Karpilov, 1960)

[1] The numbers mean the references in original article.

Decades later, one of us (K. Biel) tried to find out from Professor N. G. Doman, one of the most authoritative biochemists and physiologists of Russia in the second half of the 20th century, how it could happen that his laboratory did not detect this phenomenon. Nikolai G. Doman laughed it off, saying that his team "...just did not have enough speed to make short light expositions (5–15 sec) with maize leaves in the presence of $^{14}CO_2$" (N. Doman, per. com.).

In public, Karpilov reported his results in 1962 at the International Biochemical Congress in Moscow. Unfortunately, the reactions were clearly hostile; some foreign researchers expressed mistrust, while domestic scientists were extremely negative. It went so far that he was advised by some to stop all his scientific studies (Y. Karpilov, per. com. with K. Biel).

However, Yuri Karpilov was not from the timid. On the contrary, after such reviews, he decided to continue his studies and went away from the central Laboratories, accepting the invitation to work in Tiraspol (Moldavian SSR, USSR) from Professor P. I. Dvornikov. The President of Moldavian Institute for Research in Irrigation Farming and Vegetable Growing, Dvornikov has been a very good leader, who had a tremendous talent to apprehend the potential capacity of scientists to make scientific discoveries. In this Institute Karpilov organized the Laboratory of isotopes and biochemistry of photosynthesis and, without scientifically-bureaucratic problems, worked 10 years, from 1962 to 1972.

It should be noted that before moving to Tiraspol Karpilov published an article in the News of the USSR Academy of Sciences (Karpilov, 1961), and defended his Ph.D. thesis "The influence of reducing the area of assimilation and conditions of nitrogen-phosphorus nutrition in the beginning of vegetation on some physiological processes and yield of maize" (Kazan, 1962). In the same year, Drs. Igor A. Tarchevsky and Yuri S. Karpilov submitted an article in the Russian Journal of Plant Physiology, which was published a year later (Tarchevsky and Karpilov, 1963). In this paper the authors convincingly proved the correctness of earlier chromatographic analyses of ^{14}C-labeled compounds. They also demonstrated that the composition of photosynthetic products is not dependent on the fixation technique, as well as the light intensity and CO_2 concentration.

In his new place, *i.e.* in Tiraspol, Karpilov had a good opportunity to investigate photosynthetic carbon metabolism using a large collection of agricultural higher plants. His main aim was to find out species, which will show the phenomenon of photosynthetic CO_2 fixation similar to corn. For this study were involved practically all employees of

Karpilov's Laboratory and even colleagues from other institutions. In different years of this period in Karpilov's Laboratory worked Svetlana Neustrueva, Nina Maslova (Kotova), Vladimir Anisimov, Vladimir Medvedev, Inna Nedopekina (Karpilova), Evgeniya Zarvanskaya, Valentina Karpenko, Lelya Russu, Oleg Malishev, Karl Biel, and some students from the local Pedagogical University. Collaborations of our Team were carried out with: Clara Shchupak from the same Institute; Nikolai Pristupa from the Institute of Plant Physiology, USSR Academy of Sciences, Moscow; Pavel Brick from the Agricultural Institute, Moldavian Academy of Sciences, Kishinev; Valery Karnaukhov from the Institute of Biophysics, USSR Academy of Sciences, Pushchino Moscow Region, and Mikhail Vaskovsky from the Botanical Institute, USSR Academy of Sciences, Leningrad[2].

Thus, in Tiraspol Karpilov studied the photosynthetic carbon metabolism of almost all agriculturally important crops, which were used for research by other scientists of the Moldavian Institute for Research in Irrigation Farming and Vegetable Growing.

The accumulated experimental data, their systematization and full analysis has been done during 1963–68. This work gave rise to a series of publications (see, for example, Karpilov and Nedopekina, 1965; Karpilov and Shchupak, 1965; Karpilov and Kotova, 1969; Karpilov *et al.*, 1969; Karpilov and Vaskovsky, 1969; Karpilov and Brik, 1969; Karpilov and Zarvanskaya, 1969*a*, 1969*b*; Karpilov and Biel, 1970; and others). Most of the monocotyledonous and dicotyledonous species after the examination were classified into three groups according to the type of primary products of CO_2 photoassimilation (Karpilov, 1966; Karpilov *et al.*, 1968; Karpilov, Nedopekina *et al.*, 1968).

- The first group comprised the plants in which the metabolites of the Benson–Bassham–Calvin cycle are formed slowly but carboxylation of phosphoenolpyruvate (PEP) proceeds readily – maize, millet and sorghum.
- Plants with higher rates of reductive reactions of the pentose-phosphate cycle and low rates of PEP-carboxylation were assigned to the second group – pea, bean, alfalfa, radish, and cabbage.
- The third group included species with intermediate properties – tomato, eggplant, pepper, sugar beet, grape, and sunflower.

---

[2] The authors apologize if any colleagues of Yuri Karpilov have not been mentioned here.

12.4 From the Memories of Karl Biel

On the recommendation of my friend Oleg Malyshev (Obituary of Malishev can be seen in Bil and Nishio, 1999), the first time I came to Karpilov's Team in the summer of 1966 was as a volunteer, and in January of 1968 I became a member of the Laboratory. Joining with a friendly Group, I quickly realized that Karpilov was not just a talented scientist, but also a wonderful person. He could very quickly "infect" other people with his enthusiasm and scientific interests. His ideas seemed to me very important, and I wanted to implement them immediately.

My first thought was to analyze the accumulated biochemical data on photosynthesis relevant to our interests (see, *e.g.*, Karpilov, 1966; Karpilov *et al.*, 1968; Karpilov, Nedopekina *et al.*, 1968), considering the anatomical features of plant leaves that has not been done before. This approach to my surprise proved to be very promising. In particular, as mentioned above, according to Karpilov, there were three agriculturally important plant groups related to the character of the path of the primary CO_2 photoassimilation (see Karpilov, 1966; Karpilov *et al.*, 1968; Karpilov, Nedopekina *et al.*, 1968; see also Karpilov, 1970*a*, 2007; Karpilov and Bil', 1970; Biel, 2000; and others). The group which was of interest to us included species having a low rate of formation of the Benson-Bassham-Calvin cycle as initial products and high capacity for PEP carboxylation. The leaves of these plants have a special, so-called, "Kranz" leaf anatomy (*e.g., Zea mays, Panicum miliaceum, Sorghum bicolor*) (Figures 12.2 and 12.4; see also Karpilov, 1970*a*, 1971, 2007; Karpilov and Bil', 1970; Biel, 1976, 1988, 1993, 2000). Leaves of plants of the other two groups had high activity of the reductive pentose phosphate cycle and low capacity of PEP carboxylation. While there were some differences between the second and third groups in leaf anatomy (variation in palisade and spongy parenchyma tissues, uniform mesophyll tissue, *etc.*), none of these species had the distinctive Kranz anatomy.

Following this, the main attention of our Team (Figure 12.3) was given to representative species with "Kranz" leaf anatomy. In these years (1966–1968), Karpilov was very insightful in proposing how a large set of biochemical reactions of photosynthetic carbon metabolism may be "distributed" between the two types of phototrophic cells in leaves of maize and other species of these groups of plants.

I continued to search for such higher plants, which were later named species with the Cooperative photosynthesis, or C_4 plants. For this purpose, in addition to literature sources, I performed a simple test to study anatomical structure of leaves using leaf cross sections of cultivated and,

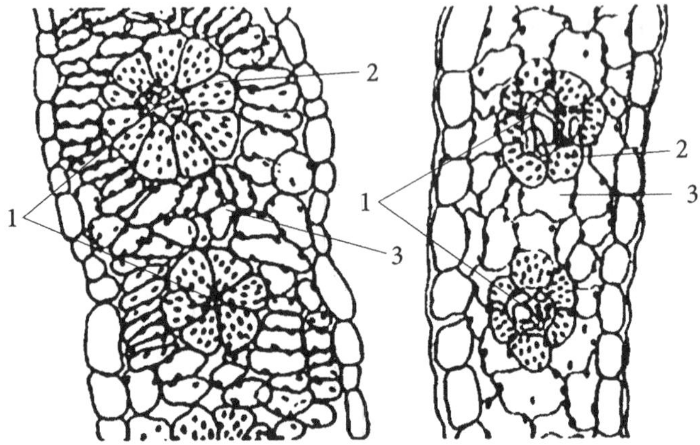

Fig. 12.2. Features of photosynthetic apparatus structure in species with "C_4-pathway" of carbon metabolism: *from the left* – cross section of *Amaranthus retroflexus* leaves; *from the right* – cross section of *Zea mays* leaves. 1 – bundles; 2 – bundle sheath cells, 3 – mesophyll cells (copy from Karpilov, 1969)

Fig. 12.3. Main participants in development of Cooperative Photosynthesis Concept in the Laboratory of Isotopes and Biochemistry of Photosynthesis, Moldavian Institute for Research in Irrigation Farming and Vegetable Growing, Tiraspol, Moldavian SSR, USSR. *From left to right*: Oleg G. Malishev, Yuri S. Karpilov and Karl Y. Biel – 1966–1972 (photo images from the private collections of Karl Biel and Inna Karpilova)

especially – wild species which was naturally growing in Moldavian SSR. Also, using a large collection of maize genotypes of Professor Valery N. Lisikov (Kishinev Agricultural Institute, Moldavian SSR, USSR) I tried to find chlorophyll mutants which lacked chlorophyll pigments in either mesophyll or bundle sheath cells (see Biel, Arkhipov *et al.*, 1974; Karpilov and Biel, 1975; Biel, 1976). A more detailed analysis of the photosynthetic apparatus in the assembled collection of species was carried out with fixed material (see, *e.g.*, Figure 12.4). Frequently, depending on the task, colored cytochemical reactions were used to study the anatomy (Karpilov, Bil *et al.*, 1970; Karpilov and Biel, 1974; Biel, 1976; and others). Additionally, many other aspects were examined in species in our collection, including the ultrastructure of photosynthetic tissue; aspects of the chloroplast photochemistry; and primary carbon metabolism during photosynthesis in plants growing in the natural environment and in greenhouse conditions (Karpilov and Malishev, 1970; Karpilov, 1971, 2007; Biel, 1976; and others).

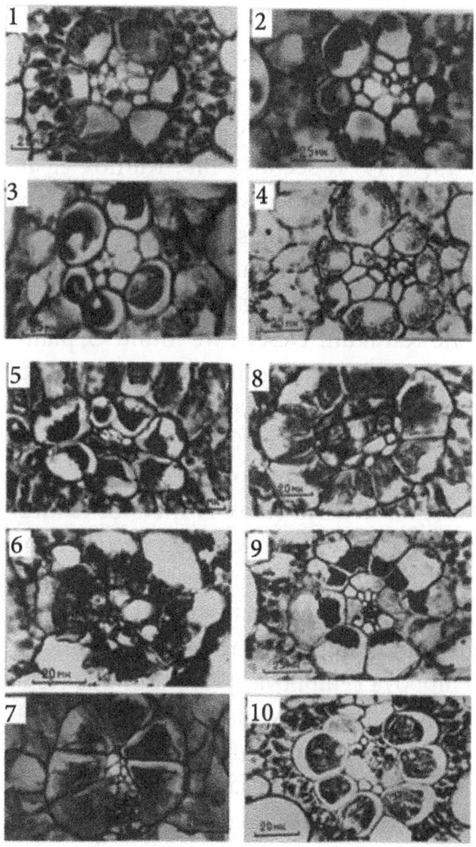

Fig. 12.4. Cross section of C_4 plants leaves with chloroplasts in bundle sheath cells (BSC) shifted to the mesophyll (1 – *Sorghum vulgare*; 2 – *Echinochloa crusgalii*; 3 – *Setaria viridis*; 4 – *Zea mays*), and with chloroplasts in BSC shifted to the bundle (5 – *Tibullus terrestris*; 6 – *Portulaca oleracea*; 7 – *Atriplex tatarica*; 8 – *Amaranthus retroflexus*; 9 – *Panicum miliaceum*; 10 – *Cynodon dactylon*) (copy from Biel, 1976). **Note**: Fixation – Navashin, coloration – hematoxylin

I remember that in those years Karpilov in discussions with scientists from other laboratories and visitors of the Institute joked about us young colleagues, saying that "... dissertations of my fellows are growing up in the dustbin". His interlocutors were often at a loss, and demanded clarification. Clarification was that in most of the natural conditions in Moldova, C_4 species were found in all sorts of wastelands, abandoned farmlands, landfills, garbage pits, *etc.*

At the end of 1967, and beginning of 1968, Karpilov systematically divided a considerable number of species with the cooperative photosynthesis into two groups (Karpilov, 1969; Cooperative photosynthesis of xerophytes, 1970; Karpilov, 2007). The first group included representative grass species with a predominant labelling by $^{14}CO_2$ into malic acid during a short light exposure. Lamellar chloroplasts of bundle sheath cells of these plants are localized in the cells centrifugally (see Figure 12.4) and are characterized by reduced photosystem II activity (Karpilov *et al.*, 1970, 1971; see also Bil', Klimov *et al.*, 1975; Karpilov, 2007). The second group included species in which aspartic acid was the primary initial C_4 product. Bundle sheath cells of these plants contained granal chloroplasts which were located in the centripetal position (Figure 12.4; see also Karpilov, 1971, 2007; Karpilov *et al.*, 1970, 1971; Biel, 1976). Later, in the Karakum Desert, we found C_4 plant with an absolute predominance of aspartate among the primary products of photosynthesis, *Suaeda arcuata* Bunge (*Chenopodiaceae)* which, in contrast to the representatives of malate group of C_4 plants, have lamellar chloroplasts with inactive photosystem II in mesophyll cells (Bil' and Gedemov, 1980).

In the late 1960's, for detailed study of the biochemical characteristics of both phototrophic tissues, we also intensively searched for methods of separation of mesophyll cells and bundle sheath cells. Initially Oleg Malishev, who became at that time a postgraduate student of Yuri Karpilov, applied the non-aquatic method of gradient centrifugation according to description of Dr. Ludmila A. Filippova (Filippova, 1967). But this technique did not give satisfactory results (Karpilov and Malishev, 1970; see also Karpilov, Malishev *et al.*, 1974). Success came unexpectedly, when I was trying to prove the absence of photosystem II activity in lamellar chloroplasts of bundle sheath cells of *Zea mays* by the method of Engelmann (see Biel, 1976).

The idea was that it was necessary to put the plant material between the object and cover glasses with one to two drops of physiological solution together with the bacteria *Proteus vulgaris* (or *Bacilius mycoides*)[3]. Isolated

---

[3] The strains of *Proteus vulgaris* and *Bacilius mycoides* we got from the Institute of Microbiology of the USSR Academy of Sciences, Moscow, USSR.

bundle sheath cells and mostly broken mesophyll cells of *Zea mays* were used. Next, after the edges of the cover glass were "glued" by liquid vaseline / paraffin, the samples were placed in the dark for a period sufficient for the bacteria to stop moving once all the oxygen was absorbed. Afterwards samples of bundle sheath cells were illuminated through a special diaphragm under an optical microscope: as a result, the bacteria started to move and concentrate in certain areas indicative of production of oxygen by photosynthesis (Figure 12.5). It should be noted that some histochemical reagents, for example, 2,3,5–thriphenyl-tetrazolium chloride, and some other tetrazolium derivatives, which we used for our purposes, proved to be more affective and beneficial (see Figure 12.10; Karpilov *et al.*, 1970; Karpilov and Biel, 1974; Biel, 1976) than bacteria.

To prepare bundle sheath cells, I prepared cross-sections of the leaf, and then separated the epidermis and mesophyll cells from the bundle sheath cells by mechanical laceration using micro-needles at low magnification of the microscope. The separation of mesophyll from bundle sheath cells was easy, but the mesophyll cells were always disrupted, releasing chloroplasts (Figure 12.5).

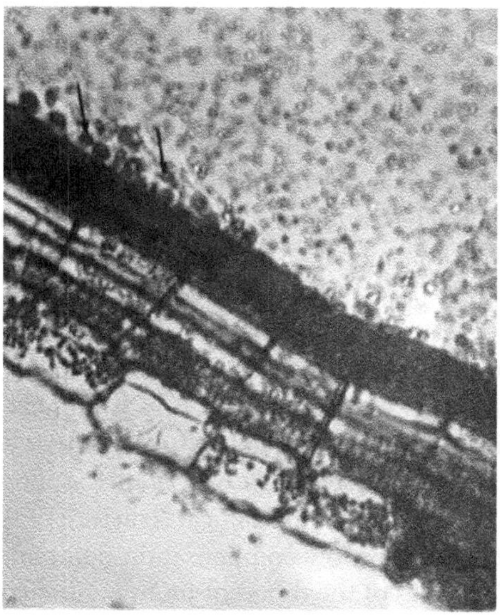

Fig. 12.5. Concentration of the bacterial strain *Proteus vulgaris* near the lightening bundle sheath cells which were isolated by mechanical method from *Zea mays* leaves. Arrows indicate the remaining mesophyll chloroplasts near the BSC after separation of phototrophic tissues (copy from Biel, 1976)
Note: in the lower part of the photo image it is possible to see the damage of some bundle sheath cells, which lost the chloroplasts

Very fast we understood that the strength of the bundle sheath cells walls in *Zea mays* is much higher than the strength of the cell walls of the mesophyll cells. After that it was a matter of technique to use. Soon enough by grinding leaves in a mortar, we obtained sequentially: the damaged mesophyll cells, then intermediate fractions, and finally substantially pure bundle sheath cells. We were delighted .., but the delight did not last long. In the midst of getting intriguing results from the separated tissues, we found that Drs. Björkman and Gauhl (1969) had already published biochemical results using mechanical methods similar to ours for separation of mesophyll chloroplasts from bundle sheath strands; we were a little late.

In the scientific journey of Karpilov, one can say with certainty that since 1968, a bright period in his creativity had begun. By this time, the research which he had performed during his postgraduate studies had been verified, and supported the hypotheses developed. This included obtaining support experimentally which showed that *Zea mays* and other species with similar anatomical features carry out carbon assimilation by spatial separation of reactions of CO_2 fixation and reduction which he called species with "Cooperative photosynthesis" (see Cooperative Photosynthesis of Xerophytes, 1970; Karpilov, 2007). The concepts for this phenomenon were published by Yuri Karpilov in 1969 in the special issue of the Proceedings of the Moldavian Institute for Research in Irrigation Farming and Vegetable Growing (Karpilov, 1969). In this work, he proposed a fundamentally new scheme of cooperative interaction between the mesophyll cells and bundle sheath cells in leaves of C_4 plants (Figure 12.6) and analyzed the possible differences between photochemical production of energy and in carbon metabolism of monocotyledonous and dicotyledonous C_4 species. Following this, in 1970 he published work on evolutionary aspects of the structural and functional organization of the photosynthetic apparatus of succulents and C_4 plants (Karpilov, 1970*b*).

These theoretical views were backed up by a number of experimental data, published in 1970 under the name "Cooperative Photosynthesis of Xerophytes". This book presents five works of Karpilov and colleagues, showing a correlation between anatomical structure of the leaves and the cooperative photosynthesis, low compensation point and high transpiration activity in C_4 species of the different mono- and dicotyledonous families. Features of leaf anatomy of representative species and the position of plastids and the occurrence of lamellar versus granal chloroplasts in bundle sheath cells are described. The hypothesis that the lamellar chloroplasts of bundle sheath cells were specialized for function of photosystem I was partially confirmed by experimental data. It was suggested in leaves of maize that some C_4 acids produced in the light may be oxidized in mitochondria in bundle sheath cells with production of pyruvate (Karpilov 1970*c*).

Based on the radio-chromatographic analysis of photosynthetic products in fractions enriched in mesophyll and bundle sheath cells chloroplasts, along with the kinetics of the photosynthesis products, possible transport of metabolites between mesophyll and bundle sheath cells was discussed (Karpilov, 1969, 1971, 2007; Cooperative Photosynthesis of Xerophytes, 1970; Karpilov and Malishev, 1970; Photosynthesis in Maize. Structural and Functional Properties of Photosynthetic Apparatus, 1974; Biel, 1976; and others).

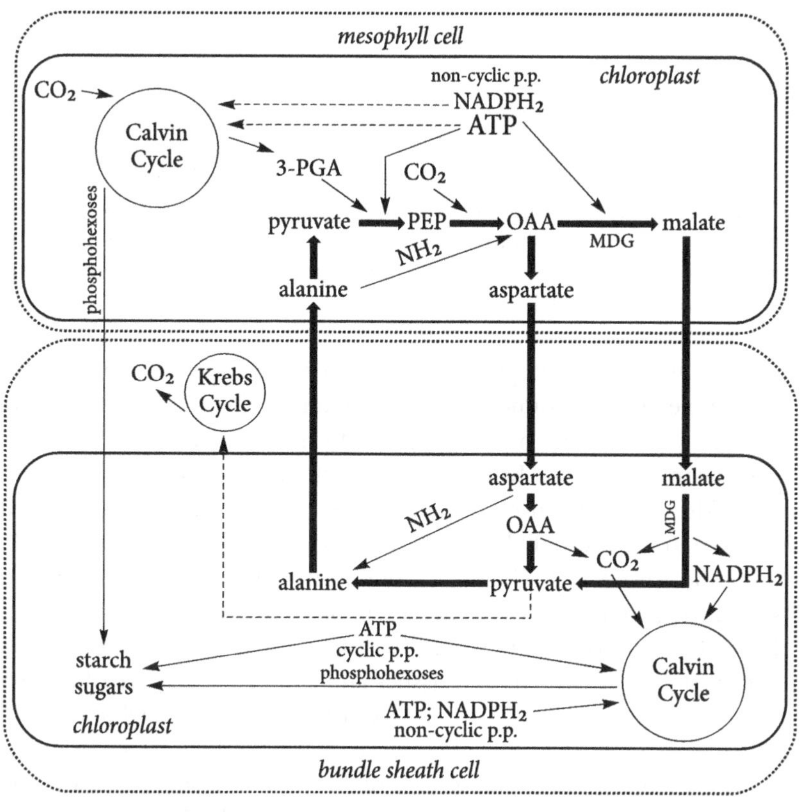

Fig. 12.6. Cooperative participation of chloroplasts of mesophyll and bundle sheath cells in photosynthesis of C_4 plants (adapted from Karpilov, 1969; see also Osmond *et al.*, 2016). Pretty soon Karpilov improved the scheme of Cooperative photosynthesis, showing that most part of the Calvin cycle (Benson–Bassham–Calvin cycle) is absent in mesophyll cells (Karpilov and Malishev, 1970; Karpilov *et al.*, 1974; see also Karpilov, 2007)
Note: p.p. – photophosphorylation

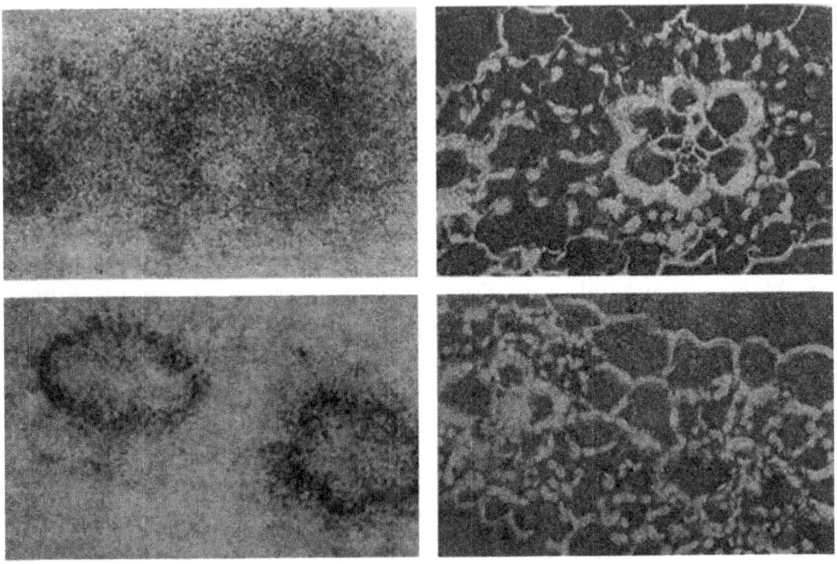

Fig. 12.7. Micro-radioautographs (left) and photos (right) of cross sections *Zea mays* leaves after 20 sec photosynthesis with $^{14}CO_2$ (top) and after additional illumination 100 sec in $^{12}CO_2$ (bottom) during pulse-chase experiment (adapted from Karpilov *et al.*, 1970; Karpilov, 2007)

Special microradioautographic experiments to identify the alcohol-water soluble ^{14}C-compounds within the photosynthetic tissues of leaves of *Zea mays* was carried out together with Dr. Nikolai A. Pristupa in Academician Andrei L. Kursanov' Laboratory (Institute of Plant Physiology, Moscow), which confirmed that the initial $^{14}CO_2$ photo-fixation takes place in the mesophyll cells, from which ^{14}C was quickly transported into bundle sheath cells (Figure 12.7; Karpilov, Bil' *et al.*, 1970; see also Pristupa, 1972; Biel *et al.*, 1975; Biel, 1976; Biel and Karpilov, 1978).

This redistribution of the fixed ^{14}C-carbon was considered to occur by the close contact of two types of tissues and a large occurrence of the pore fields and plasmodesmata in adjacent cell walls of mesophyll and bundle sheath cells (Figure 12.8; Karpilov, Bil' *et al.*, 1970; Biel *et al.*, 1975; Biel, 1976, 1993). This provides a large area of symplastic interactions (Karpilov, Bil' *et al.* 1970; see also Photosynthesis in Maize, 1974; Biel *et al.*, 1975; Biel 1976, 1988, 1993), and also by the presence of a peripheral reticulum for example, in mesophyll and bundle sheath cells chloroplasts of *Amaranthus retroflexus*, *Zea mays* and *Panicum miliaceum* leaves (see Figures 15–17 and 19 in Biel, 1976), which increases the surface area for transport by formation of invaginations and protrusions (Karpilov, Bil' *et al.*, 1970; see also Photosynthesis in Maize, 1974; Biel *et al.*, 1975; Biel, 1976, 1988, 1993).

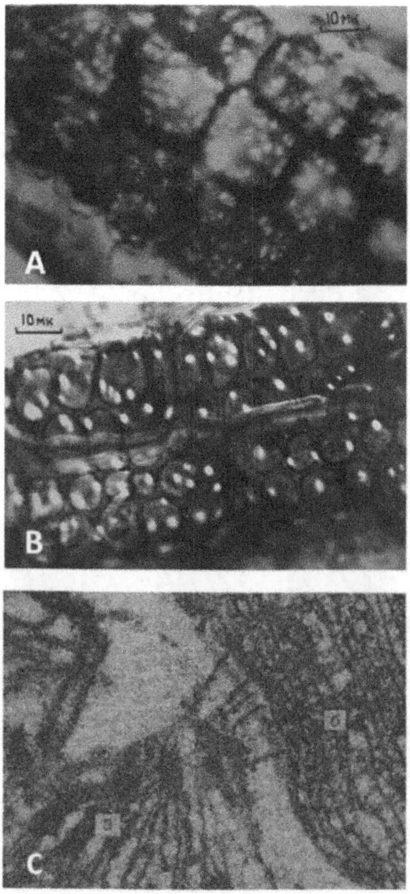

Fig. 12.8. Pores fields in adjacent cell walls of *Zea mays* leaves on (A) mesophyll : bundle sheath cell and (B) mesophyll : epidermis surfaces; (C) the plasmodesmata in pore of bundle sheath cell and mesophyll cell walls (adapted from Karpilov *et al.*, 1970; Biel, 1976)
Note: The light holes [gleams] in Figures 12.8 A and B identified in histochemical reaction with *Cl-Zn-J* correspond to the pores

As mentioned above, in the earlier studies (see Karpilov, 1969; Cooperative Photosynthesis of Xerophytes, 1970; see also Karpilov, 2007) the possibility for active symplastic transport of the photosynthetic products between mesophyll and bundle sheath cells was suggested. Later, in histochemical experiments and studies with ATPases inhibitors it was shown that an active Mg^{2+}-dependent ATPase is present in the chloroplast peripheral reticulum and envelope, and in the plasma membranes and plasmodesmata (Figure 12.9; Karpilov *et al.*, 1975*a*; Biel, 1988, 1993).

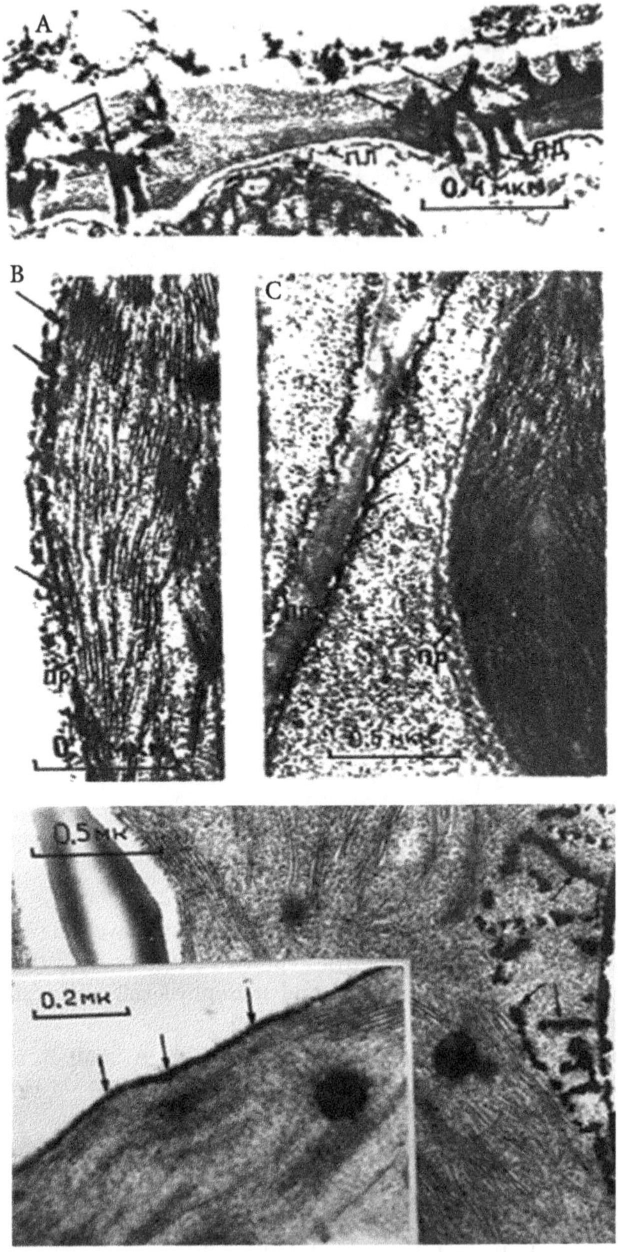

Fig. 12.9. Location of Mg^{2+}-dependent ATPase in peripheral reticulum of bundle sheath cells chloroplasts, plasmalemma and the plasmodesmata of *Amaranthus retroflexus* leaf (adapted from Biel, 1976, 1988; Bil' *et al.*, 1976)

Note: Precipitate of lead phosphate formed in the histochemical reaction and the corresponding of enzyme localization indicated by arrows

Also, the rate of the transition of ^{14}C from dicarboxylic acids into products of the reductive pentose phosphate cycle was discussed in relation to the ratio of malate / aspartate and placement of chloroplasts in bundle sheath cells (Karpilov *et al.*, 1975*b*). Evidence that the anatomical features of C_4 plants is correlated with the composition of the primary CO_2 fixation products was confirmed: in C_4 grasses malic acid dominated among the products of photosynthesis in C_4 plants with lamellar bundle sheath cells chloroplasts that are centrifugally located in the cells; while aspartic acid prevails in C_4 species with granal bundle sheath cells chloroplasts which are centripetally located in cells.

Thus, it has been experimentally shown that (a) lamellar bundle sheath cells chloroplasts of leaves of malate type C_4 plants have a high Chl *a* / *b* ratio and a high content of long-wave forms of Chl *a* (Karpilov, Bil' *et al.*, 1970; Biel, Arkhipov *et al.*, 1974; Karpilov and Biel, 1975; Biel, 1976); (b) a very low level of photo-induced changes of Chl *a* fluorescence yield (approximately 50 times less than in mesophyll chloroplasts) (Bil', Klimov *et al.*, 1975; Biel, 1976); (c) a reduced rate of $NADP^+$ reduction from water and an increased ability for cyclic photophosphorylation (Biel, 1976); (d) little or no capacity for light-dependent reduction of tetrazolium derivatives (Figure 12.10) without adding an electron donor (Karpilov *et al.*, 1970; Karpilov, Bil' *et al.*, 1971; Karpilov and Biel, 1974). It was concluded that the photosystem II activity in lamellar chloroplasts is greatly reduced in comparison with granal plastids and photochemical systems do not contribute significantly to the formation of NADPH, which is used in the reductive pentose phosphate cycle (see Biel, 1976, 1988, 1993).

12.5 Australian Scientist Professor Barry Osmond Visited Karpilov's Laboratory in 1971

Chronologically, perhaps it's time to mention an important event which occurred with Karpilov in 1969–1971.

Interest in the unique physiological and biochemical attributes of photosynthesis in certain species was growing in the 1960s. By 1969 hypothetical schemes were independently provided in several laboratories worldwide for the mechanism of photosynthesis in these species. One of these laboratories, as mentioned above, was Karpilov's Laboratory.

Naturally Karpilov made every effort to inform Russian and foreign scientists about our scientific achievements and sent them some publications of our Laboratory with cover letters to them (see, *e.g.*, Karpilov correspondence with foreign scientists, Karpilov, 2007: 314–326).

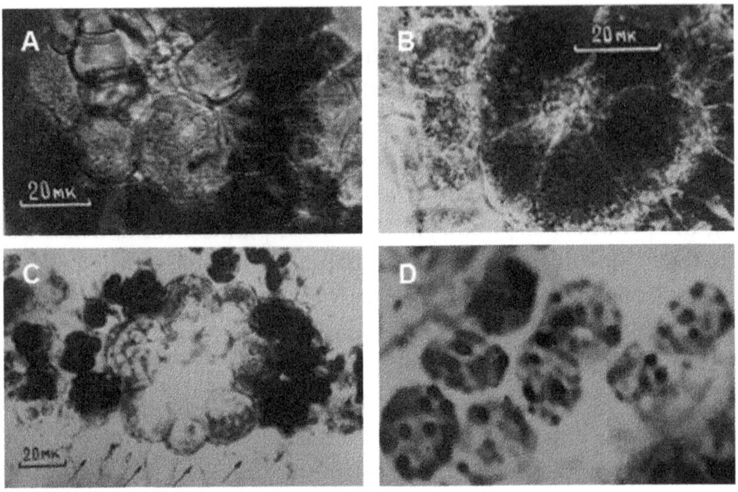

Fig. 12.10. Red formazan crystals formation during reduction of 2.3.5 – thriphenyl-tetrazolium chloride by chloroplasts during illumination of (A) *Zea mays*, (B) *Amaranthus retroflexus* and (C) pigment mutant *Zea mays* leaves (arrows show absence of chloroplasts in bottom part of mesophyll cells); D – formazan crystals in mesophyll chloroplasts of *Z. mays* leaves; (adapted from Karpilov *et al.*, 1970; Karpilov and Biel, 1974)

Foreign scientists were interested in the results of Karpilov and his colleagues. However, due to many difficulties, the trip abroad from Tiraspol was virtually impossible for Karpilov. In 1971, Professor Barry Osmond, a famous Australian scientist came to visit Karpilov in Tiraspol (Figure 12.11). In fact, the visit to our laboratory by this foreign scientist who was so well known it caused psychological "turmoil" in the scientific community of Soviet leaders in the field of photosynthesis research. And there was nothing surprising – the country's leading experts "missed the boat" for a one of the major discoveries in biochemistry and physiology of photosynthesis. It so happened that by 1970-71, the C_4 syndrome was well known to most colleagues in the world of plant physiology science. However, the early studies which were conducted in the Soviet Union only became known through the scientist, who himself contributed to the discovery of the C_4 mechanism! Truly, the Utterance of Jesus Christ: "No prophet is in his own Fatherland", – works at different levels of the social hierarchy.

As a matter of urgency, leaders in the field of physiology and biochemistry of photosynthesis began to find out about Karpilov, forgetting that about 10 years ago, tried to literally push him out of plant physiology when not believing his work on the C_4 phenomenon which is now accepted unconditionally.

Fig. 12.11. The visit of Professor Barry Osmond to Yuri Karpilov in the Moldavian Institute for Research in Irrigation Farming and Vegetable Growing, Tiraspol, USSR. *From left to right*: Isaac Y. Vinogradov (interpreter - English, Chinese), Vera A. Grigorian (interpreter - English, German), Barry Osmond, Yuri S. Karpilov, and Karl Y. Biel - 1971 (photo image from the private archive of Inna Karpilova; see also Osmond *et al.*, 2016)

12.6 Moving from Tiraspol to Pushchino, Moscow Region, to the Institute of Photosynthesis of the USSR Academy of Sciences

At the beginning of 1972 Yuri Karpilov and Karl Biel relocated to work at the Institute of Photosynthesis, USSR Academy of Sciences in Pushchino, Moscow Region (now the Institute of Basic Biological Problems, Russian Academy of Sciences), which had been recently organized and consisted of young scientists.

Before moving to Pushchino Karpilov in Tiraspol completed preparation of his Dr. Sci. Thesis "Participation of nitrogen and phosphorus in photosynthetic metabolism in relation with its specificity in various plant species" (Karpilov, 1971), and in 1972 in Kazan (Russian SSR, USSR) he very successfully defended it.

Now new opportunities and scientific perspectives were opened for Karpilov. Professor Barry Osmond was the scientific angel-associate, who "lobbied" the Russian "C_4-team" to continue researches in this chosen scientific direction.

At the Institute of Photosynthesis, Karpilov organized the Laboratory for Carbon Metabolism (Figure 12.12) and led work on the physiology and biochemistry of photosynthetic carbon metabolism. Under his leadership, various questions on photosynthesis were addressed and some of them have been resolved.

Fig. 12.12. Laboratory for Carbon Metabolism (1972–1977) of the Institute of Photosynthesis, USSR Academy of Sciences, Pushchino, Moscow Region, USSR. *From left to right:* 1st *row* – Raisa Kartasheva, Yuri Karpilov, Simon Hertz; 2nd *row* – Alexandr Maslov, Valery Lyubimov, Lyudmila Belobrodskaya (Ignatova), Liya Oparina; 3rd *row* – Alexandr Kuzmin, Raisa Karpova (Demidova), Karl Biel, and Irina Novitskaya (Maslova) – 1975 (photo image from the archive of the Institute of Basic Biological Problems, Russian Academy of Sciences, 142290 Pushchino, Moscow Region, Russia)

One of the reason for success based on the introduction into practice of the enzymatic method of cell separation, proposed by Kanai and Edwards (1973), which allowed to isolate mesophyll protoplasts and the bundle sheath cells (Figures 12.13 A, B) from the leaves of certain C_4 species. This technique gave us much more accurate results than we have got with the mechanical method used earlier (Figure 12.13 C).

Our studies have focused on identifying the mechanisms of photosynthetic carbon assimilation and the regulation of activity of the associated enzymes (Karpilov, Malishev *et al.*, 1974; Karpilov *et al.*, 1975, 1979; Karpilov, 1976, 1978; Persanov *et al.*, 1976; Persanov, Voronova *et al.*, 1976); the organization of the electron transport chain of granal and lamellar chloroplasts (Bil', Klimov *et al.*, 1975; Biel, 1976, 1988, 1993; and others); and its connection with the reactions of carbon and nitrogen metabolism (Karpilov *et al.*, 1975); mechanisms and pathways of transport of photosynthetic products between mesophyll and bundle sheath cells (Biel *et al.*, 1975; Karpilov *et al.*, 1976*a*, 1976*b*; Karpilov and Bil', 1976; Biel and Karpilov, 1978; see also Biel, 1988, 1993).

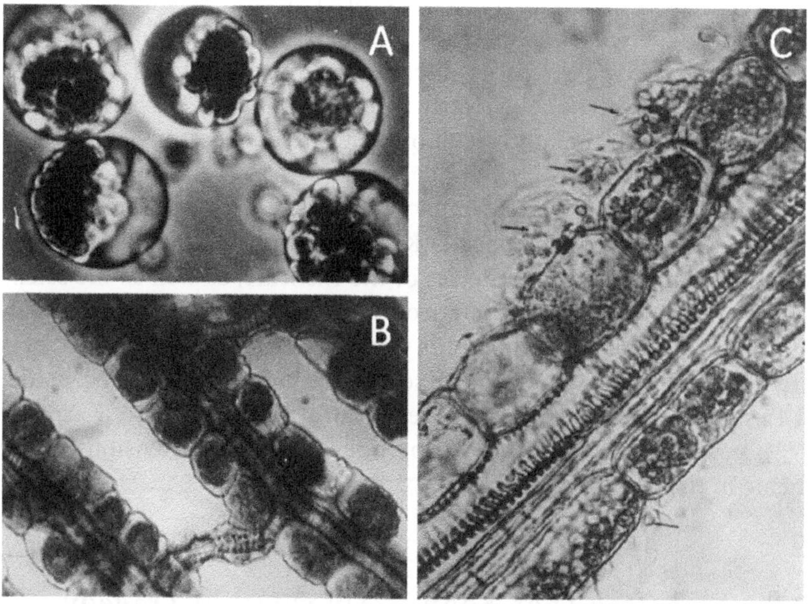

Fig. 12.13. Examples of photo images: (A) protoplasts of mesophyll cells, (B) bundle sheath cells separated by enzymatical method during treatment of *Zea mays* leaves segments by 1% of cellulase "Onozuka", and (C) bundle sheath cells obtained by mechanical method (adapted from Bil', Klimov *et al.*, 1975; Biel, 1976; both methods are described in a great detail in Biel, 1976)
Note: (C) – arrows show the remaining part of damaged mesophyll cells with chloroplasts

More attention began to be paid to the questions of the oxidative processes in the photosynthetic cell, especially to the mechanism of photorespiration in leaves of different groups of C_3 and C_4 plants (Karpilov, Kerimov *et al.*, 1978; Karpilov, Novitskaya, Belobrodskaya *et al.*, 1978; Lyubimov and Karpilov, 1977, 1978; Karpilov, Lyubimov *et al.*, 1978; Lyubimov *et al.*, 1978; Mechanism of Photorespiration and Its Specificity in Different Types of Plants, 1978; and others); and, to the evolution of structural and functional aspects of photosynthesis (Karpilov, 1974; Karpilov *et al.*, 1975; and others).

12.7 International Botanical Congress ...

In the summer of 1975 Yuri Karpilov participated in the International Botanical Congress in Leningrad. At the Congress Karpilov met with foreign colleagues, including Professors Black C. C., Hatch M. D., Edwards G. E., Laetsch W., Smillie R. M., and others. In conversations, Karpilov and foreign scientists discussed pointed questions of C_4 phenomenon that were not yet clearly resolved, mutually exchanged their unpublished results and at the end of the Congress became not only colleagues, but also friends.

After this meeting, Yuri Karpilov was on an emotional lift. His long-term research plans were huge. At Karpilov's initiative, members of his Laboratory and other Laboratories of the Institute, as well as the scientists from several scientific centers of the USSR were involved in developing joint projects; seemingly for decades ahead. Soon Karpilov prepared a "draft" of a biochemical scheme, which was reported at the IX Pushchino Reading Conference (Figure 12.14; see also Karpilov, 1976); he had planned to verify experimentally many provisions in the scheme in the coming years.

12.8 And after That, Soon ... Unexpected Tragedy

However, Karpilov was destined to work for the benefit of Plant Physiology and Biochemistry in the Institute of Photosynthesis of the USSR Academy of Sciences for a little over two years, until 21 September 1977 when a tragic traffic accident in Pushchino, Moscow Region, took the life of a brilliant scientist and a wonderful person. He was only 46 ½ years old.

The memory of Yuri Karpilov resides in the hearts of his family, friends, colleagues, students, and everyone who knew this wonderful man and great scientist. There have been many dedications and tributes to him including (Figures 12.15 and 12.16): meeting (All Soviet-Union Meeting, 1981), book (Biel 1993), article in the book (see Ilin *et al.*, 2011), recollections (see Biel K. Y., Lyubimov V. Y., Makarov A. D., Oparina L. A.,

Ignatova L. K., Pavlova E., Buntukova E. K. and Pakhomova V. M. in Karpilov, 2007; Osmond *et al.*, 2016), articles in the periodical press (Mokronosov and Karpilova, 1992; Bil', 2000), and publication of his selected works (Karpilov, 2007).

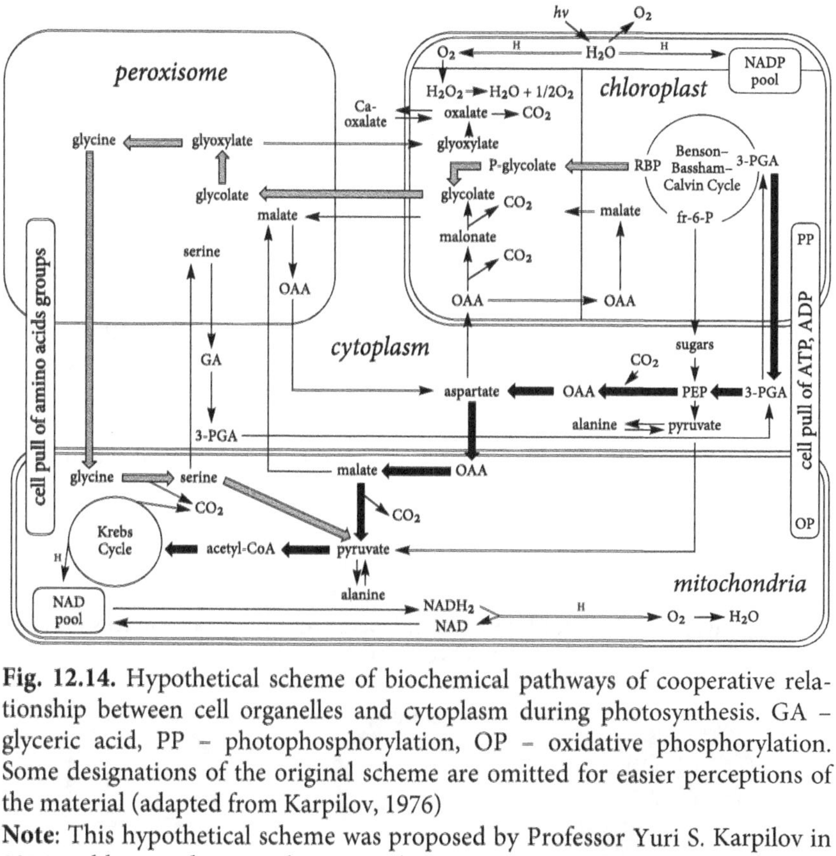

Fig. 12.14. Hypothetical scheme of biochemical pathways of cooperative relationship between cell organelles and cytoplasm during photosynthesis. GA – glyceric acid, PP – photophosphorylation, OP – oxidative phosphorylation. Some designations of the original scheme are omitted for easier perceptions of the material (adapted from Karpilov, 1976)

Note: This hypothetical scheme was proposed by Professor Yuri S. Karpilov in 1976 and has not lost its relevance today, and on some of points has been confirmed in the same years (see Karpilov *et al.*, 1978; and other articles in "Mechanism of Photorespiration and Its Specificity in Different Types of Plants"). In this series of papers, different taxa of angiosperms were studied in detail regarding the interrelationships between carbon metabolism and chloroplasts' energy production in the photosynthetic leaf cells, the participation of cell organelles in photosynthesis and photorespiration, the photoassimilates role in supplying the substrates for mitochondrial respiration, and others. It should be noted, however, that a few the provisions of the scheme have not yet been studied probably due to the complexity of the issue and the lack of methodological approaches

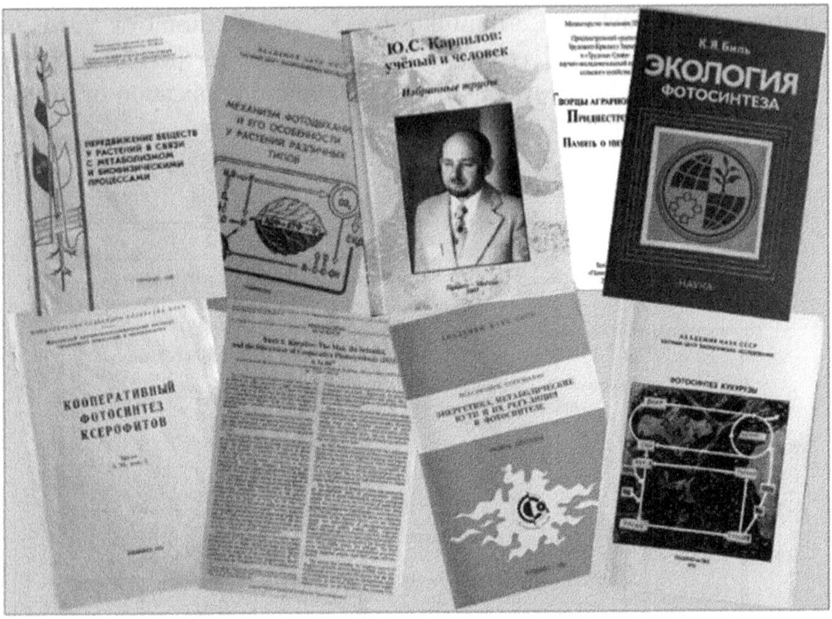

Fig. 12.15. Some of the books written or reviewed by Yuri S. Karpilov and some publications dedicated to his memory (copy from Osmond *et al.*, 2016)

Yuri S. Karpilov
[21 March 1931 – 21 September 1977]
Soviet plant physiologist, Ph.D. (1962), Doctor of Biological Sciences (1972), Professor (1977)

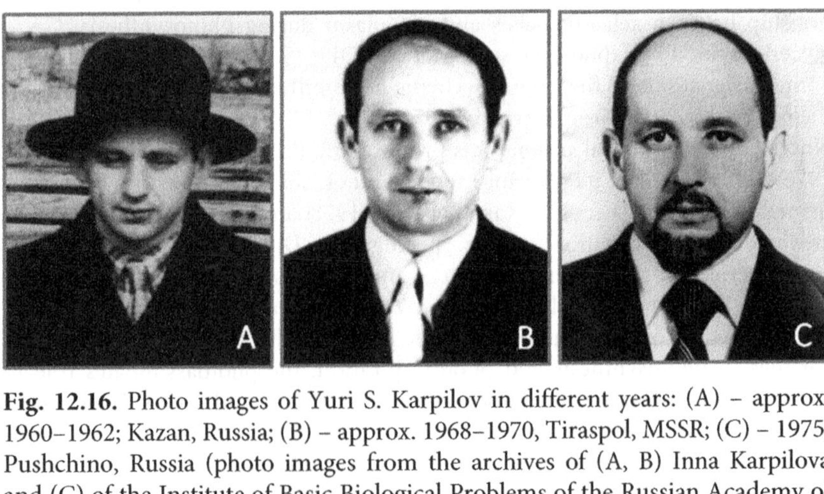

Fig. 12.16. Photo images of Yuri S. Karpilov in different years: (A) – approx. 1960–1962; Kazan, Russia; (B) – approx. 1968–1970, Tiraspol, MSSR; (C) – 1975, Pushchino, Russia (photo images from the archives of (A, B) Inna Karpilova and (C) of the Institute of Basic Biological Problems of the Russian Academy of Sciences; see also Osmond *et al.*, 2016)

12.9 Short Biography of Yuri S Karpilov

1953 – Yuri Karpilov graduated from the Kazan Agricultural Institute

1953–1954 – worked as a correspondent of the republican newspaper of the Tatar ASSR in Tetyushi

1954–1956 – Karpilov was the first secretary of the district committee of the Komsomol in Tetyushi

1957 – Karpilov entered the post-graduate school of the Kazan Agricultural Institute

1959–1960 – The first two publications of Yuri Karpilov appeared in the "Works of Young Scientists Conference in Kazan" and in the "Proceedings of the Kazan Agricultural Institute" in his studies on maize which suggested a previously unknown pathway of carbon fixation in photosynthesis, which is different from the Benson-Bassham-Calvin cycle. After finishing the post-graduate school Karpilov worked at the Institute as an assistant

1962 – Karpilov defended his Ph.D. thesis "The Influence of Reducing the Area of Assimilation and Conditions of Nitrogen-Phosphorus Nutrition in the Beginning of Vegetation on Some Physiological Processes and Yield of Maize"

1962 – Karpilov moved to Tiraspol (Moldavian SSR, USSR), where until 1972 he worked as a Head of the Laboratory of Isotopes and Biochemistry of Photosynthesis in the Moldavian Institute for Research in Irrigation Farming and Vegetable Growing

1972 – At the Kazan University Yuri S. Karpilov defended Doctor of Science thesis "Participation of Nitrogen and Phosphorus in Photosynthetic Metabolism in Relation to their Specificity in Various Plant Species", which mainly dedicated to the study of the Cooperative photosynthesis in C_4 plants

1972–1977 – Yuri S. Karpilov worked as a Head of the Laboratory for Carbon Metabolism in the Institute of Photosynthesis, USSR Academy of Sciences (now Institute of Basic Biological Problems, Russian Academy of Sciences, Pushchino Moscow Region, Russia)

1977 – Yuri S. Karpilov was awarded with the Diploma of Professor of Biology of the USSR Academy of Sciences by the Highest Attestation Committee of the USSR Academy of Sciences in Moscow (Russia)

1977 – Yuri Karpilov died tragically in a traffic accident (Pushchino Moscow Region, USSR)

Acknowledgements

We are grateful to Professor Barry Osmond (Centre for Sustainable Ecosystem Solutions, School of Biological Sciences, Faculty of Science, Medicine and Health, University of Wollongong, NSW 2522, and Plant Sciences Division Research School of Biology, College of Medicine, Biology and Environment, Australian National University, Canberra ACT 2601, Australia) for many years of support and promotion of the recognition at the international level of Karpilov's contribution to the development of the C_4 concept as one of the pioneers of the mechanism of C_4 photosynthesis and to Professor Gerald (Gerry) Edwards (School of Biological Sciences, Washington State University, Pullman, Washington 99164, USA) for the reading and editing of this recollection. Also, many thanks to Professor Valery Lyubimov (Institute of Basic Biological Problems, Russian Academy of Sciences, 142290 Pushchino Moscow Region, Russia) for the restoration of old photos, to Drs Inna Karpilova and Lidiya Kuznetsova (142290 Pushchino, Moscow Region, Russia) who provided personal recollections and some photos, and to Drs Olga Tarasova (Onoprienko) and Vladimir Medvedev (Tiraspol, Transnistria) for help in finding additional information.

Abbreviations

3-PGA – 3-phosphoglyceric acid
BSC – bundle sheath cells
Chl – chlorophyll
PEP – phosphoenolpyruvate

References

Bassham J.A., Benson A.A., Kay L.D., Harris A.Z., Wilson A.P., Calvin M. The path of carbon dioxide in photosynthesis. XXI. The cyclic regeneration of carbon dioxide acceptor, *J. Am. Chem. Soc.*, 1954, Vol. 76, p. 1760–1770.

Benson A.A. Paving the path, *Annu. Rev. Plant Biol.*, 2002, Vol. 53, p. 1-25.

Benson A.A., Bassham J.A., Calvin M., Goudate T.C., Haas U.A., Stepka W. The path of carbon in photosynthesis. 5. Paper chromatography and radioautography of the products, *J. Am. Chem. Soc.*, 1950, Vol. 72, p. 1710–1718.

Benson A.A. and Maretzki A. Constance Endicott Hartt (1900-1984) and the path of carbon in sugarcane leaf, *Advances in Photosynthesis and Respiration. C_4 Photosynthesis and Related CO_2 Concentrating Mechanism*, Raghavendra A.S., Sage R.F. eds., 2011, Vol. 32, Chapter 2, p. 13-16.

Biel K.Y. *Relationship between Structure and Function of the Photosynthetic Apparatus in Different C_4 Plants*, Ph.D. Thesis, Biol Sci, Pushchino-na-Oke, 1976, 173 p. [in Russian]

Biel K.Y. *Phototrophic Carbon Metabolism in Plants with Differential Organization of Photosynthetic Apparatus*, Dr. Sci. Thesis, Pushchino, 1988, 543 p. [in Russian]

Biel K.Y. *Ecology of Photosynthesis*, Kefely V.I. ed., Nauka, Moscow, 1993, 221 p. [in Russian]

Biel K.Y., Arkhipov V.A., Karpilova I.F., Moshkov D.A. Some characteristics of jointly and separately functioning tissues of maize leaf, *Photosynthesis in Maize. Structural and Functional Properties of Photosynthetic Apparatus*, Karpilov Y.S. ed., NCBI USSR AS, Pushchino-na-Oke, 1974, p. 7–23. [in Russian]

Biel K.Y., Karpilov Y.S. The transport mechanism of the photosynthetic products in leaves of C_4 plants, *Plant Physiology. Bulgarian Academy of Sciences*, Sofia, 1978, Vol. 4, No. 1, p. 25–40. [in Russian]

Biel K.Y., Stoilov M.A., Karpilov Y.S. Peculiarities of the transport of photosynthetic products in C_4 plants, *Transport of Substances in Plants in Connection with Metabolism and Biophysical Processes*, Gorkii State University, Anisimov A.A., Opritov B.A. eds., Gorkii, 1975, Vol. 6, p. 36–44. [in Russian]

Bil' K.Y., Nishio J.N. Oleg Georgievich Malishev, *The Newsletter of the American Society of Plant Physiologies (ASPPNEWS)*, Obituaries, July / August, 1999, 1999, Vol. 26, p. 4.

Bil' K.Y. Yuri S Karpilov: The Men, the Scientist, and the Discoverer of cooperative photosynthesis (1931–1977), *Russ. J. Plant Physiol.*, 2000, Vol. 47, No. 4, p. 565–568. [Translated from *Fiziologiya Rastenii*, 2000, Vol. 47, No. 4, p. 641–645.]

Bil' K.Y., Belobrodskaya L.K., Karpilov Y.S. The localization ATPases in cellular structures of assimilating tissues in *Amaranth* leaves, *Proceedings of the USSR Academy of Sciences*, 1976, Vol. 226, No. 5, p. 16–19. [Translated from *Dokl. Akad. Nauk SSSR*, 1976, Vol. 226, No. 5, p. 1229–1231.]

Bil' K.Y., Gedemov T. Structural and functional properties of assimilation apparatus in *Suaeda arcuata* Bunge (*Chenopodiaceae*) related to cooperative photosynthesis, *Dokl. Bot. Sci.* [A translation of selected sections of *Dokl. Akad. Nauk SSSR*], 1980, Vol. 250, No. 4, p. 9–13.

Bil' K.Y., Klimov V.V., Karapetyan N.V., Karpilov Y.S. Activity and cooperation of two photosystems in chloroplasts of bundle sheath cells of maize leaves, *Proceedings of the USSR Academy of Sciences*, 1975, Vol. 223, No. 1. [Translated from *Dokl. Akad. Nauk SSSR*, 1975, Vol. 223, No. 1, p. 225–228.]

Björkman O., Gauhl E. Carboxydismutase activity in plants with and without β-carboxylation photosynthesis, *Planta (Berl)*, 1969, Vol. 88, No. 3, p. 197–203.

Cooperative Photosynthesis of Xerophytes, Proceedings of the Moldavian Institute for Research in Irrigation Farming and Vegetable Growing, Fillipov N.A. ed., Kishinev 1970, Vol. 11, No. 3, 67 p. [in Russian]

Doman N.G. Nature of intermediate products of photosynthesis, *Dokl. Akad. Nauk. SSSR*, 1952, Vol. 84, No. 5, p. 1017–1019. [In Russian]

Doman N.G., Khadzhi-Murat L.N., Demina S.E. Unities and peculiarities of carbon assimilation pathway(s) in different plant species, *Dokl. Akad. Nauk. SSSR*, 1958, Vol. 122, p. 111–113. [in Russian]

Doman N.G., Kuzin A.M., Mamul Y.V., Khudyakova R.I. Variability of primary products of photosynthesis in various plants, *Dokl Akad Nauk SSSR*, 1952, Vol. 86, No. 2, p. 369–372. [in Russian]

Energy, Metabolic Pathways and Their Regulation in Photosynthesis, All Soviet-Union Meeting, Institute of Photosynthesis, USSR Academy of Sciences, Erokhin Y.E., Biel K.Y. eds., XI CO_2 Pushchino Reading, Pushchino 1981, 72 p. [in Russian]

Filippova L.A. About methods of isolation of chloroplasts in a non-aqueous medium, *Soviet Plant Physiol.*, 1967, Vol. 14, No. 6, p. 1107–1112. [in Russian]

Ilin I.R., Sadykin A.V., Maidurova V.E. Karpilov Yuri Solomonovich (21.03.1931–21.09.1977), *Creators of Agricultural Science in Transnistria: The Memory About Them is Alive*, Ilin I.R. ed., The Agriculture Research Institute of Transnistria, Bendery Polygraph, 2011, p. 95–100. [in Russian]

Hartt C.E., Kortschak H.P., Burr G.O. Photosynthesis by sugar cane fed radioactive carbon dioxide, *Proc. Haw. Acad. Sci.*, 1953, p. 13–14.

Kanai R., Edward G.E. Separation of mesophyll protoplasts and bundle sheath from maize leaves for photosynthetic studies, *Plant Physiol.*, 1973, Vol. 51, p. 1133–1137.

Karpilov Y.S. About change of some indicators of metabolism exchange at mechanical damage of corn leaves, *Works of Young Scientists Conference in Kazan (Section of Agricultural Chemistry, Physiology, Agriculture)*, 1959, p. 43–51. [in Russian]

Karpilov Y.S. Distribution of radioactive carbon among the products of photosynthesis of maize, *Works Kazan Agricult. Inst.*, 1960, Vol. 41, No. 1, p. 15–24. [in Russian]

Karpilov Y.S. Maize requirement to nitrogen nutrition as related to metabolic specificity during development, *News of the USSR Academy of Sciences, Series Biology*, 1961, No. 1, p. 112–116. [in Russian]

Karpilov Y.S. *The Influence of Reducing the Area of Assimilation and Conditions of Nitrogen-Phosphorus Nutrition in the Beginning of Vegetation on Some Physiological Processes and Yield of Maize*, Ph.D. Thesis, Kazan, 1962. [in Russian]

Karpilov Y.S. Formation and metabolism of amino acids during photosynthesis, *Ways to Increase of the Intensity and Efficiency of Photosynthesis*, Kiev, Naukova Dumka, 1966, p. 58-76. [in Russian]

Karpilov Y.S. Features of the functions and structure of the photosynthetic apparatus of some plant species of tropical origin, *Proceedings of the Moldavian Institute for Research in Irrigation Farming and Vegetable Growing, Kishinev*, 1969, Vol. 11, Special Issue, 35 p. [in Russian]

Karpilov Y.S. Participation of nitrogen and phosphorus in the photosynthetic reactions, *Mineral Elements and Mechanism of Photosynthesis*, Kishinev,1970*a*, p. 3–10. [in Russian]

Karpilov Y.S. *Photosynthesis of Xerophytes (Evolutionary Aspects)*, Kishinev, 1970*b*, 19 p. [in Russian]

Karpilov Y.S. Photorespiration of corn leaves, *Cooperative Photosynthesis of Xerophytes. Proceedings of the Moldavian Institute for Research in Irrigation Farming and Vegetable Growing*, Kishinev 1970*c*, Vol. 11, No. 3, p. 46–64. [in Russian]

Karpilov Y.S. 1971 *Participation of Nitrogen and Phosphorus in Photosynthetic Metabolism in Relation with Its Specificity in Various Plant Species*, Dr. Sci. Thesis, Tiraspol, 379 p. [in Russian]

Karpilov Y.S. Evolution of maize photosynthesis, *Photosynthesis in Maize. Structural and Functional Properties of Photosynthetic Appa-*

ratus, Karpilov Y.S. ed., 1974, NCBI, Pushchino-na-Oke, p. 152–165. [in Russian]

Karpilov Y.S. Organization and regulation of carbon metabolism during photosynthesis of C_3 and C_4 plants, *Results of Studies on the Mechanism of Photosynthesis.* Reports IX All-Union Conference Study of Mechanisms of Photosynthesis, Evstigneev V.B. ed., Pushchino, 1976, p. 58–62. [in Russian]

Karpilov Y.S.: *The Scientist and the Man*, Selected works, Editorial Board: Biel K.Y., Ignatova L.K., Karpilova I.F., Kuznetsova L.G., Lyubimov V.Y., Mitenko G.V., Prisyazhnaya A.A., Fomina I.R., NIA-Priroda – Fond "Infosphera", Moscow, 2007, 338 p. [in Russian]

Karpilov Y.S., Avdeeva T.A., Persanov V.M. Localization of carbon metabolism in two tissues of maize leaves, *Genetic Aspects of Photosynthesis*, Nasyrov Y.S., Sestak Z. eds., Hague, Junk BW, 1975, p. 177–185.

Karpilov Y.S., Biel K.Y. Application of cytochemical methods for the Hill reaction localization, *Methods for Investigation of Photosynthetic Electron Transport*, Akulova E.A. ed., NCBI USSR Acad. Sci., Pushchino, 1974, p. 27–33. [in Russian]

Karpilov Y.S., Biel K.Y. Quantitative, ultrastructural and functional peculiarities of chloroplasts from assimilating tissues of maize chlorophyll mutants, *Quantitative Traits of Maize Mutants*, Lisikov V.N. *et al.* eds., Kishinev, 1975, p. 96–99. [in Russian]

Karpilov Y.S., Biel K.Y., Malishev O.G., Karnaukhov V.N. 1970 Some properties of the photosystems of chloroplasts in mesophyll cells and parenchyma sheaths of conducting bundles in leaves of monocotyledonous and dicotyledonous, *Cooperative Photosynthesis of Xerophytes. Proceedings of the Moldavian Institute for Research in Irrigation Farming and Vegetable Growing*, Kishinev, 1970, Vol. 11, No. 3, p. 25–32. [in Russian]

Karpilov Y.S., Bil' K.Y. Ratio of two types of chloroplasts in maize leaves as dependent on nitrogen — phosphorus nutrition of plants, *Proceedings of the USSR Academy of Sciences*, 1970, Vol. 193, Number 5, p. 105–107. [Translated from *Doklady Akademii Nauk SSSR* 1970, Vol. 193, No. 5, p.1198–1200.]

Karpilov Y.S., Bil' K.Y. Transport of photosynthetic intermediates through the cytoplasm assimilating cells in C_4 plants, *Proceedings of the USSR Academy of Sciences*, 1976, Vol. 226, No. 6, p. 24-27. [Translated from *Doklady Akademii Nauk SSSR* 1976, Vol. 226, No. 6, p. 1469–1471.]

Karpilov Y.S., Bil' K.Y., Gukasyan I.A. Participation of ATPases in transport of intermediate photosynthesis products between assimilating tissues of C_4 plants, *Soviet Plant Physiol.*, 1976*a*, Vol. 22, No. 6, p. 974–980. [Translated from *Fiziologiya Rastenii*, 1975, Vol. 22, No. 6, p.1113–1120.]

Karpilov Y.S., Bil' K.Y., Malishev O.G. Rate of transfer of ^{14}C from dicarboxylic acids into the pentose phosphate cycle as a function of the ratio of malate and aspartate and disposal of the chloroplasts in C_4 plants, *Soviet Plant Physiol.*, 1976*b*, Vol. 22, No. 5, p. 796–803. [Translated from *Fiziologiya Rastenii*, 1975, Vol. 22, No. 5, p. 910–917.]

Karpilov Y.S., Bil' K.Y., Malishev O.G., Karnaukhov V.N. Characteristic properties of the photosystems of chloroplasts in mesophyll cells and in parenchymatous sheaths of vascular bundles in leaves of monocotyledonous and dicotyledonous plants, *Proceedings of the USSR Academy of Sciences*, 1971, Vol. 197, No. 2, p. 49–52. [Translated from *Doklady Akademii Nauk SSSR*, 1971, Vol. 197, No. 2, p. 480–483.]

Karpilov Y.S., Kerimov S.K., Maslov A.I., Belobrodskaya L.K., Karpilova I.F, Novitskaya I.L., Gerus E.V., Kartasheva R.I. Serine and glycine conversions in leaves of C_3 and C_4 plants and their involvement in mitochondrial respiration, *Mechanism of Photorespiration and Its Features in Various Plant Types*, Karpilov Y.S., Romanova A.K. eds., NCBI USSR AS, Pushchino, 1978, p. 187-224. [in Russian]

Karpilov Y.S., Kuzmin A.N., Bil' K.Y. Distribution of glycolytic enzymes in assimilating tissues of C_4 plants leaves and their link with characteristics of photosynthetic and photorespiratory reactions, *Soviet Plant Physiol.*, 1979, Vol. 25, No. 6, p. 891–897. [Translated from *Fiziologiya Rastenii*, 1978, Vol. 25, Number 6, p. 1129-1135.]

Karpilov Y.S., Lyubimov V.Y., Chermnykh R.M., Kosobryukhov A.A. The nature of light-induced oxidation of organic acids and its possible role in the regulation of pseudocyclic photorespiration in chloroplasts, *Mechanism of Photorespiration and Its Features in Various Plant Types*, Karpilov Y.S., Romanova A.K. eds., NCBI USSR AS, Pushchino,1978, p. 74–79. [in Russian]

Karpilov Y.S., Malishev O.G. Carbon metabolism of photosynthesis in the leaves of maize, *Cooperative Photosynthesis of Xerophytes. Proceedings of the Moldavian Institute for Research in Irrigation Farming and Vegetable Growing*, Kishinev, 1970, Vol. 11, No. 3, p. 33–45. [in Russian]

Karpilov Y.S., Malishev O.G., Persanov V.M., Kuznetsova L.G. Photosynthetic carbon metabolism of maize leaves, *Photosynthesis in Maize.*

Structural and Functional Properties of Photosynthetic Apparatus, Karpilov Y.S. ed., NCBI USSR AS, Pushchino-na-Oke, 1974, p. 89–130. [in Russian]

Karpilov Y.S., Maslova N.F., Russu L.N. Features of the pigment composition of leaves and photosynthesis depending on the origin and early ripening of varieties, *Physiology and Biochemistry of Varieties*, Irkutsk,1969. [in Russian]

Karpilov Y.S., Nedopekina I.F. Photosynthetic products of tomato and influence on their formation of nitrogen and phosphorus nutrition, *Proceedings of the Moldavian Institute for Research in Irrigation Farming and Vegetable Growing*, 1965, Vol. 7, No. 1, p. 35–40. [in Russian]

Karpilov Y.S., Nedopekina I.F., Kotova N.F. Some features of carbon pathways in photosynthesis of various plant species, *Proceedings of the Moldavian Institute for Research in Irrigation Farming and Vegetable Growing*, 1968, Vol. 8, p. 19–34. [in Russian]

Karpilov Y.S., Nedopekina I.F., Kotova N.F., Zarvanskaya E.I. The involvement of nitrogen and phosphorus in photosynthetic biochemical reactions in various species of vegetable plants, *Proceedings of the Moldavian Institute for Research in Irrigation Farming and Vegetable Growing*, 1968, Vol. 9, p. 181–187. [in Russian]

Karpilov Y.S., Novitskaya I.L., Belobrodskaya L.K., Biel K.Y., Maslov A.I., Kuzmin A.N., Karpilova I.F., Kerimov C.X., Popova E.I., Petrukhin Y.A., Hertz S.M. The glycolysis reactions in autotrophic cells; their role in photosynthetic metabolism and photorespiration, *Mechanism of Photorespiration and Its Specificity in Different Types of Plants*, Karpilov Y.S., Romanova A.K. eds., NCBI USSR AS, Pushchino, 1978, p. 90–187. [in Russian]

Karpilov Y.S., Pristupa N.A., Biel K.Y., Brik P.L. The structure of the photosynthetic apparatus in the leaves of plants with cooperative photosynthesis, *Cooperative Photosynthesis of Xerophytes. Proceedings of the Moldavian Institute for Research in Irrigation Farming and Vegetable Growing*, Kishinev, 1970, Vol. 11, No. 3, p. 18–24. [in Russian]

Karpilov Y.S., Shchupak C.D. Some features of photosynthesis of tomato according to the temperature conditions of cultivation, *Proceedings of the Moldavian Institute for Research in Irrigation Farming and Vegetable Growing*, 1965, Vol. 7, No. 1, p. 25–33. [in Russian]

Karpilov Y.S., Vaskovsky M.D. Dependence of the reaction of photophosphorylation from the level of the plants nutrition with nitrogen and phosphorus, *Mineral Elements and Mechanisms of Photosynthesis*,

Editing and Publishing Department, MSSR Academy of Sciences, Kishinev, USSR, 1969, p. 87–93. [in Russian]

Kortschak H.P., Hartt C.E., Burr G.O. Carbon dioxide fixation in sugarcane leaves, *Plant Physiol.*, 1965, Vol. 40, No. 1, p. 209–213.

Lyubimov V.Y., Biel K.Y., Karpilov Y.S. Peculiarities of the glycolate formation in assimilation tissues of C_4 plants, *Plant Physiology. Bulgarian Academy of Sciences*, Sofia, 1978, Vol. 4, No. 1, p. 41–56. [in Russian]

Lyubimov V.Y., Karpilov Y.S. The effect of organic acids on light-induced oxygen uptake by maize chloroplasts, *Dokl. Akad. Nauk USSR*, 1977, Vol. 237, p. 746–748. [in Russian]

Lyubimov V.Y., Karpilov Y.S. Light-dependent oxidative metabolism of organic acids in maize mesophyll chloroplasts, *Mechanism of Photorespiration and Its Features in Various Plants Types*, Karpilov Y.S., Romanova A.K. eds., NCBI USSR Acad. Sci., Pushchino, 1978, p. 58–74. [in Russian]

Magomedov I.M. On the history of the discovery C_4 photosynthesis. Current state of problem, *Advance in Current Natural Sciences (Biological Sciences)*, 2015, Vol. 1, p. 962–965. [in Russian]

Mechanism of Photorespiration and Its Specificity in Different Types of Plants, Karpilov Y.S., Romanova A.K. eds., NCBI USSR Acad. Sci., Pushchino, 1978, 225 p. [in Russian]

Mokronosov A.T., Karpilova I.F. Problems of photosynthesis in the investigations of Y.S. Karpilov (1931–1977), *Soviet Plant Physiol.*, 1992, Vol. 39, No. 4, p. 631–634. [in Russian]

Nezgovorova L.A. To the question about the products of photosynthesis, *Soviet Plant Physiol.*, 1956, Vol. 3, No. 6, p. 497–507. [in Russian]

Nezgovorova L.A. Influence of water regime of plants on the receipt and distribution of carbon in photosynthesis, *Soviet Plant Physiol.*, 1957, Vol. 4, No. 5, p. 440–449. [in Russian]

Nichiporovich A.A. Products of photosynthesis and physiological role of the photosynthetic apparatus of plants, *Proc. Inst. Plant Physiol. USSR Acad. Sci.*, 1953, Vol. 8, p. 3–41. [in Russian]

Nichiporovich A.A. Light and carbon nutrition of the plants — photosynthesis, *Publishing USSR Acad. Sci.*, Moscow, 1955, 284 p. [in Russian]

Nichiporovich A.A. About the diversity of the photosynthetic products and physiological impact, *Plant Physiology. Agrochemistry. Soil Sciences*, Publishing House USSR Acad. Sci., Moscow, 1958, p. 56–65. [in Russian]

Osmond C.B., Biel K.Y., Berry J.A. Enigmatic involvement of C_4 acids in photosynthesis and CAM before and after Hatch and Slack (1966), Oral report, *C_4 Photosynthesis Conference 2016: Past, Present and Future. Celebrating 50 Years of C_4 Photosynthesis Discovery and Innovation; Session 1: Historical Perspectives and the C_4 Renaissance*, Canberra, Australia, 2016.

Photosynthesis in Maize. Structural and Functional Properties of Photosynthetic Apparatus, Karpilov Y.S., ed., NCBI USSR Acad. Sci., Pushchino-na-Oke,1974, 170 p. [In Russian]

Persanov V.M., Voronova E.A., Karpilov Y.S. Investigation of the properties of NADP-malate dehydrogenase from the leaves of corn, *Biochemistry*, 1976, Vol. 41, No. 6, p. 1014–1021. [in Russian]

Persanov V.M., Voronova E.A., Oparina L.A., Karpilov Y.S. Isolation and purification of NADP-dependent "malic"-enzyme from the leaves of corn, *Biochemistry*, 1976, Vol. 41, No. 5, p. 921–924. [in Russian]

Pristupa N.A. Localization of short-term products of photosynthesis in the maize leaf tissues, *Soviet Plant Physiol.*, 1972, Vol. 19, No. 3, p. 536–544. [in Russian]

Tarchevsky I.A., Karpilov Y.S. On the nature of products of short-term photosynthesis, *Soviet Plant Physiol.*, 1963, Vol. 10, No. 2, p.183–184. [in Russian]

Zalensky O.V. *Plant photosynthesis in natural conditions, Questions of Botany*. Publishing House USSR Acad. Sci., Moscow, 1954, 61 p. [in Russian]

Zalensky O.V. *Ecological and Physiological Aspects of the Study of Photosynthesis, Reported at the 37th Annual Timiryazev Reading*, June 3, 1976, Nauka, Leningrad Branch, Leningrad, 1977, 57 p. [in Russian]

13
Dr. Nicholas Yensen's *Curriculum Vitae*

Karl Y. Biel[1,2,*], **Irina R. Fomina**[1,2,*]

[1]*Institute of Basic Biological Problems, Russian Academy of Sciences, Pushchino, Moscow Region 142290, Russia*
[2]*Biosphere Systems International Foundation, Tucson, Arizona 85755, USA*

Abstract
This Chapter is dedicated to the memory of Dr. Nicholas (Nick) Patrick Yensen who passed away 12 years ago on August 24, 2006. Dr. Yensen had a Ph.D. in Ecology from The University of Arizona (Tucson, Arizona, USA). During his scientific carrier, he was the founder and president of Ecosphere Associates, the Intercultural Center for the Study of Deserts and Oceans, NyPa Inc., an International Outlet for Saline Agriculture and Aquaculture Technology, and the Biosphere Systems International. He had designed various closed systems and was the holder of over 20 patents. His interest was in the remediation and restoration of natural ecosystems, saline agriculture and on the development of closed and semi closed ecological biosphere systems. Nick was also developer of photo-halosynthesis concept which may be beneficial to most terrestrial living systems as well as the concept of criticality with positive and important ramifications for our planet.

Keywords: Yensen' protocol, Saline agriculture, Closed and semi-closed biospheres, Concept of criticality

* *Corresponding authors*: karlbiel@hotmail.com; irafomi@rambler.ru

13.1 Introduction

We dedicated Chapter 13 to the memory of Dr. Nicholas (Nick) Patrick Yensen (Figure 13.1) who passed away twelve years ago on August 24, 2006, succumbing to a battle with pancreatic cancer. He was only 60 years old. During many years Nick was our colleague and very good friend. His broad knowledge in different fields of the sciences, enthusiasm in collaboration and excellent capacity to carry out laboratory and field experiments as well as the warmest friendship were so magnificent that he always was in the center of our scientific Group.

Fig. 13.1. Photo image of Dr. Nicholas Patrick Yensen six months and three days before he passed away. Nick looks a perfectly healthy person. At that time, he knew nothing about his pancreatic disease (this photo image of Nick Yensen was made by Karl Biel in February 21, 2006; photo from K. Biel archive)

Dr. Yensen had a very broad scientific interest, but in the last 25 years of his life he mostly focused his attention on the two big research directions – the remediation and restoration of natural ecosystems and saline agriculture (Chapter 4 of this book; see also Biel and Yensen, 2004–2006;

Biel, Fomina, Yensen, *et al.*, 2014; and others), and on the closed and semiclosed ecological micro- and macrobiosphere systems (Chapter 2 of this book; see also Yensen and Biel, 2005; Biel, Fomina, Yensen *et al.*, 2014; and others). Also, Nick was spending much time for design and proof of the photo-halosynthesis concept which is explaining some mechanisms of positive protections and benefits in many terrestrial living systems as well as on development of the concept of closed system criticality which may be useful instrument for predictions of important events and their ramifications on the planet Earth.

The reader can find a description of these scientific projects in Chapters 2 and 4 of this book and in the set of Nick's articles (see the list of Publications (selected) below).

In this Chapter, we would like to submit some information about Nick's scientific and social life, which is not widely known in scientific society.

13.2 Biographical Note about Dr. Nicholas Patrick Yensen

13.2.1 Education

Dr. Nick Yensen received a Bachelor's degree in Mathematics and Biology at Idaho College, Caldwell, Idaho, USA in 1968. Then he moved to Tucson, where he received his Master of Biology degree from the University of Arizona (Tucson, Arizona, USA) in 1973, and after 6 years, in 1979, Nick defended his Doctor of Philosophy degree (Ph.D.) thesis on Ecology (University of Arizona, Tucson, Arizona, USA).

13.2.2 Teaching experience

In 1967–1973 Nick Yensen was: curator of Natural History at Albertson College of Idaho (Caldwell, Idaho, USA); teacher of ecology, biology, advanced biology, geology, Earth sciences and mathematic in Boise (Idaho, USA); and an assistant of the leader of the expedition to the Great Barrier Reef (Australia, Albertson College of Idaho, Caldwell, Idaho, USA).

Then, in 1973–1979, Nick was a graduate teaching assistant in biology, marine algae, and marine ecology at the University of Arizona (Tucson, Arizona, USA). Four years, since 1986 till 1989, Yensen worked as an Adjunct professor of botany, mathematics and algebra at Pima Community College (Tucson, Arizona, USA).

In 1989–1990, he was a lecturer in saline agriculture (University of Arizona, Tucson, Arizona, USA); 2000–2003 – Adjunct Professor of the Louisiana Tech University (Ruston, Louisiana, USA); 2003–2006 – Adjunct Professor of chemistry, ecology, biology and botany at Central Arizona College, Aravaipa Campus (Arizona, USA).

He had given lectures and taught courses on closed systems at Central Arizona College, Columbia University (Oracle, Arizona, USA) and at the University of Arizona (Tucson, Arizona, USA).

13.2.3 Founder and Leader of Scientific Organizations

For seven years, from 1973 till 1979, Nick was the president of Panamic Environmental Consultants; Environmental Impact, USA and MEXICO; and in 1979 he was one of the founders of the Intercultural Center for the Study of Deserts and Oceans (CEDO) (Puerto Peñasco, MEXICO).

In 1985–1987 Dr. Nicholas Yensen was president of CEDO USA Board (Tucson, Arizona, USA); and the founder of Ecosphere Associates (Tucson, Arizona, USA).

In 1985 Yensen also became one of the founders and until his death, in 2006, remained the principal of Partner Salt Weeds; Partnership, Saline Remediation Tech (Tucson, Arizona, USA). In addition, he was the founder of the Coastal Laboratory, Inc., Tucson, Arizona, USA (1986–2006), and the founder / vice president of Chemicals and Equipment Supplies Inc., Tucson, Arizona, USA (1988-2006).

In 1988–2005 Nick organized and led, as president and chairman, NyPa Inc., (Tucson, Arizona, USA). Yensen had been closely involved in the organization of the Biosphere System International (Tucson, Arizona, USA).

13.2.4 Member of Board of Directors, Consultant, and Chairman

In 1995–1998 Yensen was a member of the board of directors of Growing Connections, Inc. (Tucson, Arizona, USA); 1997–2001 – member of the board of directors in Far West Group, Inc. (Tucson, Arizona, USA).

It is interesting that for two years, in 2002-2003, Nick was even a member of the board of directors in Tucson Girls Chorus (Tucson, Arizona, USA).

From 1997 till 2006 Yensen worked as a consultant and legal consul in the Research Center for Food and Development in Hermosillo (Sonora, MEXICO).

In 1997–2004 – member of the board of directors, Montessori Schoolhouse (Tucson, Arizona, USA);

2003–2005 – Chairman, NyPa International (Dubai, UAE), and in 2005–2006 – President, Biosphere Systems International (Tucson, Arizona, USA).

Dr. Yensen also had served as a consultant to the Biosphere 2 Center (Oracle, Arizona, USA), under various managements.

13.2.5 Languages

Nick's ability to memorize languages was amazing. In particular, he could fluently speak, read, write, and do simultaneous translation from English to Spanish and *vice versa*; Nick could minimal speak and read in Arabic, Chinese, French, German, Italian, Portuguese, Russian; also he could read in Latin.

13.2.6 Oratorical Talent

Dr. Yensen possessed phenomenal oratorical talent. He was invited and gave lectures to Royal Botanic Gardens, Kew, England, UK.; National Academy of Science, Cairo, Egypt; Secretary of Agriculture, Riyadh, Saudi Arabia; Royal Agriculture Centers of the King of Morocco, Casablanca; University of Singapore (Singapore); University of Namibia, Windhoek; University of Valencia (Spain); University of Nanjing (China); Charles and Ann Morrow Lindbergh Foundation, Little Falls, Minnesota (USA); Universidad de Iquique (Chile); United Nations – FAO Water Quality Workshop for Latin America; USA EPA – Remediation of salt affected soils, Arkansas; NATO conference on Salt-affected Soils (Uzbekistan); Williams Institute for the Production of Forage and Fodder, Moscow (Russia); Lecturer, Krasnoyarsk Scientific Center, Krasnoyarsk-Sayani (Russia) (Figure 13.2), *etc.*

13.2.7 Dr. Yensen's International Teamwork, Expeditions and Visitations

Dr. Yensen visited many countries of the world as a member of the International Expeditions, a co-author of collaborative scientific works, a professor with lectures and a participant in conferences.

Here we just name (not all!) the countries which Yensen visited and worked in Argentina, Australia, Austria, Belgium, Botswana, Brazil, Chile, China, Costa Rica, Colombia, Egypt, Ecuador, France, Germany, Hungary, Mexico, Morocco, Namibia, Niger, Pakistan, Panama, Peru,

Fig. 13.2. Dr. Nicholas Yensen (at the photo Yensen is shown by the blue arrow) together with participants of International XII Symposium "The Complex Systems under Extreme Conditions", July 27-31, 2005 in Russia (photo image from the collection of the International Scientific Centre for Organism Extreme States Research of the Krasnoyarsk Scientific Centre, Siberian Branch of the Russian Academy of Sciences, Krasnoyarsk 660036, Russia)
Note: The XII International Symposium was held at one of the tourist bases of the Western Sayan Mountains at an altitude of 1460 meters above sea level, Russia

Puerto Rico, Qatar, Russia, Saudi Arabia, Senegal, Singapore, South Africa, Spain, Thailand, Tunisia, United Arab Emirates, United Kingdom, United States, Uzbekistan.

Among the scientific organizations that he visited were: Smithsonian Institution, National Museum, Washington D.C. (USA); Lecturer, Charles and Ann Morrow Lindbergh Foundation, Washington D.C. (USA); National Academy of Science in China; Department of Forestry in Perth and Australia; Instituto Nacional de Tecnologia Agricola in Buenos Aires (Argentina); President Sam Nujoma, (Namibia, Africa); Crown Prince Agricultural Farm, Al Ain (United Arab Emirates); Lecturer, Rafan Ltd., Faiselabad (Pakistan); Department of Soils, Bangkok (Thailand); Senator Osman Nagome, Dakar (Senegal); Instituto Nacional de Fomento Agricola, Santiago (Chile); Universidad Nacional Agraria, Lima (Peru); Department of Agriculture, (Niger, Africa); Rossing Foundation and Department of Agriculture, water and Rural Affairs (Windhoek, Namibia); Universidad de Fortaleza (Brazil); Waite Agricultural

Institute (Adelaide, Australia); Trelew and Instituto de Zonas Aridas (Mendoza, Argentina); Nuclear Physics Institute C-70 (Chelyabinsk, Russia); Institute of Basic Biological Problems, Russian Academy of Sciences (Pushchino, Russia); and many others.

13.2.8 *Distinctions*

Dr. Yensen had designed and synthesized one of the world's first synthetic self-sustaining closed ecosystems of various sizes and compositions. He was a co-designer of the first Earth-analog microbiosphere closed ecosystems including the commercial EcoSphere™ system, of which many thousands units have been made. As listed in Introduction, he was also a co-developer of the Biotube (macrobiosphere closed system) and co-author of the concept of closed system criticality. The above works are described in detail in Chapter 2 of this book.

Dr. Yensen was co-developer of the salt-plant conduction remediation concept and the photo-halosynthesis concept (see Chapter 4) and awarded over 20 patents and patent registrations for the remediation of saline environments worldwide. His famous variety yensen 4a of the highly productive conductor plant *Distichlis spicata* (see Figure 13.3) is described in detail in Chapter 4 of this book. *Distichlis spicata* var. yensen 4a is very good for soil salt remediation and soil restoration, forage, turf and salt disperse (see Chapter 4 and References therein). In some countries this variety is already used.

Numerous articles and books of Dr. Yensen's (see Publications (selected)), continue to be useable around the World.

13.2.9 *Articles, Videos and Documentaries about Dr. Yensen's Work*

Dr. Nicholas Yensen was a public person. He traveled to over 40 countries, studied for a Masters of Business Administration, founded several International companies and had experience working with large corporations and governments conducting remediation work in salt-affected, impoverished regions on many continents. He had discussed in public places important topics, related to sciences and human life, how to protect our Globe from catastrophes, *etc.*

Dr. Yensen and descriptions of his work were appearing in various documentaries, magazines and newspapers, such as: National Geographic Magazine; Discovery Channel TV [Invention Program 1994–1996; Discovery Channel TV - Limits to Populations - Documentary]; Arizona

Fig. 13.3. Photo images of Dr. Nickolas Yensen (photo images from the archives of Karl Biel and Nick Yensen, 15 August 2005)
Note: *from the left* – Nick near the *Distichlis spicata* var. yensen 4a, which was planted in a priming mix of sand, clay, construction debris, and without mineral nutrition. According to special aim the plants were not watered during spring-summer time – 4 months; *from the right* – Nick in a field

Highways Magazine; Sea Frontiers Magazine; Los Angeles Times Newspaper; The West TV Program – Documentary; NBC TV Namibia – Documentary; Nieman Marcus Magazine; Farm Industry News Magazine; Cornucopia – Book; The Tumbleweed Gourmet – Book; The Charles Lindbergh Foundation – Biography; Tucson Magazine – Biography; and local TV programs, books, newspapers and magazines.

13.2.10 Skill and Avocation

Dr. Yensen had an interesting and important skill – construction of closed and semi closed ecosystems, pilot and SCUBA. Nick's avocation was in the fields of nuclear physics and cosmology.

13.3 Conclusion

Dr. Nicholas P. Yensen died calm and in a good mood on August 24, 2006, only a few months after the loss of his wife Susana.

Fig. 13.4. Nick and Susana Yensen at a young age (copy of photo image from the *Memorial Meeting* related to the ebb and flow of the Life of the Drs. Nicholas Patrick Yensen and Susana Bojorquez Yensen, November 10–11, 2006, in Puerto Peñasco, Sonora, México)

According to fulfillment of Nick and Susana Yensen's wishes, on November 11th, 2006 the members of their families returned the ashes to the sea at sunrise in the place where they first met (Figure 13.4), at Puerto Peñasco, Sonora (MEXICO), near CEDO, on the shore of the Gulf of California.

Many relatives, friends and colleagues participated in this celebration of Nick and Susana's lives.

Nick was a very intelligent, gentle and caring person who patiently explained the complexity of natural phenomena in understandable language to students, friends, and to everybody who had an interest in discussing the topic. At the same time, Dr. Nicholas Yensen was strong and persistent about ethical responsibility for environmental use, and was concerned for future of our Globe and Humanity.

Dr. Yensen had a huge professional experience and active social life. He traveled the whole world with an unquenchable interest in people and their lives in different parts of the Planet. He was working out projects to improve environmental conditions in regions with contaminated soil not for profit, but from a sincere desire to realize his plans for the benefit of mankind.

13.4 Addendum

In the archive of K. Biel, we found a note in which Nick Yensen wrote about himself long time ago. And we decided to present this note here in its original state:

"*...NyPa International, 90 Carrington Street, Adelaide SA 5000, Australia. Nicholas P. Yensen, Ph.D. Ecology, University of Arizona. My principal work is with plants / crops that grow in saline environments.*

I have spent some 25–30 years in this field and have 7 patents and am in the process of writing another 13 on various aspects of salinity, water remediation... one of which is CDI (capacitive deionization). CDI is a technology that extracts salts or any dissolved solids from water using aerogel electrodes. The process promises to be more efficient that RO and DI for many applications.

The bulk of my work, however, is in agriculture. NyPa International with NyPa Australia, Yancheng NyPa and now with NyPa Iberica is a technology clearing house to distribute salt tolerant plants and other technology for saline conditions (including aquaculture).

I spend much of my time traveling around the world [sounds great until you've gone around the world 6 times in a year].

In spare time and between flights I have initiated a bulletin called SALT BULLETIN to report on the developments in salinity and we have a few subscribers and to whom I must address, if there are any of you brave souls out there, that another issue is due and I am still committed to publishing.

Hence, for those out there who wish to publish: articles of a half-page to a page will be accepted for publication, if they meet editorial scrutiny. e-mail to NyPa@aol.com or NickYensen@aol.com" – *Nick P. Yensen.*

Acknowledgements

Authors' thanks to Bob Yensen, Jim Bedell, Sasha Alexander, Lloyd Findley, Linda Leigh, Elisabeth Poscher, Nariman and Zebry Shamsutdinov who shared with us information about the life and professional achievements of talented scientist and brilliant person Dr. Nickolas Patrick Yensen.

Publications (selected)

Biel K.Y., Fomina I.R., Yensen N.P., Nishio J.N., Matichenkov V.V., Nazarova G.N., Soukhovolsky V.G., Khlebopros R.G. *Complex Biological Systems: Adaptation and Tolerance to Extreme Environments*, Collective Monograph, Khlebopros R.G., Biel K.Y. eds., Gorod, Pushchino-Krasnoyarsk, 2014, 343 p.

Biel K.Y., Yensen N.P. *The Hypothesis of Halosynthesis, Photoprotection, and Soil Remediation via Salt-Conduction*, Offprint, Khlebopros R.G., Soukhovolsky V.G. eds., 2004, 36 p.

Biel K.Y., Yensen N.P. The hypothesis of halosynthesis, photoprotection, soil remediation via salt-conduction, and potential medical benefits, *Complex Systems under Extreme Conditions*, Proc. XII Int. Symp., Khlebopros R.G., Soukhovolsky V.G. eds., Sayani, July 27–31, 2004, Krasnoyarsk, Russia, 2005, p. 86–115.

Biel K.Y., Yensen N.P. The hypothesis of photo-halosynthesis and its application for some agricultural and medical problems, *Functional Foods for Cardiovascular Diseases. Functional Foods Can Help Reduce the Risks of Cardiovascular Diseases*, Proc. Int. Conf., Martirosyan, D.M. ed., Richardson, Texas, USA, 2005, p. 215–227.

Biel K.Y., Yensen N.P. ICBA associates propose revolutionary new concept for halophytes: Photohalosynthesis, *Biosalinity News* (www.biosaline.org), 2006, Vol. 7, No. 1, p. 6–7.

Biel K.Y., Yensen N.P. *Salt Removal by Conductor Plants*, Patent Pending, Submitted to the United States Patent and Trademark Office, 2006, 26 p.

Bohnert H.J., Yensen N.P. *Alternative Crops. Arizona, Land and People*, University of Arizona, College of Agriculture, USA, 1987, Vol. 38

Efremov A.V., Antonov A.R., Litvinova V.A., Novoselov J.B., Veterina O.A., Kostina N.A., Martirosyan D.M., Yensen N.P. Variability of biometals and electrolytes under myocardial infarction and methods of its correction, *Functional Foods for Cardiovascular Diseases. Functional Foods Can Help Reduce the Risks of Cardiovascular Diseases*, Martirosyan D.M. ed., Richardson, Texas, USA, 2005, p. 80-88.

El Shourbagy N., Yensen N.P. Direct seeding of *Distichlis palmeri* into soil irrigated with sea water, *Proc. VI Symp. Sobre el Medio Ambiente del Gulfo de California*, Loreto, Baja California, México, 1982.

Leake J.E., Barrett-Lennard E., Sargeant M., Yensen N.P., Prefumo J. *NyPa Distichlis Cultivars: Rehabilitations of Highly Saline Areas for*

Forage, Turf and Grain, Rural Industries Research and Dev. Corp., Australian Fed. Gov., 2002, 440 kb, www.ridic.gov.au

Ortiz M., Vender G., Adami M., Leigh L.S., Biel K.Y., Yensen N.P. Earth-analog microbiospheres for evaluating biomass and other criticality conditions of global-system models, Poster, *VII International Carbon Dioxide Conference*, Boulder, Colorado, USA, 2005.

Sheppard C., Yensen N.P., Vaught C. Capacitive deionization (CDI) with carbon aerogel: A new technology for purifying contaminated waters of salts and pollutants, *International Workshop Regarding Water Quality and Control of Contamination in Latin America; Food and Agriculture Organization of the United Nations*, 30 September – 4 October 1999 (English and Spanish), 1998, 5 p.

Yensen N.P. Complimentary nutrition for cancer patients using functional foods: a review, *Functional Foods for Chronic Diseases. The Modern-Day Cure without the Side Effects of Traditional Treatments*, Martirosyan D.M. ed., Richardson, Texas, USA, 2006, p. 174–200.

Yensen N.P. *The Success of NyPa International as a Saline Agriculture Outlet: Origin, Development and Philosophy*, Kluwer Academic Publications, Netherlands, 2004.

Yensen N.P. *Salt-Tolerant Plants of the World and Their Uses*, Co publishers: International Center for Biosaline Agriculture (ICBA), Dubai and the Center for the Investigation of Food and Development (CIAD), Hermosillo, México, 2004, 600 p.

Yensen N.P. New developments in the world of saline agriculture, *Prospects of Saline Agriculture*, Ahmad R., Malik K.A. eds., Kluwer Academic Publications, Netherlands, 2002, p. 321–332.

Yensen N.P. *Halófitas del Golfo de California y sus Usos (Halophytes of the Gulf of California and their Uses)*, Traducción Julio Rodriguez-Casas, Universidad de Sonora, Hermosillo, Sonora, México (University of Sonora Press, Hermosillo), 29 illustrations of intertidal species, 2001, 296 p.

Yensen N.P. New horizons in the use of saline resources, *Natural Resource Management, Special Issue* (June), 2000, p. 4–8.

Yensen N.P. International symposium on high salinity tolerant plants summary of papers presented, *Biology of Salt Tolerant Plants*, Khan M.A., Ungar I.A. eds., Ohio University, Athens, 1995, p.1–12.

Yensen N.P., Meling A.E., Ortega P., Yensen S.B., Bedell J.L., Vázquez J. Forraje de la acuicultura, *Proc. IV Congreso Nacional de Acuicultura*, University of Sonora, Hermosillo, México, 1990.

Yensen N.P. (and other editors) *Saline Agriculture: Salt-tolerant Plants for Developing Countries*, National Acad. Science, National Acad. Press, Washington, D.C., 1990, 143 p.

Yensen N.P. Plants for salty soils, *Arid Lands Newsletter*, 1988, Vol. 27, p. 3-10.

Yensen N.P. Halófitas de América Latina y el Mundo: Sus Usos con Agua Salina y Desechos y Suelos Marginales (Halophytes of Latin America and the World: Their Use with saline and waste waters and marginal soils), *Taller Internacional sobre Gestión de la Calidad del Agua y Control de la Contaminación en América Latina (International Workshop Regarding Water Quality and Control of Contamination in Latin America)*, Food and Agriculture Organization of the United Nations, 30 September – 4 October 1998, Arica, Chile, 1998, 48 p.

Yensen N.P. Saline agriculture and new halophyte crops, *Proc. 61st Ann. Meeting Southwestern and Rocky Mountain Division, Amer. Assoc. Advancement of Science*, 19-23 March 1985, 1985, p. 27

Yensen N.P. The use of plastics in saline sea water irrigation, *Mem. Congresso Nacional Sobre el uso de Plastico in Agricultura*, Guadalajara, Jalisco, México, 1983.

Yensen N.P., Bedell J.L., Yensen S.B. Domestication of *Distichlis* as a grain and forage, *Biology of Salt Tolerant Plants*, Khan M.A., Ungar I.A. eds., Ohio University, Athens, 1995, p. 388-392.

Yensen N.P., Bedell J.L., Yensen S.B. The Use of agricultural drain water for forage production in coastal ecosystems, *Proc. Int. Symp. Salt-Affected Lagoon Ecosystems*, Batlle-Sales J. ed., Universidad de Valencia, Spain, 1995.

Yensen N.P., Bedell J.L. Consideration for the selection, adaptation, and application of halophyte crops to highly saline desert environments as exemplified by the long-term development of cereal and forage cultivars of *Distichlis spp.* (POACEAE), *Toward the Rational Use of Highly Salinity Tolerant Plants*, Lieth H., Al Masoom A. eds., Kluwer Acad. Publ., Netherlands, 1993, Vol. 2, p. 305–313.

Yensen N.P., Bedell J.L. Man's salt legacy and some possible solutions, *Proc. Int. Symp. Strategies for Utilizing Salt Affected Lands*, February17–25, 1992, Bangkok, Thailand, 1992.

Yensen N.P., Bedell J.L., Yensen S.B. The history of the development of *Distichlis* cultivars for grain, forage, and sod, *Proc. Mexican National Conf. Halophytes*, Ciudad Obregon, June 23–25, 1988, Sonora, México, 1988.

Yensen N.P., Biel K.Y. Soil remediation via salt-conduction and the hypothesis of halosynthesis and photoprotection, *Ecophysiology of High Salinity Tolerant Plants,* Khan M.A., Weber D.J. eds, published by Springer, 2006, p. 313–344.

Yensen N.P., Biel K.Y. The concept of criticality and some principles for sustainability in closed biological systems, biospheres, and their classification, *Complex Systems under Extreme Conditions,* Proc. XII Int. Symp., Khlebopros R.G., Soukhovolsky V.G. eds., Sayani, July 27–31, 2004, Krasnoyarsk, Russia, 2005, p. 134-158.

Yensen N.P., Carty D.J., Martin R., Rudder K., Todderick K., Shamsutdinov N.Z., Shamsutdinov Z. *Remediation of Oil, Gas and Mine Sites of Salts and Hydrocarbon Contaminants via Halophytes for Rangeland Use,* Kluwer Academic Publications, Netherlands, 2003.

Yensen N.P., Clark W.H., Francoeur A. A checklist of Idaho ants (Hymenoptera: Formicidae), *Pan-Pac. Entomol.,* PDF, 1977, Vol. 53, p. 181–187.

Yensen N.P., Hinchman R.R., Negri M.C., Mollock G.N., Settle T., Keiffer C.S., Carty D.J., Rodgers B., Martin R., Erickson R. Halophytes to manage oilfield saltwater: disposal by irrigation / evapotranspiration and remediation of spills, *Proc. 6th Int. Petroleum Environ. Conf., Environmental Issues and Solutions in Petroleum Exploration, Production and Refining,* Sublette K.L. ed., November 16-18, 1999, Houston, Texas, 1999.

Yensen N.P., Fontes M.R., Glenn E.P., Felger R.S. Nuevos cultivos tolerantes a la salinidad para el dieserto Sonora, *Proc. VI Symposio Sobre el Medio Ambiente del Gulfo de California,* Mexicali, Baja California Norte, México, 1981.

Yensen N.P., Fontes M.R., Glenn E.P., Felger R.S. New salt tolerant crops for the Sonoran desert. *Desert Plants,* 1981, Vol. 3, p.111–118.

Yensen N.P., El Shourbagy N. La Siembra directa de *Distichlis palmeri* en suelos riegadas con agua del Mar, *Proc. VI Symposio Sobre el Medio Ambiente del Gulfo de California,* Loreto, Baja California, México, 1982.

Yensen N.P., Yensen S.B. Development of a rare halophyte grain: prospects for reclamation of salt-ruined lands, *J. Wash. Acad. Sc.,* 1987, Vol. 77 p. 209–214

Yensen N.P., Yensen S.B., Weber C.W. A review of *Distichlis* spp. for production and nutritional values, *Arid Lands Today and Tomorrow,* Whitehead E.E., Hutchinson C.F., Timmermann B.N., Varady R.G. eds., West view Press, Boulder, Colorado, 1985, p. 809–822.

14
Rem Khlebopros: Life in Science

Vladislav G. Soukhovolsky[1,2,*], **Irina R. Fomina**[3,4,*]

[1]*International Scientific Centre for Organism Extreme States Research, Krasnoyarsk Scientific Centre, Siberian Branch of the Russian Academy of Sciences, Krasnoyarsk 660036, Russia*
[2]*Institute of Forest, Siberian Branch of the Russian Academy of Sciences, Krasnoyarsk 660036, Russia*
[3]*Institute of Basic Biological Problems, Russian Academy of Sciences, Pushchino, Moscow Region 142290, Russia*
[4]*Biosphere Systems International Foundation, Tucson, Arizona 85755, USA*

Abstract
This chapter, like the entire second edition of the book, is devoted to the ideological leader of the book, a friend and a colleague to most of its authors (and to my beloved father, I. R. Fomina), who died on July 21, 2017. Since the student years at Kiev State University and until the last day of his life, the life of Rem G. Khlebopros was inextricably intertwined with science. "Life in Science", so we called this small chapter.

Keywords: Biophysics, Ecology, Education, Mathematical modeling, Physics

[*] *Corresponding authors*: soukhovolsky@nm.ru; irafomi@rambler.ru

14.1 Introduction

A wonderful scientist, teacher, tutor, comrade, a fine man, Rem Grigorievich Khlebopros (Photo 14.1) died in the eighty-eighth year of his life on July 21, 2017. Rem lived a long life, he dedicated to serve people; the landmarks of his life were set by the history of the country.

Photo 14.1. Professor Rem Grigorievich Khlebopros in the Hall of V.A. Trapeznikov Institute of Control Sciences, of the Russian Academy of Sciences, during a coffee-break at the VII International Conference "System Identification and Control Problems" SISPRO '08, Moscow, January 28–31, 2008 (photo image from the private collection of Karl Biel)

He left a bright mark in science. His work in the field of magnetism and biological physics is admired for its non-standard thinking and incredible scientific intuition. The flow of his thought was daring, it seemed there were no obstacles on its way. Until the last day of his life he had been striving for the knowledge of the unknown.

14.2 Life in Science

Rem G. Khlebopros graduated from the famous Physics Faculty of Kiev State University, the Department of Theoretical Physics, in the difficult post-war years and subsequently managed to instill in his numerous students the classical Kiev school of physics.

He worked in the Theoretical Department of the Institute of Physics of the Siberian Branch of the USSR Academy of Sciences for many years (Figure 14.1).

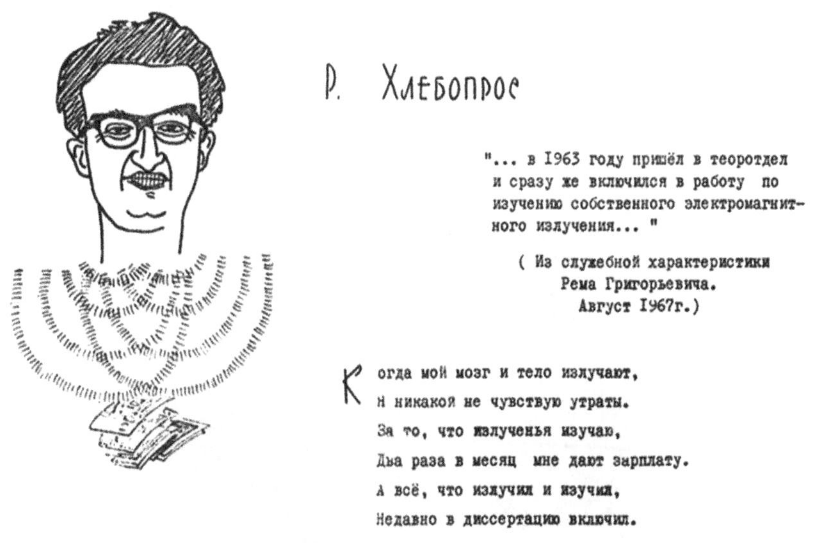

Fig. 14.1. A friendly cartoon from the cartoon book "The Museum of Friends" the first part of which was published in 1967. Its authors O. A. Bayukov and N. S. Chistyakov are the researchers for the Institute of Physics of the Siberian Branch of the USSR Academy of Sciences. The book came out in 1967 in a small edition (100 copies in all) as a gift to the Institute anniversary and instantly spread among the characters of the book and their co-workers (copied from http://photo.kirensky.ru/personalii/hlebopros-rem-grigorevich).

Note: Humorous translation: R. Khlebopros "...came to the Theoretical Department in 1963 and immediately engaged in the study of its own electromagnetic radiation..." (from the Rem Grigorievich's personal file, August 1967)

When my brain and body radiate
I do not feel any loss.
For the fact that I'm investigating radiation,
Twice a month they give me a salary.
And all that I radiated and studied,
I recently included in my Ph.D. thesis.

It was in those years that he (together with academician Alexander S. Isaev) created a phenomenological theory of the dynamic in forest insects population. To continue his work in this field, he was invited to the V. N. Sukachev Institute of Forest and Wood of the Siberian Branch of the USSR Academy of Sciences (Krasnoyarsk, Russian Federation, USSR), where he organized the Laboratory of Mathematical Methods of Research in 1975. The further step of his scientific career was the creation, in 1989, and management of the Theoretical Department of the Institute of Biophysics of the Siberian Branch of the Russian Academy of Sciences (Krasnoyarsk, Russia). And then for many years until his death he headed the International Scientific Centre for Organism Extreme States Research of the Krasnoyarsk Scientific Centre of the Siberian Branch of the Russian Academy of Sciences (Krasnoyarsk, Russia).

Rem's contribution to scientific and technological progress is over two hundred scientific works: articles, monographs and remarkable rationalization proposals (Photo 14.2).

Photo 14.2. Professor Vladislav G. Soukhovolsky (left) and Professor Rem G. Khlebopros (right) discuss some scientific hypothesis and the celebration of the 85th anniversary of Rem Khlebopros; the International Scientific Centre for Organism Extreme States Research of the Krasnoyarsk Scientific Centre of the Siberian Branch of the Russian Academy of Sciences, Krasnoyarsk, 2015 (photo image from the collection of the International Scientific Centre for Organism Extreme States Research)

In the second part of twentieth century a strange tendency was observed in the world of science: numerous scientists recognized or even famous in their fields suddenly began to work in biology in general and ecology in particular. Chemist Professor Alfred Lotka and outstanding mathematician Professor Vito Volterra laid the foundations for modern mathematical ecology. In the 1930s, Volterra's work on models of population dynamics was generalized by the great Russian mathematician Academician Andrei N. Kolmogorov, the author of the "strict" probability theory and other important theories such as computational complexity, turbulence, *etc.* Professor Alexander A. Witt (the main specialist in the sustainability theory) and his colleague and friend since their student years at Moscow State University Professor Georgy F. Gause (whose name is known for the Gause principle!), became, in fact, the first to analyze the real series of time dynamics of bacterial populations.

Mathematician Academician Andrei M. Lyapunov and engineer Dr. Igor A. Poletayev developed methods for describing biocenosis... The strange thing about this brilliant company of scientists who came to the ecology was there were engineers, chemists, mathematicians (and so many!), but surprisingly few physicists. Of course, we must not forget about one of the creators of quantum mechanics, Nobel Prize-winning Professor Erwin Schrödinger[1], the initiator of work on atomic bomb creation in the US, Professor Leo Szilard (creator of the microorganism cultures theory), the creator of the DNA model Professor Francis Crick (though not very successful in physics). Of course, it's not that physicists do not like living systems, "it's not about that, and yet, yet, yet..."

There is a feeling that in the last century, when ecology was young and pleasant in all respects science, which was looking for a partner for a long life together, she opted for mathematics, not physics. On the other hand, of course, physics at that time was not up to some bugs: the fascinating epic unfolded from the mid-1920s on the creation of quantum mechanics, the study of the atomic nucleus and elementary particles; and the mathematics of that time was still relatively free science.

What mathematics has brought to the environment is well known. Much less has been told about what the physics could bring to the ecology of that time. For example, the idea of the nonuniqueness of models of studied objects (quantum objects can be considered both as particles and as waves), which only now penetrates into the theoretical ecology; the understanding of the impossibility of obtaining exact solutions for models of complex objects and the development of approximate models with an estimate of the degree of reliability of this approximation (this actually does not yet exist in ecology!); taking into account the fact that exact

---

[1] Schrödinger E. *What Is Life? The Physical Aspect of the Living Cell*, Cambridge University Press, United Kingdom, 1944, 194 p.

measurements of all characteristics of the studied object are impossible: until now many ecologists and mathematical-ecologists behave as if everything in the ecosystem can be measured…

The scientific success of Rem Khlebopros was determined by the ability to use methods of describing complex systems developed in physics, with the understanding that to describe the system it is not necessary to increase the number of variables in the model, sometimes only two variables are enough: population density and its multiplication factor, knowledge (from his teacher at the Kiev State University Academician Nikolai N. Bogolyubov) that it is necessary to choose what characteristic time to consider the object and, most importantly, the talent that allowed all these physical principles to be applied to biological, medical and social objects.

Coming into ecology of Rem Khlebopros, a successful theoretical physicist, a specialist in ferromagnetism, a disciple of the great scientists Nikolai N. Bogolyubov, was a matter of chance: a neighbor, who was not very familiar at the time, asked whether insect mass outbreaks really resembled a chain reaction in uranium. However, success in phenomenological theory of the dynamics of the number of forest insects, which was created jointly with this neighbor Professor Alexander S. Isaev – the future Academician, the Director of the V. N. Sukachev Institute of Forest and Wood of the Siberian Branch of the USSR Academy of Sciences and the member of the Government of the USSR, was not accidental.

It is impossible to list in a small chapter all the works of Rem Khlebopros even in the field of ecology (not to mention physics, biophysics and other fields of science), but we would like to mention one of the aspects of his scientific activity (Photo 14.3).

Rem G. Khlebopros brought up the scientific school of specialists in the physics of magnetic phenomena, the dynamics of the number of animals, forest ecology, the economy of natural resources, and theoretical medicine. For active work on the training of scientists Rem G. Khlebopros awarded the title of Professor of Ecology by the Highest Attestation Committee of the USSR Academy of Sciences in Moscow (USSR) in 1982.

His pedagogical talent was not inferior to the talent of the scientist and he spared neither time nor energy in training new generations of scientists[2]. He was a Professor at the Department of Biophysics of Krasnoyarsk

[2]"…Ira, do you remember that the student will come to your lab for joint research?" These were the last words that I heard from my father already awaiting the surgical operation. Until the last minute of his life he continued to think about scientific projects and about his students. (Irina R. Fomina, pers. recol.)

State University (KSU) in 1978-1996; the Department of Applied Mathematics of KSU in 1996–2003; the Department of Social and Economic Planning of the Faculty of Economics of KSU in 2003–2009; the Department of Nature Management of the Institute of Economics, Management and Nature Management of the Siberian Federal University from 2009 until his death.

Photo 14.3. Successful defense of the Ph.D. thesis of Olga Vshivkova under supervision of Rem G. Khlebopros, Institute of Experimental and Theoretical Biophysics of the Russian Academy of Sciences, Pushchino, Moscow Region, 2012 (photo image from the collection of the International Scientific Centre for Organism Extreme States Research)
Note: *From left to right*: Rem Khlebopros, Nadezhda Lanina, Rostislav Trashcheev, Olga Vshivkova, Alexander Medvinsky, Yuri Tyutyunov

Abroad, he lectured at Boston University, Massachusetts Institute of Technology, Harvard University (USA) in 1990 and 1991; University of Toronto (Canada) in 1989, 1990 and 1997; Courant Institute, New York (USA) in 2002; Institute of Higher Scientific Research in Paris (France) in 2002 and 2004.

Rem Khlebopros is a winner of the National Environmental Award in the category "Education for Sustainable Development". Twelve Doctor of Sciences theses and fifty-eight Philosophy Degree theses were defended by his students under his supervision.

Who in physics is considered a good scientist? Of course, someone who has obtained nontrivial new results (there are many such scientists) or someone who is able to explain nontrivially the new results of other authors and to teach how to obtain such results (for example, Professor Paul Ehrenfest and few except him). The top of the top is a physicist who knows how to get new results and creates a scientific school whose students are making progress.

Of course, when Rem Khlebopros started to count on his fingers his students, who became famous scientists, and said that he lacks fingers, he boasted a little. But behind this account are real, well-known persons and important results (Photo 14.4).

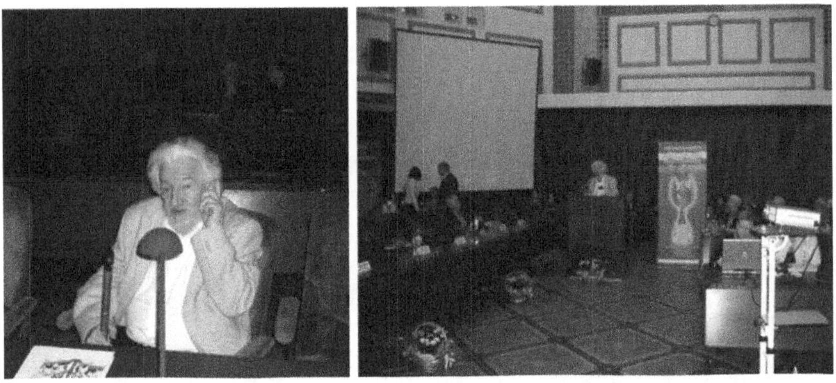

Photo 14.4. National Environmental Award of the Vernadsky Foundation, nomination "Education for Sustainable Development", Moscow 2009. The first prize was awarded to the project of Rem G. Khlebopros with co-authors (photo images from the private collection of Irina R. Fomina)

Ecology is a science more complex than physics, and it is not possible to obtain new non-trivial results, as on the conveyor belt. Nevertheless, the School of Rem Khlebopros made outstanding works - sometimes with a very strange appearance in unexpected fields of science (for example, the creation of methods for identifying falsifications in elections on the basis of models of forest insect populations dynamics; phenomenological theories of mass reproduction outbreaks and cancer tumors as phase transitions of the second kind; the theory of an optical microscope with superresolution; the theory of computational neural networks and a whole bunch of others).

Rem G. Khlebopros was a man whose destiny is an example of extremely severe trials. Still quite young, he almost completely lost his vision as a result of the explosion of the bulb during the training experiment. And he carried through these tests surprisingly sensitive, open to people's heart. He was a true ascetic. The subtlest tactfulness was inherent in him in communicating with colleagues and friends. Rem Khlebopros was a man of a living mind and an amazing sense of humor, understanding and appreciating poetry. He treated heartfelt, with a tender love for Jewish history and culture.

Rem Khlebopros was a man of exceptional honesty. He was always able to defend his position on any fundamental issue, whether it was attitude towards Stalinism or incorrect behavior of colleagues.

Especially principally and tirelessly, he fought for the ecology of Krasnoyarsk, the beloved city in which he lived and worked for more than 50 years, not only criticizing the irresponsibility of industrial enterprises and the disorganization of city development, but also offering concrete ways, approaches and techniques for solving the problem of "black sky" (see Khlebopros, TV-interviews).

He could still have much to do... His life is the highest example for all of us.

The bright memory of Rem Grigorievich Khlebopros will always remain in our hearts (Photos 14.5 and 14.6).

Photo 14.5. Professor Rem Khlebopros in the Scientific Campus (Akadem Gorodok), Krasnoyarsk 2008 (photo image from the collection of the International Scientific Centre for Organism Extreme States Research)

Photo 14.6. Professor Rem Khlebopros on the banks of the Yenisei in the Scientific Campus (Akadem Gorodok), Krasnoyarsk, May 13, 2009 (photo image from the private collection of Lyudmila Petrova)

14.3 Selected Scientific Publications and Speeches by Rem G. Khlebopros

14.3.1 Video-Interviews about Ecology in Krasnoyarsk

Open Society – https://www.youtube.com/watch?v=BtXPESEUoOw
Time over the Edge – https://www.youtube.com/watch?v=RFrR7IS0KXI
News – https://www.youtube.com/watch?v=nQHD2udhoyU

14.3.2 Books

Khlebopros R.G., Biel K.Y. eds., Biel K.Y., Fomina I.R., Yensen N.P., Nishio J.N., Matichenkov V.V., Nazarova G.N., Soukhovolsky V.G., Khlebopros R.G. *Complex Biological Systems: Adaptation and Toler-*

ance to Expreme Environments, Gorod, Pushchino-Krasnoyarsk, 2014, 344 p.

Khlebopros R.G., Okhonin V.A., Fet A.I. *Catastrophes in Nature and Society: Mathematical Modeling of Complex Systems*, World Scientific Publishing Co, Singapore, 2007, 336 p.

14.3.3 Articles

Khlebopros R.G., Zakhvataev V.E., Gabuda S.P., S. Kozlova G., Slepkov V.A. Possible mantle phase transitions by the formation of SiO_2 peroxides: Implications for mantle convection, *Doklady Earth Sciences*, 2017, Vol. 473, No. 2, p. 416–418.

Kryukova O.V., Pyankov V.F., Khlebopros R.G., Kopylov A.F. Effect of electromagnetic microwave radiation on the growth of Ehrlich ascites carcinoma, *Doklady Biological Sciences*, 2016, Vol. 470, No. 1, p. 237–239.

Gavrikov V.L., Sharafutdinov R.A., Knorre A.A., Pakharkova N.V., O. M. Shabalina O.M., Bezkorovaynaya I.N., Borisova I.V., Erunova M.G., Khlebopros R.G. How much carbon can the Siberian boreal taiga store: a case study of partitioning among the above-ground and soil pools, *Journal Forestry Research*, 2016, Vol. 27. No. 4, p. 907–912.

Khlebopros R.G., Zakhvataev V.E., Slepkov V.P., Kuzmin M.I. On the possibility of phase transitions with the formation of SiO_2 peroxide forms in the earth mantle and their effect on mantle convection, *Journal of Structural Chemistry*, 2016, Vol. 57, No. 2, p. 417–421.

Shadrin K.V., Morgulis I.I., Pakhomova V.G., Rupenko A.P., Kryukova O.V., Khlebopros R.G. Characteristics of oxygen transport through the surface of the isolated perfused rat liver, *Doklady Biochemistry and Biophysics*, 2015, Vol. 464, No. 1, p. 298–300.

Isaev A.S., Pal'nikova E.N., Ovchinnikova T.M., Soukhovolsky V.G., Tarasova O.V., Khlebopros R.G. Population dynamics and stability of forest insects with low density (using the example of populations of the bordered white *Bupalus piniarius* L.), *Contemporary Problems of Ecology*, 2014, Vol. 7, No. 7, p. 733–742.

Klyuchevskii A.V., Khlebopros R.G. Large earthquakes in the Baikal Region as response to bifurcations in nonlinear resonance hysteresis, *Doklady Earth Sciences*, 2014, Vol. 457, No. 1, p. 845–849.

Gavrikov V.L., Khlebopros R.G. The problem of carbon forest management: Much or fast?, *Herald of the Russian Academy of Sciences*, 2014, Vol. 84, No. 3, p. 211–216.

Gavrikov V.L., Khlebopros R.G. Phenomenological model of carbon-sequestering forest management, *Contemporary Problems of Ecology*, 2013, Vol. 6, No. 7, p. 708–713.

Gurevich Yu.L., Mankov Yu.I., Khlebopros R.G. Self-assembly of superparamagnetic ferrihydrite nanoparticles, *Doklady Physics*, 2013, Vol. 58, No. 11, p. 478–481.

Klyuchevskii A.V., Khlebopros R.G. Coupled large earthquakes in the Baikal rift system: Response to bifurcations in nonlinear resonance hysteresis, *Geoscience Frontiers*, 2013, Vol. 4, No. 6, p. 66–68.

Kruglik O.V., Morgulis I.I., Khlebopros R.G. Effect of microwave electromagnetic radiation (UHF EMR) on tumor cell viability in experiment, *Doklady Biochemistry and Biophysics*, 2013. Vol. 449, No. 1, p. 66–68.

Mankov Yu.I., Khlebopros R.G. Effect of the shape of an antiferromagnetic nanoparticle on its magnetization, *Technical Physics*, 2012, Vol. 57, No. 5, p. 733–734.

Isaev A.S., Soukhovolsky V.G., Khlebopros R.G. Model approaches to description of critical phenomena in forest ecosystems, *Contemporary Problems of Ecology*, 2011, Vol. 4, No. 7, p. 699–705.

Biel K.Y., Fomina I.R., Nazarova G.N., Soukhovolsky V.G., Khlebopros R.G., Nishio J.N. Untangling metabolic and spatial interactions of stress tolerance in plants. 1. Patterns of carbon metabolism within leaves, *Protoplasma*, 2010, Vol. 245, No. 1, p. 49–73.

Shishatskaya E.I., Goreva A.V., Voinova O.N., Inzhevatkin E.V., Khlebopros R. G., Volova T.G. Evaluation of antitumor activity of rubomycin deposited in absorbable polymeric microparticles, *Bulletin of Experimental Biology and Medicine*, 2008, Vol. 145, No. 3, p. 358–361.

Slepkov V.A., Sukhovolsky V.G., Khlebopros R.G. Population dynamics in tumor growth modeling, *Biophysics*, 2007, Vol. 52, No. 4, p. 426–431.

Slepkov V.A., Sukhovolskii V.G., Khlebopros R.G. Modeling the growth and morphogenesis of malignant tumors, *Doklady Biochemistry and Biophysics*, 2006, Vol. 411, p. 342–345.

Semenov D.A., Khlebopros R.G. Influence of biota on global climate, *Biophysics*, 2005. Vol. 50, No. 4, p. 659–662.

Slepkov V.A., Khlebopros R.G. Chain reaction as mechanism of explosive crystallization in amorphous films, *Metallofizika i Noveishie Technologii*, 2005, Vol. 27, No. 1, p. 11–18.

Isaev A.S., Sukhovol'skii V.G., Khlebopros R.G., Buzikin A.I., Ovchinnikova T.M. The phenomenological model of forest formation, *Doklady Biological Sciences*, 2004, Vol. 398, p. 406–409.

Sukhovol'skii V.G., Khlebopros R.G., Iskhakov T.R. An optimization model of competition between species for resource, *Doklady Biochemistry and Biophysics*, 2003. Vol. 390. P: 171–173.

Semenov D.A., Khlebopros R.G. Biophysical aspects of the biosphere impact on global climate, *Doklady Biochemistry and Biophysics*, 2002. Vol. 387. P: 338–339.

Semenov D.A., Sudachkova N.E., Khlebopros R.G. A new scheme of lignin biosynthesis and the mechanism of its regulation of functional properties, *Doklady Biochemistry and Biophysics*, 2002. Vol. 382. P: 50–52.

Soukhovolsky V.G., Fomina I.R., Bil K.Y., Nishio J.N., Khlebopros R.G. An optimization model of the photosynthetic leaf: the model of optimal photosynthetic CO_2 fixation within leaves of mesophytic C_3 plants, *Doklady Biological Sciences*, 2002, Vol. 382, p. 28–30.

Ovchinnikova T.M., Savchenko A.A., Sukhovol'skii V.G., Khlebopros R.G. Structural and temporal links between the components of humoral immunity, *Doklady Biochemistry*, 2000, Vol. 374, No. 1, p. 186–188.

Berezovskaya F.S., Isaev A.S., Karev G.P., Khlebopros R.G. The role of taxis in the population dynamics of forest insect, *Doklady Biological Sciences*, 1999, Vol. 365, No. 3, p. 148–151.

Agapov I.A., Kashkin V.B., Khlebopros R.G. Identification of random fields of points, *Signal Processing-Image Communication*, 1998. Vol. 13, No. 1, p. 21–43.

Bazykin A.D., Berezovskaya F.S., Isaev A.S., Khlebopros R.G. Dynamics of forest insect density: Bifurcation approach, *Journal of Theoretical Biology*, 1997, Vol. 186, No. 3, p. 267–278.

Berezovskaya F.S., Karev G.P., Kisliuk O.S., Khlebopros R.G., Tselniker Y.L. A fractal approach to computer-analytical modelling of tree crowns, *Trees – Structure and Function*, 1997, Vol. 11, No. 6, p. 323–327.

Gitelzon I.I., Bartsev S.I., Okhonin V.A., Soukhovolsky V.G., Khlebopros R.G. The strategy of development: What kind?, *Herald of the Russian Academy of Sciences*, 1997, Vol. 67, No. 5, p. 415–420.

Sudachkova N.E., Sukhovolskii V.G., Khlebopros R.G. Competitive type of free amino-acids distribution in morphological structures of tree, *Doklady Akademii Nauk*, 1996, Vol. 348, No. 4, p. 556–559.

Bazykin A.D., Berezovskaya F.S., Isaev A.S., Khlebopros R.G. An analysis of stereotypes of forest insect dynamics, *Zhurnal Obshchei Biologii (Journal of General Biology)*, 1995, Vol. 56, No. 2, p. 191–199.

Bazykin A.D., Berezovskaya F.S., Isaev A.S., Khlebopros R.G. Parametric substantiation of the principle of dynamic stability of the phytophage-entomophage system, *Doklady Biological Sciences*, 1993, Vol. 333, No. 5, p. 464–467.

Borisov A.N., Kashkin V.B., Khlebopros R.G. Method for indication of horizontal structure of tree stands. *Doklady Biological Sciences*, 1987, Vol. 293, No. 1-6, p. 132–133.

Nedorezov L.V., Khlebopros R.G. Peculiarities in the dynamics of population numbers in strongly inertial regulation and regulation without inertia, *Zhurnal Obshchei Biologii (Journal of General Biology)*, 1985, Vol. 46, No. 3, p. 345–348.

Iskhakov R.S., Fish G.I., Maltsev V.K., Khlebopros R.G. Determination of the nearest-surrounding symmetry in amorphous Co-P alloys, *Fizika Metallov i Metallovedenie (The Physics of Metals and Metallography*, 1984, Vol. 58, No. 6, p. 1214–1215.

Kozhevnikova N.N., Sudachkova N.E., Khlebopros R.G., Cherkashin V.P. Rhythms of peroxidase activity in tissues of Scotch pine needles, *Soviet Plant Physiology*, 1984, Vol. 31, No. 4, p. 527–534.

Iskhakov R.S., Fish G.I., Khlebopros R.G. Temperature-dependence of exchange inhomogeneity characteristics in amorphous CoP alloys, Solid State Communications, 1983, Vol. 45, No. 11, p. 961–962.

Iskhakov R.S., Fish G.I., Khlebopros R.G. Temperature-dependence of exchange interaction fluctuations in amorphous co-p alloys, *Fizika Tverdogo Tela (Solid State Physics)*, 1983, Vol. 25, No. 2, p. 568–570.

Mankov Y.I., Gorenko L.M., Khlebopros R.G. Slowly decaying waves near cyclotron-resonance in metallic ferromagnets located in an inclined magnetic-field, *Zhurnal Eksperimentalnoi i Teoreticheskoi Fiziki (Journal of Experimental and Theoretical Physics)*, 1981, Vol. 81, No. 6, p. 2180–2191.

Mankov Y.I., Gorenko L.M., Khlebopros R.G. Electromagnetic-waves in a metal near cyclotron-resonance in an inclined magnetic-field, *Zhurnal Eksperimentalnoi i Teoreticheskoi Fiziki (Journal of Experimental and Theoretical Physics)*, 1979, Vol. 77, No. 5, p. 2042–2052.

Kofman G.B., Kuzmichev V.V., Khlebopros R.G. Use of the period of intensive growth of wood plants for the construction of phylogenetic series, *Zhurnal Obshchei Biologii (Journal of General Biology)*, 1979, Vol. 40, No. 5, p. 766–771.

Chekanova L.A., Iskhakov R.S., Khlebopros R.G., Chistyakov N.S., Fish G.I. Some peculiarities of crystallization of Co_{1-X} P metallic glasses, *Fizika Tverdogo Tela (Solid State Physics)*, 1978, Vol. 20, No. 11, p. 3501–3503.

Isaev A.S., Khlebopros R.G. Delay effects in regulation of forest insect quantity, *Doklady Akademii Nauk SSSR*, 1977, Vol. 232, Iss. 6, p. 1448–1451.

Gorenko L.M., Mankov Y.I., Khlebopros R.G. Magnetostatic waver with discrete spectrum, *Fizika Tverdogo Tela (Solid State Physics)*, 1977, Vol. 19, No. 1, p.56–58.

Gorenko L.M., Mankov Y.I., Khlebopros R.G. Electromagnetic-waves with a discrete spectrum in metallic ferromagnets, *Zhurnal Eksperimentalnoi i Teoreticheskoi Fiziki (Journal of Experimental and Theoretical Physics)*, 1976, Vol. 71, No. 8, p. 627–634.

Chekanova L.A., Iskhakov R.S., Fish G.I., Khlebopros R.G, Chistyakov N.S. Spin-wave resonance and structural transformations in amorphous films, *Fizika Metallov i Metallovedenie (The Physics of Metals and Metallography)*, 1976, Vol. 41, No. 3, p. 536–541.

Mankov Y.I., Khlebopros R.G., Zakharov Y.V. Domain-wall energy in metallic ferromagnet, *Fizika Metallov i Metallovedenie (The Physics of Metals and Metallography)*, 1975, Vol. 39, No. 3, p. 461–465.

Mankov Y.I., Khlebopros R.G., Zakharov Y.V. Effect of conductive electrons on domain boundary energy, *Fizika Tverdogo Tela (Solid State Physics)*, 1975, Vol. 17, p. 352–353.

Isaev A.S., Khlebopros R.G. Analysis of number dynamics of forest insects based on principle of stability of mobile ecological systems, *Zhurnal Obshchei Biologii (Journal of General Biology)*, 1974, Vol. 35, No. 5, p. 737–745.

Isaev A.S., Khlebopros R.G. Population Ecology: Fighting a forest pest, *Nature*, 1973, Vol. 241, No. 5391, p. 504.

Acknowledgments

The authors are grateful to Professor Karl Biel, Dr. Olga Vshivkova and Dr. Lyudmila Petrova for kindly providing photos and to Master Tatyana Khlebopros for editorial and stylistic corrections.

Index

Printed and bound by CPI Group (UK) Ltd, Croydon, CR0 4YY
17/09/2026
14964245-0006